HABITATS OF AFRICA

HABITATS OF AFRICA

A Field Guide for Birders, Naturalists, and Ecologists

Ken Behrens | Keith Barnes | Iain Campbell

Special contributors
Gray Tappan, Giselle Velastegui, and Pablo Cervantes

PRINCETON UNIVERSITY PRESS
PRINCETON AND OXFORD

Copyright © 2025 by Ken Behrens, Keith Barnes, and Iain Campbell

Princeton University Press is committed to the protection of copyright and the intellectual property our authors entrust to us. Copyright promotes the progress and integrity of knowledge created by humans. Thank you for supporting free speech and the global exchange of ideas by purchasing an authorized edition of this book. If you wish to reproduce or distribute any part of it in any form, please obtain permission.

Requests for permission to reproduce material from this work should be sent to permissions@press.princeton.edu

Published by Princeton University Press
41 William Street, Princeton, New Jersey 08540
99 Banbury Road, Oxford OX2 6JX

press.princeton.edu

All Rights Reserved
ISBN (pbk.) 9780691244761
ISBN (e-book) 9780691244778

British Library Cataloging-in-Publication Data is available

Editorial: Robert Kirk and Megan Mendonça
Production Editorial: Mark Bellis
Cover Design: Wanda España
Production: Steve Sears
Publicity: William Pagdatoon and Caitlyn Robson-Iszatt
Copyeditor: Eva Silverfine
Typesetting and Design: D & N Publishing, Wiltshire, UK

Cover Credit: Christine Elder

This book has been composed in Cambay Devanagari

Printed in China

10 9 8 7 6 5 4 3 2 1

CONTENTS

Figures and Sidebars	7
INTRODUCTION	**9**
Genesis of the Book	9
What We Cover as Distinct Habitats	9
Mapping Methodology	10
"Habitats of the World" as a Complement to Other Global Habitat Classification Systems	10
Types of Habitat Boundaries and Ecotones	13
Area Covered by This Book	15
Common Canopy Leaf Types and the Forests Where You May Find Them	16
Climate Descriptions and Graphs	16
Köppen Climate Explained	17
Biomes of Africa	21
Taxonomy	24
Endemic Bird Areas	24
Useful Habitat Jargon	24
About This Book	25
Acknowledgments	26
Abbreviations	27
Selected Bibliography	27
REGIONAL INTRODUCTIONS	**33**
Northeast Africa	33
East Africa	35
Central and North-Central Africa	39
Indian Ocean	41
Greater Southern Africa	44
North Africa	47
West Africa	51
CONIFERS	**54**
■ Af1A Maghreb Fir and Cedar Forest	54
■ Af1B Maghreb Juniper Open Woodland	57
■ Af1C Afrotropical Montane Dry Mixed Woodland	61
■ Af1D Maghreb Pine Forest	67
DESERTS AND ARID LANDS	**71**
■ Af2A Saharan Reg Desert	71
■ Af2B Namib Rock Desert	77
■ Af2C Saharan Erg Desert	82
■ Af2D Namib Sand Desert	88
■ Af2E Rocky Hamada and Massif	94
■ Af2F Nama Karoo	101
■ Af2G Succulent Karoo	107
■ Af2H Maghreb Hot Shrub Desert	115
■ Af2I Afrotropical Hot Shrub Desert	121
■ Af2J Spiny Forest	127
■ Af2L Dragon's Blood Tree Semi-desert	133
TEMPERATE BROADLEAF FORESTS	**138**
■ Af3A Laurel Forest	138
WARM HUMID BROADLEAF FORESTS	**143**
■ Af4A Afrotropical Lowland Rainforest	143
■ Af4C Monsoon Forest	151
■ Af4D East Coast Forest Matrix	157
■ Af4E South Coast Forest Matrix	164
■ Af4F Swamp Forest	169
■ Af4G Moist Montane Forest	174
■ Af4J Indian Ocean Lowland Rainforest	180
■ Af4K Indian Ocean Montane Rainforest	189
■ Af4L Seychelles Granite Forest	197
■ Af4M Mavunda	203
TROPICAL DECIDUOUS FORESTS	**209**
■ Af5A Malagasy Deciduous Forest	209
■ Af5B Angolan Deciduous Forest	215
SAVANNAS	**221**
■ Af6A Mopane	221
■ Af6B Northern Dry Thorn Savanna	227
■ Af6C Kalahari Dry Thorn Savanna	235
■ Af6D Moist Mixed Savanna	243
■ Af6E Miombo	252
■ Af6F Guinea Savanna	260

CONTENTS

- Af6G Inselbergs and Koppies — 268
- Af6H Gusu — 275

GRASSLANDS — 280
- Af7A Afrotropical Grassland — 280
- Af7E Malagasy Grassland and Savanna — 287
- Af7F Montane Grassland — 293

MEDITERRANEAN FORESTS, WOODLANDS, AND SCRUBS — 300
- Af8A Fynbos — 300
- Af8B Strandveld — 309
- Af8C Renosterveld — 315
- Af8D Maghreb Maquis — 321
- Af8E Maghreb Garrigue — 326
- Af8F Maghreb Broadleaf Woodland — 328
- Af8G Albany Thicket — 332
- Af8H Tapia — 338

ALPINE TUNDRAS AND MONTANE HEATHS — 343
- Af10A Afroparamo — 343
- Af10B High Atlas Alpine Meadow — 350
- Af10C Montane Heath — 354

FRESHWATER HABITATS — 358
- Af11A Afrotropical Deep Freshwater Marsh — 358
- Af11B Afrotropical Shallow Freshwater Marsh — 365
- Af11C North African Temperate Wetland — 370
- Af11D South African Temperate Wetland — 374
- Af11E Freshwater Lakes and Ponds — 379
- Af11F Rivers — 382

SALINE HABITATS — 386
- Af12A Afrotropical Mangrove — 386
- Af12B Salt Pans and Lakes — 390
- Af12C Tidal Mudflats and Estuaries — 393
- Af12D Salt Marsh — 395
- Af12E Sandy Beach and Dunes — 397
- Af12F Rocky Shoreline — 400
- Af12G Offshore Islands — 403
- Af12H Pelagic Waters — 406

ANTHROPOGENIC — 409
- Af13A Humid Lowland Cultivation — 409
- Af13B Savanna Cultivation — 412
- Af13C Tropical Montane Cultivation — 415
- Af13D South African Temperate Cultivation — 418
- Af13E North African Temperate Cultivation — 419
- Af13F Tree Plantations — 422
- Af13G Cities and Villages — 424
- Af13H Grazing Land — 427

INDEX — 429

FIGURES AND SIDEBARS

FIGURES
Fig. 1. Hierarchical coverage of Habitats of the World   11
Fig. 2. Organism taxonomy versus habitat typology   12
Fig. 3. Different types of boundaries between habitats   13
Fig. 4. Habitat intruders and outliers   13
Fig. 5. Political map of Africa, Macaronesia and the Atlantic and Indian Ocean Islands   14
Fig. 6. Topographic map of the region   15
Fig. 7. Dominant leaf types and their global habitats   16
Fig. 8. Köppen Climate Maps 2080 and 2020   18
Fig. 9. Sample climate graphs   19
Fig. 10. The transition from the Sahara to the rainforests of West Africa   20
Fig. 11. Biomes of Africa   21

SIDEBARS
Sidebar 1. African Endemic Bird Families   29
Sidebar 2. Indian Ocean Islands Endemic Bird Families   32
Sidebar 3. The Sahara—The Realm Dictator   53
Sidebar 4. Thriving on Moderation   62
Sidebar 5. What Makes a Desert? The Saharan Example   74
Sidebar 6. Migration Madness—Trans-Saharan Sojourn   76
Sidebar 7. Extinction of People, Plants, and Animals of the Sahara   85
Sidebar 8. The Effects of World Ocean Currents on Climate   93
Sidebar 9. Wadis: Not Just an Oasis in the Desert   100
Sidebar 10. Photosynthesis—Beating Drought through Biochemistry   111
Sidebar 11. Duricrusts and Desertification   126
Sidebar 12. Owlnigmas   149
Sidebar 13. Itigi Thicket   153
Sidebar 14. Weavers, a Megadiverse Afrotropical Family   163
Sidebar 15. Galagos—the Primate Nightshift   173
Sidebar 16. Moist Montane Forest Subhabitats   177
Sidebar 17. Habitat Helpers—Dispersers   188
Sidebar 18. Sambirano Rainforest   193
Sidebar 19. Dung Beetles, Like a Rolling Stone?   226
Sidebar 20. Paleoclimate, Habitat Fluctuations, Biogeography, and Refugia   234
Sidebar 21. Water in Dry Country   236
Sidebar 22. Do Giraffes Have Toxic Neck Syndrome?   239
Sidebar 23. Moist Mixed Savanna Microhabitats   246
Sidebar 24. Edge Case Study: Highland Acacia Woodland   247
Sidebar 25. Miombo Etymology   252
Sidebar 26. Savanna Icebergs   259
Sidebar 27. West African Dry versus Wet Season Savanna Transformations   263
Sidebar 28. Africa's Vanishing Mammals   265

FIGURES AND SIDEBARS

Sidebar 29. South to North Transition from Sahel Thornscrub to Guinea Savanna in Senegal   267
Sidebar 30. Hyrax   272
Sidebar 31. A Tale of Two Seasons   286
Sidebar 32. Edge Case Study: Highland Ouhout Shrubland   299
Sidebar 33. Fire: Friend or Foe?   303
Sidebar 34. Habitat Helpers—Pollinators   306
Sidebar 35. Is It a Mole or Is It a Rat?   312
Sidebar 36. Where Did the Southern Ungulates Roam?   318
Sidebar 37. What's in a Thicket?   335
Sidebar 38. Malagasy Tapia, a Global Classification Enigma   341
Sidebar 39. Africa's Great Rift Valley   349
Sidebar 40. Habitat Architects   373
Sidebar 41. Intra-African Migrants   378
Sidebar 42. In Denial: An African Corridor to the Med   383

INTRODUCTION

GENESIS OF THE BOOK

All the authors of this field guide have had a lifelong fascination with biogeography and wildlife habitats. Like the vast majority of other passionate traveling naturalists, we are most consistently interested in birds and larger mammals, while also paying some attention to reptiles, amphibians, butterflies, and other groups, especially in places where they are conspicuous. We have all been frustrated by the approach to habitat classification used in most books and the complete absence of habitat information in many vertebrate field guides. What we are hoping to do in this book is present our view of African wildlife habitats and, via online resources, their relationship with other vegetation mapping systems in order to allow others to understand them far more readily than we were able to.

There are innumerable lenses through which planet Earth's habitats can be assessed. Geology, geography, and botany are all critically important. But we don't view any of them as the final word on habitats, and much of what these single-discipline models prioritize is of little immediate relevance to traveling naturalists. A specialist in entomology or herpetology will also apply a different, and fascinating, lens to the world. Our reason for prioritizing a mammal and bird "lens" is that we look at the world primarily through this lens, and so do the vast majority of the world's traveling naturalists. A handful of specialists seek out moths in Tanzania's Eastern Arc Mountains, whereas millions of tourists visit the Serengeti and Masai Mara to see big mammals and glamorous birds. A few people venture to the Congo to seek out its incredible diversity of insects, but masses visit mountain lodges in search of Mountain Gorillas. So our approach in this book might lack the apparent clarity of a botanical approach to the world's habitats, but we think at the scale we present habitats, our book has far greater utility to most world travelers and conservationists than any other previous perspective on habitats.

In our attempt to cover the wildlife habitats of the entire continent of Africa, this is an ambitious book, in which hard decisions had to be made about what to include and exclude. We freely admit that habitats like wetlands, or the oceans, are worthy of far more detailed coverage. People who know their local area well may be frustrated by a lack of information about "their patch." Please remember that this book is about giving people a broad overview, especially of places they have never been. It also offers a sort of "virtual travel"; the first thing all the authors do when they find out they are headed somewhere new on the globe is conduct a bit of research about the local wildlife habitats. Deciding to "split" or "lump" some habitats was very tricky, and some of our decisions could be argued endlessly. But condensing a huge amount of research into a finite number of pages, and simply finishing this project, required a certain brutality. Our approach is certain to alienate some, but we firmly believe it will be both enjoyable and useful to other global naturalists like ourselves.

WHAT WE COVER AS DISTINCT HABITATS

We evaluate habitats based on two main criteria: (1) their visual distinctiveness, which can be easily assessed by a casual observer and usually relates to the types of structure (forest vs. shrubland) and species of vegetation present (conifers vs. grasses); and (2) their assemblage of wildlife, primarily mammals and birds, but we also considered other vertebrates. Most of the listed habitats are easily validated by a moderate score in both categories. Each habitat is described with a suite of obligate and indicator bird species available online at www.habitatsoftheworld.org/bird assemblages. An example is African MIOMBO woodland, which is quite distinct in appearance from adjacent savanna habitats and supports a fairly distinctive set of wildlife, including quite a few species restricted to this habitat. But in some cases, one or the other criterion is of predominant importance. Except to the eye of a trained botanist, Indian Ocean rainforest is not very different from other humid broadleaf forests around the world. But it has almost a continent's worth of diversity for many wildlife groups and virtually no overlap with any other place on earth, so it is considered a distinct habitat. An example of the opposite case is African MOPANE savanna. This habitat is characterized by the dominance of the Mopane tree, which is highly distinctive and easily recognizable. So the Mopane savanna qualifies strongly for the first criterion, even as it lacks a cohesive set of wildlife, having, rather, a subset of the wildlife of surrounding habitats.

MAPPING METHODOLOGY

This book is a description of habitats and not a mapping project, but we did make serious attempts to provide accurate maps of habitat distribution. The whole continent of Africa is covered by two good vegetation mapping systems: *The Vegetation of Africa* by White and the International Vegetation Classification (IVC) by NatureServe. Further coverage of West Africa is provided by *Landscapes of West Africa* from the US Geological Survey; of South Africa by the South African VEGMAP from the South African National Biodiversity Institute (SANBI); and of East Africa by the Vegetation and Climate Change in Eastern Africa (VECEA) project, now called the vegetationmap4Africa. When one of the habitats covered in this book aligned well with one or a mix of units from one of these systems, we used their boundaries to generate the distribution maps for this habitat. In South Africa we primarily used the SANBI mapping; in East Africa we primarily used the VECEA mapping; and in the remainder of sub-Saharan Africa we primarily used Natureserve Macrogroup mapping. The exceptions to these sources were West Africa and North Africa where, because the three systems covering these regions were predictive, based on plots and extrapolation, they frequently failed to predict habitat distribution adequately. So we resorted to syncretizing our habitats with the US Geological Survey mapping and manually remapped these regions by the interpretation of aerial photography rather than extrapolation. We then merged the resulting individual habitat distribution maps to generate the regional vegetation maps. Note that the regional maps show the dominant habitats down to the finest scale that is visible in a map of this size. However, no map showing a vast area can fully encompass the complexity of habitat distribution, and many habitats overlap at a small scale. If superimposed, the individual habitat maps included in each habitat chapter show the broad interlocking and overlapping between habitats that is the reality on the ground.

"HABITATS OF THE WORLD" AS A COMPLEMENT TO OTHER GLOBAL HABITAT CLASSIFICATION SYSTEMS

The understanding and correct classification of habitats is crucial to the development of useful and viable nature reserve systems, as is knowing what wildlife occurs in threatened habitats within them. The problem is that as of now, there is no system to classify all the world's habitats at a level that is appreciable to most casual naturalists, birders, conservationists, and ecologists. There are reasons for this, such as, but not limited to, systems and typologies being overly hierarchical by design and the difficulty of trying to syncretize different national mapping systems. The Global Ecosystem Typology (GET) and the International Vegetation Classification (IVC, mainly developed through Natureserve) both have excellent ecosystem classifications that aim to define and protect ecological communities. These systems are incredibly useful in principle, but neither works globally at the scale that conservation groups, birders, or ecologists can use because they do not yet have all habitats described at a level that is convenient to use and understand (fig.1). The *Habitats of the*

Fig. 1. The hierarchical coverage of Habitats of the World system compared with the typologies of the Global Ecosystem Typology (GET) system and NatureServe's International Vegetation Classification (IVC) system.

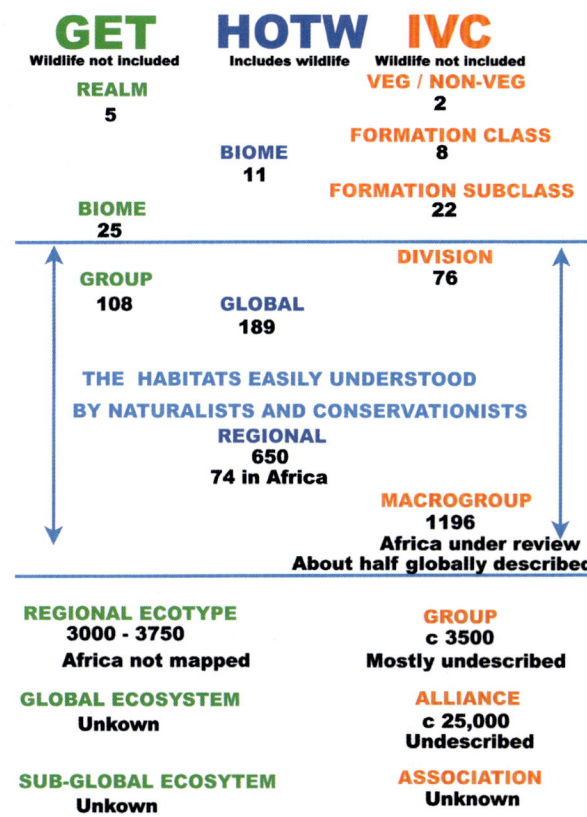

World (HotW) system works as the "Rosetta Stone" for applying GET and the IVC habitat classification systems at a global scale. A complete walkthrough from the HotW system to the IVC and GET systems is available online at www.habitatsoftheworld.org/intotheweedstypology/africa.

The GET system jumps from 108 (described) units at the "Group" level up to 3750 (as yet undescribed) units at the "Regional Ecotype" level. Similarly, the IVC system jumps from 76 (described) units at "Division" level to around 3500 (mostly undescribed) units at "Group" level. Neither system describes the animals living in the habitats. The IVC system is currently being revised to align with the GET system and has macrogroups for North America, South America, and Africa, but the revised system is still a few years away from being global and uniform. The HotW system has global coverage and includes 650 mainly terrestrial habitats, covering almost all the world's birds and much of the other wildlife. At this level, it becomes much easier for non-botanists to discern one habitat from another, understand how they differ, and develop the understanding and criteria to be able to understand the varying ecologies of these systems. Habitats understood at this level of detail will be much easier to use in conservation planning, mapping, and ecological work. The HotW system is in the "Goldilocks zone" that makes it "just right"; detailed enough to be valuable but easy enough for anyone to understand.

Habitats (ecosystems, but also referred to as "vegetation types") are assemblages of organisms (among which animals and plants are the focus of this book), under specific environmental conditions,

Organism Evolution Taxonomy

Kindom: Plant	**Kingdom:** Animalia
Phylum: Tracheaophytes	Phylum: Chordata
Class: Angiosperm	Class: Aves
Order: Poales	Order: Pelecaniformes
Family: Cyperaceae	Family: Balaenicipitidae
Genus: Cyperus	Genus: Balaeniceps
Species: *Cyperus papyrus*	Species: *Balaeniceps rex*
Papyrus	**Shoebill**

Habitat Typology

Humid Forest	**Freshwater Wetland**
African Lowland Humid Forest	African Tropical Wetland
Swamp Forest	**Afrotropical Deep Freshwater Wetland**

Fig. 2. Organism taxonomy versus habitat typology.

that may have artificially human-imposed boundaries delineating them from adjacent similar habitats in what is almost always a gradual boundary at some scale. They are not taxonomic entities the way animals and plants are. Organisms have a distinct genetic code and evolutionary history, which infers relatedness that can be detailed in a strict dendrogram form (see fig. 2). All current habitat classification systems use a similar hierarchical typology. But as you will see in the breakdowns of different biomes in the habitat sections, the dendrograms are helpful in showing relationships, but many habitats are crossovers or can easily be classified in multiple biomes. The relationships are not linear but more often web-like. Examples include SWAMP FOREST, which can be classified as forests *or* wetlands, and MAVUNDA, which has characteristics of both a dry deciduous forest *and* a humid evergreen forest; not either/ors but both/ands.

A major reason why there are so many more birders than enthusiasts for other groups, such as bats, rats, or beetles, is that birds are readily identifiable animals with a limited number of species. And the reason we use bird assemblages as the main tool to check ecosystem classification is that it is easy to identify them and to catalog them (keep lists), as so many birders are prone to do. This listing then becomes a meaningful contribution to citizen science projects like Cornell's eBird, which can be used to analyze global trends in species distribution and occurrence. To compare, beetles have roughly 400,000 known species, a number that defies easy identification or the ability to discern relationships between what has been identified. Bats and rats suffer the problem of being mostly nocturnal, often cryptic, and extremely difficult to identify, which results in a dearth of bat and rat enthusiasts and, consequently, much less knowledge on these groups.

TYPES OF HABITAT BOUNDARIES AND ECOTONES

An ecotone is a place where two or more biomes or habitats meet, usually possessing traits of both. So for example, AFROTROPICAL MANGROVE is a ecotone between marine systems and terrestrial systems, savannas are an ecotone between tropical rainforests and deserts, and monsoon forests are ecotones between savannas and rainforests. So most boundaries are ecotones at some scale. We try to avoid use of this confusing term ecotone in the book and prefer to treat transition zones as boundaries, where the ends of the transitions are different habitats with different bird assemblages. Sometimes the transitions are extremely **sharp**, especially where you have natural forests and anthropogenic farmlands, or wetland systems in arid terrains. There can be mosaics, where there are distinct patches of different habitats in the one area. This is very common in coastal forests and savanna edges. **Mélange** occurs where you have distinct systems that intertwine in a complex manner across a broad zone. This pattern is very common in mountainous areas with complex geomorphology and geology, and with complex microclimates. The fourth kind of transition is **nebulous**, where the changes are so gradual that it is difficult to determine which habitat you are in. This is an ecotone where the bird and animal assemblages mix and it can be difficult to determine the habitat based on either plant or animal assemblages, and so in a finer-scale habitat typology, these nebulous transition zones might be split out as a different habitat.

There are also regions where habitats **intrude** into one another, such as moist forests that extend far into arid terrains or terrains that have strong fire regimes. The opposite exists where

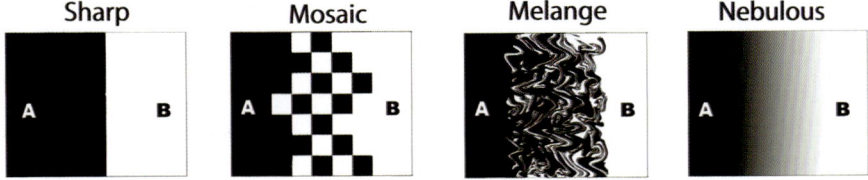

Fig. 3. Different types of boundaries between habitats.

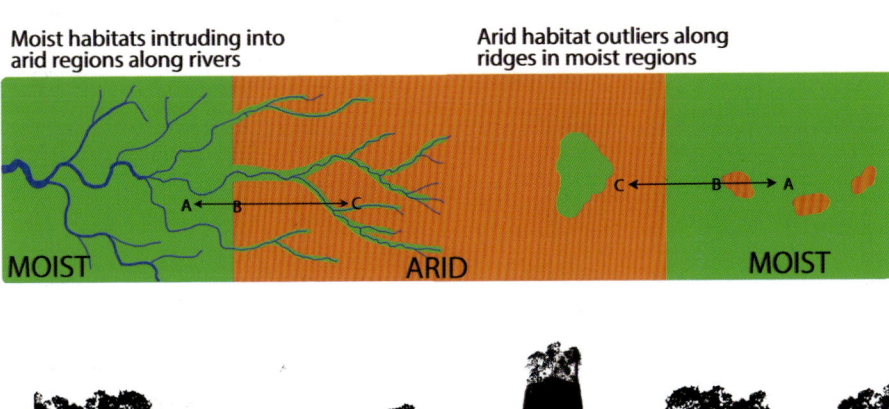

Fig. 4. Outliers of habitats where humid habitats extend into arid zones and where arid habitats extend into humid zones. Orange silhouette of human for scale.

dry and heath-type habitats can extend as **outliers** in very humid environments along ridgelines with nutrient-deficient rocks such as granites or on extremely inert weathered soils, such as lateritic plateaus. If these outliers are very high, they can become cold and/or attract orographic rainfall; there the system flips, and wetter forests can occur in mountains within dry terrains, such as on the Saharan Hamada massifs (p. 94).

Fig. 5. Political map of Africa, Macaronesia and the Atlantic and Indian Ocean islands covered in the book.

AREA COVERED BY THIS BOOK

This book covers all continental Africa, with our northeastern cutoff at the Gulf of Suez. We also cover Macaronesia (the Azores, Madeira, the Canary Islands, and Cape Verde), the Gulf of Guinea Islands, the remote islands off the South Atlantic, and all the w. Indian Ocean islands (Madagascar, the Seychelles, the Comoros, and the Mascarenes).

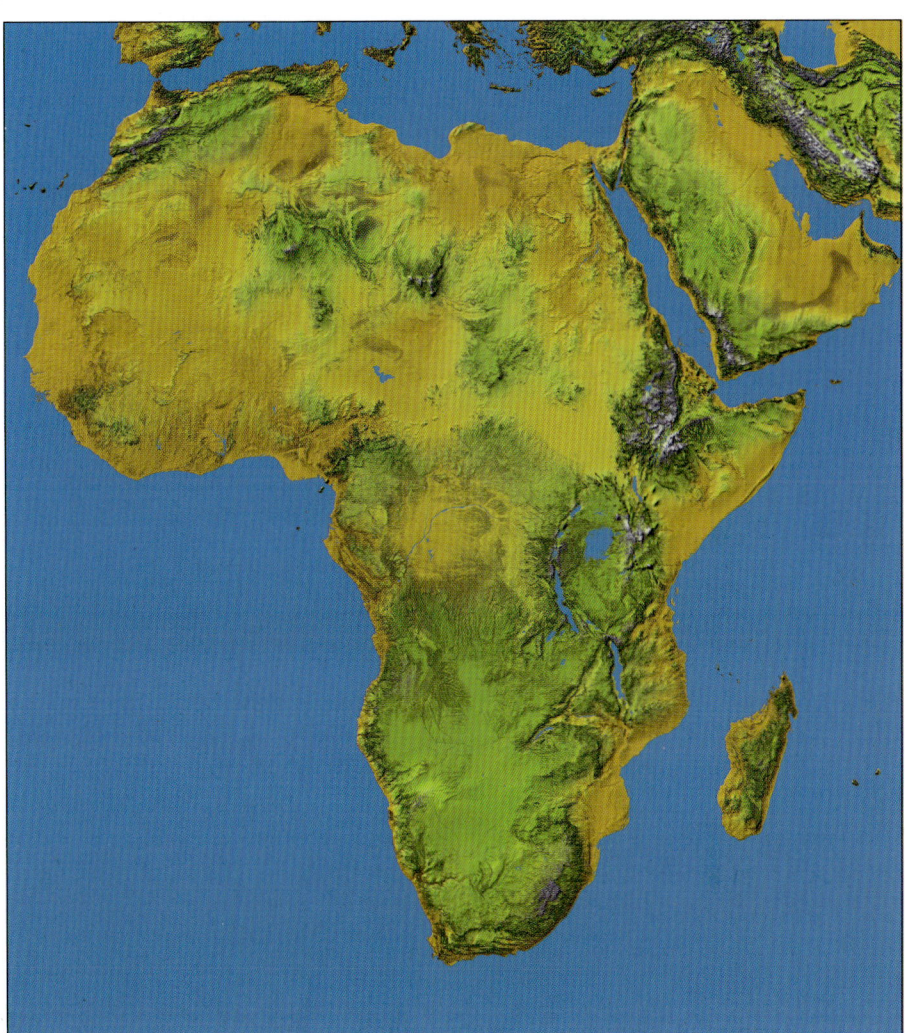

Fig. 6. Topographic map of the region.

COMMON CANOPY LEAF TYPES AND THE FORESTS WHERE YOU MAY FIND THEM

Figure 7 presents the most common leaf types from around the world used in describing different types of forest canopy and some of the habitats where they are prominent. This figure does not take into account the many types of leaves of understory plants such as grasses, sedges, ferns, forbs, and succulents.

CLIMATE DESCRIPTIONS AND GRAPHS

Throughout the book, the habitat descriptions include a brief overview of the climate with the Köppen Climate code and Walter-Lieth climate graphs, which are a powerful tool when combined. When using the climate graphs, the reader has to understand that one climate does NOT mean one vegetation. Areas often contain a mosaic of different habitat types and even biomes within the one climate classification; also, some vegetation types can exist over a range of rainfalls and temperatures. That said, an understanding of the exemplar climate in which a habitat occurs does help the reader better understand the habitat.

In sidebars sprinkled throughout the book, we provide a deeper look into the relationships between vegetation and latitude and climate, the interaction of general climate with fire, the influence of drought on vegetation, and anthropogenic effects on habitat modification. Looking at habitat distributions and their relationships to not only temperature and rainfall but also the distribution of rainfall through the year, it became apparent that this annual rainfall distribution is often a more important factor in vegetation type than average precipitation alone. To help illustrate these variations through the year, we have created climate graphs for each habitat based on the original work of Walter and Lieth, though we have heavily modified them to make them easier to read and interpret.

Reading these graphs may seem intimidating at first, but when their relevance is explained, they become more scrutable (see next section for examples). When temperature and precipitation are plotted together, and where each 0.8 in. (20 mm) of precipitation is compared with each 50°F (10°C), some really interesting patterns emerge. When the precipitation plot drops below the temperature plot, the area is in a period of water stress (drought) because transpiration rates (the rate at which

LEAF SHAPE		LEAF NAME	HABITATS
G Y M N O S P E R M S		Conifer Lobe Flat, lobed, evergreen	Temperate forests, mixed conifer/broadleaf forests
		Conifer Needle Thin linear leaves. Usually evergreen.	Boreal conifer forests, dry conifer forests
A N G I O S P E R M S		Deciduous Broadleaf Broad, thin leaves that grow quickly and last one season.	Temperate deciduous forests, wet/dry deciduous forests
		Evergreen Broadleaf Broad, thin, often with drip tips. They last a long time.	Rainforests, cloud forests
		Sclerophyllous Evergreen Thick, leathery leaves resist transpiration and fires.	Eucalypt forests, sclerophyll forests, heathlands, maquis, fynbos, mallee, mulga, matorral, cerrado
		Microphyllous Small Leaves that resist transpiration.	Acacia savanna, thornscrub, Chaco seco, desert scrubs

Fig. 7. Dominant leaf types and their global habitats. Note that most leaf types can be found in almost all habitats in very small numbers.

plants lose water) are higher than precipitation levels. We have colored these drought periods in orange. When the precipitation plot lifts above the temperature line, the area has a surplus of water, and plant growth is strong; these periods are colored light blue. However, once the precipitation exceeds 4 in. (100 mm) a month, there is an extreme surplus of water, and regardless of the temperature, most water runs off and is not used by plants; we have colored these periods in dark blue. Because the whole method makes sense only when used with the metric system, we have included temperature only in Celsius and rainfall in millimeters on the graphs.

KÖPPEN CLIMATE EXPLAINED

The Köppen system is the most widely used global method to classify and categorize different climatic regions. It is a simple three-letter code.

The first letter denotes the average temperature (B is the exception):

A: Tropical climates, with year-round average temperatures above 64°F (18°C)
B: Arid climates with low precipitation
C: Midlatitude climates with mild to cool temperatures
D: Midlatitude climates with cold winters and mild to cool summers
E: Polar or alpine climates with extremely cold temperatures

The second letter denotes when most precipitation occurs:

f: Year-round rainfall pattern, with precipitation evenly distributed throughout the year
m: Monsoonal, with a pronounced wet season and a dry season
w: Dominant dry winter season
s: Dominant dry summer season
t: Lacks a true summer

The third letter denotes maximum and minimum temperatures or hot/cold desert:

a: Hot summers, with the warmest month having an average temperature above 71.6°F (22°C)
b: Mild summers, with the warmest month averaging below 71.6°F (22°C) but above 50°F (10°C)
c: Cool summers, with the warmest month averaging below 50°F (10°C) but above 32°F (0°C)
d: Very cold winters, with the coldest month averaging below 32°F (0°C)
e: Cold summers, with the warmest month averaging below 50°F (10°C)
h: Hot Desert
k: Cold Desert

Africa encompasses a wide range of temperature zones, from the temperate area north of the Sahara (in the Northern Hemisphere) through the tropical zones of central and east Africa to the southern temperate zone of South Africa's Western Cape, with its Southern Hemisphere Mediterranean climate. Simultaneously, there are also varying rainfall regimes, resulting in everything from cold deserts to hot rainforests, and pretty much everything in between. When used in conjunction with the climate graphs explained earlier, these Köppen codes can explain why most habitats occur where they do.

1. Tropical Humid (**Afa**, **Afb**): These areas have high rainfall and high temperatures throughout the year. Habitats are moist forests. Mainly in the equatorial regions of West Africa, the Congo Basin, and the Indian Ocean seaboard and islands.
2. Tropical Monsoonal (**Awa**, **Awb**): These regions have distinct wet summer and dry winter seasons. Some locations would have sufficient precipitation to support rainforest if the rain were distributed more evenly throughout the year. The main habitats here are moist

savannas, tropical grasslands, and monsoon forests. They are prevalent in East Africa, West Africa, and the subtropical parts of Southern Africa.

3. Mediterranean (**Csa, Csb**): These habitats are both boreal and austral, occurring in temperate zones on either extremity of the continent, in the Maghreb of nw. Africa and the Cape provinces of South Africa. They are categorized by hot, dry summers and mild, wet winters. The main habitats in n. Africa are wooded heathlands and dry forest, such as MAGHREB MAQUIS, oak forest, and MAGHREB GARRIGUE. In Southern Africa there are heath habitats such as FYNBOS, RENOSTERVELD, and STRANDVELD.
4. Desert (**Bwh, Bwk**): These regions are characterized by extremely low annual precipitation, high temperatures, and little vegetation. They include the barren deserts such as Reg (rock), Hamada, and Erg (sand) deserts and shrub deserts. They are mainly found in the Sahara, Horn, and Namib.
5. Semiarid (**Bsh, Bsk**): These areas receive more rain than deserts but not enough rain in the wet season to allow the development of lush savannas. The main habitats are shrubland, dry grasslands, and thorn savannas. They are found in the Sahel region south of the Sahara, in patches through East Africa and in sw. Africa.

In figure 8 the Köppen climate for the recent past is presented against the projected Köppen climate for Africa and Europe late this century. Although climate change is of great concern, it is beyond the scope of this book to predict future habitats related to these changes in Africa. What this prediction does show is that Africa will be less affected than Europe.

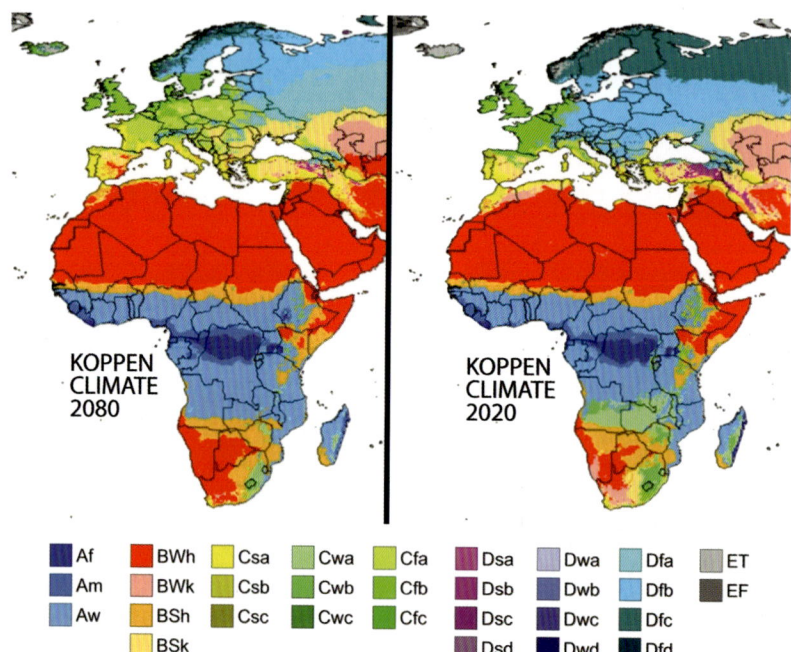

Fig. 8. Köppen Climate Maps 2080 and 2020.

KÖPPEN CLIMATE EXPLAINED   19

MONSOON FOREST: Awa

> Temperature hot much of the year
> Dry conditions during an extensive dry season
> Very intense monsoonal rains
> Significant overlap with savanna climate

Fig. 9. Sample climate graphs.

SAVANNA: Awa, Awb

> Temperature hot throughout the year
> Long dry season
> Very intense monsoonal rains

LOWLAND RAINFOREST: Afa, Afb

> Temperature hot throughout the year
> Abundant precipitation throughout the year

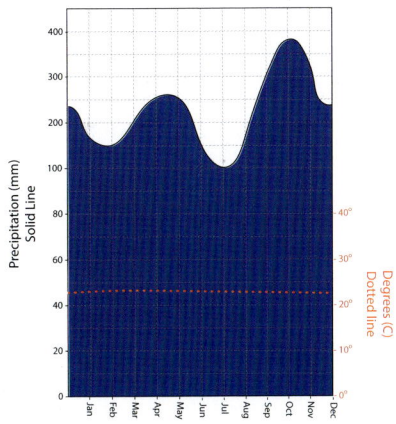

INTRODUCTION

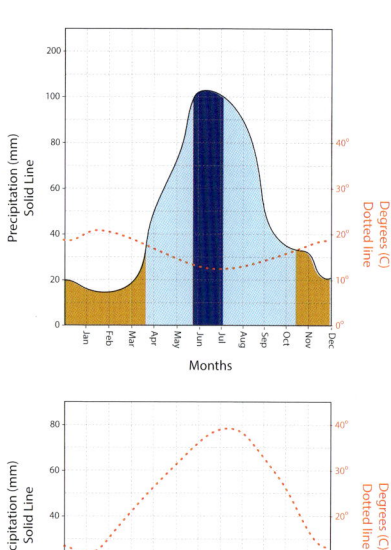

Fig. 9. Sample climate graphs *(continued)*

MEDITERRANEAN SCRUB (INCLUDING FYNBOS): Csa, Csb

> Cold winters, warm summers
> Wet winters
> Moderately dry summers

WARM DESERT: Bwh

> Dry and warm or hot throughout the year

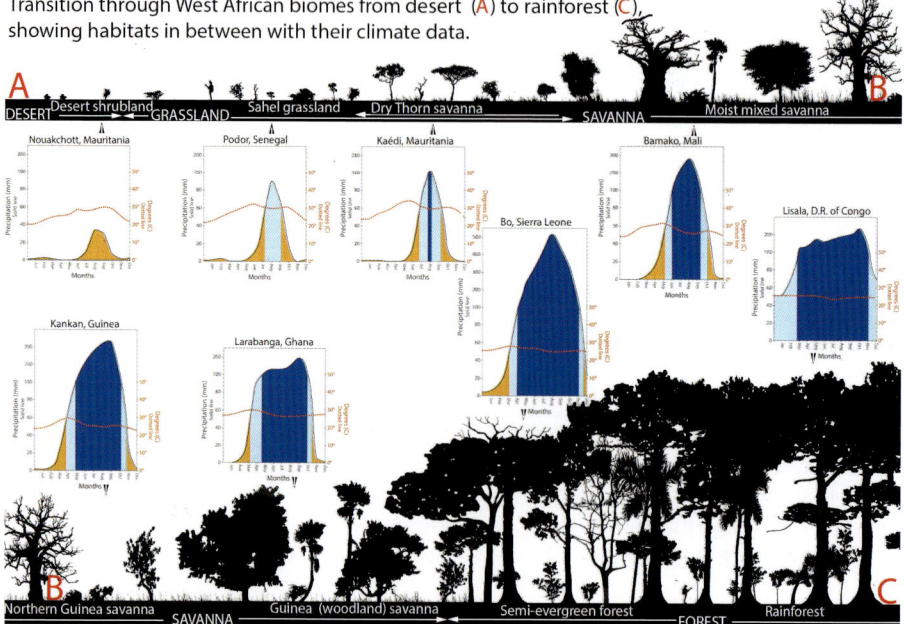

Transition through West African biomes from desert (A) to rainforest (C), showing habitats in between with their climate data.

Fig. 10. The transition from the Sahara to the rainforests of West Africa. Habitats can be predicted from the climate graphs at the extremes, but the different habitats in the intervening savanna regions have similar climates, suggesting that here other factors such as soil, fire, and herbivory determine habitat types.

BIOMES OF AFRICA   21

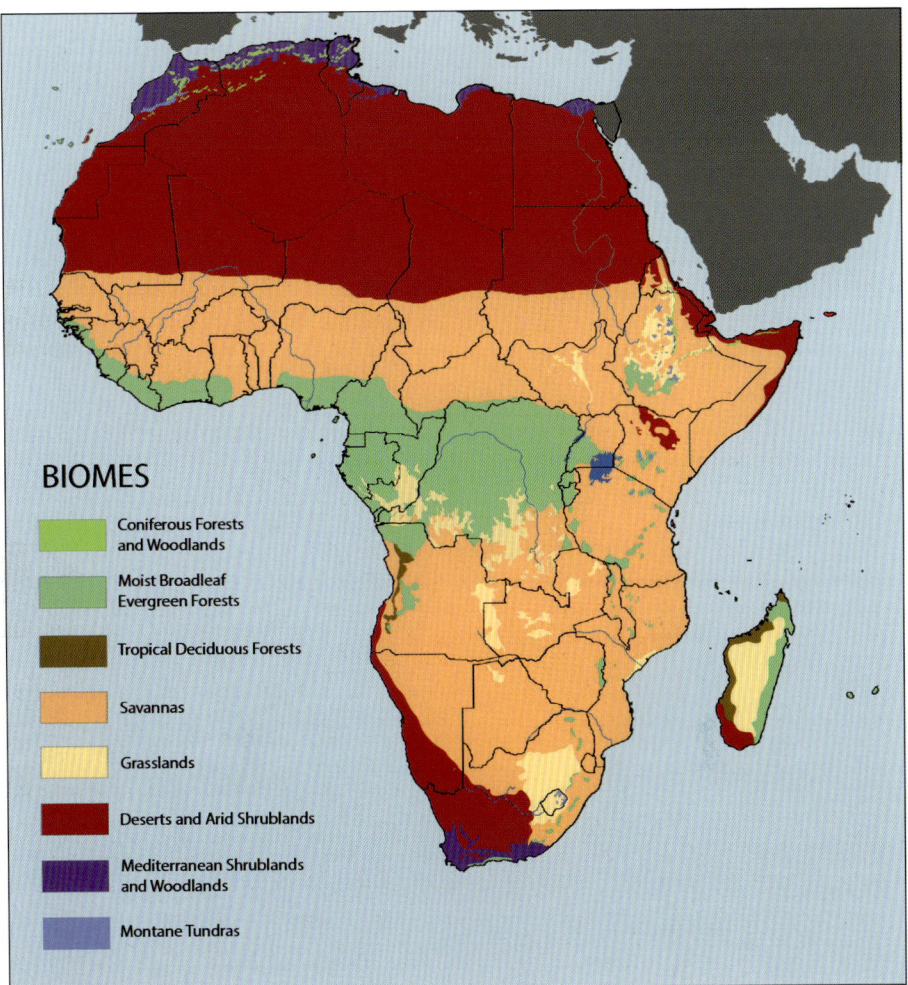

Fig. 11. Biomes of Africa.

BIOMES OF AFRICA

The broad habitat categories and subcategories used in this book are briefly explained in this section; the example habitats listed can be located in the table of contents. Refer to Useful Habitat Jargon, below, for further explanation of most of the terms used here. Note that the color coding used here corresponds with that used in the maps throughout the book. These descriptions broadly correspond to the major ecological community types known as "biomes."

CONIFER FORESTS: Forests made up of coniferous trees (which generally don't seasonally lose their leaves, with the exception of larches and a few others).

- **Dry Conifer Forest:** Forests and woodlands in dry temperate and tropical areas.
  Example: **Maghreb Pine Forest**

DESERTS AND ARID SCRUBS: Arid areas with little plant growth.

- **Barren Desolate Desert:** Harshest of deserts, where little grows, with areas of rock and sand dunes.
  Example: **Saharan Erg Desert**
- **Hot Desert:** Tropical/subtropical deserts with hot summers and cool or warm winters.
  Example: **Saharan Reg Desert**
- **Desert and Semidesert Shrubland:** Open arid areas with small shrubs with generally small leaves; cacti or euphorbias can be present and may be large.
  Example: **Maghreb Hot Shrub Desert**
- **Desert and Semidesert Thornscrub:** Arid areas with thickets of tall thorn bushes and little grass growth.
  Example: **Afrotropical Hot Shrub Desert**

TROPICAL HUMID FORESTS: Quintessential warm and wet rainforest-type environments.

- **Lowland Rainforest:** Wet, evergreen tall forest with thick full canopy cover and open undergrowth.
  Example: **Afrotropical Lowland Rainforest**
- **Semi-Evergreen Forest:** Generally humid forests with near-complete canopy cover in which a minority of the trees lose their leaves.
  Example: **East Coast Forest Matrix**
- **Montane/Subtropical Evergreen Forest:** Warm, wet forests in montane and/or subtropical areas, with almost closed canopy of evergreen or partially deciduous trees.
  Example: **Moist Montane Forest**

DRY DECIDUOUS FORESTS: Warm forests that lose most of their leaves in dry periods.

- **Closed Deciduous Forest:** Closed-canopy, dry deciduous forests that in summer appear lush, but many trees lose their leaves in winter. Fire-intolerant.
  Example: **Malagasy Deciduous Forest**

SAVANNAS AND SEASONALLY MOIST SHRUBLANDS: Habitats with an open canopy, lots of grass or shrubbery, and a strongly seasonal (usually wet-summer/dry-winter) climate. Most habitats in this category are heavily influenced by fire.

- **Open Broadleaf Woodland:** Non-sclerophyllous woodlands that can have tall trees and open canopy cover. Can have many deciduous trees. Fire-tolerant.
  Example: **Miombo**
- **Thorn Savanna:** Open (tall or short) woodlands with lots of grass cover. Trees are often dominated by acacias (of various genera), many of them with thorns.
  Example: **Kalahari Dry Thorn Savanna**

GRASSLANDS AND STEPPES: Habitats that are dominated by grasses with or without shrubs and flowers, and few or no trees. Fire-dependent.

- **Tropical Grassland:** Grasslands that remain warm or hot year-round. Grasses can grow throughout the year but are dependent on rain. Fire-tolerant.
  Example: **Afrotropical Grassland**
- **Montane Grassland:** Grasslands in highlands, often receiving orographic rainfall. Fire-tolerant.
  Example: **Montane Grassland (e.g., Highveld)**

- **Flooded Grassland:** Grasslands that spend most of the year as lush grasslands but turn into huge wetlands during the wet season.
  Example: **parts of Afrotropical Grassland**

MEDITERRANEAN FORESTS, WOODLANDS, AND SHRUBLANDS: Thick scrub in areas with climates with cold, wet winters and hot, dry summers.

- **Maquis, Chaparral, and Matorral:** Low shrubland that can be either closed or open. Dominated by fire and grazing. Plants are often similar to those of nearby forests.
  Example: **Maghreb Maquis**
- **Mediterranean Heathland:** Low heathlands with nearly 100% ground cover of sclerophyllous bushes and forbs. Fire-dependent.
  Example: **Fynbos**

TUNDRAS: Very low vegetation dominated by mosses and lichens. Found at extreme latitudes or elevations, where temperatures, snow cover, or exposure to wind prohibit the growth of trees.

Example: **Afroparamo**

FRESHWATER HABITATS: Habitats whose most important aspect is their inundation with fresh water.

- **Swamp Forest:** Forested habitats that are seasonally or permanently inundated.
  Example: **Swamp Forest**
- **Freshwater Wetland:** Nonforested habitats whose most important aspect is that they are seasonally or permanently flooded.
  Example: **Afrotropical Deep Freshwater Marsh**

SALT-DOMINATED HABITATS: Habitats where the dominant force is the presence of high levels of salt in the water or soil.

- **Salt Pan:** Areas in which evaporation or volcanic activity have produced extremely high salt concentrations in the soil. Mostly unvegetated, though algae grows quickly when floods occur.
  Example: **Salt Pans and Lakes**
- **Mangrove:** A specialized forest that grows in tidally flooded coastal areas.
  Example: **Afrotropical Mangrove**
- **Salt Marsh:** Salt-tolerant marsh vegetation that grows in sheltered coastal areas that are periodically flooded with seawater.
  Example: **Salt Marsh**
- **Tidal Mudflat:** Nutrient-rich areas of mud that are frequently flooded with seawater, usually in estuaries.
  Example: **Tidal Mudflats and Estuaries**
- **Rocky Coastline and Sandy Beach:** Nutrient-poor sandy and rocky beaches, cliffs, and other coastline types.
  Example: **Sandy Beach and Dunes**
- **Pelagic Waters:** Marine environments with deep water.
  Example: **Pelagic Waters**
- **Offshore Islands:** Small islands that are well offshore and that support a low growth of grass and/or shrubs.
  Example: **Offshore Islands**

ANTHROPOGENIC HABITATS: The primary force shaping these habitats is the presence of humans.
- **Grazing Lands:** Areas that are heavily grazed by domestic animals.
- **Cultivated Lands:** Areas cultivated by humans for the production of crops.
- **Human Habitation:** Areas directly inhabited by humans.

TAXONOMY

For birds we follow the eBird-Clements taxonomy. It is up to date and carefully maintained and is the most popular global taxonomy for birders. For mammals, reptiles, amphibians, insects, and plants, we mainly follow Wikipedia and/or iNaturalist. This is sure to shock some scientists and purists, but in writing this book, we found these sources to be accurate and up to date for the groups that we know intimately well, giving us confidence that other groups are covered similarly. In an effort to keep the text flowing, we include scientific names for only plants and some herps, since this is how plants are normally discussed even by amateur naturalists.

ENDEMIC BIRD AREAS

Endemic Bird Areas (EBAs), first identified in 1987 by Birdlife International, are defined as areas that contain two or more bird species with restricted ranges. A secondary EBA is one that contains the range of a single species. Range-restricted species are defined as those with a breeding range of less than 19,300 mi.^{2} (50,000 km^2), as recorded historically, that is, since 1800. If at some point after 1800 the species had a breeding range larger than this, it is not considered a range-restricted species. The size of each EBA is flexible and is dictated by the ranges of the species contained therein.

While the identification of an EBA is a valuable tool in pinpointing areas of endemism, it doesn't show the whole picture, especially for the habitat-based approach used in this book. In the EBA designation, there is a natural bias toward island species, the ranges of which are intrinsically restricted. The same holds true for continental species restricted to montane environments. Despite a bird species being restricted to a single continental lowland habitat, it can have a large distribution that is more extensive than the measurable threshold of a range-restricted species, and the area in which it lives may not count as an EBA, even though it is a major area of endemism. Examples of such excluded habitats include the Karoo of Southern Africa and the Miombo woodland savanna belt. In addition, this book uses a dual bird and larger mammal lens, which obviously is different from that of the purely bird-driven EBA concept. The new Key Biodiversity Area (KBA) concept is one that combines the EBA with other biota to create a system of delineating conservation priorities. The KBA system ties in very well with the HotW system to highlight crucial areas in need of protection and create a uniform language for conservationists and ecologists.

USEFUL HABITAT JARGON

This section includes definitions that will help the reader to understand some of the most important terms used in naming and defining habitats. These are terms that appear over and over in the book.

- **Austral.** Southern Hemisphere.
- **Azonal.** A habitat whose occurrence is not directly related to climate, but is determined by local factors such as soil type or fire regime.
- **Boreal.** Northern Hemisphere.

- **Desert.** Very dry and either unvegetated or sparsely vegetated habitat.
- **Edaphic.** Relating to the soil. Usually used in a case where a habitat (most often grassland) is produced primarily by the lack of nutrients in the soil of a certain area.
- **Grassland.** Habitat dominated by grasses, with few shrubs or trees.
- **Halophytic.** Refers to a plant that can grow in highly saline environments.
- **Heath/Heathland.** Shrubland that is dominated by fine-leaved evergreen members of the erica family (Ericaceae). Moorland is a moist type of heath often at high elevation.
- **Rainforest.** Lush forest that receives abundant moisture.
- **Savanna.** A variably wooded tropical grassland with prominent wet and dry seasons.
- **Sclerophyll.** Plants with hard desiccation-resistant leaves, often small leaves, but not exclusively.
- **Semi-evergreen forest.** Forest that has rainforest structure, but in which some trees lose at least some of their leaves at some point during the year.
- **Wetland.** Habitat that is frequently or permanently flooded.
- **Woodland.** Habitat with abundant trees, forming a nearly interlocking canopy, but in which sun still reaches the ground, allowing the growth of grass. Broadly overlaps with Savanna.
- **Xeric.** Refers to a dry environment with little moisture.
- **Xerophytic.** Refers to a plant that needs very little water and can grow in xeric conditions.
- **Zonal.** A habitat that is limited to one climatic zone.

ABOUT THIS BOOK

Each of the habitat accounts includes the following sections, which are briefly explained here:

In a Nutshell: A succinct explanation of what makes a habitat distinctive and worthy of separation from other habitats.

Global Habitat Affinities: Habitats from other continents that are structurally and climatically similar. This is a way of cross-referencing similar habitats on other continents. Perhaps one of the habitats mentioned will be familiar to you, helping you to understand the unfamiliar habitat covered.

Continental Habitat Affinities: Habitats from elsewhere in Africa that are structurally and climatically similar.

Species Overlap: The habitats that have the most similar assemblages of mammals and birds. These are ranked from the most similar habitat down. The vast majority of these are habitats within the same zoogeographic region as the habitat covered.

Habitat Silhouette: These silhouettes are designed to give a quick visual snapshot of a habitat, showing some of its distinctive plant shapes and its overall height and structure. They include a human silhouette for scale. These diagrams are obviously simplifications, especially in the case of habitats with a huge range of variation, such as Africa's dry thorn savanna.

Range Map: These are visual representations of a habitat's occurrence. Dark shading is used for areas where the habitat is the predominant habitat or one of the predominant habitats. In some maps, pale shading is used to indicate areas where the habitat is found only locally or predicted to occur in regions that are poorly known, for example, Sudan.

Description: This section explains what makes a habitat distinctive and how that habitat works. Some of the commonly included information is the height and composition of the various layers of vegetation, the overall "feel" and accessibility, and local temperature and rainfall. The accompanying climate graphs are discussed in Climate Descriptions and Graphs, above. In these descriptions, we have purposefully chosen not to always include exactly the same information or to put it in the same order. This approach both allows us to stress what is most important about a given habitat and simply to vary these sections to keep them interesting for readers.

Conservation: This section provides a quick summary of the conservation status of the habitat and major issues it is facing.

Wildlife: This section may be the most interesting for a typical reader. Beyond the nuts and bolts of what makes a habitat distinctive, and what makes it work, most visitors are keen to learn about and to find its wildlife. Throughout this book, when considering wildlife, larger mammals and birds are our primary focus, but in many accounts we go well beyond this to feature a broad array of vertebrates. Species that are restricted to a certain habitat (endemics) are given special weight, as finding these will be the priority for many visitors. A species that is an indicator species for that habitat has "(IS)" beside its name.

Endemism: When included, this section talks about a habitat's degree of endemism, as well as endemic hotspots *within* a given habitat. Some habitats, especially montane ones or those occurring on islands, host many distinct nodes of endemism and are given their own Endemism section. For many habitats, if there are no major zones of endemism, the details are described in the main Description section or are incorporated into the Wildlife section.

Distribution: This section, and the accompanying range map, indicate where the habitat occurs within a given zoogeographic region. The elevations at which it is found are sometimes mentioned, though this information may also be in the Description.

Where to See: These are places that you can visit to experience a given habitat. In general, these are the most readily or frequently visited places, in the most accessible country or countries.

Photos: Photos are included that illustrate both the habitat itself and some of its charismatic wildlife. Some photos are chosen since they effectively show both the habitat and some of its wildlife.

Sidebars: Throughout the book there are sidebars that discuss aspects of a habitat, biome, or region—in some cases these discussions are slightly more tangential, in others more in-depth. Many of these are about geology, ecology, and climate. We have chosen to place this sort of information in sidebars to make it more accessible and relevant (rather than in long and dry introductory sections that are likely to be ignored by most readers!).

ACKNOWLEDGMENTS

Many thanks to our review readers Gray Tappan and Odette Curtis-Scott. Gray also went well "above the call of duty" in becoming a book contributor by reworking our mapping for North and West Africa, as well as by finalizing the formatting of the regional maps. Many thanks to Pablo Cervantes who created the climate diagrams, dendrograms, and some other graphics that greatly enhanced the visual impact of this book. Giselle Velastegui and Pablo Cervantes played key roles in the creation of the habitat range maps. The following folks helped clarify our understanding of various African habitats: Frank Willems, Odette Curtis-Scott, Nik Borrow, Adam Scott Kennedy, Kristal Maze, Phillip Desmet, Cliff Dorse, Ross Wanless, Don Faber-Langendoen, and Dale Forbes. Dayne Braine provided timely and targeted information about African herps, which allowed us to do more justice to this wonderful group of animals. The team at Princeton University Press were excellent as always; thanks to Robert Kirk, Lisa Black, Megan Mendonça, Mark Bellis, and our copyeditor Eva Silverfine. Thanks also to our excellent typesetter, D & N Publishing. All three authors thank their families for their patience and understanding during the long hours, days, and months that it took to bring this book to fruition. Keith would like to thank Professor William Bond for teaching him how to think more critically.

ABBREVIATIONS

Directions (north, south, east, west, central) are abbreviated only when they directly precede a geographical place name.

c.	central	IS	indicator species	oz.	ounce
cm	centimeter	KBA	Key Biodiversity Area	p./pp.	page/pages
e.	east/eastern	kg	kilogram	s.	south/southern
DRC	Democratic Republic of the Congo	km	kilometer	sc.	south-central
		lb.	pound	se.	southeastern
EBA	Endemic Bird Area	m	meter	spp.	species (plural)
ft.	foot/feet	mi.	mile	sw.	southwestern
g	gram	mm	millimeter	vs.	versus
ha	hectare	MYA	million years ago	w.	west/western
HotW	Habitats of the World	n.	north/northern	YBP	years before present
hr	hour	ne.	northeastern		
in.	inch/inches	nw.	northwestern		

SELECTED BIBLIOGRAPHY

We are indebted to a great many wonderful books and papers on the habitats, plants, and wildlife of Africa and the world. Although we thought about referencing these individually, in academic fashion, we decided that doing so would unduly disrupt the flow of a book, which is intended to be accessible to all. Below we mention some of our most important reference material, but note that this is not intended as a comprehensive bibliography.

Acocks, J.P.H. 1988. *Veld Types of South Africa*. 3rd ed. Pretoria: Botanical Research Institute.
Alexander, G., and J. Marais. 2007. *A Guide to the Reptiles of Southern Africa*. Cape Town: Struik.
Barnes, K., T. Stevenson, and J. Fanshawe. 2024. *Field Guide to Birds of Greater Southern Africa*. London: Bloomsbury.
Behrens, K., and K. Barnes. 2016. *Wildlife of Madagascar*. Princeton, NJ: Princeton University Press.
Birds of Africa. 1986–2004. *Helm Field Guides*. 7 vol. Dallas, TX: Helm.
Bond, W. J. 2019. *Open Ecosystems: Ecology and Evolution beyond the Forest Edge*. Oxford: Oxford University Press.
Branch, B. 1998. *Field Guide to Snakes and Other Reptiles of Southern Africa*. 3rd ed. Cape Town: Struik.
Campbell, I., K. Behrens, C. Hesse, and P. Choan. 2021. *Habitats of the World: A Field Guide for Birders, Naturalists and Ecologists*. Princeton, NJ: Princeton University Press.

Clements, J. F., P. C. Rasmussen, T. S. Schulenberg, M. J. Iliff, T. A. Fredericks, J. A. Gerbracht, D. Lepage, et al. 2023. *The eBird/Clements Checklist of Birds of the World*, v2023b. https://www.birds.cornell.edu/clementschecklist/download/.

du Preez, L., and V. Carruthers. 2018. *Frogs of Southern Africa: A Complete Guide*. 2nd ed. New York: Penguin Random House South Africa.

Faber-Langendoen, D., T. Keeler, D. Meidinger, C. Josse, A. Weakley, D. Tart, G. Navarro, et al. 2016. *Classification and Description of World Formation Types*. General Technical Report RMRS-GTR-346. Fort Collins, CO: US Department of Agriculture, Forest Service, Rocky Mountain Research Station.

Glaw, F., and M. Vences. 2007. *A Field Guide to the Amphibians and Reptiles of Madagascar*. 3rd ed. Germany: Vences and Glaw Verlag GbR.

Goodman, S. M., and J. P. Benstead, eds. 2003. *The Natural History of Madagascar*. 1st ed. Chicago: University of Chicago Press.

Jenkins, M. D., ed. 1987. *Madagascar: An Environmental Profile*. Gland, Switzerland: International Union for Conservation of Nature.

Kingdon, J. 2007. *The Kingdon Field Guide to African Mammals*. London: Bloomsbury.

Levin, S., ed. 2000. *Encyclopedia of Biodiversity*. 7 vol. Cambridge, MA: Academic Press.

Low, A. B., and A. G. Rebelo. 1998. *Vegetation of South Africa, Lesotho and Swaziland*. Pretoria: Department of Environmental Affairs and Tourism.

Mucina, L., and M. C. Rutherford. 2006. *The Vegetation of South Africa, Lesotho and Swaziland*. Strelitzia 19. Pretoria: South African National Botanical Institute.

NatureServe. "Map of Potential Distribution of Vegetation Macrogroups of Africa for Ecosystem Assessment, Planning, Management, and Monitoring." https://www.natureserve.org/projects/map-potential-distribution-vegetation-macrogroups-africa.

OneEarth. "Afrotropics." https://www.oneearth.org/realms/afrotropics/.

SANBI (South African National Biodiversity Institute). "National Vegetation Map." https://www.sanbi.org/biodiversity/foundations/national-vegetation-map/.

Sayre, R., P. Comer, J. Hak, C. Josse, J. Bow, H. Warner, M. Larwanou, et al. 2013. *A New Map of Standardized Terrestrial Ecosystems of Africa*. 2013. Washington, DC: Association of American Geographers.

Shugart, H., P. White, S. Saatchi, and J. Chave. 2022. *The World Atlas of Trees and Forests*. Princeton, NY: Princeton University Press.

Sinclair, I., and P. Ryan. 2010. *Birds of Africa South of the Sahara*. 2nd ed. New York: Penguin Random House South Africa.

Spawls, S., K. Howell, R. C. Drewes, and J. Ashe. 2001. *Field Guide to the Reptiles of East Africa: All the Reptiles of Kenya, Tanzania, Uganda, Rwanda and Burundi*. Princeton, NJ: Princeton University Press.

Tappan, G., W. M. Cushing, S. E. Cotillon, J. A. Hutchinson, B. Pengra, I. Alfari, E. Botoni, A. Soulé, and S. M. Herrmann. 2016. *Landscapes of West Africa: A Window on a Changing World*. Garretson, SD: US Geological Survey.

Walter, H. K., E. Harnickell, F.H.H. Lieth, and H. Rehder. 1967. *Klimadiagramm-weltatlas*. Jena: Fischer.

White, F. 1983. *The Vegetation of Africa*. Paris: United Nations Educational, Scientific, and Cultural Organization.

WWF (World Wildlife Fund). "Ecoregions." https://www.worldwildlife.org/biomes.

SIDEBAR 1 — AFRICAN ENDEMIC BIRD FAMILIES

There are 25 bird families endemic to the continent and associated islands. Four of these are found only on Madagascar and one more on Madagascar and the Comoros. This leaves 20 families endemic to continental Africa.

Left: **Guineafowl, a small family of chunky chicken-like birds. Helmeted Guineafowl.** © KEN BEHRENS, TROPICAL BIRDING TOURS

Right: **Turacos, colorful and highly vocal, a bit like the guans and chachalacas of the Neotropics. Knysna Turaco.** © KEN BEHRENS, TROPICAL BIRDING TOURS

Below: **Flufftails are small and incredibly secretive rail-like birds. Madagascar Flufftail.** © KEN BEHRENS, TROPICAL BIRDING TOURS

Left: **Egyptian Plover, a monotypic family of n. sub-Saharan Africa, is not a plover and is no longer found in Egypt!** © KEN BEHRENS, TROPICAL BIRDING TOURS

Right: **Mousebirds, a small family of highly social, puffy, long-tailed birds. White-backed Mousebird.** © KEITH BARNES, TROPICAL BIRDING TOURS

Left: **Secretarybird, a well-known monotypic family, distantly related to raptors.** © KEN BEHRENS, TROPICAL BIRDING TOURS

(continued overleaf)

| SIDEBAR 1 | *AFRICAN ENDEMIC BIRD FAMILIES* (continued) |

Left: **Woodhoopoes and scimitarbills,** dark, long-tailed, long-billed birds with iridescent highlights. Green Woodhoopoe. © KEN BEHRENS, TROPICAL BIRDING TOURS

Above: **Ground hornbills,** a family of two species; massive ground-striding birds. Southern Ground Hornbill. © KEN BEHRENS, TROPICAL BIRDING TOURS

Left: **African barbets** vary from tiny to hefty and are boldly marked. Pied Barbet. © KEN BEHRENS, TROPICAL BIRDING TOURS

Right: **Bushshrikes** are a large family that includes some of the most delightful African birds, strikingly colored and with huge vocal repertoires. Gray-headed Bushshrike. © KEN BEHRENS, TROPICAL BIRDING TOURS

Above: **Wattle-eyes and batises** are small, mostly strikingly black-and-white birds with piping vocalizations. White-tailed Shrike. © KEN BEHRENS, TROPICAL BIRDING TOURS

Rockjumpers, a small family of colorful birds with two members, both endemic to South Africa and Lesotho. Drakensberg Rockjumper. © KEN BEHRENS, TROPICAL BIRDING TOURS

Left: **Rockfowl,** a family with two members. Secretive, mysterious, cave- and boulder-nesting rainforest birds. White-necked Rockfowl. © KEITH BARNES, TROPICAL BIRDING TOURS

Hyliotas are a small family of somewhat generic warbler-like birds. Southern Hyliota. © KEN BEHRENS, TROPICAL BIRDING TOURS

Nicators are a small family of greenbul-like birds with loud ringing songs. Yellow-throated Nicator. © KEN BEHRENS, TROPICAL BIRDING TOURS

Left: **African warblers** comprise a small but remarkably diverse set of birds, ranging from tiny crombecs to chunky robin-like species. Rockrunner. © KEN BEHRENS, TROPICAL BIRDING TOURS

Right: **Oxpeckers** are a small family with two members; well-known safari birds. Red-billed Oxpecker. © KEN BEHRENS, TROPICAL BIRDING TOURS

Left: **Modulatricidae** is a small, recently recognized family of incredibly secretive birds that live in the understory of montane forest. Spot-throat. © ADAM SCOTT KENNEDY

Right: **Sugarbirds**, another small family, has two members, endemic to Southern Africa. Long-tailed, nectar-eating birds. Cape Sugarbird. © KEN BEHRENS, TROPICAL BIRDING TOURS

Left: **Indigobirds and whydahs** are odd birds that lay their eggs in the nests of other birds. Indigobirds are blackish and generic, whereas male whydahs have long tails and bright plumage. Eastern Paradise-Whydah. © KEITH BARNES, TROPICAL BIRDING TOURS

SIDEBAR 2 — *INDIAN OCEAN ISLANDS ENDEMIC BIRD FAMILIES*

There are 25 bird families endemic to Africa and associated islands. Four of these are found only on Madagascar and one more on Madagascar and the Comoros.

Mesites are rail-like forest floor birds. Distinct and ancient, they constitute their own order; their closest relatives remain mysterious. This is a Brown Mesite. © KEN BEHRENS, TROPICAL BIRDING TOURS

Cuckoo-Roller is a large forest bird that gives a haunting call in a swooping display flight. Exact taxonomic affinities are mysterious, though it makes up its own order and qualifies as a sort of "living fossil"! Found on Madagascar and the Comoros.
© KEN BEHRENS, TROPICAL BIRDING TOURS

Above: Ground-Rollers are beautiful birds that are the ultimate prize for visiting birders. You can easily make a case for any of the five species as the "best of the lot"! This is a Rufous-headed Ground-Roller.
© KEN BEHRENS, TROPICAL BIRDING TOURS

Above right: Asities form an endemic family with ties to the African Grauer's Broadbill and the Asian broadbills. Four species in two very different genera. This is a Common Sunbird-Asity. © KEN BEHRENS, TROPICAL BIRDING TOURS

Right: Tetrakas long "hid" within widespread families, but recent DNA analyses revealed their status as an endemic family. Diverse members range from greenbul-like understory birds to warbler-like canopy species. This is Yellow-browed Oxylabes. © KEN BEHRENS, TROPICAL BIRDING TOURS

REGIONAL INTRODUCTIONS

NORTHEAST AFRICA

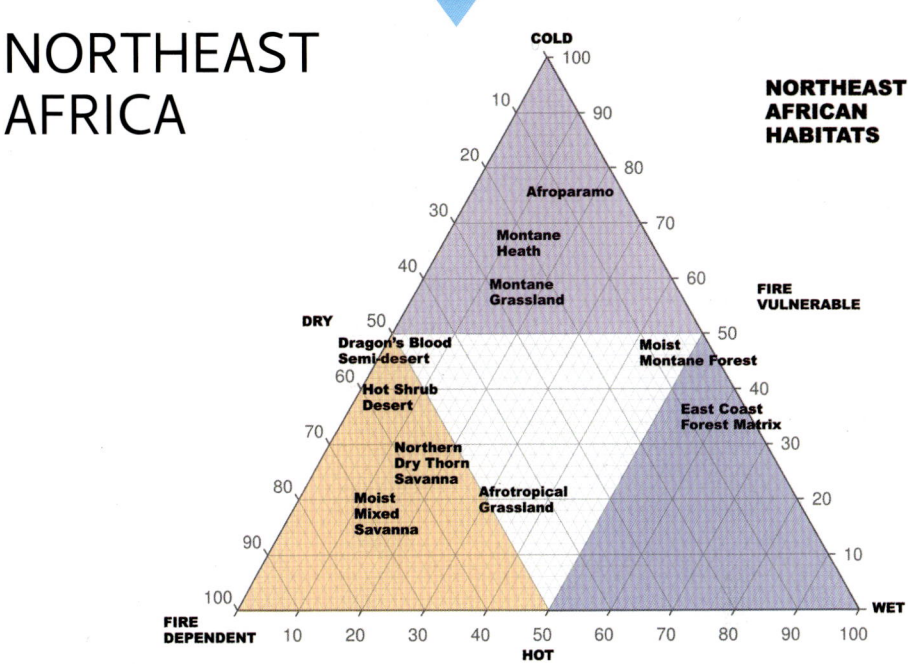

NORTHEAST AFRICAN HABITATS

This area, often referred to as the "Horn of Africa," is one of the smallest mainland regions, comprising Ethiopia, Eritrea, Djibouti, Somalia, and the Yemeni island of Socotra. At 750,000 mi.^{2} (1,950,000 km^2) this area is slightly larger than Alaska, or Spain and France combined. This region contains 31 habitats, of which one is endemic. In general, this is a very dry region (Köppen **Bwh**), in which desert and semi-desert climates are widespread. Most rain falls during the boreal summer, though some areas of the south experience two short annual rainy periods. Most of the eastern and northern part of this region, from n. Eritrea through Djibouti, well into Ethiopia's Great Rift Valley, and most of Somalia, is covered in a mix of dry thorny savanna and shrub desert. The Shebelle and Juba Rivers flow through s. Somalia, and support bands of lusher vegetation, which resemble MOIST MIXED SAVANNA. The long band of coastal forest running all the way north from s. Mozambique terminates in s. Somalia.

The Abyssinian highlands are this region's outstanding geographical feature. These rise in c. Ethiopia and extend north into Eritrea, with another long spur projecting east into n. Somalia. Remarkably, the central mountain block is completely split by the Great Rift Valley. This valley holds many lakes, including both freshwater lakes lined with rich wetlands as well as saline ones. The mountains are the main driver of diversity in ne. Africa. Historically there was much more forest and woodland, but intense human pressure has destroyed or degraded much of this. The most common treed habitat is dry montane woodland, in which conifers are common. The southern slopes also hold areas of MOIST MONTANE FOREST. The dominant montane habitats today are

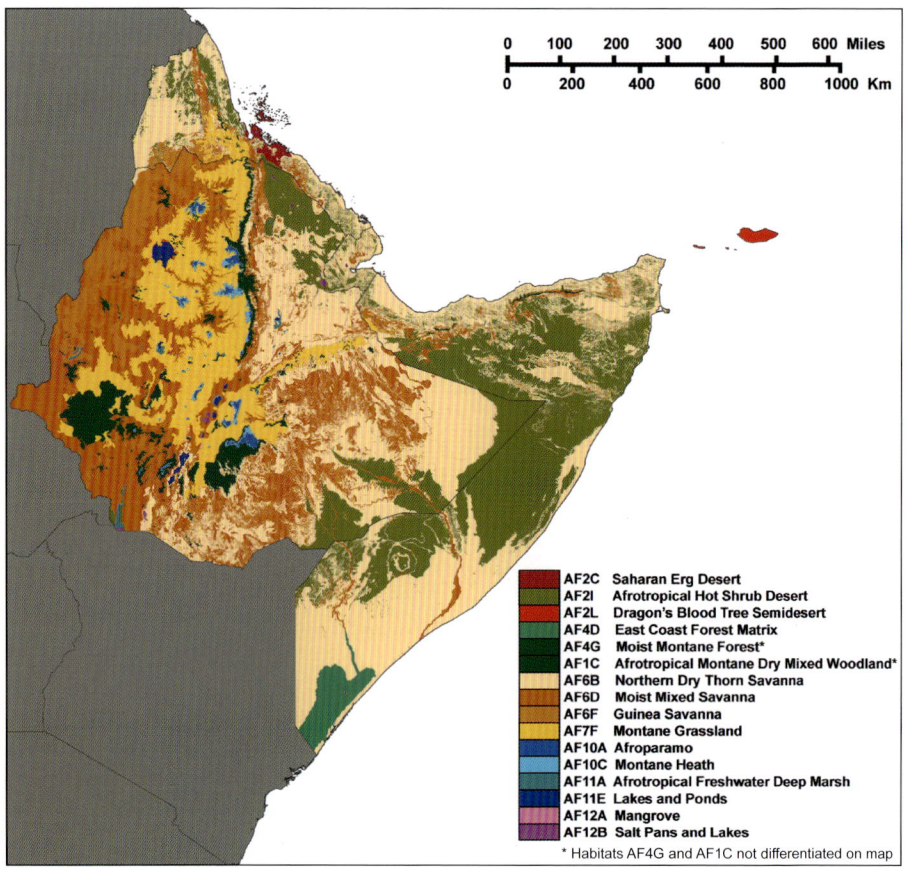

MONTANE GRASSLAND and cultivation, fields of wheat, barley, and tef that support Ethiopia's huge human population. The highest mountains in Ethiopia, which tower up to 14,930 ft. (4550 m) have a cold climate (Köppen **Cfc**) and usually have a band of heath vegetation, while their heights support AFROPARAMO, a sparsely vegetated alpine habitat. The mountains of n. Somalia have dry juniper-dominated woodland. The western lowlands of Ethiopia and Eritrea hold the easternmost tracts of GUINEA SAVANNA, a broadleaf woodland that strongly resembles the MIOMBO found farther south in the continent. Ethiopia's remote w. Gambela region holds vast freshwater wetlands.

Socotra is politically part of Yemen but biogeographically African. It has a distinctive type of dry habitat, which in this book we have called DRAGON'S BLOOD TREE SEMI-DESERT, named for its characteristic umbrella-shaped euphorbia trees.

Northeast Africa has two main endemic hotspots: the arid lowlands and the mountains. Despite their inhospitality, the lowlands have a rich set of birds, most of which are shared with Kenya and n. Tanzania, though some are endemic. The highlands are even richer. There are dozens of endemic birds, such as the Blue-winged Goose and Rouget's Rail, and some spectacular mammals like Ethiopian Wolf, Walia Ibex, and Mountain Nyala.

NORTHEAST AFRICAN HABITATS

CODE	CATEGORY/NAME	PAGES	CODE	CATEGORY/NAME	PAGES
	DRY CONIFERS			**FRESHWATER HABITATS**	
Af1C	Afrotropical Montane Dry Mixed Woodland	61–66	Af11A	Afrotropical Deep Freshwater Marsh	358–364
	DESERTS AND XERIC SHRUBLANDS		Af11B	Afrotropical Shallow Freshwater Marsh	365–369
Af2E	Saharan Erg Desert	94–100	Af11E	Freshwater Lakes and Ponds	379–381
Af2I	Afrotropical Hot Shrub Desert	121–126	Af11F	Rivers	382–385
Af2L	Dragon's Blood Tree Semi-Desert (Endemic)	133–137		**SALINE HABITATS**	
	TROPICAL AND SUBTROPICAL MOIST BROADLEAF FORESTS		Af12A	Afrotropical Mangrove	386–389
			Af12B	Salt Pans and Lakes	390–392
Af4D	East Coast Forest Matrix	157–163	Af12C	Tidal Mudflats and Estuaries	393–394
Af4G	Moist Montane Forest	174–179	Af12D	Salt Marsh	395–396
	SAVANNAS		Af12E	Sandy Beach and Dunes	397–399
Af6B	Northern Dry Thorn Savanna	227–233	Af12F	Rocky Shoreline	400–402
Af6D	Moist Mixed Savanna	243–251	Af12G	Offshore Islands	403–406
Af6F	Guinea Savanna	260–267	Af12H	Pelagic Waters	406–408
Af6G	Inselbergs and Koppies	268–274		**ANTHROPOGENIC**	
	GRASSLANDS		Af13B	Savanna Cultivation	412–415
Af7A	Afrotropical Grassland	280–286	Af13C	Tropical Montane Cultivation	415–417
Af7F	Montane Grassland	293–299	Af13F	Tree Plantations	422–423
	ALPINE TUNDRAS AND MONTANE HEATHS		Af13G	Cities and Villages	424–426
Af10A	Afroparamo	343–349	Af13H	Grazing Land	427–428
Af10C	Montane Heath	354–357			

EAST AFRICA

Although one of the smallest regions, made up of the countries of Uganda, Rwanda, Burundi, Kenya, and Tanzania, East Africa is incredibly rich in habitats and wildlife. This region is essentially the crossroads where all the other regions intersect. If you plan to visit Africa only once, this is the place to see the maximum diversity in a small area. At 700,000 mi.² (1,800,000 km²) this area is about the size of Alaska, or Spain and France combined. There are 34 habitats, none endemic, ranging from shrub desert in n. Kenya to lowland rainforest in w. Uganda. Part of what makes this region special is the complexity of its geography. There are several mountain ranges and isolated stand-alone mountains, as well as two branches of the Great Rift Valley. In combination with the complex dynamics of a tropical climate, this geography produces a complicated matrix of different habitats, often within a very small area. Mount Kilimanjaro is a good example—within a few hundred square miles, habitats range from semi-desert thornscrub up through MOIST MONTANE FOREST to extremely dry (Köppen **Cfc**) and nearly barren AFROPARAMO at the summit (19,341 ft./5895 m).

The east coast is covered in a complex matrix of moist savanna, semi-evergreen forest, and anthropogenic habitats. This coastal forest mix continues south to s. Mozambique and north to n. Somalia. A rich set of wildlife is restricted to this coastal belt, though many species are shared with adjacent regions.

There is a wide arid zone in ne. Tanzania and most of eastern and northern Kenya. The primary habitat here is arid thorn savanna, which is tied to that of n. Africa via Ethiopia. The driest parts of

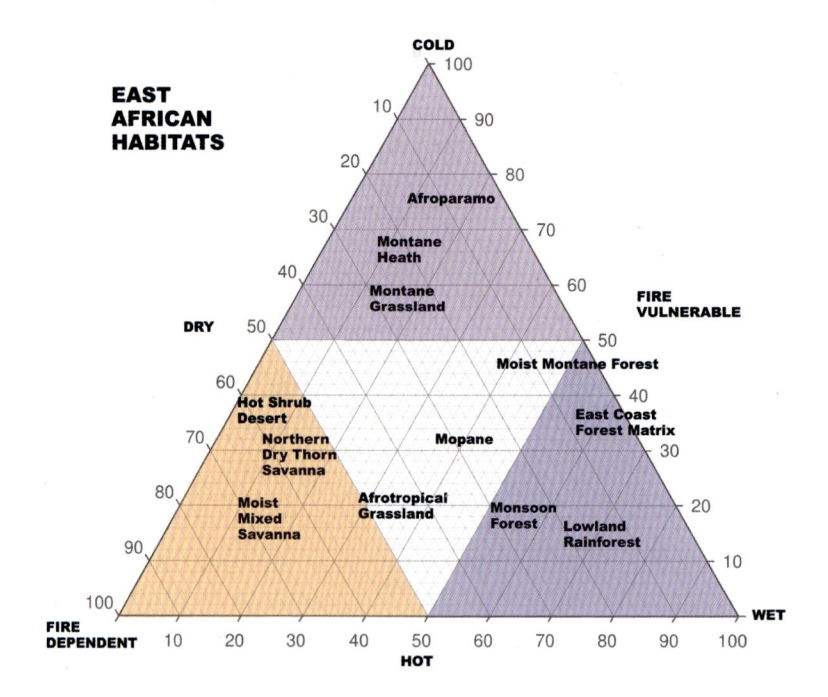

EAST AFRICAN HABITATS

[Ternary diagram with axes COLD (top), DRY (left) / FIRE DEPENDENT (bottom left), HOT (bottom), WET (bottom right) / FIRE VULNERABLE (right). Habitats labeled: Afroparamo, Montane Heath, Montane Grassland, Moist Montane Forest, Hot Shrub Desert, Northern Dry Thorn Savanna, Mopane, East Coast Forest Matrix, Moist Mixed Savanna, Afrotropical Grassland, Monsoon Forest, Lowland Rainforest.]

n. Kenya support hot shrub desert. Despite their superficial inhospitality, these areas are the most biologically rich arid areas of the continent: rich in mammals and birds that are shared with similar habitats in Northeast Africa.

Moister savannas cover much of Tanzania, Uganda, and sw. Kenya. There are two broadleaf savanna types: MIOMBO in Tanzania and GUINEA SAVANNA in Uganda, the farthest southeastern reaches of this primarily n. African habitat. Within Tanzania's "Miombo zone" moister areas such as drainage lines and areas with abundant groundwater support MONSOON FOREST, which resembles coastal forest, and provides a bridge between that habitat and the Congo Basin forests to the west. MOIST MIXED SAVANNA is interspersed throughout the region and is part of a complicated savanna mélange in places such as n. Tanzania, where Miombo and dry thorn savanna intermingle. East African savannas generally lack localized endemics but are incredibly rich in species shared with adjacent regions. The global universality of the Swahili word "safari" hints at the world-class nature of the mammal watching here. This part of the world includes one of the world's greatest mammal migrations: the annual movements of Blue Wildebeest and Plains Zebra through the Masai Mara and Serengeti of Kenya and Tanzania. The abundant savanna herbivores are naturally stalked by a full range of predators, including Cheetah, Lion, Leopard, and Wild Dog.

The moderate elevations of the northwestern portion of the region hold true rainforests, which are connected to the vast forests of the Congo Basin. The easternmost outpost of rainforest is Kakamega Forest in w. Kenya. Rainforests in Uganda hold tantalizing rainforest wildlife such as Chimpanzee and Green-breasted Pitta.

The most extensive mountain ranges are the Eastern Arc Mountains of c. Tanzania and the mountains along the Albertine Rift, from Uganda south to nw. Tanzania. There are also isolated mountains elsewhere in Uganda, Kenya, and n. Tanzania, including Kilimanjaro, the continent's tallest. Drier montane areas such as rain shadow slopes, often have dry mixed montane woodland,

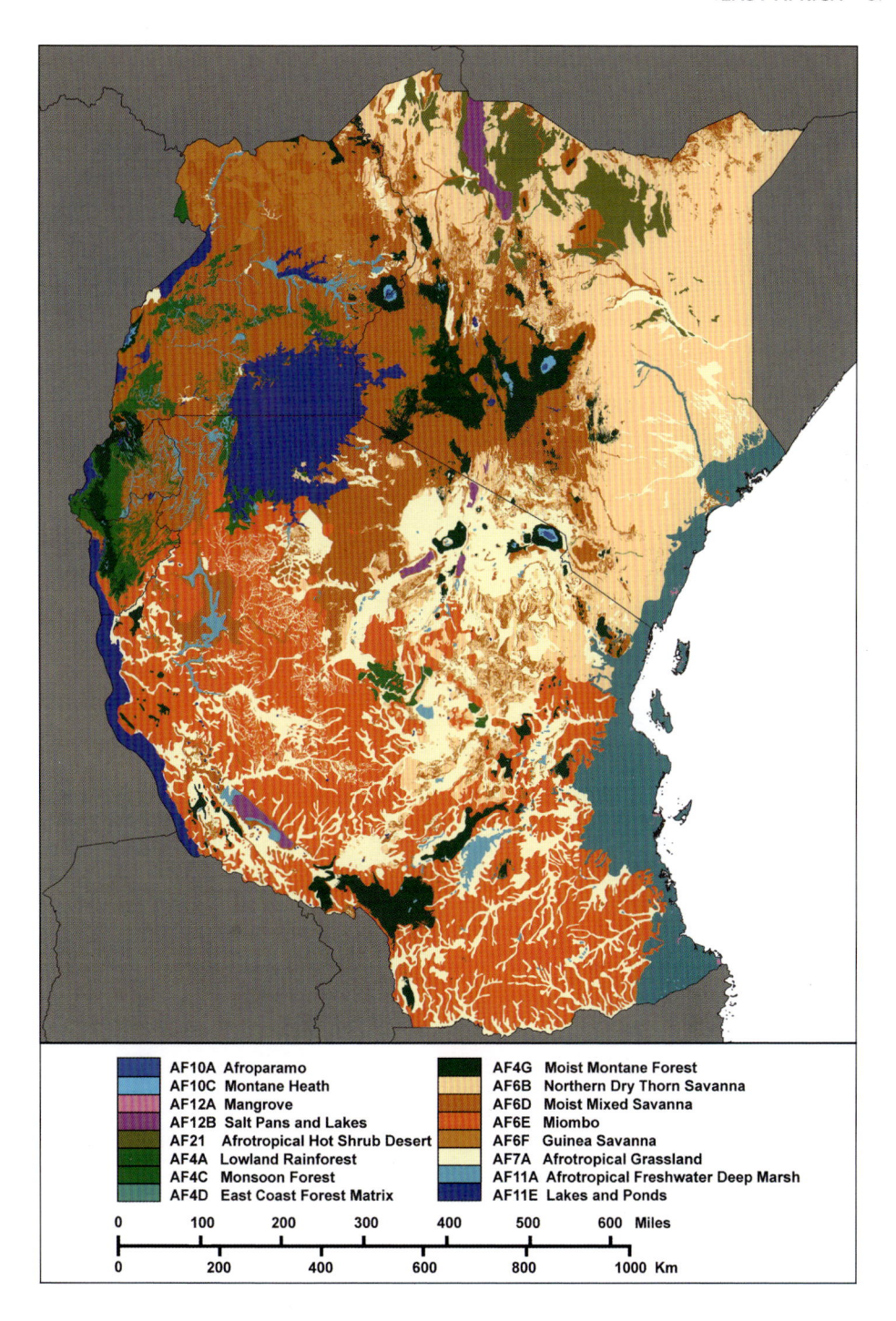

EAST AFRICA 37

AF10A	Afroparamo	AF4G	Moist Montane Forest
AF10C	Montane Heath	AF6B	Northern Dry Thorn Savanna
AF12A	Mangrove	AF6D	Moist Mixed Savanna
AF12B	Salt Pans and Lakes	AF6E	Miombo
AF21	Afrotropical Hot Shrub Desert	AF6F	Guinea Savanna
AF4A	Lowland Rainforest	AF7A	Afrotropical Grassland
AF4C	Monsoon Forest	AF11A	Afrotropical Freshwater Deep Marsh
AF4D	East Coast Forest Matrix	AF11E	Lakes and Ponds

0 100 200 300 400 500 600 Miles
0 200 400 600 800 1000 Km

though too locally to be mapped on the regional map on page 37. **MOIST MONTANE FOREST** is found widely. **MONTANE GRASSLAND** occurs in a similar elevational zone to forest and is often the result of human disturbance. Above the forest and grassland belt is a narrow band of **MONTANE HEATH**, while the tops of the highest mountains support Afroparamo. Both the Eastern Arc and Albertine Rift mountains are very rich overall and blessed with many endemic species. The Eastern Arc holds more chameleons than any place outside Madagascar. The Albertine Rift endemic species are shared with e. Democratic Republic of the Congo (DRC), though they are virtually inaccessible in that troubled country. For wildlife watching, the mountains of Rwanda and Uganda are renowned for holding accessible populations of "Mountain" Eastern Gorilla.

East Africa's complex matrix of habitats includes an abundance of freshwater wetlands. These are especially prominent in the Lake Victoria basin, which holds vast papyrus swamps. Papyrus is the key habitat component of arguably Africa's most remarkable bird, the massive Shoebill, a stork-like bird with a huge craggy bill.

EAST AFRICAN HABITATS

CODE	CATEGORY/NAME	PAGES
	DRY CONIFERS	
Af1C	Afrotropical Montane Dry Mixed Woodland	61–66
	DESERTS AND XERIC SHRUBLANDS	
Af2I	Afrotropical Hot Shrub Desert	121–126
	TROPICAL AND SUBTROPICAL MOIST BROADLEAF FORESTS	
Af4A	Afrotropical Lowland Rainforest	143–150
Af4C	Monsoon Forest	151–156
Af4D	East Coast Forest Matrix	157–163
AF4F	Swamp Forest	169–173
AF4G	Moist Montane Forest	174–179
	SAVANNAS	
Af6B	Northern Dry Thorn Savanna	227–233
Af6D	Moist Mixed Savanna	243–251
Af6E	Miombo	252–259
Af6F	Guinea Savanna	260–267
Af6G	Inselbergs and Koppies	268–274
	GRASSLANDS	
Af7A	Afrotropical Grassland	280–286
Af7F	Montane Grassland	293–299
	ALPINE TUNDRAS AND MONTANE HEATHS	
Af10A	Afroparamo	343–349
Af10C	Montane Heath	354–357

CODE	CATEGORY/NAME	PAGES
	FRESHWATER HABITATS	
Af11A	Afrotropical Deep Freshwater Marsh	358–364
Af11B	Afrotropical Shallow Freshwater Marsh	365–369
Af11E	Freshwater Lakes and Ponds	379–381
Af11F	Rivers	382–385
	SALINE HABITATS	
Af12A	Afrotropical Mangrove	386–389
Af12B	Salt Pans and Lakes	390–392
Af12C	Tidal Mudflats and Estuaries	393–394
Af12D	Salt Marsh	395–396
Af12E	Sandy Beach and Dunes	397–399
Af12F	Rocky Shoreline	400–402
Af12G	Offshore Islands	403–406
Af12H	Pelagic Waters	406–408
	ANTHROPOGENIC	
Af13A	Humid Lowland Cultivation	409–411
Af13B	Savanna Cultivation	412–415
Af13C	Tropical Montane Cultivation	415–417
Af13F	Tree Plantations	422–423
Af13G	Cities and Villages	424–426
Af13H	Grazing Land	427–428

CENTRAL AND NORTH-CENTRAL AFRICA

CENTRAL & NORTH-CENTRAL AFRICAN HABITATS

Habitats shown on ternary diagram (COLD–HOT–WET axes, with FIRE DEPENDENT and FIRE VULNERABLE sides):
- Afroparamo
- Montane Heath
- Montane Grassland
- Moist Montane Forest
- Hot Shrub Desert
- Northern Thorn Savanna
- Guinea Savanna
- Monsoon Forest
- Moist Mixed Savanna
- Afrotropical Grassland
- Lowland Rainforest
- Swamp Forest

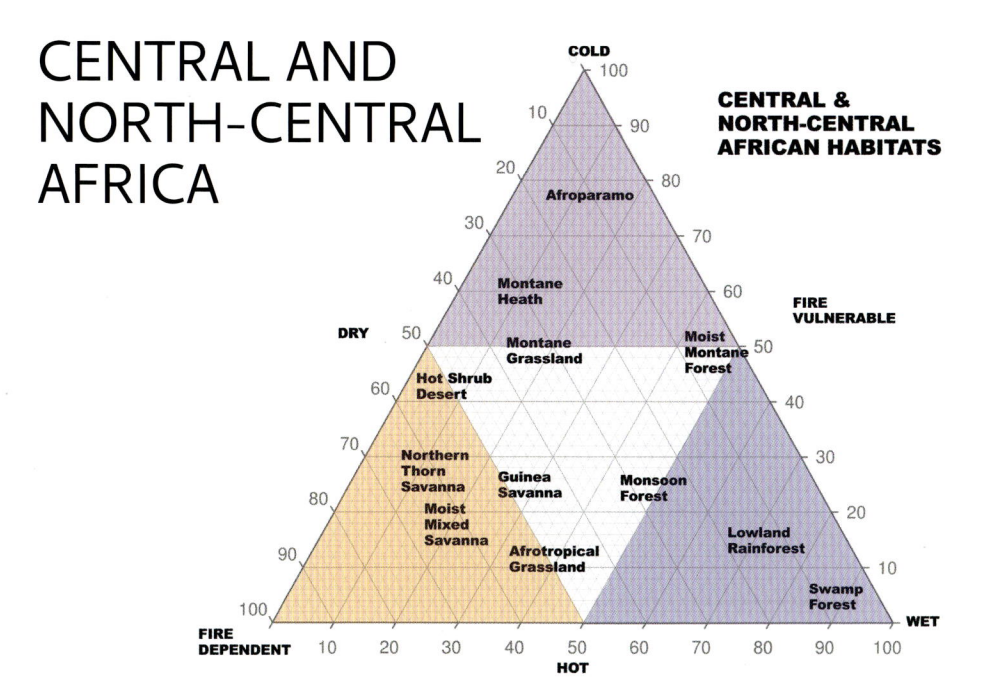

This region is by far Africa's least frequently visited and most poorly known. It comprises two different zones, the first being mostly forest: the vast country of the DRC, along with the Angolan enclave of Cabinda, the Republic of Congo, Gabon, Equatorial Guinea, São Tomé and Príncipe, and s. Cameroon. The second zone is North-Central Africa, from s. Chad in the west, through n. Central African Republic to South Sudan in the east. This latter area is so rarely visited and poorly studied that exact affinities are very hard to define. At 2.4 million square miles (6.2 million square kilometers) this area is slightly larger than Western Europe and slightly smaller than Australia. Parts of it are similar to West Africa, while the east has affinities with the Horn of Africa. There are 34 habitats, none of which is endemic. The heart of this region is the great swath of forest that grows along the Congo River and its tributaries. This is the world's second-largest expanse of rainforest, after the Amazon. There are large areas of SWAMP FOREST and vast tracts of lowland rainforest. Surrounding the humid core are drier forests, in which many trees lose leaves when water stressed, though in this book these are still considered as rainforests. This region is unusually poor in savannas, at least by African standards, but large tracts of MIOMBO occur in the south and there is also some MOIST MIXED SAVANNA, mainly near the coast. The Miombo zone also holds MONSOON FOREST, a semi-evergreen forest that grows mostly along watercourses. Atlantic coastal areas have the typical mix of saline habitats, including abundant AFROTROPICAL MANGROVE. Heading north into North-Central Africa the rainforests give way to GUINEA SAVANNA, MOIST MIXED SAVANNA, and the dry thorn savanna and tropical grasslands of the Sahel, which abuts the Sahara and where the Afrotropics transition to the Palearctic realm. Huge flooded grasslands occur in Chad and South Sudan. There are two mountainous areas in the core area: the Albertine Rift of the far east and the arc of mountains along the border between Cameroon and Nigeria. The tallest mountain is Mt. Stanley in the Ruwenzoris, at 16,762 ft. (5109 m). The highest western peak is Mt. Cameroon at 13,250 ft. (4040 m). These mountains support islands of MOIST

REGIONAL INTRODUCTIONS

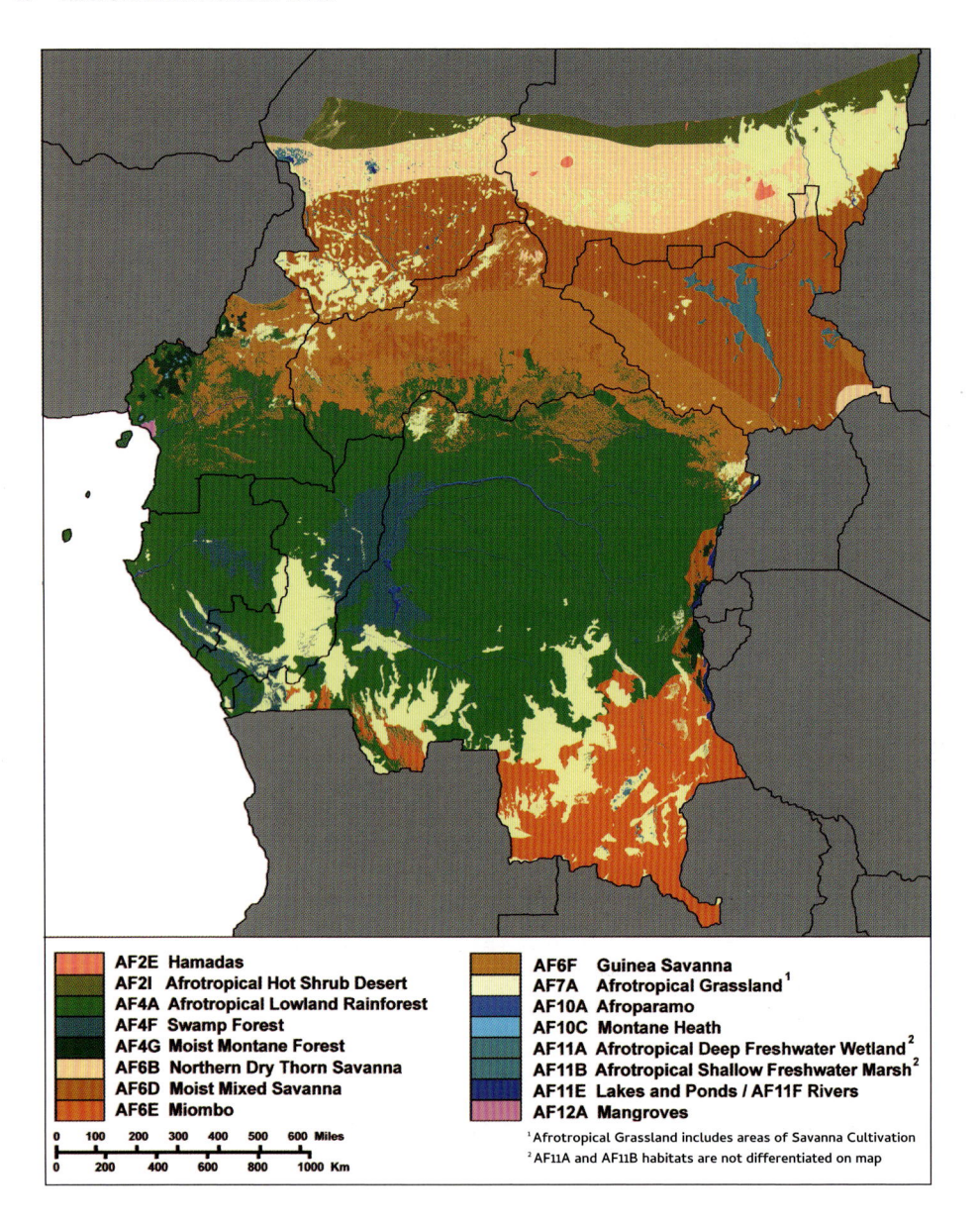

	AF2E Hamadas		AF6F Guinea Savanna
	AF2I Afrotropical Hot Shrub Desert		AF7A Afrotropical Grassland [1]
	AF4A Afrotropical Lowland Rainforest		AF10A Afroparamo
	AF4F Swamp Forest		AF10C Montane Heath
	AF4G Moist Montane Forest		AF11A Afrotropical Deep Freshwater Wetland [2]
	AF6B Northern Dry Thorn Savanna		AF11B Afrotropical Shallow Freshwater Marsh [2]
	AF6D Moist Mixed Savanna		AF11E Lakes and Ponds / AF11F Rivers
	AF6E Miombo		AF12A Mangroves

[1] Afrotropical Grassland includes areas of Savanna Cultivation
[2] AF11A and AF11B habitats are not differentiated on map

MONTANE FOREST, **MONTANE HEATH**, and **MONTANE GRASSLAND**. The Ruwenzoris and Mt. Cameroon are tall enough to also have **AFROPARAMO** on their summits. As would be expected of a great rainforest, this region has remarkable biodiversity. It is the heart of the range of a rich set of rainforest mammals and birds, though many of these are very difficult to see. There are enigmatic creatures like Okapi, Bonobo, and Congo Peacock, plus some more frequently sighted ones like Western and Eastern Gorillas, and Mandrill. Both of the mountainous areas are rich areas of

INDIAN OCEAN

CENTRAL AND NORTH-CENTRAL AFRICAN HABITATS

CODE	CATEGORY/NAME	PAGES	CODE	CATEGORY/NAME	PAGES
	DRY CONIFERS			**FRESHWATER HABITATS**	
Af1C	Afrotropical Montane Dry Mixed Woodland	61–66	Af11A	Afrotropical Deep Freshwater Marsh	358–364
	DESERTS AND XERIC SHRUBLANDS		Af11B	Afrotropical Shallow Freshwater Marsh	365–369
Af2E	Hamada	94–100	Af11E	Freshwater Lakes and Ponds	379–381
Af2I	Afrotropical Hot Shrub Desert	121–126	Af11F	Rivers	382–385
	TROPICAL AND SUBTROPICAL MOIST BROADLEAF FORESTS			**SALINE HABITATS**	
Af4A	Afrotropical Lowland Rainforest	143–150	Af12A	Afrotropical Mangrove	386–389
Af4C	Monsoon Forest	151–156	Af12B	Salt Pans and Lakes	390–392
AF4F	Swamp Forest	169–173	Af12C	Tidal Mudflats and Estuaries	393–394
AF4G	Moist Montane Forest	174–179	Af12D	Salt Marsh	395–396
	SAVANNAS		Af12E	Sandy Beach and Dunes	397–399
Af6B	Northern Dry Thorn Savanna	227–233	Af12F	Rocky Shoreline	400–402
Af6D	Moist Mixed Savanna	243–251	Af12G	Offshore Islands	403–406
Af6E	Miombo	252–259	Af12H	Pelagic Waters	406–408
Af6F	Guinea Savanna	260–267		**ANTHROPOGENIC**	
Af6G	Inselbergs and Koppies	268–274	Af13A	Humid Lowland Cultivation	409–411
	GRASSLANDS		Af13B	Savanna Cultivation	412–415
Af7A	Afrotropical Grassland	280–286	Af13C	Tropical Montane Cultivation	415–417
Af7F	Montane Grassland	293–299	Af13F	Tree Plantations	422–423
	ALPINE TUNDRAS AND MONTANE HEATHS		Af13G	Cities and Villages	424–426
Af10A	Afroparamo	343–349	Af13H	Grazing Land	427–428
Af10C	Montane Heath	354–357			

endemism. The Albertine Rift montane forests are Africa's most diverse and hold many endemic species, though these are shared with several East African countries. The western mountains also have a tantalizing set of endemics, including birds like Bannerman's Turaco and Mount Kupe Bushshrike. In Sudan there are isolated low mountains where dry broadleaf and conifer forest occur. There is huge scope for investigation with the likelihood of new species to be discovered there.

INDIAN OCEAN

The w. Indian Ocean contains four main island groups: Madagascar, the Seychelles, the Comoros, and the Mascarenes. At 230,000 mi.^{2} (595,700 km^2) this area is slightly smaller than France or Texas. Madagascar and the Seychelles are ancient fragments of Gondwana with some rocks over 3 billion years old. Around 180 MYA, while still joined with India, they separated from Africa and started heading north. At that time there were still dinosaurs, and birds had not evolved! Around 80 MYA Madagascar split from India and started heading back toward Africa. With Madagascar's split and resulting ocean creation, some of the oldest rocks in the Comoros formed through uplift of the ocean floor. But most of the Comoros and Mascarenes are much younger islands, formed through hotspot volcanism and formation of shield volcanoes (similar to the Galápagos and Hawaiian Islands) that is ongoing today.

REGIONAL INTRODUCTIONS

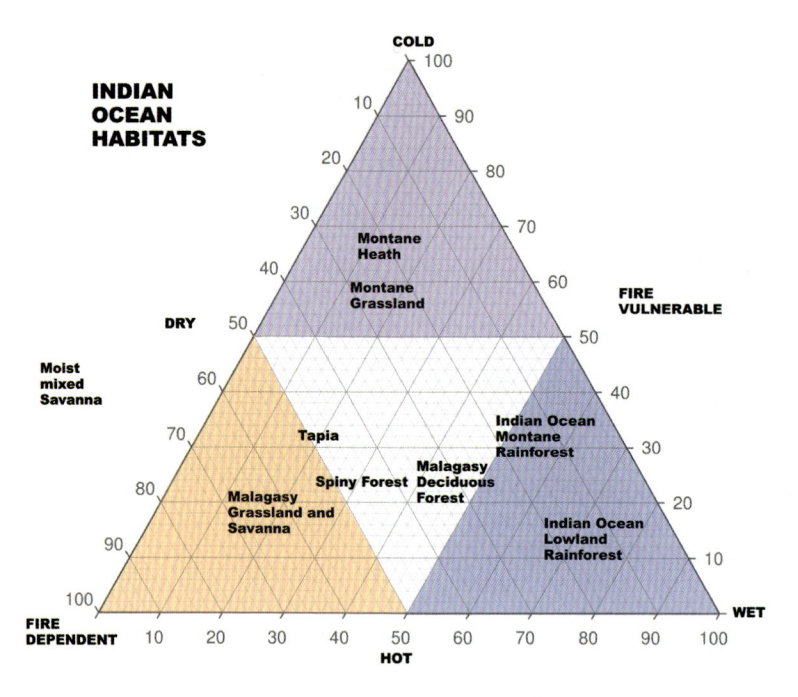

INDIAN OCEAN HABITATS

Triangular diagram with axes COLD (top), DRY (left), WET (right), HOT (bottom), FIRE DEPENDENT (bottom-left) and FIRE VULNERABLE (right). Habitats plotted: Montane Heath, Montane Grassland, Moist mixed Savanna, Tapia, Spiny Forest, Malagasy Deciduous Forest, Malagasy Grassland and Savanna, Indian Ocean Montane Rainforest, Indian Ocean Lowland Rainforest.

INDIAN OCEAN HABITATS

CODE	CATEGORY/NAME	PAGES
	DESERTS AND XERIC SHRUBLANDS	
Af2J	Spiny Forest (endemic)	127–132
	TROPICAL AND SUBTROPICAL MOIST BROADLEAF FORESTS	
Af4J	Indian Ocean Lowland Rainforest (endemic)	180–188
Af4K	Indian Ocean Montane Rainforest (endemic)	189–196
Af4L	Seychelles Granite Forest (endemic)	197–202
	TROPICAL AND SUBTROPICAL DRY BROADLEAF FORESTS	
Af5A	Malagasy Deciduous Forest (endemic)	209–214
	GRASSLANDS	
Af7E	Malagasy Grassland and Savanna (endemic)	287–292
Af7F	Montane Grassland	293–299
	MEDITERRANEAN FORESTS, WOODLANDS AND SHRUBS	
Af8H	Tapia (endemic)	338–341
	ALPINE TUNDRAS AND MONTANE HEATHS	
Af10C	Montane Heath	354–357

CODE	CATEGORY/NAME	PAGES
	FRESHWATER HABITATS	
Af11A	Afrotropical Deep Freshwater Marsh	358–364
Af11B	Afrotropical Shallow Freshwater Marsh	365–369
Af11E	Freshwater Lakes and Ponds	379–381
Af11F	Rivers	382–385
	SALINE HABITATS	
Af12A	Afrotropical Mangrove	386–389
Af12B	Salt Pans and Lakes	390–392
Af12C	Tidal Mudflats and Estuaries	393–394
Af12D	Salt Marsh	395–396
Af12E	Sandy Beach and Dunes	397–399
Af12F	Rocky Shoreline	400–402
Af12G	Offshore Islands	403–406
Af12H	Pelagic Waters	406–408
	ANTHROPOGENIC	
Af13A	Humid Lowland Cultivation	409–411
Af13B	Savanna Cultivation	412–415
Af13C	Tropical Montane Cultivation	415–417
Af13F	Tree Plantations	422–423
Af13G	Cities and Villages	424–426
Af13H	Grazing Land	427–428

INDIAN OCEAN

This region contains 27 of the habitats covered in this book, and 7 of these are endemic. Madagascar is by far the largest island and the hotspot of biological diversity. This is partially explained by its possessing nearly a whole continent's worth of diversity in its natural environments. The eastern plain and escarpment is covered in lush rainforest, which changes character above 2600 ft. (800 m) elevation. On the highest mountains, there is heath habitat as well as some grassland. The middle of Madagascar is the "High Plateau," which has been ravaged by fire, and today is covered in biologically poor grassland but which originally held a mix of forest, woodland, and savanna. The northern two-thirds of the west is also largely deforested, but the original habitat was drier deciduous forest—shorter than eastern rainforest and with many trees that lose their leaves during the dry season. The southwestern third is the driest part of Madagascar, and the vegetation is the

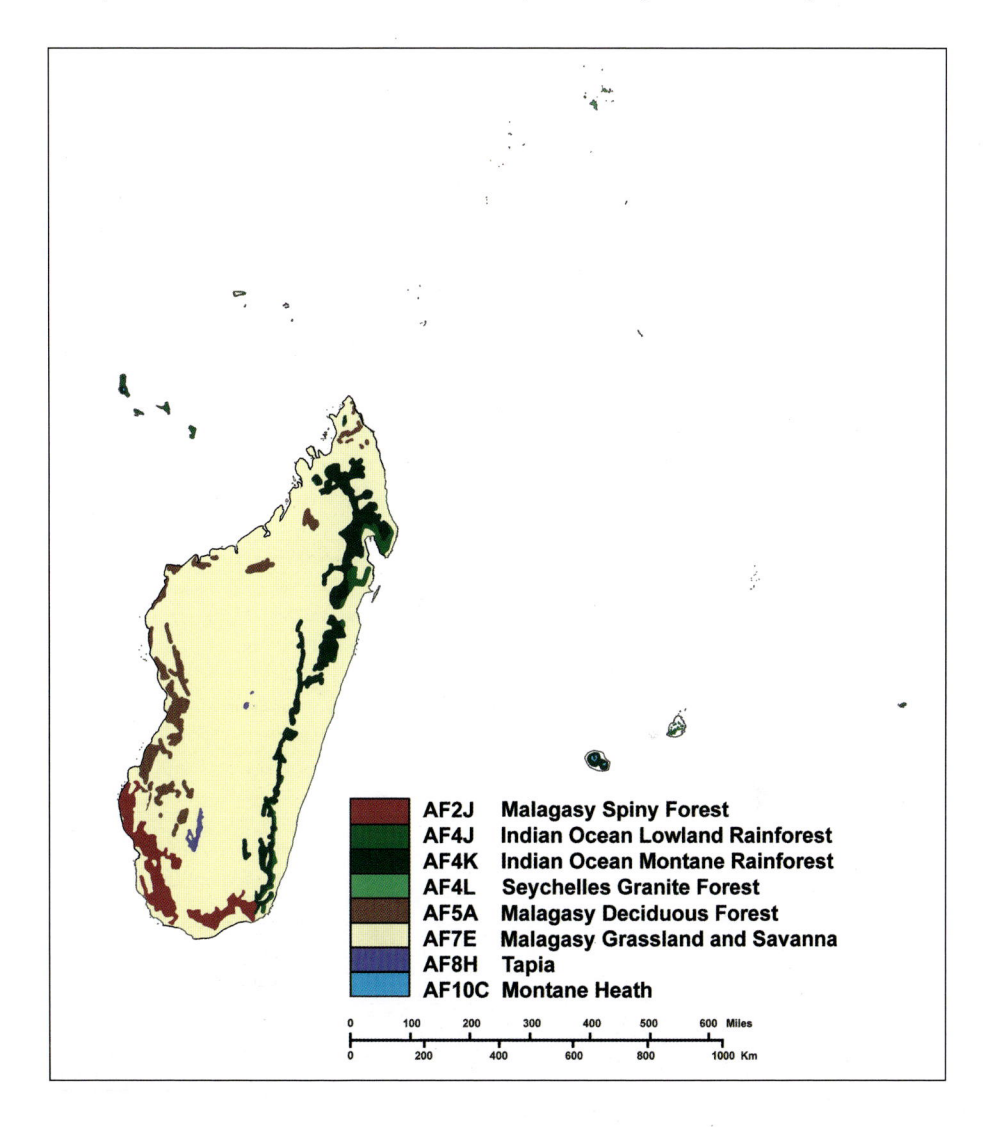

	AF2J	Malagasy Spiny Forest
	AF4J	Indian Ocean Lowland Rainforest
	AF4K	Indian Ocean Montane Rainforest
	AF4L	Seychelles Granite Forest
	AF5A	Malagasy Deciduous Forest
	AF7E	Malagasy Grassland and Savanna
	AF8H	Tapia
	AF10C	Montane Heath

semi-desert SPINY FOREST, an other-worldly landscape of baobab trees, succulent plants, and endemic many-armed "octopus trees."

Madagascar is the land of lemurs—there are over 100 species of these delightful primates, found in all types of forest. Other prominent groups of mammals include tenrecs, Malagasy carnivorans, rodents, and bats. The birdlife is low in diversity but high in endemism—there are over 100 endemic birds and five endemic families. Many common groups of birds from Africa, such as woodpeckers and bushshrikes, are lacking in Madagascar. Reptile and amphibian diversity is high, and there are hundreds of endemic species.

The default habitat on all the other Indian Ocean islands is rainforest, but most islands have been largely deforested. The two islands with the tallest and most rugged mountains, Réunion and Grande Comore, are also the two that have retained the most forest, mostly at higher elevations. The highest parts of these islands are above tree line and support heath habitat. The Inner Islands of the Seychelles are granite, and the remnant rainforest growing there has a distinctive character. The Outer Islands of the Seychelles, including Aldabra, are coral islands, where the natural vegetation is drier scrub and woodland. With a few exceptions, the wildlife on the other Indian Ocean island groups is similar to, but much poorer than, Madagascar. There are very few mammals, but huge flying foxes are conspicuous. The endemic bird families of Madagascar are lacking, save for the Cuckoo-Roller, which has colonized the Comoros.

Mangroves and mudflats are found throughout the region but are most common along the sheltered west coast of Madagascar. This is a generally poor sector of ocean for marine life, though sea turtles and tropical seabirds nest widely, and Round Island off Mauritius has a notable assemblage of breeding seabirds.

GREATER SOUTHERN AFRICA

Greater Southern Africa encompasses 10 nations stretching from temperate South Africa (34°S) to the tropics of Mozambique and across to Angola (5°S). At 2.25 million square miles (5.8 million square kilometers) this area is slightly larger than Western Europe and slightly smaller than Australia. The region's extraordinary biodiversity is linked to the varied habitats it supports, including 46 habitats (15 of which are endemic) covered in this book, making this region the African champion for both overall habitat diversity and endemism. These habitats range from some of Earth's driest deserts (Köppen **Bwk**), to lowland rainforest (Köppen **Afa**), and everything in between! This diversity is driven by complex landscapes, topography, and surrounding oceans, which affect climate. The region's interior consists mostly of a high-elevation Central Plateau lying above 3280 ft. (1000 m), surrounded by steep-edged mountains called the Great Escarpment, which fall away to a coastal plain mostly 50–125 mi. (80–200 km) in width but much broader in Mozambique. The Central Plateau is incised by several low-lying broad river valleys, notably the Orange, Limpopo, Zambezi, and Kafue, which introduce hot coastal conditions deep into the continental interior. Africa's Great Rift Valley enters the northern part of the region, forming Lake Malawi and terminating in the Zambezi Valley to the south.

The dominant habitats of the region (particularly in the north) are a variety of savannas varying from sparsely treed grassland to dense woodland. Rainfall and seasonality is strongly influenced by the zone where the trade winds of the Northern and Southern Hemispheres collide (the Intertropical Convergence Zone), creating a subtropical climate (Köppen **Amb**) characterized by three seasons: hot dry (Sep–Nov); wet rainy (Dec–Apr); and cool dry (May–Aug). The savannas

GREATER SOUTHERN AFRICA 45

GREATER SOUTHERN AFRICA HABITATS

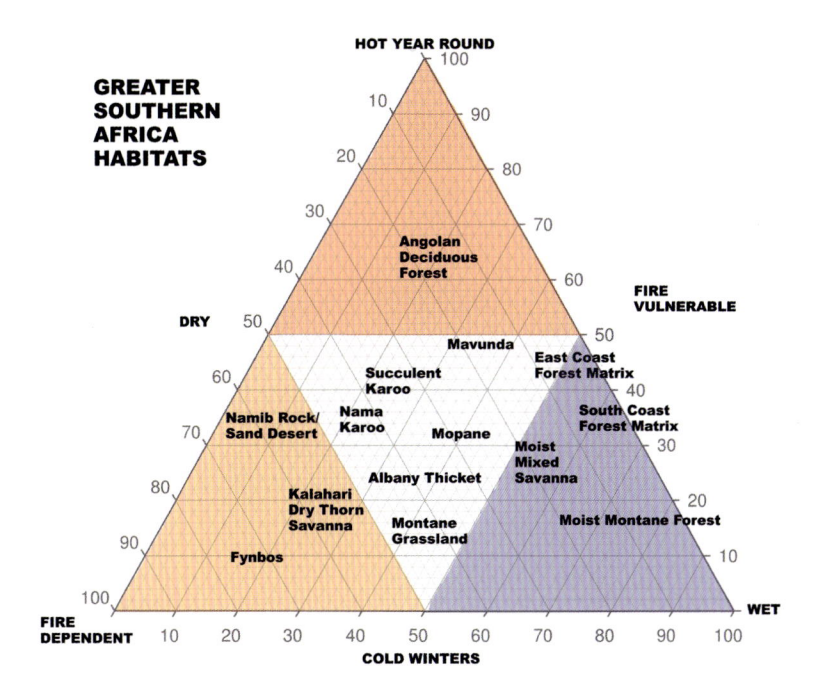

Ternary diagram with axes: HOT YEAR ROUND (top), COLD WINTERS (bottom), DRY (left), WET (right), FIRE DEPENDENT (bottom-left), FIRE VULNERABLE (right).

Habitats plotted: Angolan Deciduous Forest, Mavunda, East Coast Forest Matrix, Succulent Karoo, Namib Rock/Sand Desert, Nama Karoo, Mopane, South Coast Forest Matrix, Moist Mixed Savanna, Albany Thicket, Kalahari Dry Thorn Savanna, Montane Grassland, Moist Montane Forest, Fynbos.

vary from more closed broadleaf-type woodlands like MOPANE growing at slightly lower elevations and MIOMBO on higher plateaus to drier GUSU on well-drained Kalahari sands. These are interspersed in a complex mosaic with MOIST MIXED SAVANNA and open AFROTROPICAL GRASSLANDS. In the more tropical reaches of nw. Zambia, Angola, and elsewhere, the rainfall increases but remains highly seasonal, giving rise to deciduous and monsoon forests. In localized patches in Angola, where rainfall is both heavy and less seasonal, patches of true lowland rainforest develop.

The coast has a dramatic and underappreciated influence on habitats. Two oceanic currents, the cold-water Benguela (Atlantic Ocean) and warm-water Agulhas (Indian Ocean) both radically influence the climate. The icy waters of the Benguela create a variety of arid habitats in the west, including the Namib Desert and Karoo. The rain fronts from the Southern Ocean that buffet the Cape Fold Mountains of s. South Africa during the austral winter (May–Aug) create a collection of Mediterranean-type heathlands. The east coast, in contrast, is influenced by the warm-water Agulhas current. This creates a humid environment resulting in the formation of coastal forest types. In se. South Africa, where the Agulhas starts cooling, and the drier influence of the Central Plateau becomes prevalent, the unique succulent ALBANY THICKET develops.

There are also plenty of mountains that emerge above the Central Plateau. Thaba Ntlenyana (Lesotho) is the tallest at 11,423 ft. (3482 m). Although peaks occur along the Namibian-Angolan escarpment and the Cape Fold Mountains in South Africa, mountains are tallest in the east, running from the Drakensberg (South Africa) north to the Nyika Plateau (Malawi-Zambia). Generally, montane habitats like MOIST MONTANE FOREST and MONTANE GRASSLAND are restricted to elevations above 4600 ft. (1400 m) on isolated tropical massifs, but farther south in temperate zones these "montane" habitats occur at progressively lower elevations until they reach the coast in South Africa. Only the region's tallest alpine peaks support small and isolated fragments of

46 REGIONAL INTRODUCTIONS

AFROPARAMO and MONTANE HEATH. Freshwater wetlands are divided between temperate ones in the cooler south and both deep and shallow freshwater marshes in the warmer north.

The wildlife of this region is legendary. It is home to almost every iconic African mammal and much more. There are tens of thousands of plants (many of which are endemic), over 1300 bird species, of which over 250 are endemic or near-endemic, and similar levels of diversity for reptiles, amphibians, fish, and insects. The details of this stunning array of biodiversity are explored in the accounts.

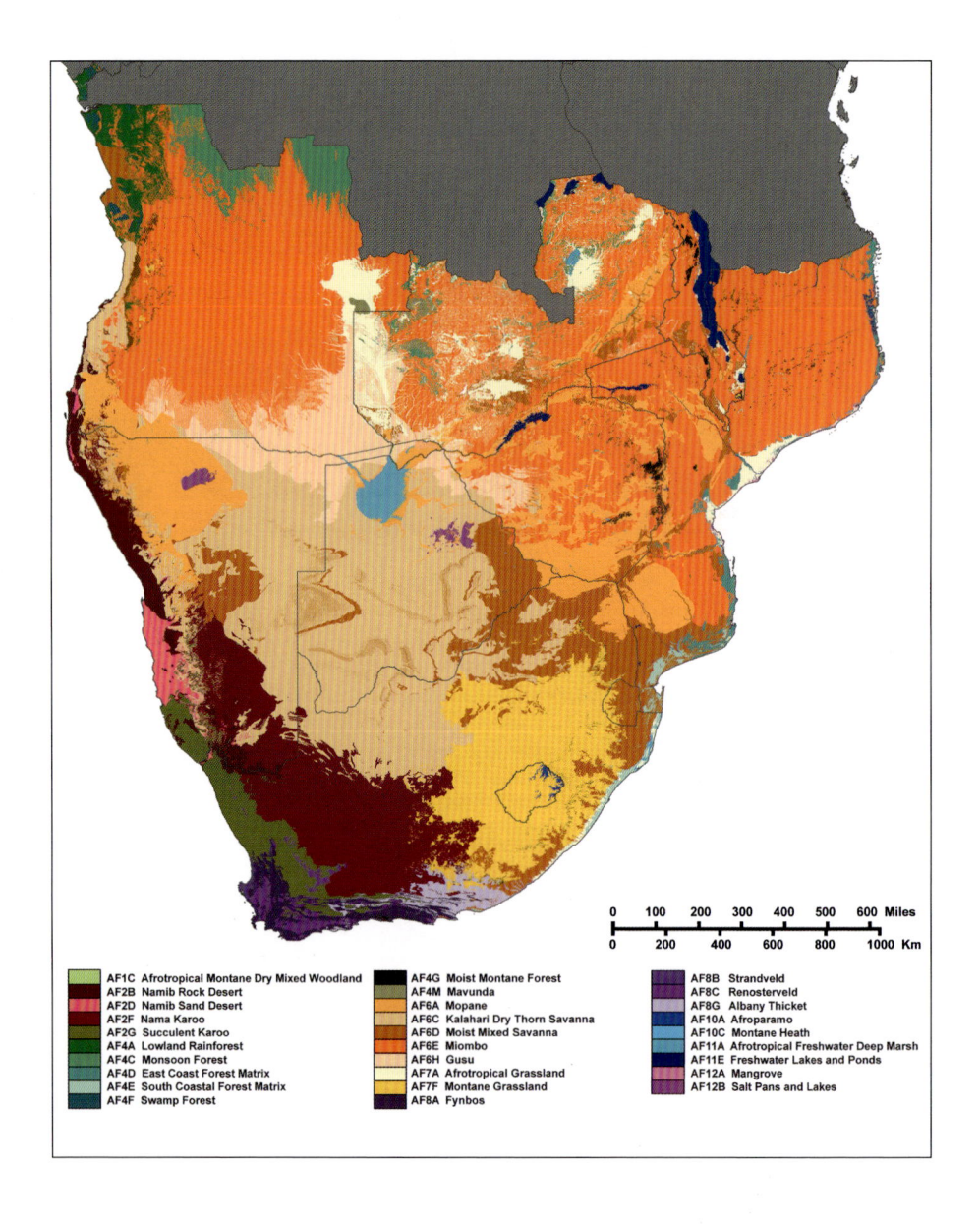

AF1C	Afrotropical Montane Dry Mixed Woodland
AF2B	Namib Rock Desert
AF2D	Namib Sand Desert
AF2F	Nama Karoo
AF2G	Succulent Karoo
AF4A	Lowland Rainforest
AF4C	Monsoon Forest
AF4D	East Coast Forest Matrix
AF4E	South Coastal Forest Matrix
AF4F	Swamp Forest
AF4G	Moist Montane Forest
AF4M	Mavunda
AF6A	Mopane
AF6C	Kalahari Dry Thorn Savanna
AF6D	Moist Mixed Savanna
AF6E	Miombo
AF6H	Gusu
AF7A	Afrotropical Grassland
AF7F	Montane Grassland
AF8A	Fynbos
AF8B	Strandveld
AF8C	Renosterveld
AF8G	Albany Thicket
AF10A	Afroparamo
AF10C	Montane Heath
AF11A	Afrotropical Freshwater Deep Marsh
AF11E	Freshwater Lakes and Ponds
AF12A	Mangrove
AF12B	Salt Pans and Lakes

GREATER SOUTHERN AFRICAN HABITATS

CODE	CATEGORY/NAME	PAGES
	DESERTS AND XERIC SHRUBLANDS	
Af2B	Namib Rock Desert (endemic)	77–81
Af2D	Namib Sand Desert (endemic)	88–93
Af2F	Nama Karoo (endemic)	101–106
Af2G	Succulent Karoo (endemic)	107–114
	TROPICAL AND SUBTROPICAL MOIST BROADLEAF FORESTS	
Af4A	Afrotropical Lowland Rainforest	143–150
Af4C	Monsoon Forest	151–156
Af4D	East Coast Forest Matrix	157–163
Af4E	South Coast Forest Matrix (endemic)	164–168
AF4F	Swamp Forest	169–173
AF4G	Moist Montane Forest	174–179
AF4M	Mavunda (endemic)	203–207
	TROPICAL AND SUBTROPICAL DRY BROADLEAF FORESTS	
Af5B	Angolan Deciduous Forest (endemic)	215–219
	SAVANNAS	
Af6A	Mopane	221–226
Af6C	Kalahari Dry Thorn Savanna (endemic)	235–242
Af6D	Moist Mixed Savanna	243–251
Af6E	Miombo	252–259
Af6G	Inselbergs and Koppies	268–274
Af6H	Gusu (endemic)	275–279
	GRASSLANDS	
Af7A	Afrotropical Grassland	280–286
Af7F	Montane Grassland	293–299
	MEDITERRANEAN FORESTS, WOODLANDS AND SHRUBS	
Af8A	Fynbos (endemic)	300–308
Af8B	Strandveld (endemic)	309–314
Af8C	Renosterveld (endemic)	315–320
Af8G	Albany Thicket (endemic)	332–337

CODE	CATEGORY/NAME	PAGES
	ALPINE TUNDRAS AND MONTANE HEATHS	
Af10A	Afroparamo	343–349
Af10C	Montane Heath	354–357
	FRESHWATER HABITATS	
Af11A	Afrotropical Deep Freshwater Marsh	358–364
Af11B	Afrotropical Shallow Freshwater Marsh	365–369
Af11D	South African Temperate Wetland (endemic)	374–378
Af11E	Freshwater Lakes and Ponds	379–381
Af11F	Rivers	382–385
	SALINE HABITATS	
Af12A	Afrotropical Mangrove	386–389
Af12B	Salt Pans and Lakes	390–392
Af12C	Tidal Mudflats and Estuaries	393–394
Af12D	Salt Marsh	395–396
Af12E	Sandy Beach and Dunes	397–399
Af12F	Rocky Shoreline	400–402
Af12G	Offshore Islands	403–406
Af12H	Pelagic Waters	406–408
	ANTHROPOGENIC	
Af13A	Humid Lowland Cultivation	409–411
Af13B	Savanna Cultivation	412–415
Af13C	Tropical Montane Cultivation	415–417
Af13D	South African Temperate Cultivation (endemic)	418–419
Af13F	Tree Plantations	422–423
Af13G	Cities and Villages	424–426
Af13H	Grazing Land	427–428

NORTH AFRICA

North Africa is massive; at 3.5 million square miles (9.9 million square kilometers) it is pretty close to the size of the United States. It stretches from the Atlantic Ocean in the west to the Red Sea in the east, from the Mediterranean Sea in the north to the Sahel of Mali, Niger, Chad, and Sudan in the south. The northwest is referred to as the Maghreb (Mauritania, Western Sahara, Morocco, Algeria, Tunisia, and Libya), and has a distinctly temperate Mediterranean climate (Köppen **Csa**, **Csb**) with hot dry summers and wetter winters. But the Maghreb also includes the mighty Atlas Mountains, which reach heights of 13,671 ft. (4167 m). The remainder of North Africa is part of the Saharan region with a subtropical Mediterranean desert (Köppen **Bsa**) in the north and a summer

REGIONAL INTRODUCTIONS

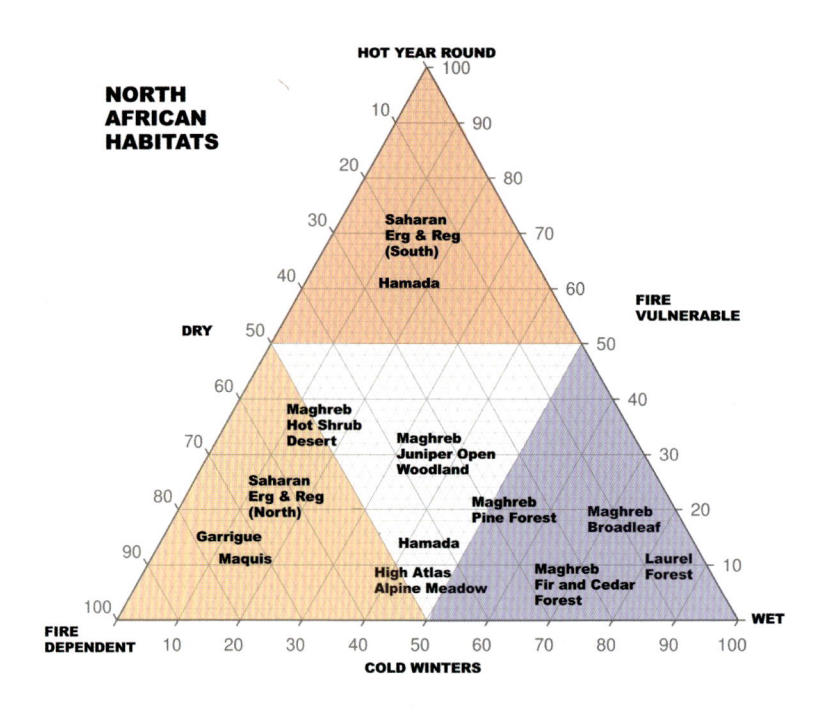

NORTH AFRICAN HABITATS

Ternary diagram axes:
- HOT YEAR ROUND (top)
- FIRE VULNERABLE (right)
- COLD WINTERS (bottom)
- FIRE DEPENDENT (bottom-left)
- WET (right-bottom)
- DRY (left)

Habitats plotted:
- Saharan Erg & Reg (South)
- Hamada
- Maghreb Hot Shrub Desert
- Maghreb Juniper Open Woodland
- Saharan Erg & Reg (North)
- Garrigue
- Maquis
- Maghreb Pine Forest
- Maghreb Broadleaf
- Hamada
- High Atlas Alpine Meadow
- Maghreb Fir and Cedar Forest
- Laurel Forest

NORTH AFRICAN HABITATS

CODE	CATEGORY/NAME	PAGES
	DRY CONIFERS	
Af1A	Maghreb Fir and Cedar Forest (endemic)	54–57
Af1B	Maghreb Juniper Open Woodland (endemic)	57–60
Af1D	Maghreb Pine Forest (endemic)	67–69
	DESERTS AND XERIC SHRUBLANDS	
Af2A	Saharan Reg Desert (endemic)	71–76
Af2C	Saharan Erg Desert (endemic)	82–87
Af2E	Rocky Hamada and Massif (endemic)	94–100
Af2H	Maghreb Hot Shrub Desert (endemic)	115–120
	TEMPERATE BROADLEAF AND MIXED FORESTS	
Af3A	Laurel Forest (endemic)	138–141
	MEDITERRANEAN FORESTS, WOODLANDS AND SHRUBS	
Af8D	Maghreb Maquis (endemic)	321–325
Af8E	Maghreb Garrigue (endemic)	326–327
Af8F	Maghreb Broadleaf Woodland (endemic)	328–331
	ALPINE TUNDRAS AND MONTANE HEATHS	
Af10B	High Atlas Alpine Meadow (endemic)	350–353
Af10C	Montane Heath	354–357

CODE	CATEGORY/NAME	PAGES
	FRESHWATER HABITATS	
Af11C	North African Temperate Wetland (endemic)	370–373
Af11E	Freshwater Lakes and Ponds	379–381
Af11F	Rivers	382–385
	SALINE HABITATS	
Af12A	Afrotropical Mangrove	386–389
Af12B	Salt Pans and Lakes	390–392
Af12C	Tidal Mudflats and Estuaries	393–394
Af12D	Salt Marsh	395–396
Af12E	Sandy Beach and Dunes	397–399
Af12F	Rocky Shoreline	400–402
Af12G	Offshore Islands	403–406
Af12H	Pelagic Waters	406–408
	ANTHROPOGENIC	
Af13E	North African Temperate Cultivation (endemic)	419–421
Af13F	Tree Plantations	422–423
Af13G	Cities and Villages	424–426
Af13H	Grazing Land	427–428

rainfall tropical-influenced desert (Köppen **Bwa**) in the south. This region contains 28 of the habitats covered in this book, and 14 of these are endemic.

The Atlas Mountains are three distinct ranges, with the tallest being the High Atlas in the center, dominated by sedimentary rocks. Counterintuitively, the Middle (in height) Atlas lie farther north, closer to the Mediterranean Sea, and comprise both sedimentary and volcanic rocks, and the Anti-Atlas are an older outlying range of predominantly much older Precambrian metamorphic rocks such as gneisses and quartzites. These lie to the south of the High Atlas and border the Sahara Desert.

The Sahara is so massive that any summary of the geology will always be an oversimplification. But in brief, it is dominated by Precambrian metamorphic complexes (4.6 billion to 500 million years old) and overlain by younger sedimentary deposits. Although these deposits range over 500 million years in age, they have mostly been deposited in shallow environments and have not been subject to much alteration since formation. The distinct landscapes of the Sahara have been produced by geomorphological processes that move this material around through wind (aeolian erosion and deposition), which creates the Reg (stony plains), Hamada (rocky terrain), and Erg (sand surfaces and seas).

Between 5.9 and 5.3 MYA North Africa was attached to Europe by land, and what is now the bottom of the Mediterranean Sea was a basin covered in low-lying savanna and a series of smaller hypersaline seas about 6000 ft. (1800 m) lower than current sea level. That came to a sudden end with a flood that poured the equivalent of 500 times the flow of today's Amazon through the Straits of Gibraltar and into the basin to create the Mediterranean Sea. With the sea level rising at 30 ft. (10 m) per day it is hard to conceive how any terrestrial animals could have escaped that flood. The expansion of the Sahara in the east, in conjunction with the creation of the Mediterranean Sea, has split the region into two nodes of speciation: on the north side of the Mediterranean Iberian Lynx and Savi's Pine Vole evolved in Pleistocene (2.6 million–11,700 YBP) European refugia, while on the African side animals such as Barbary Ground Squirrel, Egyptian Greater Jerboa, and Cuvier's Gazelle evolved. Birds and bats are mobile and have mostly colonized both sides of the Mediterranean.

Along North Africa's Mediterranean fringe, the habitats are very similar to those occurring in s. Europe and the Middle East. Coastal areas are dominated by MAGHREB GARRIGUE heathland, with larger patches of dry sclerophyllous woodland called MAGHREB MAQUIS, coastal juniper woodlands, and oak forests dominated by Cork Oak. On the slopes of the Atlas Mountains in Morocco, Algeria, and Tunisia (Köppen **Csb**), the oak forests merge into forests dominated by cedars and pines. Above the cedar in wet areas, and also on the lower slopes in the rain shadows, junipers, surprisingly of the same species as in the littoral zone, form shrublands and low forests.

To the south of the Atlas Mountains there is hot shrub desert, where cultivation is still possible along the northern fringes of the dry Sahara. Farther east, in Libya and Egypt, the North African coast is much lower in elevation, and the Mediterranean climate continues across most of coastal Libya, petering out before reaching Egypt, where the Sahara runs all the way to the coast. Half of the Sahara is covered with small pebbles, forming a desert pavement called Reg, where most of the sand has been eroded away by wind, leaving shiny cobbles. The aforementioned sand is deposited in dunes and seas of sand called Erg, which are nearly devoid of life until stabilized by date palms and scattered vegetation. Within the Erg and Reg landscapes there are areas where freshwater comes to the surface as springs, forming oases. These areas formerly supported marshes and dense stands of palms and other trees, teaming with life. Unfortunately, the vast majority of oases have been transformed by cultivation and turned into towns with little room left for wildlife. There are also Hamada areas of barren, rocky plateau, some flat and some mountainous. Hamadas often contain wadis—valleys where the water flowing off barren rocky plateaus feeds a water table close to the surface, allowing lusher growth than would normally grow in these areas. Though few people realize this, there are rocky mountain ranges scattered across the Sahara that support pockets of montane shrubland in areas above 6000 ft. (1800 m), such as in the Tibesti range.

The Macaronesia Islands contain the last major vestiges of LAUREL FOREST, a habitat that was once widespread across Europe but over time has been excluded by other habitats and climates. Several endemic birds and other fauna survive only here.

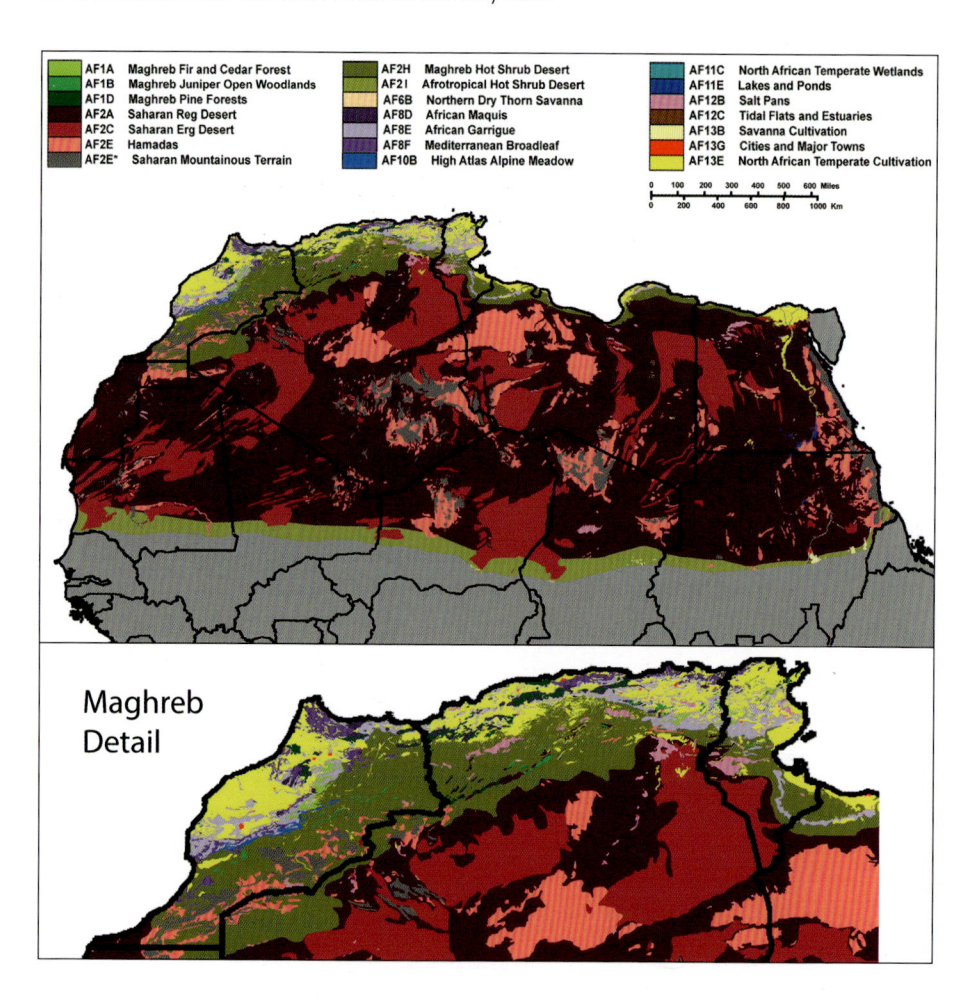

AF1A	Maghreb Fir and Cedar Forest	AF2H	Maghreb Hot Shrub Desert	AF11C	North African Temperate Wetlands
AF1B	Maghreb Juniper Open Woodlands	AF2I	Afrotropical Hot Shrub Desert	AF11E	Lakes and Ponds
AF1D	Maghreb Pine Forests	AF6B	Northern Dry Thorn Savanna	AF12B	Salt Pans
AF2A	Saharan Reg Desert	AF8D	African Maquis	AF12C	Tidal Flats and Estuaries
AF2C	Saharan Erg Desert	AF8E	African Garrigue	AF13B	Savanna Cultivation
AF2E	Hamadas	AF8F	Mediterranean Broadleaf	AF13G	Cities and Major Towns
AF2E*	Saharan Mountainous Terrain	AF10B	High Atlas Alpine Meadow	AF13E	North African Temperate Cultivation

Maghreb Detail

WEST AFRICA

WEST AFRICAN HABITATS

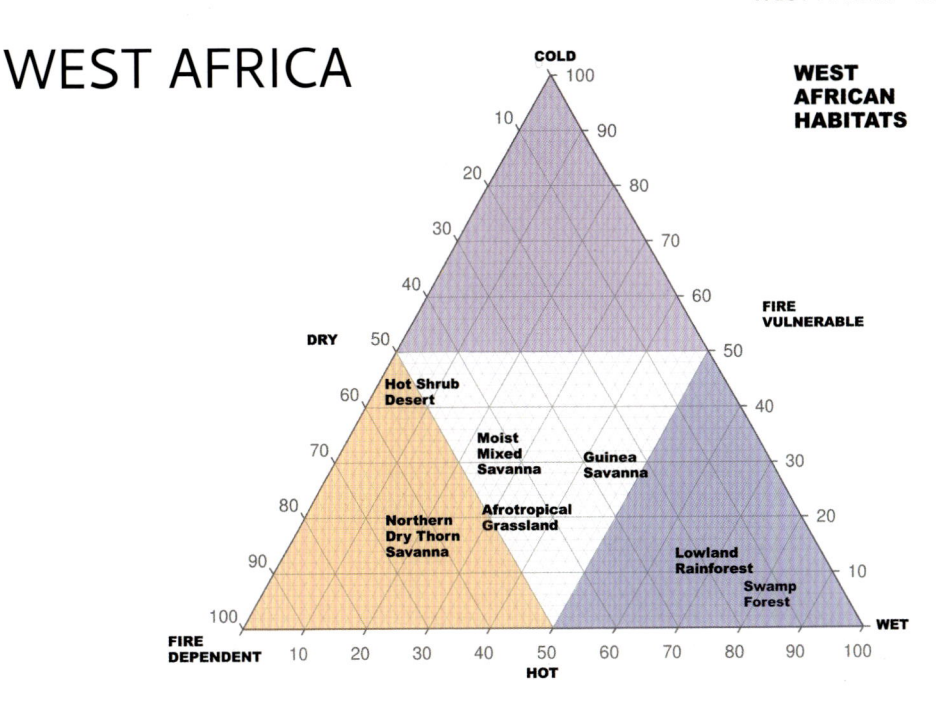

Ternary diagram of West African habitats plotted on axes of Cold–Hot, Dry–Wet, and Fire Dependent–Fire Vulnerable. Habitats shown include: Hot Shrub Desert, Northern Dry Thorn Savanna, Moist Mixed Savanna, Afrotropical Grassland, Guinea Savanna, Lowland Rainforest, and Swamp Forest.

West Africa lies between the Sahara and the Gulf of Guinea, stretching east to Chad. At 1.4 million square miles (3.6 million square kilometers) it is the size of the US Eastern Seaboard across to the Great Plains, or slightly larger than India. It covers the northern half of Cameroon and the southern parts of Niger, Burkina Faso, Mali, and Mauritania. From east to west along the West African coast it includes Nigeria, Benin, Togo, Ghana, Ivory Coast, Côte d'Ivoire, Liberia, Sierra Leone, Guinea, Guinea-Bissau, Senegal, and The Gambia. It contains 26 habitats, none of which are endemic.

The savanna belt comprises GUINEA SAVANNA, MOIST MIXED SAVANNA, and dry thorn savanna. Guinea Savanna occurs adjacent the rainforest belt, then slowly transitions into moist mixed then dry thorn savanna to the north as the climate becomes drier (see diagram of this transition on page 20). This savanna belt extends from the Senegal coastline to Chad, and then extends eastward, gradually changing and becoming a minor habitat through north-central Africa. Guinea Savanna is structurally a woodland savanna though botanically has affinities with the rainforests to the south. The climate has a very strong seasonality (Köppen **Ama**) with extremely humid wet seasons and very dry seasons. Rainfall is sufficient to support rainforest, but the prolonged dry season combined with a fairly regular fire regime maintains the savanna.

The once extensive Upper Guinea lowland rainforests to the south of the savanna belt are separated from the similar Lower Guinea Central African lowland rainforests by the Dahomey Gap, where the Guinea Savanna reaches the coast in Togo and Benin. Upper Guinea forests contain endemics such as White-breasted Guineafowl, Brown-cheeked Hornbill, White-necked Rockfowl, and Sharpe's Apalis. These forests, however, have been decimated, turned into secondary rainthicket to the point at which it is difficult to find any real "primary" rainforest, and even our understanding of what is a primary forest may need to change.

The habitats to the north of the Guinea Savanna gradually become dryer and shorter in stature, with thorn savanna and grassland of the Sahel merging with the shrub deserts of the Sahara. Much

REGIONAL INTRODUCTIONS

WEST AFRICAN HABITATS

CODE	CATEGORY/NAME	PAGES
	DESERTS AND XERIC SHRUBLANDS	
Af2I	Afrotropical Hot Shrub Desert	121–126
	TROPICAL AND SUBTROPICAL MOIST BROADLEAF FORESTS	
Af4A	Afrotropical Lowland Rainforest	143–150
Af4F	Swamp Forest	169–173
	SAVANNAS	
Af6B	Northern Dry Thorn Savanna	227–233
Af6D	Moist Mixed Savanna	243–251
Af6F	Guinea Savanna	260–267
Af6G	Inselbergs and Koppies	268–274
	GRASSLANDS	
Af7A	Afrotropical Grassland	280–286
Af7F	Montane Grassland	293–299
	FRESHWATER HABITATS	
Af11A	Afrotropical Deep Freshwater Marsh	358–364
Af11B	Afrotropical Shallow Freshwater Marsh	365–369
Af11E	Freshwater Lakes and Ponds	379–381
Af11F	Rivers	382–385

CODE	CATEGORY/NAME	PAGES
	SALINE HABITATS	
Af12A	Afrotropical Mangrove	386–389
Af12B	Salt Pans and Lakes	390–392
Af12C	Tidal Mudflats and Estuaries	393–394
Af12D	Salt Marsh	395–396
Af12E	Sandy Beach and Dunes	397–399
Af12F	Rocky Shoreline	400–402
Af12G	Offshore Islands	403–406
Af12H	Pelagic Waters	406–408
	ANTHROPOGENIC	
Af13A	Humid Lowland Cultivation	409–411
Af13B	Savanna Cultivation	412–415
Af13F	Tree Plantations	422–423
Af13G	Cities and Villages	424–426
Af13H	Grazing Land	427–428

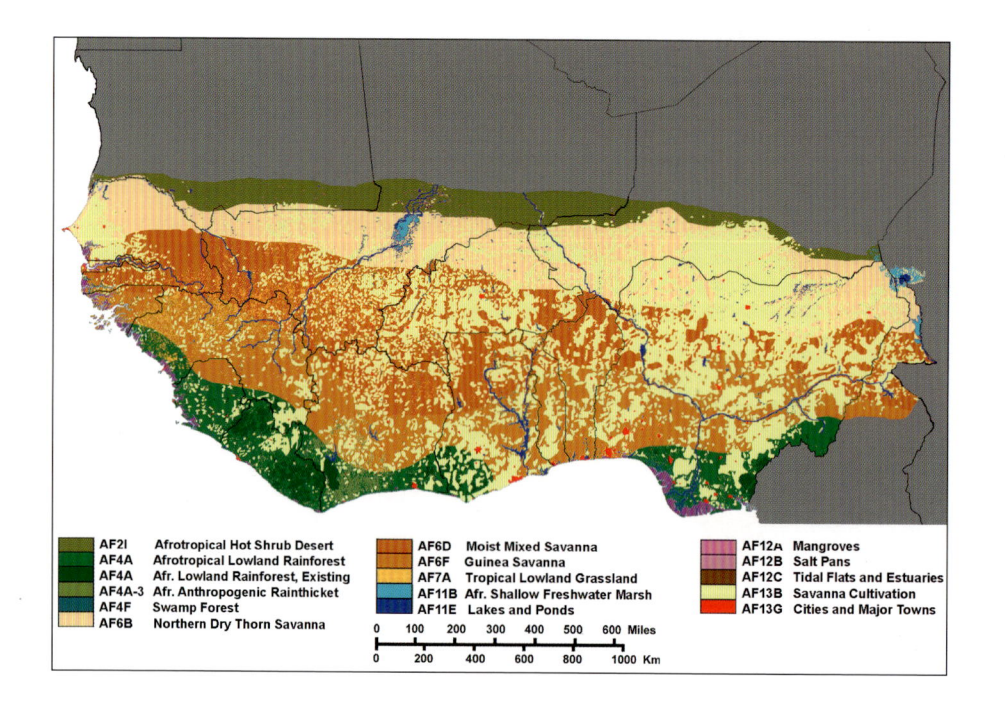

of the birdlife of the East African savannas wraps around the northern half of West Africa, with familiar examples such as Gray-headed Kingfisher, Lappet-faced Vulture, and Red-billed Firefinch.

In contrast to all other regions of Africa, there are no extensive or very tall mountain ranges in West Africa. There are elevated areas such as the Guinea Highlands, but not to a point where they have orographic rainfall and develop true montane rainforests or AFROPARAMO. There are some montane grasslands, but these have likely been produced by human-caused fire. Geologically, this zone is centered on a geological craton (extremely old and highly stable parts of the world), with some ancient rock over 2 billion years old at the surface. There are some soils that developed at least 14 MYA yet have remained at the surface unaltered, having received no additional weathering. Overall, this region has an extremely stable soil environment that has remained constant despite climate and vegetation changes; the soils tend to be iron oxide and silica-rich clay, which are highly inert and contain few other minerals and nutrients for plants to use. In some areas, these ancient soils form laterite caps and, in more recent soils, ferricretes (see sidebar 11, p. 126), which prevent most woody vegetation from growing but allows the growth of an edaphic grassland.

SIDEBAR 3 — THE SAHARA—THE REALM DICTATOR

The continent of Africa comprises two very distinct biogeographic realms separated by the Sahara (hence "sub-Saharan" Africa). To the south lies the Afrotropics with bird groups such as guineafowl, turacos, African barbets, mousebirds, African warblers, and hyliotas, and to the north is a rapid change to the Palearctic where loons, gulls, nuthatches, grouse, redstarts, buntings, finches, and *Curruca* warblers predominate. This pattern follows with other wildlife too. The Sahara is mostly linked to the desert realms of Asia, stretching through the Middle East and Indian subcontinent all the way to Mongolia's Gobi! Extending from the west coast of Mauritania through to the Red Sea and occupying an area the size of Brazil, some 25% of Africa, the Sahara is the world's largest "hot" desert, being eclipsed only by both frozen poles (cold deserts) in size. The massive inhospitable expanse helps to keep the fair-weather biota of the two realms isolated, as the desert forms a formidable barrier: a mixture of bare rock mountains, barren plateaus called Hamadas, bleak stony plains called Reg, windblown sand deserts called Erg, and the rare oasis. But some wildlife is able to find a way to cross the desert (see sidebars 6, p. 76 and 42, p. 383).

African Warblers like this Cape Crombec *(left)* are found only south of the Sahara, while *Curruca* warblers like this Tristram's Warbler *(right)* are common north of the Sahara.
© KEITH BARNES (LEFT); © KEN BEHRENS (RIGHT), TROPICAL BIRDING TOURS

CONIFERS

Af1A MAGHREB FIR AND CEDAR FOREST

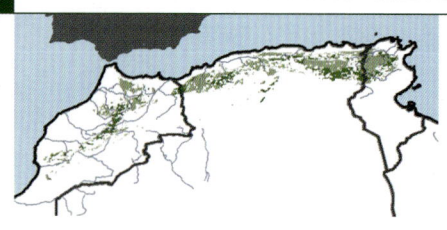

IN A NUTSHELL: Conifer forests dominated by fir and cedar trees that will be familiar to people from the Northern Hemisphere. **Global Habitat Affinities:** MEDITERRANEAN JUNIPER AND CYPRESS FOREST; NEARCTIC MONTANE MIXED-CONIFER FOREST. **Continental Habitat Affinities:** No similar habitats in Africa. **Species Overlap:** MAGHREB JUNIPER OPEN WOODLAND; MAGHREB PINE FOREST; MEDITERRANEAN PINE FOREST; MAGHREB BROADLEAF WOODLAND.

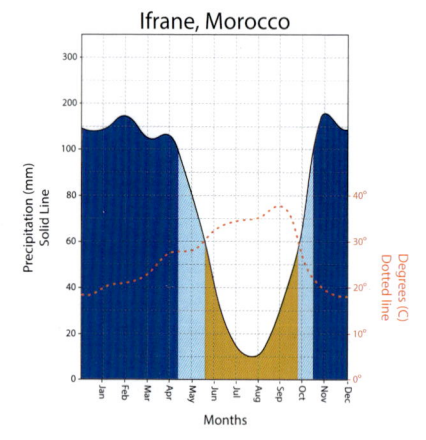

DESCRIPTION: Cedar and fir forests of North Africa are concentrated in the Atlas Mountains from Morocco to Algeria. They form an extension of the cedar and fir forests that surround the Mediterranean and even extend into the Caucasus Mountains between the Black and Caspian Seas. Because the Atlas Cedar (*Cedrus atlantica*) is almost identical to the Lebanese Cedar (*Cedrus libani*), these forests will feel very familiar to people who have visited Israel, the Iberian Peninsula, and the Greater Caucasus Mountains.

Along with some of the deciduous and evergreen oak forests of the Maghreb oak forests, the cedar and fir forests form in the wettest part of the northern slopes of the mountain ranges, where the moisture-laden winds of the Mediterranean Sea and Atlantic Ocean rise up the mountain ranges and orographic precipitation generates between 19 and 39 in. (500–1000 mm), rarely up to 83 in. (2100 mm), of rain and snow. Because this region mainly has a Mediterranean-type climate (hot dry summer/cold wet winter), at the higher elevations the early snowfalls can result in snow ground cover persisting from October through to April. Although some periods of the year have abundant moisture, other periods are very dry, so the plant communities of this habitat are xerophytic (adapted to low-water environments).

Atlas Cedar forest is confined to the mountains of Algeria, Morocco, and Tunisia and mainly occurs between 3280 and 9180 ft. (1000–2800 m). The Moroccan Fir (*Abies pinsapo marocana*, a subspecies of Spanish Fir) occurs between 4200 and 6890 ft. (1300–2100 m), though it is most common between 4920 and 6200 ft. (1500–1900 m). These forests almost always contain a component of Atlas Cedar, so they can be seen as clinal, where one extreme contains mainly fir and some cedar and the other is cedar with little fir. Atlas Cedar trees can be towering giants reaching 200 ft. (60 m) tall, although most are in the 130–160 ft (40–50 m) range. The very conical shape and lower height (65–80 ft./20–25 m) of the fir contrasts with the dome crown of the cedars, though very tall and old Moroccan Firs can become irregularly shaped and be less obvious in the canopy. The Algerian Fir (*Abies numidica*), which looks very similar to the Moroccan Fir, also exists in a mosaic within the cedar forests but prefers alkaline soils and tends to associate with clusters of oaks and another conifer, the English Yew (*Taxus baccata*).

Where the forests are thickest and in the hardest-to-access areas, they are dominated by Atlas Cedar, which grows on a variety of soils. The general structure of the forest is open, and the crowns of the Atlas Cedar rarely touch, though when fir trees are present, they grow much closer together and form a thicker canopy. There is also a component of broadleaf trees such as Holly Oak (*Quercus ilex*) and Montpellier Maple (*Acer monspessulanum*), with bushes of Common Holly (*Ilex aquifolium*), Wild Service Tree (*Sorbus torminalis*), and Cade Juniper (*Juniperus oxycedrus*). The Holly Oaks generally form the succession tree when a large cedar falls, but are replaced over time by cedars, and thus the oak spends years waiting for a chance to grow quickly, only to be replaced. Very little light penetrates to the ground around the firs, and the winter snowpack takes many months to melt; subsequently ground cover is limited to a few grasses under the firs and shrubs and orchids such as Broad-leaved Helleborine (*Epipactis helleborine*) and Red Helleborine (*Cephalanthera rubra*).

At the lower elevations, where orographic rainfall is less prevalent, these forests merge with, and are eventually replaced by, Aleppo Pine (*Pinus halepensis*), with some Maritime Pine (*Pinus pinaster*), Holly Oak, and Phoenician Juniper (*Juniperus phoenicea*). At these transitions, the Atlas Cedar canopy exists as emergents reaching 130 ft. (40 m) above the more uniform oaks, pines, and junipers, which exist as a subcanopy at around 50–65 ft. (15–20 m). At their upper limit where the climate is colder and drier, the cedar and fir forests become shorter and thin out to where they become a woodland and merge into the Spanish Juniper (*Juniperus thurifera*) of the MAGHREB JUNIPER OPEN WOODLAND or have well-defined boundaries with the HIGH ATLAS ALPINE MEADOW. Over much of their distribution, tree lopping and overgrazing has made these forests become much shrubbier and take on an almost MAGHREB MAQUIS nature, dominated by oak and juniper shrubs, with many climbers and grasses; eventually they look like an anthropogenic form of MAGHREB JUNIPER OPEN WOODLAND with a few remnant cedars and fir.

Also on the massifs in the Sahara to the south are scattered patches of cypress woodlands (which resemble the upper limits of the Maghreb Fir and Cedar Forest), though these areas have a distinctly summer rainfall, a much more open canopy, and shorter structure. These cypresses are evolved from trees that may be remnants of a much broader coniferous woodland that existed during the Mesozoic. This is fascinating, because it implies that the Sahara had a climate that allowed the existence of broadleaf

Maghreb Fir and Cedar Forest can occur up to 9000 ft. (2800 m) elevation. Igunane, Morocco.
© NICK ATHANAS, TROPICAL BIRDING TOURS

forest, but not surprising, considering that angiosperms (flowering plants) only evolved in the Cretaceous (the later part of the Mesozoic), so as long as the climate was conducive to plant growth, conifers had little competition from Angiosperms.

WILDLIFE: Mammals are few in these forests. Several endemic races of megafauna that formerly inhabited these mountains have gone extinct in historical times, including Atlas Brown Bear, "North African" Savanna Elephant (the famous war elephants of Carthage), Bubal Red Hartebeest, Red Gazelle, and Atlas Wild Ass. The Roman Empire may have been responsible for the extinction of a couple of these species. Barbary Lion went extinct in the early 1900s, and "Barbary" Leopard (a subspecies of African Leopard) may still occur in these forests, although it is unlikely that a viable population persists. Barbary Macaque still exists here and relies on cedars in winter, though they seem to be doing much better in Gibraltar than in the Atlas Mountains. African Wolf (formerly known as Golden Jackal) occurs in small numbers, although it is much rarer than farther south in the continent. Lataste's Gerbil survives in both natural stands of cedar and areas turned over to cultivation along windbreaks of cedar and hedgerows. Black-tailed Garden Dormouse and Wood Mouse occur both in cedar forests and juniper woodlands.

Birdlife in these forests is remarkably similar to that of coniferous forests of s. Europe, with species such as Common Cuckoo, Great Spotted Woodpecker, Eurasian Wren, Short-toed Treecreeper, European Robin, Common Redstart, Mistle Thrush, Spotted Flycatcher, Great Tit, Eurasian Nuthatch, Common Firecrest, Eurasian Golden Oriole, Eurasian Jay, and Hawfinch, as well as a swath of warblers such as Common Whitethroat, Eurasian Blackcap, Melodious, Western Subalpine, and Western Bonelli's Warblers. Birds found here but not in Europe include Barbary Partridge, Levaillant's Woodpecker, and Atlas Flycatcher. The rare endemic Algerian Nuthatch prefers forest dominated by Atlas Oak (*Quercus tlemcenensis*) and Atlas Cedar. Reptiles of these forests include Kolliker's Glass Lizard (*Hyalosaurus koellikeri*), Atlas Mountain Viper (*Vipera monticola*), and Lataste's Lizard (*Ophiomorus latastii*).

CONSERVATION: These forests are mere remnants of their former massive extent across the Maghreb mountain ranges. Extensive clearing for agriculture has turned lower elevations into anthropogenic farmland. Increased fire frequency and intensity has caused some cedar and fir forests to be replaced by the far more fire-resistant Spanish Juniper, while overgrazing and tree lopping has caused the upper limit to be converted to alpine grasslands and shrublands. Most of the large mammals have been extirpated, but this habitat remains important for migrating birds.

These forests combined with the **MAGHREB BROADLEAF WOODLAND** are the most important habitats for many Palearctic migrants that winter in Africa, as these are the most significant "wet" habitats between the forests of the Alps and the rainforest and moist savanna of West Africa.

Maghreb Fir and Cedar Forest mainly has plant and animal species that are shared with Europe, like Common Firecrest.
© KEN BEHRENS, TROPICAL BIRDING TOURS

Af1B MAGHREB JUNIPER OPEN WOODLAND

Algerian Nuthatch is one of the few local and restricted-range endemics of this habitat.
© ANDREW SPENCER

Degradation of these forests is placing increasing stress on these migrants, which are already enduring harsher migration conditions across the Sahara because of climate change.

DISTRIBUTION: In Algeria Maghreb Fir and Cedar Forest is common on the northern slopes of Aurès, and in Morocco it is extensive in the northern slopes of the Middle Atlas and Rif Mountains. Small patches exist in the Kroumirie and Mogod Mountains of Tunisia. Although cedar and fir forests still cover hundreds of thousands of hectares, they are much reduced, especially on lower slopes.

WHERE TO SEE: Ifrane, Morocco; Talassemtane National Park, Morocco; Foret des Cedres de Tizi Ouzou, Algeria.

Af1B MAGHREB JUNIPER OPEN WOODLAND

IN A NUTSHELL: Open, generally low, conifer woodlands dominated by juniper, with a grassy understory. **Global Habitat Affinities:** MIDDLE EASTERN JUNIPER FOREST; PINYON-JUNIPER WOODLAND. **Continental Habitat Affinities:** AFROTROPICAL MONTANE DRY MIXED WOODLAND. **Species Overlap:** MAGHREB PINE FOREST; MAGHREB BROADLEAF WOODLAND.

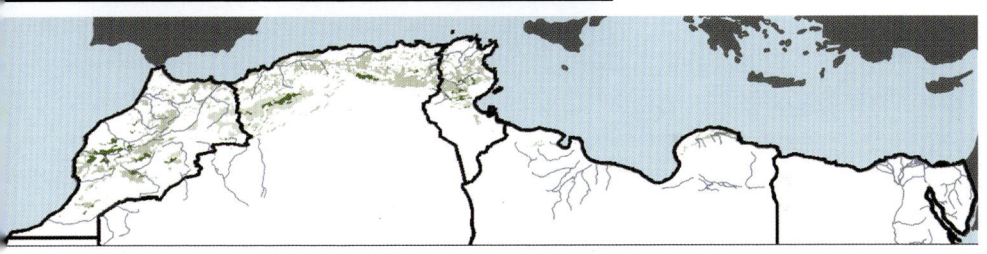

DESCRIPTION: The Atlas Mountains tower above the Mediterranean Sea and Sahara Desert, stretching from s. Morocco to Tunisia. Part of the mix of habitats in these mountains and the surrounding lowlands is a dry, open, juniper-dominated woodland. This type of habitat is also found in many other parts of the Mediterranean, such as the Iberian Peninsula and Israel. The north-facing slopes of the Atlas Mountains have a typical Mediterranean mild wet winter–hot dry summer climate (Köppen **Csb**) and annual precipitation between 20 and 39 in. (500–1000 mm). The south-facing slopes adjoining the Sahara receive much lower annual precipitation of 10–27 in. (250–700 mm) and are subject to frequent desiccating winds from the Sahara.

Djelfa, Algeria

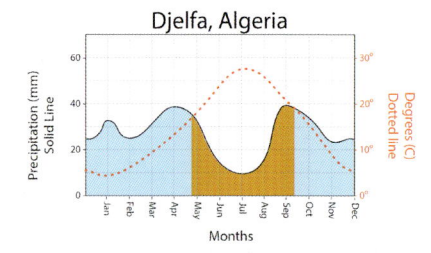

(Precipitation (mm), Solid Line; Degrees (C), Dotted line)

Left: **High-elevation juniper woodland at Oukaïmeden in the Atlas Mountains, Morocco. Juniper often grows in harsh areas where other trees cannot survive.**
© NICK ATHANAS, TROPICAL BIRDING TOURS

Below: **This is one of the habitats favored by the North African endemic Moussier's Redstart.**
© DANIELE ARDIZZONE, FLUYENDO PHOTOGRAPHY

Af1B MAGHREB JUNIPER OPEN WOODLAND

Throughout the semiarid parts of the Maghreb, juniper woodlands occur both as near-monotypic stands and as minor elements mixed with pines and oaks, in the case of Phoenician Juniper (*Juniperus phoenicea*), and with fir/cedar forests, in the case of Spanish Juniper (*J. thurifera*). As a general rule, juniper grows in a wide variety of areas where other trees struggle to grow. At high elevations, above the "forest" tree line and along the boundary with the Sahara, the landscape becomes dominated by junipers with their own faunal assemblage. Although the juniper forests and woodlands of the Maghreb have been seriously degraded by humans, they still form some extensive tracts. On the northern slopes, the Phoenician Juniper forests are replaced by the Spanish Juniper groves at higher elevations, and they are in turn replaced by Common Juniper (*J. communis*) shrublands in the highest areas.

Phoenician Juniper again forms large stands on the south slope bordering the Sahara. This tree usually grows in open stands as a large shrub to small rounded tree, with a height of 13–23 ft. (4–7 m) and an overall pear shape, though individuals of just under 32 ft. (10 m) in height with 23 in. (60 cm) diameter trunks have been recorded, and some larger trees take on a lopsided appearance. Spanish Juniper is found over a much wider variety of soils and seems more constrained by climate than by soil type. This species can grow very large, up to 50 ft. (15 m) tall with a 6.5 ft. (2 m) diameter trunk, though they are usually around 32–40 ft. (10–12 m) tall. The canopy is usually domed although some individual trees take on a Christmas tree-like appearance.

These juniper woodlands (though often referred to as forests) are open with little canopy overlap, and plenty of light reaches the understory and ground, though it is probable that extensive closed-canopy Spanish and Phoenician Juniper forests existed within Roman times. In most Phoenician Juniper stands the canopy is shared with species such as Algerian Ash (*Fraxinus xanthoxyloides*), Wild Olive (*Olea europaea*) and Cade Juniper (*Juniperus oxycedrus*), with a well-developed shrub layer including Balearic Boxwood (*Buxus balearica*), Mediterranean Buckhorn (*Rhamnus alaternus*), Egyptian Lavender (*Lavandula multifida*), and Scorpion Broom (*Genista scorpius*). In more open areas where the junipers cover less than half the area, White Wormwood (*Artemisia herba-alba*) becomes prevalent in the shrub layer. Perennial grasses such as Halfah (*Stipa tenacissima*) cover the ground in drier areas.

The lower reaches of the Spanish Juniper woodlands merge into MAGHREB FIR AND CEDAR FOREST, with their typical understory of Holly Oak (*Quercus ilex*) and Montpellier Maple (*Acer monspessulanum*), alongside bushes of Common Holly (*Ilex aquifolium*), Wild Service Tree (*Sorbus torminalis*), and Cade Juniper. In higher, more open and shrubbier woodlands the juniper share the subcanopy with shrubby trees like European Box (*Buxus sempervirens*), the deciduous hawthorn species *Crataegus laciniata*, and honeysuckles (*Lonicera*). The smaller shrub layer becomes more important with Mountain Cherry (*Prunus prostrata*) and Spurge Laurel (*Daphne laureola*), that gradually gets replaced by small, very shrubby and sometimes krummholzed (windblown in one direction) Spanish Juniper. Although fire-resistant, and thus often left as a remnant after the death of surrounding Atlas Cedars (*Cedrus atlantica*) and angiosperms (flowering plants), Spanish Juniper woodland does not easily regenerate because of intense overgrazing over most of its distribution. However, if grazing pressure were reduced, the juniper woodlands would quickly regenerate. The existing coastal juniper forests of Phoenician Juniper are limited to stunted coastal thickets associated with dunes and oaks, so are best treated as a form of MAGHREB MAQUIS.

CONSERVATION: Juniper woodlands have been heavily degraded over most of their range. On the northern side of the Atlas and along the coast, they have mainly been turned over to cultivation or overgrazed to the degree that they have changed to MAGHREB MAQUIS and MAGHREB GARRIGUE. On the southern side of North African mountain ranges, juniper woodlands have been severely degraded to shrub desert and even into outright desert through extensive clearing and overgrazing.

Mixed juniper and olive woodland. Most of North Africa has been heavily populated for a very long time. Middle Atlas, Morocco. © CARLOS N. G. BOCOS

WILDLIFE: For a discussion of the now-extinct megafauna of North Africa see the **MAGHREB FIR AND CEDAR FOREST** account. Where Maghreb Juniper Open Woodland is surrounded by very arid lands, it has a distinct suite of wildlife that is very different from the surrounding areas of shrub desert and dry thornscrub. Rock Dove, Stock Dove, Common Woodpigeon, and European Turtle-Dove all occur. Blue Rock-Thrush and Common Rock-Thrush breed, and Ring Ouzel is a particularly common winter migrant. Tristram's Warbler breeds in the coastal juniper woodlands, alongside Greater Whitethroat, especially where these woodlands have been degraded slightly. Barbary Partridge; Golden Eagle; Short-toed, Booted, and Bonelli's Eagles; Long-legged Buzzard; Black and Black-winged Kites; Eurasian Sparrowhawk; Common Kestrel; Little Bustard; Moussier's Redstart; and Black-eared Wheatear occur in grassy openings within Maghreb Juniper Open Woodland, where this habitat almost takes on a wooded steppe appearance.

DISTRIBUTION: Phoenician Juniper is found throughout the northern Maghreb. Related patches of juniper are dotted along the northern fringe of the Sahara all the way to Somalia.

WHERE TO SEE: Souss-Massa National Park, Morocco; Taza National Park, Algeria; Djebel Orbata Nature Reserve, Tunisia.

Left: **Spanish Sparrow is a habitat generalist that can be found in juniper woodland.**
© KEITH BARNES, TROPICAL BIRDING TOURS

Below: **Although the mammal population of this habitat has been vastly reduced, it retains some mammals—mostly typical of the Palearctic—such as Red Fox.**
© KEITH BARNES, TROPICAL BIRDING TOURS

Af1C AFROTROPICAL MONTANE DRY MIXED WOODLAND

IN A NUTSHELL: Dry woodland and forest found at middle and high elevations of sub-Saharan Africa, which can be pure broadleaf, pure conifer, or more often a mix of both. **Global Habitat Affinities:** SUBHUMID YUNGAS. **Continental Habitat Affinities:** MAGHREB JUNIPER OPEN WOODLAND; MOIST MONTANE FOREST; MIOMBO; EAST COAST FOREST MATRIX. **Species Overlap:** MOIST MONTANE FOREST; MOIST MIXED SAVANNA; MONTANE HEATH; MONTANE GRASSLAND; MIOMBO; EAST COAST FOREST MATRIX.

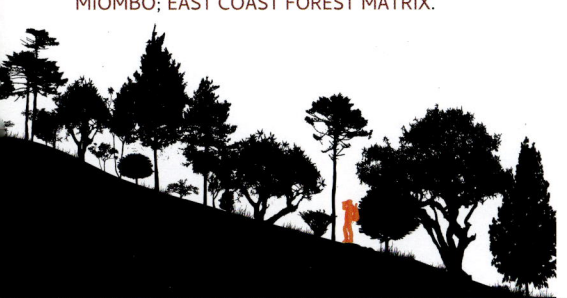

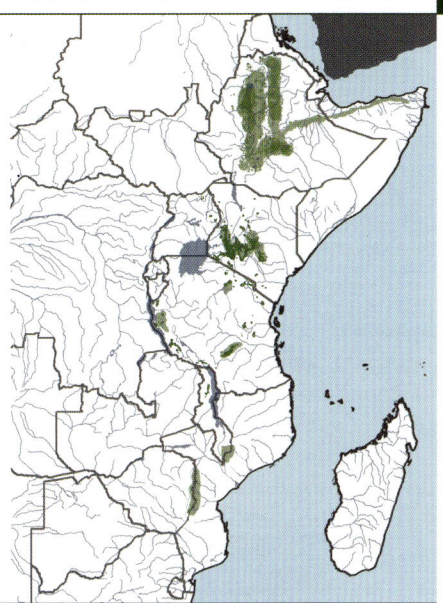

DESCRIPTION: There is great internal complexity within Afrotropical montane forests. Much of this diversity is driven by geography—different mountain blocks have very different biodiversity—but some is linked to climatic differences. While these climatic changes often occur gradually, and incrementally, there is a striking difference between classic mossy MOIST MONTANE FOREST and the drier montane forests and woodlands. Teasing these apart can be difficult, but there is enough difference both in terms of drier woodlands' wildlife assemblage and their plant composition to qualify it as distinctive.

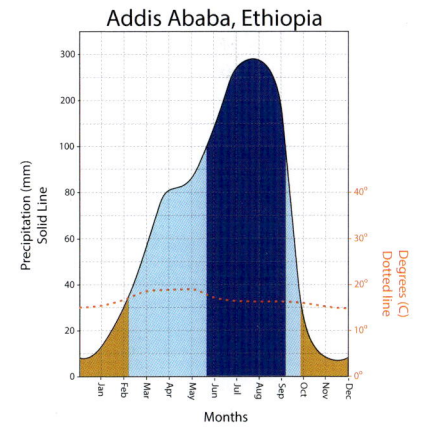

Moist Montane Forest (Köppen **Csb**) can be found in areas with annual precipitation as low as 40 in. (1000 mm), while dry montane woodland and forest grows in a narrow precipitation band just below and slightly overlapping this. The driest juniper-dominated woodland (Köppen **Bwa**) in Somalia receives around 27 in. (700 mm) of rain, while the upper limit for dry montane forest is around 45 in. (1150 mm). Dry woodland and forest represent a transition from Moist Montane Forest to other habitats. They also seem to create a sort of buffer around montane forest; fascinatingly, dry montane forest is not bound to a specific elevation but can, in some places, occur both above and below Moist Montane Forest. This odd pattern of distribution mirrors that of North Africa's MAGHREB JUNIPER OPEN WOODLAND. What makes this habitat a good buffer is that it is far more tolerant of fire and long dry periods than moist forest. Dry montane woodland is generally found from 5900 to 9800 ft. (1800–3000 m) but locally can be

CONIFERS

> **SIDEBAR 4 — THRIVING ON MODERATION**
>
> Every continent and region has some generalists. Even a place like Madagascar that was likely dominated by forest before humans arrived has a few generalists that use a wide variety of habitats. But something notable about sub-Saharan Africa is the huge suite of birds that use a similarly broad set of habitats, all of which have a moderately open structure. The extremes of this broad niche run from quite arid to quite moist: from the type of MOIST MIXED SAVANNA found along watercourses running through dry thorn savanna all the way to the edges of MOIST MONTANE FOREST and AFROTROPICAL LOWLAND RAINFOREST. These habitats include the likes of MIOMBO, GUINEA SAVANNA, GUSU, MOPANE, AFROTROPICAL MONTANE DRY MIXED WOODLAND, and portions of the EAST and SOUTH COAST FOREST MATRICES. This is obviously an expansive niche, stretching across a huge diversity of climatic conditions. It seems counterintuitive that so many birds would use this whole range of habitats. Perhaps it is best explained by Africa's biogeographic history, in which there was always abundant open habitat but in which wet and dry periods alternated, expanding and contracting the extent of the wettest and driest habitats. While species specialized in desert or the interior of moist forest would suffer boom or bust in this cycle, species using the intermediate, moderately open habitats, would have always found niches in which to thrive.

found as low as 3300 ft. (1000 m). Since this habitat is not strictly linked to an elevation range, it occurs in a wide variety of temperature regimens, from moderate/warm up to areas with near-freezing nighttime temperatures.

The most typical trees are African Juniper (*Juniperus procera*), Hagenia (*Hagenia abyssinica*), and yellowwoods (*Podocarpus* and *Afrocarpus*). Juniper sometimes dominates this habitat, especially in its driest forms, as in n. Somalia and Eritrea, or where it burns frequently. Juniper-dominated woodland seems to depend on fire for its propagation and has benefited from human-caused fires, the same dynamic that has permitted the expansion and proliferation of montane grasslands. But even juniper cannot absorb repeated intense fires and can eventually be replaced by grassland. The juniper woodland of Eritrea and Sudan is closely allied with outliers occurring on Saharan mountains, but those are treated under ROCKY HAMADA AND MASSIF. Hagenia is a broadleaf tree with large compound leaves, usually growing in a dense round shape. It can occur within Moist Montane Forest, though usually as an edge or successional tree. In both Ethiopia and East Africa, Hagenia is also typical of the high-elevation transitional zone from Moist Montane Forest to MONTANE HEATH. It occurs there in an odd mix alongside giant heath plants, often draped in picturesque epiphytic moss. In some dry areas, especially in Ethiopia, virtually monotypic stands of Hagenia occur. These may also be products of human disturbance. In some places, Hagenia-dominated forest may also be a product of heavy browsing by ungulates, such as African Buffalo. More typically, Hagenia is mixed with juniper and yellowwood, with the latter also being a typical tree of Moist Montane Forest. Mixed woodland and forest is the most common forest type in Ethiopia, where true Moist Montane Forest occurs in only the south. Like juniper, Hagenia is somewhat fire-resistant but can succumb after repeated intense fires.

Sub-Saharan Africa is poor in conifers, but those that do occur are fascinating in various ways. African Juniper is likely a recent colonist and is closely related to the Greek Juniper (*Juniperus excelsa*) of Eurasia. Yellowwoods are remnants of the flora of Gondwanaland and are mainly found in the southern two-thirds of the globe. The only other genus of coniferous tree in the Afrotropics is the *Widdringtonia* cedars. There are several species, distributed from the Western Cape of South Africa

Af1C AFROTROPICAL MONTANE DRY MIXED WOODLAND

north and east to s. Malawi. The Clanwilliam Cedar (*Widdringtonia wallichii*) is namesake of and endemic to the Cederberg Mountains of sw. South Africa. The closest relatives of these African cedars are two genera of cedars from Australia. While fascinating, these southern cedars occur essentially as isolated bushes and trees within other habitats, most frequently montane FYNBOS. The only place where they form a cohesive woodland is on Mt. Mulanje in s. Malawi, where two species are found along with a mix of broadleaf species.

In addition to the Hagenia, Juniper, and/or Cedar formations mentioned above, there are other types of drier montane woodland and forest that are not easily categorized. These largely or completely lack these characteristic trees but instead comprise a wide variety of broadleaf species. They often resemble moist MIOMBO, and, in terms of wildlife, have a mix of savanna and Moist Montane Forest species. Most of this dry woodland and forest was destroyed long ago, as it lies in a zone that is optimal for human cultivation. Remnants are preserved in some of the lower elevations of Arusha National Park, in Tanzania, and the western parts of Nairobi National Park in Kenya. Some tree species typical of this dry montane broadleaf formation include yellowwood (*Podocarpus*), Olive (*Olea europaea*), Forest Elder (*Nuxia floribunda*), African Cherry (*Prunus africana*), Brittlewood (*Nuxia congesta*), Wild Peach (*Kiggelaria africana*), Cape Holly (*Ilex mitis*), Stinkwood (*Ocotea bullata*), and Cape Beech (*Rapanea melanophloeos*). Note that all these are also found in some versions of Moist Montane Forest, and many are listed in this book as typical trees of that habitat! Some are also found in MOIST MIXED SAVANNA and other moist savanna habitats. As explained in

Juniper woodland is unevenly scattered at the lower elevations of Bale Mountains National Park, Ethiopia. © KEN BEHRENS, TROPICAL BIRDING TOURS

A dry montane woodland with a mix of juniper and broadleaf elements. This habitat forms a transition between the savanna habitats of lower elevations and Moist Montane Forest at higher elevations. Arusha National Park, Tanzania. © KEN BEHRENS, TROPICAL BIRDING TOURS

the Wildlife section below, this is a habitat made up of a mix of elements from adjacent habitats. In parts of s. Tanzania, e. Zimbabwe, Malawi, Mozambique, and e. Zimbabwe, where Moist Montane Forest above transitions to Miombo broadleaf woodland below, a similar type of transitional dry forest is found, again with a mix of Moist Montane Forest and Miombo wildlife.

Another type of drier montane habitat with woody vegetation is found in Southern Africa: the Ouhout (*Leucosidea sericea*) shrublands that occur along the edges of the Highveld MONTANE GRASSLAND (see sidebar 32, p. 299). The wildlife of this shrubland is essentially an extremely reduced subset of what is found in South African Moist Montane Forest. Yet another formation that could be placed in the dry montane woodland category is the higher-elevation *Acacia* woodlands that are found all the way from Abyssinia south to Malawi. These also form a transition habitat from Moist Montane Forest to grassland and savanna habitats, but given the dominance of Acacia, we consider them a high-elevation form of Moist Mixed Savanna.

The exact structure of dry montane woodland and forest is highly variable. At the most extreme end of aridity and openness, they can form a savanna-type environment with scattered juniper trees at a height of around 20 ft. (6 m). A more typical formation is a matrix of mixed open woodland and closed forest, with a canopy of 30–65 ft. (9–20 m). The dry forest on Malawi's Mt. Mulanje has a canopy around 90 ft. (27 m), with emergent cedars as tall as 130 ft. (40 m). The n. Somali mountains can have junipers alongside a mix of evergreen shrubs that structurally resemble the MAGHREB MAQUIS of North Africa. Tree and shrub species in this Somali formation include *Pistacia aethiopica* and *Buxus hildebrandtii*.

CONSERVATION: Across its range, this habitat is highly threatened. Most of the lower-lying dry montane woodland and forest in East Africa has already been destroyed. The higher-elevation versions, occurring above montane forest, have been less affected. This type of woodland is prone to fire and can quickly be replaced by grassland, especially in areas where humans increase fire frequency. The Mulanje Cedar (*Widdringtonia whytei*) has been heavily logged and the extent of its woodlands reduced by around half.

WILDLIFE: Dry montane woodland is not particularly rich or distinctive in mammals. Most of the large mammals are widespread and use a broad range of habitats, including adjacent savanna and/or

The dry woodlands of the Abyssinian highlands are rich in localized birds such as the stunning White-cheeked Turaco.
© AYUWAT JEARWATTANAKANOK

MOIST MONTANE FOREST. These include Angola and Guereza Colobuses, African Savanna Elephant, Leopard, Slender Mongoose, African Buffalo, Blue Duiker, Suni, and Bushbuck. Dry woodland is one of the key habitats for the Ethiopian-endemic Mountain Nyala and is also sometimes used by Ethiopian Wolf and Walia Ibex. All are considered Threatened by International Union for Conservation of Nature.

Dry montane woodland isn't a premier habitat for birding but does hold a good variety of birds, probably due to the prevalence in Africa of species that key into moderately open habitats (see sidebar 4, p. 62). Some typical and widespread birds of dry woodland and forest are Rameron Pigeon, Red-eyed Dove, Dusky Turtle-Dove, Mountain Gray Woodpecker, Red-fronted Tinkerbird, Scaly-throated Honeyguide, White-eyed Slaty-Flycatcher, African Dusky Flycatcher, Brown Woodland-Warbler, several species of white-eyes, Northern and Black-backed Puffbacks, Common Bulbul, Black Sawwing, and Tacazze and Bronze Sunbirds. This is one of the preferred habitats for several raptors, including Augur Buzzard, African Cuckoo-Hawk, and African Goshawk. Although this dry habitat isn't as lush as Moist Montane Forest, there is still often a dense understory, which provides refuge for skulking species such as Abyssinian Ground-Thrush; Abyssinian Thrush; Rüppell's, White-browed, and Cape Robin-Chats; Cinnamon Bracken Warbler; Tropical Boubou; and Green-backed Camaroptera. As in adjacent savanna, there is a good variety of seedeaters, including Baglafecht Weaver, African and Southern Citrils, Streaky Seedeater, Bronze Mannikin, and Yellow-bellied Waxbill. Dry woodland, especially that dominated by conifers, is favored by owls, notably Abyssinian Owl, African Wood-Owl, and Cape Eagle-Owl. At dusk, this habitat is often serenaded by the haunting song of Montane Nightjar. This isn't as rich a habitat for Palearctic migrants as lower-lying savanna but is favored by Tree Pipit, Eurasian Blackcap, and Spotted Flycatcher.

Endemism: Africa's most exciting dry montane woodlands for birding are those in the Abyssinian highlands, where this habitat is widespread and common. It supports more Abyssinian endemic birds than any other habitat, including Yellow-fronted Parrot, White-cheeked Turaco, Abyssinian Woodpecker, White-backed Tit, and Abyssinian Catbird. It is also used by several other endemics, such as White-collared Pigeon, Black-winged Lovebird, Prince Ruspoli's Turaco, and Banded Barbet.

The mountains of n. Somalia form another node of endemism. They share some of the characteristic birds of the Abyssinian highlands and also have a couple pure endemics, namely Somali Thrush and Warsangli Linnet.

Southern Malawi is a minor area of endemism, though it has a mix of moist and dry forest. Species found on Mt. Mulanje include the widespread Crowned Hornbill plus local endemics Black-browed Mountain Greenbul and Yellow-throated Apalis.

Banded Barbet is an Abyssinian endemic that uses both montane Moist Mixed Savanna and dry woodland habitats.
© KEITH BARNES, TROPICAL BIRDING TOURS

DISTRIBUTION: Found from the hills of se. Sudan, across into Djibouti, then in a narrow band across the mountains of n. Somalia. This is the most common montane forest formation throughout Ethiopia. In slightly moister East Africa, the broadleaf-dominated versions of this habitat are found locally, blending gradually into MOIST MONTANE FOREST, from which they are not easily distinguished. Mt. Mulanje supports an outlier dry montane cedar woodland. Drier montane woodland is found in the eastern part of the MIOMBO zone, at the transition from Moist Montane Forest to Miombo savanna.

WHERE TO SEE: Bale Mountains National Park, Ethiopia; Debre Libanos, Ethiopia; Nairobi National Park, Kenya; Arusha National Park, Tanzania.

Top: **Guereza Colobus uses a wide variety of woodland and forest types, though dry montane forest is one of its favored habitats.** © KEN BEHRENS, TROPICAL BIRDING TOURS

Left: **Some widespread Afrotropical savanna species such as Levaillant's Cuckoo will also use dry montane woodland.** © KEITH BARNES, TROPICAL BIRDING TOURS

Af1D MAGHREB PINE FOREST

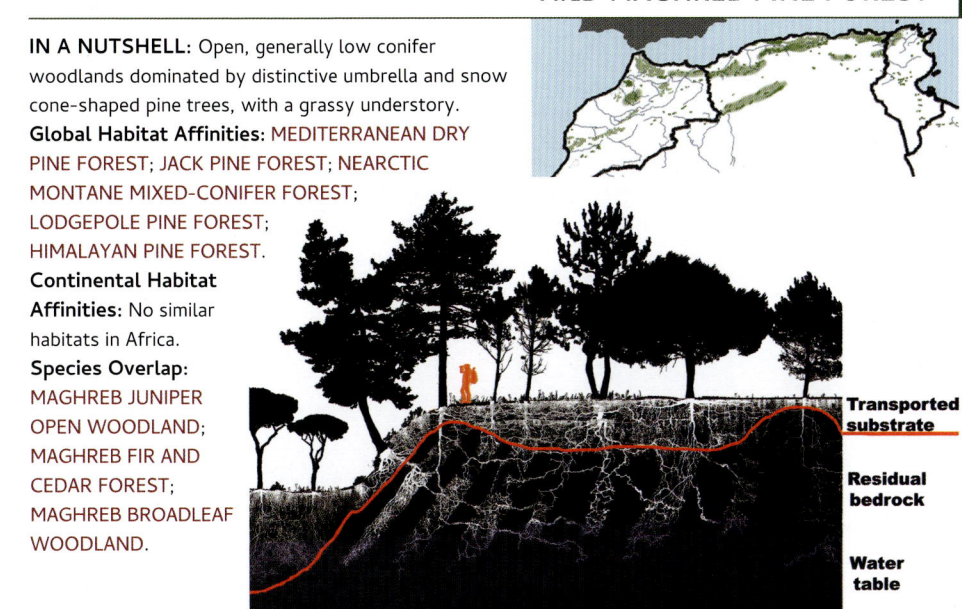

IN A NUTSHELL: Open, generally low conifer woodlands dominated by distinctive umbrella and snow cone-shaped pine trees, with a grassy understory.
Global Habitat Affinities: MEDITERRANEAN DRY PINE FOREST; JACK PINE FOREST; NEARCTIC MONTANE MIXED-CONIFER FOREST; LODGEPOLE PINE FOREST; HIMALAYAN PINE FOREST.
Continental Habitat Affinities: No similar habitats in Africa.
Species Overlap: MAGHREB JUNIPER OPEN WOODLAND; MAGHREB FIR AND CEDAR FOREST; MAGHREB BROADLEAF WOODLAND.

DESCRIPTION: Mixed or monotypic pine forests are the most common and widespread forests in the Maghreb (Western Mediterranean) region of Africa. The dominant pines are medium-sized (40–80 ft./15–25 m) and have an open growth form, red-stained trunk, and long needles. Although many of the individual plant species are endemic, this habitat is an extension of the widespread xerophytic (tolerant of dry conditions) conifer forests that surround the Mediterranean from the Iberian Peninsula all the way to the Middle East in regions with the hot summer-dry summer (Köppen **Csa**) Mediterranean climate.

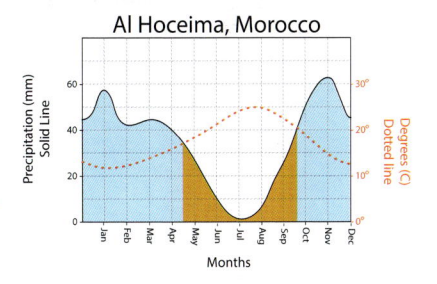

Aleppo Pine (*Pinus halepensis*) is the most prevalent of the pines in North Africa, having both a limited native distribution as well as being planted over large parts of the Mediterranean coastal plain. Arartree, a *Tetraclinis* conifer that looks similar to a small juniper or even *Callitris* of Australia, is common in the west of the region. The very distinctive Umbrella or Stone Pine (*Pinus pinea*), with its distinct umbrella-shaped growth form when young and below 40 ft. (12 m) in height, existed in n. Africa in prehistoric times, was eradicated, but recently has naturally recolonized wild areas as well as being cultivated. Pine forests usually have a canopy between 40 and 60 ft. (12–18 m) in height. They are associated with Phoenician Juniper (*Juniperus phoenicea*) and broadleaf trees such as Algerian Oak (*Quercus canariensis*), Cork Oak (*Quercus suber*), Holm Oak (*Quercus ilex*), and Wild Olive (*Olea oleaster*) The shrub layer is often composed of Turbith (*Athamanta turbith*), Egyptian Lavender (*Lavandula multifida*), and multiple rosemary species.

Above: **Pine forest on the Canary Islands, comprising the endemic Canary Island Pine (*Pinus canariensis*).**
© VINCENT LEGRAND

Right: **Maghreb Pine Forest is one of several treed habitats used by the North African endemic Levaillant's Woodpecker.** © KEN BEHRENS, TROPICAL BIRDING TOURS

The plants of this habitat are highly resistant to drought, desiccating winds, and fire but require lots of light, so they cannot establish as an understory in the shade of forests dominated by preexisting larger conifers. When fires become too frequent, Aleppo Pines are replaced, and Arartree becomes the locally dominant species due to its ability to regenerate faster than other species. Pines occur on a wide range of soils in drier and hotter areas, but in higher rainfall and higher elevation areas, pines are restricted to less productive calcareous soils and are outcompeted by firs and cedars in better soils. In the sandiest soils, Umbrella Pine dominates, and it has been planted along coastlines. The lower pine forests are also restricted by temperature and can survive to about −4°F (−20°C), though Arartree is far less tolerant of the cold. An exception to the above description are the stands of Maritime Pine (*Pinus pinaster*) that occur in the subhumid zones, where precipitation is between 31.5 in. (800 mm) and 47 in. (1200 mm), and are able to withstand long periods of cold weather. This species often grows in transitional areas from other pine forests to the fir and cedar forests of higher elevations, and they share most of their shrub layer plants with oak forests and the degraded heathlands of the MAGHREB MAQUIS. The moister w. Canary Islands support montane pine forest and woodland of the endemic Canary Island Pine (*Pinus canariensis*).

Af1D MAGHREB PINE FOREST

This habitat has been extensively modified by grazing, logging, and firewood gathering, and few if any pristine examples remain. Although these forests can have an understory, ungrazed examples are so rare that almost all the existing forests have an open understory and are easy to walk through. During the moist winter, the ground is covered with grasses, forbes, and ferns. In summer, most plants die off and there are extensive areas of bare sandy ground.

CONSERVATION: Most of the original pine forests have been destroyed, and most existing pine forests are managed to some degree for cones and timber.

WILDLIFE: For a discussion of the now-extinct megafauna of North Africa, see the MAGHREB FIR AND CEDAR FOREST account. The Cuvier's Gazelle, endemic to the Maghreb, persists in open pine forests, which it prefers, especially in the Tunisian Atlas, although it will also use shrubby desert edges. This species underwent a dramatic decline due to overhunting and was thought extinct, but numbers have increased through captive breeding and release programs. Lataste's Gerbil occurs in the upper elevations of the Aleppo Pine forests.

There are several resident birds that are common throughout Maghreb Pine Forest, such as Great Spotted Woodpecker, Short-toed Treecreeper, Eurasian Jay, European Robin, Eurasian Blackbird, Mistle Thrush, Common Bulbul, Common Nightingale, Black Redstart, Coal and Great Tits, and Eurasian Nuthatch. Many species in this habitat are summer-breeding migrants, such as Booted Eagle, Common Nightingale, and Spotted Flycatcher. Stonechat and Blue and Common Rock-Thrushes breed in rocky areas. Barbary Partridge is a resident. Golden Eagle; Short-toed, Booted, and Bonelli's Eagles; Long-legged Buzzard; Black Kite and Black-winged Kite; Eurasian Sparrowhawk; and Common Kestrel all migrate through. Red-necked Nightjar is common in more open pine woodland with few shrubs and abundant grass. Common, Alpine, and Little Swift; Plain, House, and Crag Martins; and European and Red-rumped Swallows all breed, and they feed in the sky above pine forest. Some species are pine specialists, the most spectacular of which is the North African endemic African Blue Tit. Both Hawfinch and Red Crossbill prefer Aleppo Pine, where they can be found alongside the localized Atlas Flycatcher. Other localized specialties that use these woodlands include Maghreb Owl, Maghreb Magpie, and Levaillant's Woodpecker, which uses a variety of dry open woodlands including Aleppo Pine. Of the reptiles occurring in this habitat, perhaps the most charismatic is the Mediterranean Chameleon.

DISTRIBUTION: Maghreb pine habitat occurs widely across North Africa. Planted pine forests, particularly Aleppo Pine, can be found in various parts of Morocco from the coastal areas of Western Sahara right across the North African coastline to Egypt. Natural pine forests are also found on the coast, though are more common in the Middle Atlas and Riff Mountains. They naturally occur in the lower regions from sea level to around 6000 ft. (2000 m). Much of the area is dominated by Aleppo Pine, with tracts of tens of thousands of hectares. On the Canary Islands, pine forest occurs from 4000 to 6200 ft. (1200–1900 m).

WHERE TO SEE: Natural forests around Ifrane National Park, Morocco; planted forests around Tangiers, Morocco; Oran, Algeria; Cap Bon, Tunisia.

African Blue Tit, the "marquee bird" of North African pine forest. © DANIELE ARDIZZONE, FLUYENDO PHOTOGRAPHY

African Deserts Dendrogram

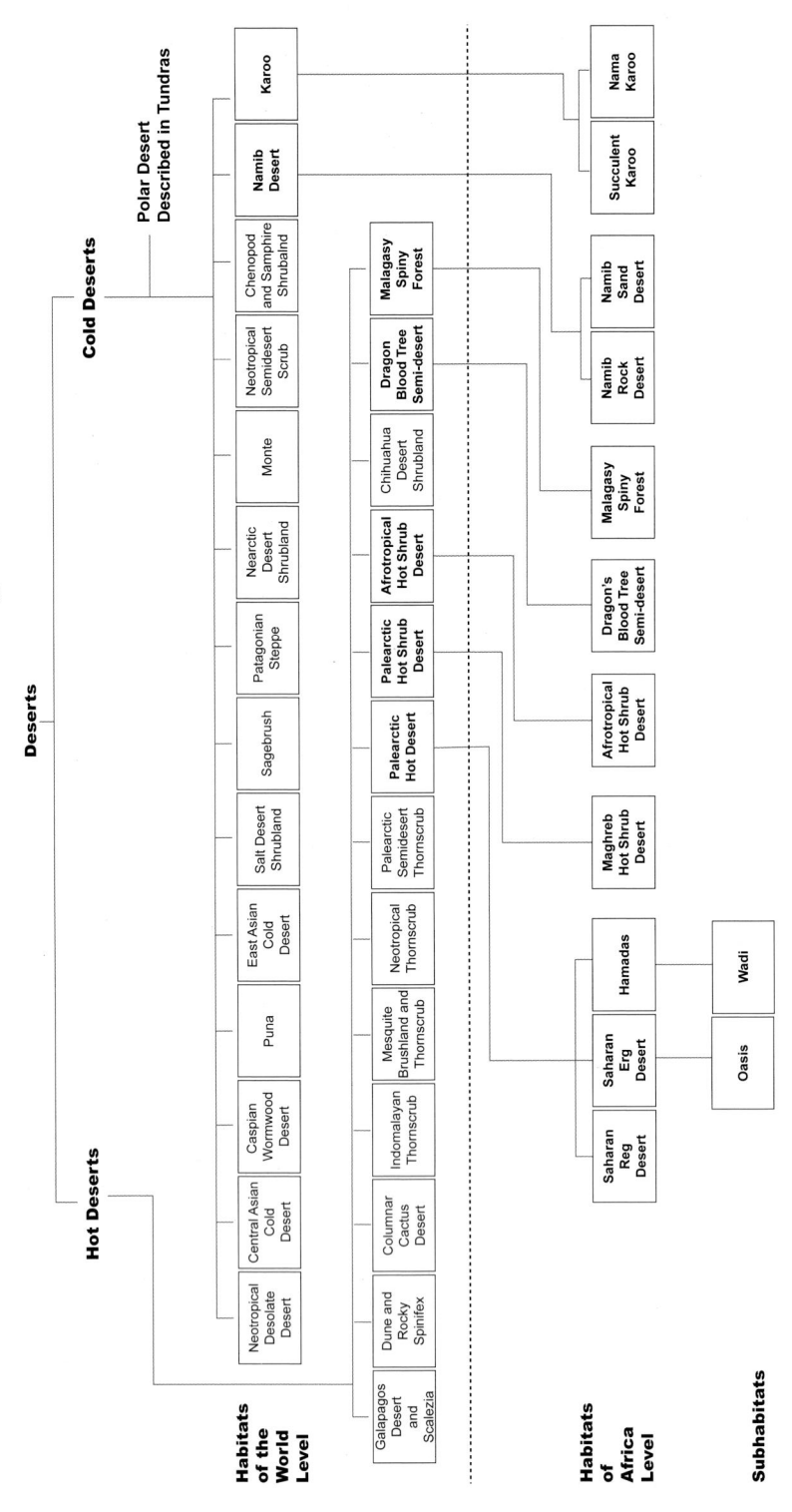

Deserts

Hot Deserts

Habitats of the World Level:
- Galapagos Desert and Scalezia
- Dune and Rocky Spinifex
- Neotropical Desolate Desert
- Columnar Cactus Desert
- Central Asian Cold Desert
- Caspian Wormwood Desert
- Indomalayan Thornscrub
- Puna
- Mesquite Brushland and Thornscrub
- East Asian Cold Desert
- Neotropical Thornscrub
- Salt Desert Shrubland
- Palearctic Semidesert Thornscrub
- Sagebrush
- **Palearctic Hot Desert**
- Patagonian Steppe
- **Palearctic Hot Shrub Desert**
- Nearctic Desert Shrubland
- **Afrotropical Hot Shrub Desert**
- Monte
- Chihuahua Desert Shrubland
- Neotropical Semidesert Scrub
- **Dragon Blood Tree Semi-desert**
- Chenopod and Samphire Shrubland
- **Malagasy Spiny Forest**

Cold Deserts
- **Namib Desert**
- **Karoo**

Polar Desert Described in Tundras

Habitats of Africa Level:
- **Maghreb Hot Shrub Desert**
- Afrotropical Hot Shrub Desert
- Dragon's Blood Tree Semi-desert
- Malagasy Spiny Forest
- Namib Rock Desert
- Namib Sand Desert
- Succulent Karoo
- Nama Karoo

Subhabitats:
- Saharan Reg Desert
- Saharan Erg Desert
- Hamadas
- Oasis
- Wadi

DESERTS AND ARID LANDS

Af2A SAHARAN REG DESERT

IN A NUTSHELL: Desolate stony deserts with almost no vegetation that are hot in summer and warm in winter. **Global Habitat Affinities:** GIBBER CHENOPODLANDS; TURANIAN REG AND HAMADA. **Continental Habitat Affinities:** NAMIB ROCK DESERT. **Species Overlap:** SAHARAN ERG DESERT; MAGHREB HOT SHRUB DESERT; ROCKY HAMADA AND MASSIF; CAUCASIAN SHRUB DESERT.

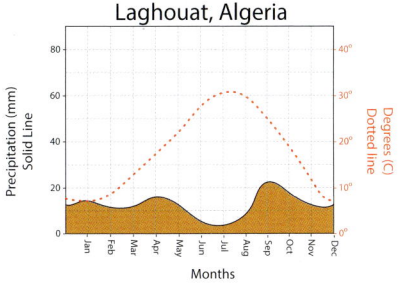

Laghouat, Algeria

DESCRIPTION: Reg is the most widespread Saharan landform, making up more than half of the desert. Comprising shimmering stony plains, Reg is one of the most intimidating environments on the planet, even less hospitable than Erg sand seas because of the blinding glare from the polished rock surfaces. Reg mostly comprises areas of colluvial outwash (loose sediment deposited downslope) with stony desert and desert pavement. Reg/Hamada and Erg are like Yin and Yang. Erg exists only because almost all the surface material from Reg and Hamada has been blown away and deposited elsewhere. Reg forms when colluvium is winnowed so that the finest material, including clay and silt, is entirely removed and the coarser sand accumulates in the surrounding Erg dunes and sand sea. During the winnowing process, heavier materials are concentrated in the Reg until they form a hard carapace protecting the underlying material from erosion. Simultaneously, hypersaline waters transport calcium, silica, and sulfur from other areas to form accumulations of calcite, gypsum, and silcrete in the powdery layer just below the protected gravel carapace. The solutes deposit a patina (a thin film) of silica mixed with iron oxides (magnetite and hematite), clays, and manganese on the cobbles. The abrasion from the silt and sand polishes the larger cobbles with this chemical veneer, giving them a varnished appearance approaching a glassy black sheen.

Places where Reg occurs are some of the hottest on Earth (Köppen **Bsh**, **Bfh**) and, away from the moderating effect of the coast, have average summer temperatures up to 113°F (45°C), with maxima topping a phenomenal 130°F (50°C). The daily temperature range can be extreme, with fluctuations

Relatively "lush" Reg in Morocco supporting a sparse growth of low shrubs. © MIGUEL ROUCO

of 70°F (21°C). Many areas receive less than 1 in. (25 mm) of precipitation per year and sometimes even go years without rain. The average annual rainfall is 4 in. (100 mm), which, when combined with evapotranspiration rates as high as 15 ft. (4.5 m) annually, makes for hyperarid conditions. Sometimes the Reg carapace is so compact and chemically bound together that it becomes a completely desolate surface where no plants can grow, but elsewhere, although the landscape is barren most of the time, vegetation sprouts when erratic rains come, generally forming 1–5% ground cover of shrubs and grasses. Common plants include *Artemesia* shrubs, chenopods, saltbushes (*Haloxylon salicornicum* and *H. scoparium*), and an ephemeral (but sometimes perennial) grass Habob (*Stipagrostis obtusa*). Where the carapace is more diffuse or a sandy component persists above it, these plants are joined by grasses such as Desert Bunchgrass (*Panicum turgidum*), Sewan Grass (*Lasiurus scindicus*), *Stipagrostis ciliata* (called Tall Bushman Grass in the Kalahari), and *S. plumosa*. After even sparse rains, the plains can burst into life with annuals such as *Daucus* of the celery family, Dahma (*Monsonia nivea*), QuraiTah (*Plantago ciliata*), which is an annual that looks a little like cabbage, Wooly Cumin (*Ammodaucus leucotrichus*), and bulb geophytes like Beyjuje (*Androcymbium gramineum*). In patches where water is more regular, denser groves can form, composed of Randonia (*Randonia africana*), Shabram (*Zilla spinosa*), and Had (*Cornulaca monacantha*), which is an important food source for camels. The groves reach 2–3 ft. (60–90 cm) tall and can thicken to cover 30% of the ground and look remarkably similar to NAMA KAROO, Succulent Puna of South America, or Gibber Chenopodlands of c. Australia.

The remarkable Addax was formerly found in all the habitats of the Sahara and its fringes. It came perilously close to extinction, but reintroduction programs in Chad have been successful. © VLADIMIR DINETS

Af2A SAHARAN REG DESERT

Cream-colored Courser is a classic cryptic Reg desert species that is partially nomadic, able to exploit meager resources in time and space. MAIN PHOTO © DANIELE ARDIZZONE, FLUYENDO PHOTOGRAPHY; INSET © KEN BEHRENS, TROPICAL BIRDING TOURS

In all the variations of Reg, seasonal or annual precipitation fluctuations can be huge, and rainfall can result in a burst of grasses mentioned above to a degree that the landscape is reminiscent of a grassland for a very short period. Some people may treat this as an ephemeral desert grassland, but for the vast majority of the time it is a desolate desert.

CONSERVATION: The situation throughout the Sahara (including all its main habitats) is poor, partially due to long-term political instability. The situation is especially grim for large mammals, most of which have been hunted out, counting among the continent's greatest conservation tragedies. The North African deserts have been heavily used by humans for a very long time, making it hard to ascertain their natural state. Another problem across Africa's deserts is off-road driving, which can have a long-lasting effect on this fragile habitat.

WILDLIFE: Saharan Reg Desert, SAHARAN ERG DESERT, ROCKY HAMADA AND MASSIF, and MAGHREB HOT SHRUB DESERT all support a subset of arid-adapted wildlife that is widespread across these habitats. Although extreme Reg is the most inhospitable environment, we deal with widespread Saharan wildlife in this account as it is presented first; we mostly consider only habitat-specialist wildlife in subsequent Saharan accounts. Larger wildlife naturally tends to be thin on the ground due to the scarcity of food. Nonetheless, deserts are home to a fair number of mammals and birds and are an excellent habitat for reptiles. Classic desert animals include gazelles, oryxes, foxes, bustards, sandgrouse, and larks. North Africa's deserts were once one of the continent's great mammal ranges, supporting vast herds of gazelles, Addax, and Scimitar-horned Oryx. An aggregation of some 10,000 oryx in Chad, witnessed in 1936, boggles the mind. Typical of desert environments, these mammals migrated or were facultative nomads around their vast range, following seasonal flushes of abundant food. What remains today is a dim shadow of this past glory. Addax, Scimitar-horned Oryx, and Dama Gazelle have sadly all but vanished from the

SIDEBAR 5 — WHAT MAKES A DESERT? THE SAHARAN EXAMPLE

The definition of true warm desert varies markedly, from environments with less than 1 in. (25 mm) of rain/month to absolute desert with less than 1 in. of rain/year, where significant plant growth is only possible with the addition of external water from either groundwater springs (e.g., oases) or accumulated runoff (e.g., wadis). These principles apply to Saharan and Namib deserts (pp. 71–100). As dry as the Sahara is, there is usually some rainfall each year, with the west being wetter than the east. But it is not all about rainfall. The interaction of precipitation with temperature and evaporation, and the timing of rainfall, are more important factors determining what types of plants and animals can live there. By analyzing climate graphs it is clear that periods of plant stress differ between regions. The n. Sahara (such as Algeria) has a subtropical Mediterranean climate with winter rains, whereas the south (such as Niger) has a climate influenced by summer rains from the tropics. The average temperature and evaporation rate are much greater in the south, so despite receiving more rain, the southern fringe of the Sahara is more barren than areas receiving less rain near the North African coastline with a more temperate climate. However, the presence of these zones is not static. During the Last Glacial Maximum (around 18,500 YBP) the northern part of the Sahara was colder, and possibly wetter, than at present. The southern fringe of the Sahara was 1100 mi. (1800 km) farther south in a much more tropical environment and was cooler and drier than at present; dune systems were moving 300 mi. (450 km) to the south of where they are today.

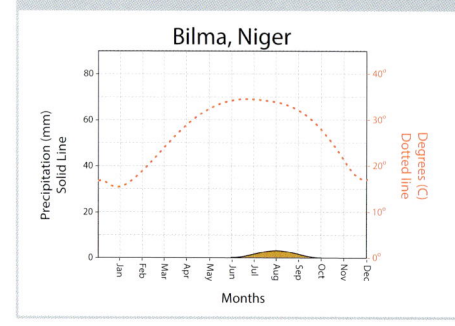

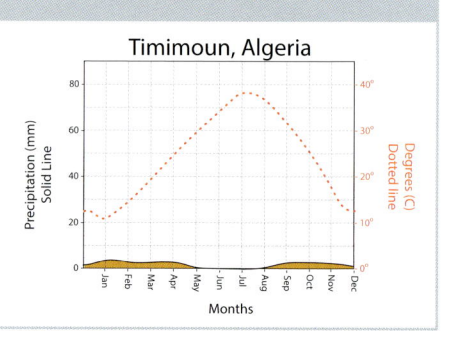

Crowned Sandgrouse is an adaptable desert nomad that quickly appears in Reg after rains have produced a burst of vegetation. © CARLOS N. G. BOCOS

Sahara, as have their main predators: Lions, Spotted Hyena, and African Wild Dogs. Captive breeding and reintroduction programs in Chad hold the last hope for these spectacular, once widespread and abundant, antelopes. A few big mammals persist, including Rhim and Dorcas Gazelles. But the Saharan mammalian food chain is driven primarily by small mammals, starting with Mauritanian, Whittaker's, and Tarfaya Shrews and the bizarre mini-kangaroo-like Lesser Egyptian Jerboa, which can leap 10 ft. (3 m) in a single bound, 25× its own body length, helping it travel in excess of 6 mi. (10 km) each night while foraging! There is also a slew of gerbils and jirds that occupy every niche

Smaller predators like Sand Cat feed on Reg's small mammal bounty of jerboas, shrews, jirds, and gerbils. © CARLOS N. G. BOCOS

possible. These in turn support a collection of predators, including the nocturnal Sand Cat, alongside Rüppell's and Fennec Foxes, and the infamously irritable Honey Badger. The cute Desert Hedgehog patrols looking for invertebrates. The Sahara is fantastic for herps, including the generalist Manuel's (*Chalcides manueli*), Moorish (*C. polylepis*), and Ocellated Skinks (*C. ocellatus*); Bosc's Fringe-toed Lizard (*Acanthodactylus boskianus*); North African Mastigure (*Uromastyx acanthinura*); and Small-spotted Lizard (*Mesalina guttulata*), while North African Rock Agama (*Agama bibronii*) and Oudri's Fan-footed Gecko (*Ptyodactylus oudrii*) like to hide in crevices and under rocks. Widespread reptilian predators include Desert Monitor (*Varanus griseus*), Egyptian Cobra (*Naja haje*), Saw-scaled Viper (*Echis pyramidum*), Saharan Horned Viper (*Cerastes cerastes*), False Cobra (*Rhagerhis moilensis*), and Schokari Sand Racer (*Psammophis schokari*).

True Reg is hostile, especially away from wadis (see sidebar 9, p. 100) or when rain hasn't fallen for a long time. All animals need food, and most need a little water, and so Reg devoid of basic elements holds little wildlife. It is surprising that any birds use it at all, but the ultimate desert survivors can. Larks lower their metabolic rate and can almost eliminate evaporative cooling, allowing Bar-tailed and Desert Larks to inhabit even Reg devoid of water or relief from the sun! In areas that support a few scattered succulents Thick-billed and Temmick's Larks miraculously appear. A range of wheatears survive in Reg, including Desert, White-crowned, Hooded, and Mourning Wheatears. Reg springs to life after rains, and nomads move in very rapidly, especially strong-flying species like Cream-colored Courser, Houbara Bustard, and Crowned, Spotted, and Lichtenstein's Sandgrouse. And wherever just a tiny little bit of vegetation clings on, like drainage lines or wadi fringes, all sorts of animals appear, including Agadir Gecko (*Saurodactylus brosseti*), which can withstand ambient temperatures exceeding 104°F (40°C), Northern Elegant Gecko (*Stenodactylus mauretanicus*), Boehm's Agama (*Trapelus boehmei*), Desert (*Tarentola deserti*) and African Wall Geckos (*T. hoggarensis*). As you are driving through seemingly endless stony plains watch out for the fat yellow-red-orange male Moroccan Mistigure (*Uromastyx nigriventris*) sitting sentry atop roadside rocks. One of the rarest Reg inhabitants is Kleinmann's Tortoise (*Testudo kleinmanni*). Once widespread from Libya to Israel, the smallest tortoise in the Northern Hemisphere is now Critically Endangered. Along wadi edges and slightly sandier Reg, small geckos like Tripoli

Trumpeter Finch has a bizarre high-pitched song, like a tiny tin trumpet, giving this species its name.
© KEN BEHRENS, TROPICAL BIRDING TOURS

Gecko (*Tropiocolotes tripolitanus*) and Oliver's Sand Lizard (*Mesalina olivieri*) are found. There are of course predators here, including the fantastic North African Cat Snake (*Telescopus tripolitanus*).

DISTRIBUTION: Reg covers about half of the Sahara from the Atlantic coast of Western Sahara and Mauritania to the Red Sea, a distance of 3300 mi. (5300 km). North to south, it runs from the southern edge of the Atlas Mountains in Morocco and Algeria and the Mediterranean coast of Libya, 1200 mi. (1900 km) south to the Sahel (**NORTHERN DRY THORN SAVANNA**).

WHERE TO SEE: Tagdilt Track, Morocco; Ténéré Desert, Niger; Tanezrouft, Algeria.

SIDEBAR 6 — MIGRATION MADNESS—TRANS-SAHARAN SOJOURN

Some 126 bird species that breed in the Palearctic (both Europe and Asia, as far east as Beijing!) escape the cooling temperate north to spend more than half of the year (generally Sep–Apr) in tropical Africa. Warmer climes and abundant food tempt some 2.1–5 billion individual birds to traverse what is arguably the most inhospitable stretch of land on Earth—the Sahara. Most Palearctic breeders have pretty specific nesting requirements. Wood Warbler, for example, likes cool and wet glades within European Mixed Broadleaf Conifer Forest, with beech, birch, and oak. Their nonbreeding quarters are the canopy of lowland forests in Central and West Africa, where they may join the more exotic Blue Cuckooshrike or Red-billed Helmetshrike in a feeding flock. Some migrants can undertake nonstop flights of 600+ mi. (1000 km), and most Wood Warblers do attempt something on this scale, crossing the Sahara in a single epic night! But they do stop, rest, and feed occasionally. It is hard to imagine a Wood Warbler feeling comfortable sitting on a tamarisk bush in the middle of the Sahara while traveling between these two wildly different canopy habitats, but most migrants are generalists by necessity. They will use just about any grove of dates, tree-lined wadi, or random bit of scrub to leapfrog across the challenging abyss that is the Sahara. Desert oases are crucial for migrants, as the tiny bit of water, fruit, and insect life that they hold may prove to be the difference between life and death for a 0.3 oz. (8 g) warbler. Hitting a Saharan oasis on the right day is the Old World equivalent of a North American "fallout." A scrubby 1 acre grove of palms and bushes can hold hundreds of birds of 20 or more species!

Lest we forget, migrants' penchant for moving, combined with their ability to tolerate vastly different habitats, makes them regular vagrants that can turn up in strange places, creating major excitement for birders. Palearctic migrant groups include ducks and other waterbirds, harriers, honey-buzzards, small falcons, crakes, cranes, shorebirds, cuckoos, swifts, bee-eaters and rollers, swallows and martins, pipits and wagtails, nightingales, robins, rock-thrushes, a smorgasbord of warblers, including genera *Sylvia*, *Acrocephalus*, *Iduna*, *Phylloscopus*, and *Hippolais*, flycatchers, shrikes, orioles, and buntings.

Red-backed Shrike braves either the vast Sahara or the deserts of the Arabian Peninsula on its migration south to its mostly Southern African wintering grounds. © KEITH BARNES, TROPICAL BIRDING TOURS

Af2B NAMIB ROCK DESERT

IN A NUTSHELL: Arid, rocky, and mountainous, mostly sparsely vegetated or unvegetated habitat with erratic grass cover. **Global Habitat Affinities:** CAUCASIAN SHRUB DESERT; ROCKY CANYON; PRE-PUNA SEMI-DESERT SCRUB. **Continental Habitat Affinities:** MAGHREB HOT SHRUB DESERT; AFROTROPICAL HOT SHRUB DESERT; SAHARAN REG DESERT; ROCKY HAMADA AND MASSIF. **Species Overlap:** NAMIB SAND DESERT; KALAHARI DRY THORN SAVANNA; NAMA KAROO; SUCCULENT KAROO.

DESCRIPTION: Deserts are very dry areas in which most of the ground is bare most of the time, even in areas with some vegetation. Despite being a fairly dry continent overall, Africa has only two sub-Saharan areas of desert, both relatively small: the Namib in the southwest and the deserts of the Horn of Africa and Kenya in the northeast. The Namib, over 50 million years old, is thought to be the world's oldest desert. This has resulted in a flora and fauna that are highly adapted to the harsh environment. Despite its desolate appearance, this is one of the world's richest deserts, a biological wonder.

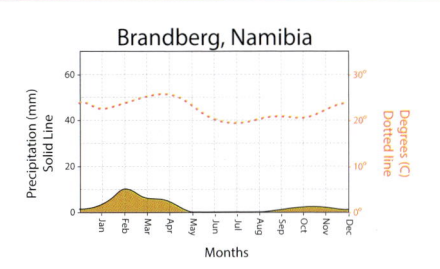

Rock desert habitat is where geology and geography can be enjoyed in their purest form, unobscured by concealing vegetation. The landforms range from rocky, eroded mountains with high plateaus and valleys of the Richtersveld (4469 ft./1362 m) and Brandberg (8432 ft./2570 m) to gravel plains with inselbergs of granite, schist, and limestone. All are open to exploration, on foot or by vehicle in areas with roads.

Deserts (Köppen **Bfk**) receive less than 10 in. (250 mm) of rain annually, but the Namib is classified as hyperarid, where evaporation is four times greater than rainfall—and most areas are much drier. The Namib Rock Desert generally receives a little more rain than NAMIB SAND DESERT, mostly in the range of 2–3 in. (50–80 mm) annually. If rain falls at all, the southern part receives rainfall during the austral winter (Jun–Aug) and the northern part during the austral summer (Dec–Feb). The inland mountainous portion of the Namib is generally hot, though it cools off at night. High temperatures average 77–95°F (25–35°C), with extremes of 113°F (45°C), and minima average 46–68°F (8–20°C), with extremes of 28°F (-2°C). Coastal areas of Namib Rock Desert receive 300–180 days of precipitation in the form of fog, but this precipitation quickly tapers off to less than 25 days circa 30–38 mi. (50–60 km) inland.

Rüppell's Bustard is a quintessential Namib Rock Desert bird and an indicator species for this habitat. Welwitschia (*Welwitschia mirabilis*) *(inset)* **makes up its own plant family and is the Namib's most remarkable plant.** © KEN BEHRENS, TROPICAL BIRDING TOURS

Some areas are virtually free of vegetation, though a close examination will usually reveal the presence of lichens, especially on gravel plains and rocky areas. The Namib Rock Desert supports at least 250 lichen species, and given they have no roots, all are exclusively supported by fog and dew. The most remarkable are the *Wanderflechten*, or wandering lichens, which are unattached and blow in the wind, accumulating in shallow depressions. However, the Namib Rock Desert's most dominant plants go largely unnoticed by visitors and occur primarily as dormant seeds (therophytes), lying in wait. After rains of anything more than 0.8 in. (20 mm), grasses will spring up, especially bushman grass (*Stipagrostis*), transforming the pebble plains into something more akin to a prairie, with swaying grasses stretching to the horizon, before producing a crop of seeds to be blown away by the wind and to lay in wait for up to 10 years before seeding en mass again, a crucial component of Namib desert ecology. One of the main bases of the entire food chain is wind-blown detritus, broken grasses, and seed bank that accumulate on plains where they are eaten for years by beetles and fishmoths.

Nonephemeral plants can survive only above subterranean watercourses or along drainage lines like fissures in rock outcrops or by using specialized root systems to obtain water. Succulents are rare because rainfall is not predictable enough. The most famous plant of the Namib Rock Desert, and indeed one of the strangest plants on Earth, is the Welwitschia (*Welwitschia mirabilis*), its scientific name meaning wondrous, a gymnosperm that composes its own plant family. It has two leathery leaves that emerge from a short trunk and grow continuously. The oldest specimens are thought to be over 2000 years old. Outcrops and hills support the fat-stemmed Desert Moringa (*Moringa ovalifolia*) and the cactus-like spurge, the Gifboom (*Euphorbia virosa*), meaning poison tree in Afrikaans; the San tribespeople used to dip their arrow tips in this plant's toxic latex. Trees grow very locally in areas with groundwater, such as valley bottoms, or at the desert fringes, forming Western Riparian Woodlands (subhabitat Af2–1). Typical groups include *Commiphora*, euphorbias, African acacias, Desert Dates (*Balanites aegyptiaca*), and golds (*Rhigozum*). Slightly moister areas permit the growth of sparse, shrubby vegetation. The interior eastern fringes of the Namib Rock Desert essentially have a nebulous transition to a narrow band of NAMA KAROO, though with far less diverse and sparser shrub growth than is found in the heart of that habitat's range.

CONSERVATION: The Namib Rock Desert is remarkably well protected by large national parks and huge diamond-mining concessions. The conservation situation in the Angolan north is much worse, mainly due to long-term political instability. One tourism-related problem across Africa's deserts is off-road driving, which can have a long-lasting effect on this fragile habitat. Just don't do it!

WILDLIFE: Larger wildlife naturally tends to be thin on the ground in deserts owing to the scarcity of food. Nonetheless, the Namib has not been heavily affected by human activities and is home to a fair number of mammals and birds and is an excellent habitat for reptiles.

Gemsbok and Springbok, the two most desert-adapted big mammals, are fairly common, especially at the desert fringes. The Gemsbok has an amazing "radiator" system in its head that allows the body's temperature to rise to 109°F (43°C), but blood is instantly cooled to 104°F (40°C) mere seconds before entering the brain. This tiny but crucial difference prevents brain damage and death, allowing the Gemsbok to retain crucial water that permits desert survival. "Hartmann's" Mountain Zebra and Angolan Giraffe are found in the adjacent habitats along the Namib Escarpment and will sometimes descend into the Namib. The major predators are Brown and Spotted Hyenas, Bat-eared Fox, Cheetah, and Black-backed Jackal. Lion, Black Rhinoceros, and African Savanna Elephant were formerly widespread but now are localized, found mainly in remote parts of nw. Namibia. Rocky outcrops support a diverse collection of small mammals, including many near-endemics such as the exceptionally cute Namib Round-eared Sengi, Damara Ground Squirrel, Stone Dormouse, Dassie Rat, Pygmy and Barbour's Rock Mice, Setzer's Hairy-footed Gerbil, and Kaokoveld Slender Mongoose.

There are a few birds endemic to the Namib Rock Desert: Rüppell's Bustard (IS), Benguela Lark, and Gray's Lark (IS), the latter a remarkable species that hardly ever has to drink. Other typical birds of gravel flats include Burchell's Courser, Namaqua Sandgrouse, Stark's Lark, and Tractrac Chat. Ostriches are remarkably able to cope with the open gravel desert, partially because they are one of only two vertebrates that can exhale air that is not fully saturated with water vapor. Rocky outcrops

Gemsbok possesses a remarkable physiological adaptation that allows it to cool the blood flowing into its brain, equipping it to survive even in shelterless rock desert. © DORIAN ANDERSON, TROPICAL BIRDING TOURS

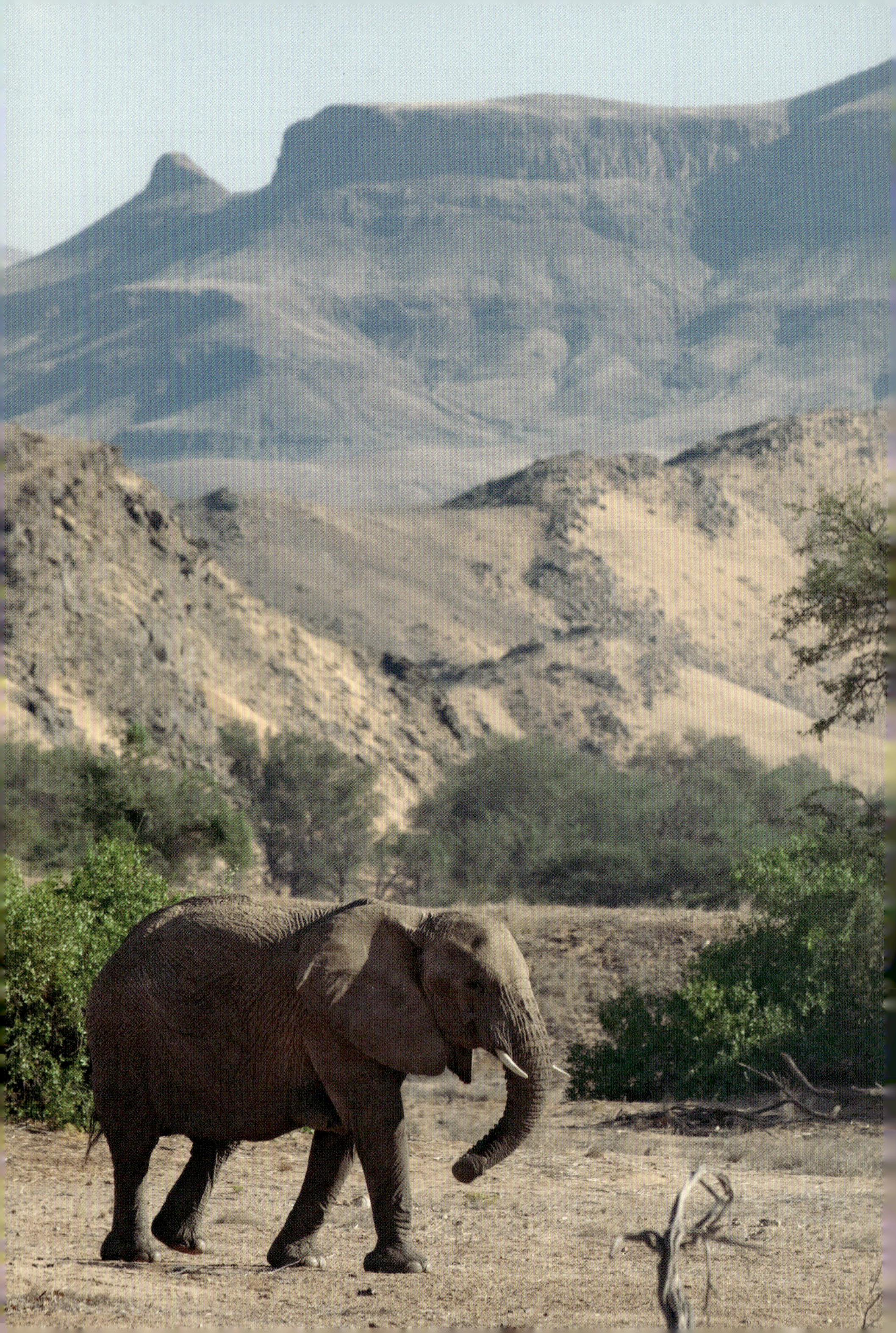

also support Mountain and Familiar Chats. After rains, and the subsequent flush of grasses, rain-tracking granivorous nomads that can be totally absent are suddenly omnipresent a few weeks later. Clouds of Gray-backed Sparrowlarks and Lark-like Buntings can be joined by the local Black-eared Sparrowlark and Black-headed Canary.

The reptile assemblage of the Namib Rock Desert is exceptional, comprising dozens of species, many of which are endemic. The Nama Padloper (*Chersobius solus*) is a near-endemic tortoise and Anchieta's Dwarf Python (*Python anchietae*), Desert Mountain Adder (*Bitis xeropaga*), Western Bark Snake (*Hemirhagerrhis viperina*), and Western Keeled Snake (*Pythonodipsas carinata*) are all localized, preferring rocky mountain outcrops. Other reptiles include legless lizards, desert lizards (*Meroles*), sand lizards (*Pedioplanis*), barking geckos (*Ptenopus*), Namib day geckos (*Rhoptropus*), and Namaqua Chameleon (*Chamaeleo namaquensis*). Despite this desert being one of the most arid on Earth, it supports frogs, including a few endemics like Dombe Pygmy Toad (*Poyntonophrynus dombensis*), Paradise Toad (*Vandijkophrynus robinsoni*), and Marbled Rubber Frog (*Phrynomantis annectens*). Most shelter in rocky crevasses, emerging almost exclusively after rain.

Endemism: As one of the world's oldest deserts, the Namib may be home to more endemic species than any other hyperarid system on Earth, with a rich set of endemic plants, birds, and reptiles.

DISTRIBUTION: The Namib Rock Desert extends from the Richtersveld of nw. South Africa, all the way along the Namibian coast and into s. Angola. Although it reaches the coast in many places, most of the gravel plains, rock desert, and mountains lie on the interior of the NAMIB SAND DESERT and extend around 125 mi. (200 km) inland, reaching the base of the Namibian escarpment.

WHERE TO SEE: Namib-Naukluft National Park, Namibia.

Opposite: **"Desert" African Savanna Elephant has exceptionally large feet, allowing it to walk through thick Namib sands.** © KEN BEHRENS, TROPICAL BIRDING TOURS

Below: **The Meerkats of the Namib Desert are ghostly pale and quite different looking from those found to the east and south in moister habitats.** © KEN BEHRENS, TROPICAL BIRDING TOURS

Af2C SAHARAN ERG DESERT

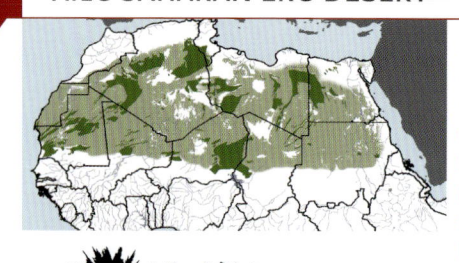

IN A NUTSHELL: Desolate sand deserts with dunes and sand seas and with sparse vegetation; hot in summer, warm in winter. **Global Habitat Affinities:** CANEGRASS DUNES; TURANIAN COLD ERG DESERT; NEARCTIC DESOLATE DESERT; ATACAMA DESOLATE DESERT. **Continental Habitat Affinities:** NAMIB SAND DESERT. **Species Overlap:** SAHARAN REG DESERT; MAGHREB HOT SHRUB DESERT; ROCKY HAMADA AND MASSIF; CAUCASIAN SHRUB DESERT.

DESCRIPTION: When most people imagine the Sahara, they think of Erg seas (moving sand). But actually, Erg and sand dunes are the least common of the desert habitats, making up only 20% of the Sahara (admittedly still a massive area!). Erg is most prevalent in Algeria and Libya and parts of Mali and Niger. At present, the Saharan sand seas, such as Erg Chech of Algeria and Mali and the Great Sand Sea of Egypt, average about 46,000 mi.^{2} (120,000 km^2), and although they encompass massive areas that reach up to 400 mi. (650 km) long and 190 mi. (300 km) wide, they are much smaller than the Ergs of Rub' al Khali of the Arabian Peninsula, which at 250,000 mi.^{2} (650,000 km^2) contains around half the sand of all the Sahara.

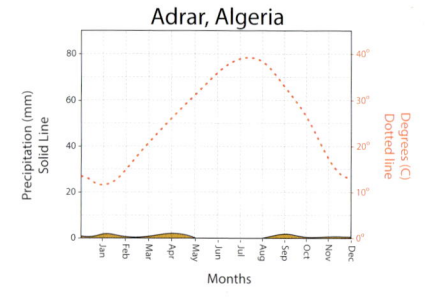

Reg/Hamada and Erg are like Yin and Yang. Erg exists only because almost all the surface sand from Reg and Hamada has been blown away and deposited elsewhere. The development of Erg requires enough wind to winnow (blow away) the finest materials, such as silts (<0.05 mm) and clays (<0.002 mm), from the Sahara and transport them out of the desert, often to far-flung destinations like South America. The remaining courser material (sand) is then accumulated in the Erg dunes and sand seas. Where the dune systems are too dry for much plant development, the dunes continually shift through aeolian (wind driven) saltation (bouncing) whereby the sand grains bounce along the ground surface and prevent plant growth. Various dune types develop in the Sahara, with the most common being Barchan, crescent-shaped sand dunes where wind direction changes often, and longitudinal dunes, where wind direction is more constant.

Places dominated by Erg are some of the hottest on Earth (Köppen **Bfh**) and, away from the moderating effect of the coast, have average summer temperatures up to 113°F (45°C), with maxima topping a phenomenal 130°F (50°C). Daily temperature range can be extreme, with fluctuations of 70°F (21°C). Many areas receive less than 1 in. (25 mm) per year and sometimes even go years without rain. The average annual rainfall is 2 in. (50 mm), which, when combined with evapotranspiration rates as high as 15 ft. (4.5 m) annually, makes for hyperarid conditions.

Dunes can reach spectacular heights, up to 820 ft. (250 m). When they are moving, they cannot sustain much perennial vegetation, and little soil development occurs, with soils remaining entisols (undifferentiated mineral soils) with no humus (organic matter) accumulation or horizon development. Vegetation is limited to robust ephemeral plants that eke out an existence after the rare rains, and during the driest years it is possible to travel tens of miles without seeing a single plant (or animal). Less mobile dunes can have small plants growing as close as 10 ft. (3 m) apart, such as Drinn Grass (*Stipagrostis pungens*) and Desert Bunchgrass (*Panicum turgidum*), which resembles Canegrass (*Zygochloa paradoxa*) from the sand dunes of the Simpson Desert in Australia. When the dunes do stabilize, water slowly accumulates beneath them over hundreds of years, enough to sustain larger shrubs, such as Christ's Thorn Jujube (*Ziziphus spina-christi*), Gum Acacia (*Senegalia senegal*), Thorn Mimosa (*Vachellia nilotica*), Desert Date (*Balanites aegyptiaca*), and even the occasional Umbrella Thorn (*Vachellia tortilis*), a common tree in Afrotropical savannas farther south. Aridisols (desert soils), with some humus accumulation and soil horizon development, form in the swales between dunes, and plant life becomes dominated by halophytic (salt tolerant) species. This includes Bedouin's Soapbush (*Anabasis articulata*), which is 1–2 ft. (30–60 cm) tall, *Arthrocnemum glaucum*, which grows to 3 ft. (1 m) and resembles Spiny Griswood (*Glossopetalon spinescens*) from the United States, tumbleweeds *Salsola*, *Tamarix* spp., a group of very tall wispy bushes, and Had (*Cornulaca monacantha*), which resembles chenopods from Australia.

Oases (subhabitat Af2-2) also occur in SAHARAN REG DESERT, ROCKY HAMADA AND MASSIF, and MAGHREB HOT SHRUB DESERT but are more influential in Erg. These are areas where fresh spring water with low salt content accumulates. Oases originally constituted a minuscule

Ténéré Sand Sea, with an interdunal wadi supporting some scattered trees and shrubs. Northern Niger. © MICHIEL KUPERS

Desert Sparrow is remarkably at home in barren dunes. © KEN BEHRENS, TROPICAL BIRDING TOURS

DESERTS AND ARID LANDS

percentage of the Saharan land surface area, and almost all of these have been converted to agriculture. When the oases had sitting water, the marshes were dominated by typical Palearctic plants, such as Broadleaf Cattail (*Typha latifolia*), Roundhead Bulrush (*Scirpus holoschoenus*), and Common Reed (*Phragmites australis*). The fringes were bound by various *Tamarix*, such as the French Tamarisk (*T. gallica*). Where the water table was close to the surface, but not open water, Nile Tamarisk (*T. nilotica*) grew to trees of 30 ft. (10 m) tall alongside Doum Palms (*Hyphaene thebaica*) and various acacias. The understory included Sodom Apple (*Calotropis procera*), and *Maerua* species such as Meru (*Maerua crassifolia*), and caper shrubs (*Capparis* spp.). The ground was covered with vines with large gourds, such as Vine of Sodom (*Colocynthis vulgari*). Oases were the only locations for permanent human settlements and now almost all are surrounded by towns and cities, supporting 75% of the Sahara's human population. In some areas, Oases have not been covered in concrete and are still farmed with Date Palm (*Phoenix dactylifera*) and small areas of natural vegetation persist between plantations and as undergrowth. Although a severely human-altered habitat, oases remain vitally important locations for migrating birds, drinking sandgrouse, and frogs that cannot exist without the water they provide.

CONSERVATION: See Conservation section for the SAHARAN REG DESERT account (p. 82).

WILDLIFE: For wildlife widespread in all Saharan habitats, see the Wildlife section in the SAHARAN REG DESERT account (p. 82). Dunes are a tough environment. The large sand dune seas

Greater Hoopoe-Lark is comfortable in both Erg and Reg and is at its most spectacular in flight! © KEN BEHRENS, TROPICAL BIRDING TOURS

support precious little vegetation, but in the swales where grasses grow and shrubs hold on, and in smaller patches of sandy desert near water, a few species can survive. Most that do are dune specialists that seldom leave Erg, like Rhim Gazelle, which has unfortunately declined massively and now borders on extinction. This gazelle is able to survive without water, obtaining moisture from dew and its plant diet while permitting its body temperature to fluctuate more than other warm-blooded mammals, which enables it to avoid evaporative cooling. Small mammals do well in Ergs as they can burrow to escape the heat of the day. Examples include dune-specialist Pygmy,

SIDEBAR 7 — EXTINCTION OF PEOPLE, PLANTS, AND ANIMALS OF THE SAHARA

In the driest and coldest recent period of the late Pleistocene glaciation (around 20,000–15,000 YBP), the Sahara was much more extensive than it is today. Moving seas of sand extended around 300 mi. (450 km) south of where they are now, and the Palearctic biogeographic region extended far into Ghana and Nigeria. The rewarming began at the end the Younger Dryas mini-glaciation period (around 11,700 YBP), and by 9000 YBP, almost all the Sahara west of the Nile became lush with savanna; wet GUINEA SAVANNA, which is actually an evergreen woodland, occurred from Ghana right up through Senegal; and moist savanna in Mauritania merged with drier Mediterranean-type shrubland in what is now hyperarid SAHARAN REG DESERT. Drainage of most of n. Africa was internal, and where vast lake systems existed, people rapidly colonized this new lush land, adapting to an aquatic diet of fish and snails but also hunting the Giraffe, Hippos, and antelopes of the wet savanna. The desert plants, and animals such as Dorcas Gazelle, found refuge in the Egyptian desert and the Arabian Peninsula. From 7000 to 5700 YBP, much of the Sahara changed to dry thorn savanna and thornscrub, and the lakes were drastically reduced. By 5000 YBP, the game was over, and most of the region was the desolate place it is now, and all previous settlements were abandoned. Whether the people of the aquatic/hunting culture died out, moved out, or became cattle herders remains unknown.

A caravan of camels crosses the dunes at Erg Chebbi, Morocco.
© KEN BEHRENS, TROPICAL BIRDING TOURS

Greater Egyptian, Pale, Lesser Egyptian, and Anderson's Gerbils alongside Lesser Sand Rat and their predators, such as Sand Cat and various foxes, especially Fennec.

Few birds are sand specialists, but one that certainly enjoys it is the courser-like Greater Hoopoe-Lark, although it needs some small shrubs or grasses to be able to find food. Desert Sparrow prefers scattered shrubs and trees, and Brown-necked and Fan-tailed Ravens often soar overhead. Fortunately, Erg is a much better environment for reptiles. On the dune fringes with scattered bushes we find Desert Agama (*Trapelus mutabilis*) and White-banded Sandfish (*Scincus albifasciatus*), along with predators like Awl-headed Snake (*Lytorhynchus diadema*). Closely related fringe-toed lizards survive side by side by occupying different niches. Long-fingered Fringe-toed Lizard (*Acanthodactylus longipes*) lives in high dunes that hardly have any vegetation, whereas Dumeril's Fringe-toed Lizard (*A. dumerili*) uses the dune edges with bushes and Halfa Grass (*Stipa tenacissima*). Here they ambush wayward Saharan Silver Ants (*Cataglyphis bombycina*), which can clock speeds of 1.9 mi./hr (3 km/hr), as fast as a walking human, making them the fastest of Earth's 12,000 ant species. Because soldier ants have saber-like mandibles, the lizards avoid ant burrows. Instead, the lizards assault worker ants that have strayed along the periphery. In turn, the ants have scouts that watch the lizards, and when the reptiles eventually retreat underground to avoid the sun, the scouts give the colony the all-clear, and the ants swarm out to forage on dunes that can reach a blistering 158°F (70°C)! Other sand specialists are Wedge-snouted (*Chalcides sphenopsiformis*) and Boulenger's Sand Skinks (*C. boulengeri*), which apparently perceive the vibrations of their prey in the sand, mainly beetle larvae, ants, or termites. At night, the cast changes, and we start seeing Erg-specialist herps like Dune Gecko (*Stenodactylus petrii*) and the amazing side-winding Saharan Sand Viper (*Cerastes vipera*). The viper buries 99% of itself, keeping its eyes and black tail tip exposed, wiggling the tail to lure hungry lizards or rodents to their death.

Where there is water in the oases, African Green Toads (*Bufo boulengeri*) can appear. Oases attract many species of bird but are probably the most vital for Palearctic migrants that use them on passage (see sidebar 6, p. 76). Sandgrouse will often come to drink water. Oases are also attractive to small mammals like Pleasant Gerbil and aerial insect hunters like Trident Leaf-nosed Bat, Christie's Long-eared Bat, and Rüppell's and Desert Pipstrelles.

DISTRIBUTION: The Erg dune habitat of the Sahara is sprinkled across North Africa where it mostly forms a mosaic with Reg and Hamada, from the Atlantic coast of Western Sahara and Mauritania to the Red Sea, a distance of 3300 mi. (5300 km). North to south, it runs from the

Fulvous Chatterer, a classic bird of Saharan oases. © DANIELE ARDIZZONE, FLUYENDO PHOTOGRAPHY

Fennec Fox is a widespread small predator of the Sahara. © MIGUEL ROUCO

Af2C SAHARAN ERG DESERT

southern edge of the Atlas Mountains in Morocco and Algeria and the Mediterranean coast of Libya 1200 mi. (1900 km) south to the Sahel (NORTHERN DRY THORN SAVANNA).

WHERE TO SEE: Douz-Redjim Maaloug, Tunisia; Erg Chebbi, Morocco; Great Sand Sea, Egypt.

Above: **The Saharan Erg is surprisingly rich in reptiles, supporting stunners such as Dune Gecko (***Stenodactylus petrii***,** *left***) and Saharan Horned Viper (***Cerastes cerastes***,** *right***).** © ZSOMBOR KÁROLYI

Below: **An oasis near Laayoune, Western Sahara.** *Cistanche phelypaea* **(inset) parasitizes chenopods (Amaranthaceae) and is often found growing on sand dunes.** © CARLOS N. G. BOCOS; INSET © ZSOMBOR KÁROLYI

Af2D NAMIB SAND DESERT

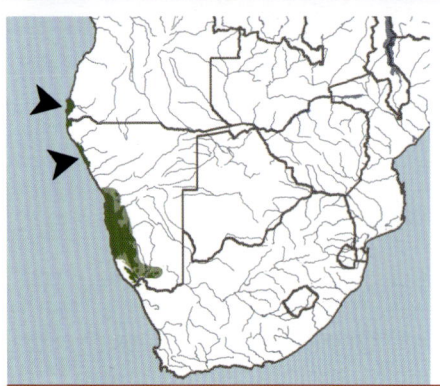

IN A NUTSHELL: Arid; linear and crescent-shaped sand dunes that form a giant sand sea and sandy coastal belt that are unvegetated or very sparsely vegetated. **Global Habitat Affinities:** ASIAN COLD ERG DESERT; CANEGRASS DUNES; TURANIAN COLD ERG DESERT; NEARCTIC DESOLATE DESERT; ATACAMA DESOLATE DESERT. **Continental Habitat Affinities:** SAHARAN ERG DESERT. **Species Overlap:** NAMIB ROCK DESERT; NAMA KAROO; KALAHARI DRY THORN SAVANNA; SUCCULENT KAROO.

DESCRIPTION: Spectacular, mostly vivid pink-to-orange or pale-brown sandy landscape that resembles the set of the *Dune* films. The vibrant Namib sand sea is one of Earth's largest, with many dunes longer than 20 mi. (32 km) and taller than 650 ft. (200 m); the tallest dune reaches 1273 ft. (388 m) in height. Most of the ground is bare most of the time, even in areas with some vegetation. This habitat is wide open, although the loose sand is difficult to walk through.

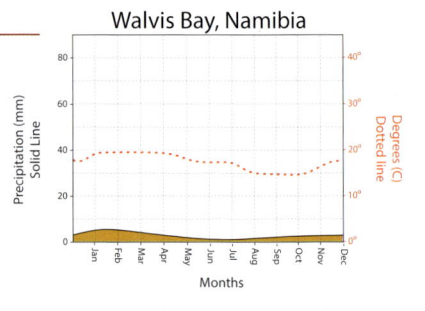

Cold deserts receive less than 10 in. (250 mm) of rain annually, but the Namib is classified as hyperarid (Köppen **Bfk**), as evaporation is typically four times greater than rainfall, and most areas are much drier. The Namib Sand Desert generally receives less rain than the NAMIB ROCK DESERT, mostly in the range of 0.2–0.8 in. (5–20 mm). However, most of this habitat's life-sustaining moisture comes in the form of thick coastal fog, which occurs 180–300 days a year, and this has a profound impact on the wildlife here. This fog is created by warm inland air meeting the cold air above the upwelling waters of the Benguela Current, which is also what keeps the coastal temperatures remarkably moderate, with Swakopmund, Namibia, receiving maximum temperatures ranging from 75–85°F (23–29°C) and minima from 45–60°F (7–15°C). The fog has caused over a thousand shipwrecks on the n. Namibian coast, leading to the grim moniker "Skeleton Coast." Farther inland, temperatures are more extreme, up to 105°F (41°C), with the sand surface temperature reaching a scorching 158°F (70°C) on the hottest days.

The dunes are mobile and constantly shifting, and the sand on wind-exposed crests is ever moving and bare; nothing grows here. However, at dune bases and in dune slacks, where a little more water accumulates, there is a meager sprinkle of plants. There are occasional tufts of perennial grasses like Namib Dune Bushman Grass (*Stipagrostis sabulicola*), *S. lutescens*, and spiny lovegrass (*Cladoraphis*). Another common plant is Herero Purslane (*Trianthema hereroensis*), a succulent in the fig marigold family that amazingly absorbs moisture from fog through its leaves! The prestigious journal *Nature* heralded this discovery as "unlikely and not very believable" when presented with a paper describing this in the 1970s. Gemsbok graze on this plant, stimulating vigorous regrowth. The base of the food chain in this dune system is essentially wind-blown detritus, broken grasses, and seeds that accumulate on dune slipfaces, where it is eaten by beetles and fishmoths. One classic plant of the Namib Sand Desert is the spiny Nara Melon (*Acanthosicyos horridus*), which

Above: **The Namib Dune Sea stretches for hundreds of miles along the coast of s. Namibia and appears desolate, but it conceals much specialized wildlife.** © DORIAN ANDERSON, TROPICAL BIRDING TOURS

Right: **The base of the dune sea has some lusher vegetation, owing to seeping groundwater.** © KEN BEHRENS, TROPICAL BIRDING TOURS

lacks leaves (thereby avoiding water loss) and instead photosynthesizes with its stems and thorns. It grows on dune edges but also along watercourses. Slightly moist areas, including those at the base of sand dunes, permit the growth of sparse, shrubby vegetation. Where dunes are adjacent drainage lines, one sees scattered Camel Thorn (*Vachellia erioloba*) trees. These have an impressive tap root system that reaches as far as 200 ft. (60 m) below the sand surface to access water. They are often accompanied by Ana Tree (*Faiderbia albida*), Wild Tamarisk (*Tamarix usneoides*), and Mustard Tree (*Salvadora persica*).

Some parts of the Namib Sand Desert have fairy circles, large circular formations 10–65 ft. (3–20 m) in diameter lacking any vegetation, often surrounded by grass. Theories abound as to what these are and how they came to be. They are somewhat analogous to Heuweltjies in the Karoo (see pp. 110–111, 114), and many think they are formed by termites that have selectively eaten grass roots. Another theory suggests that the grass plants themselves compete for water by growing longer roots to deprive adjacent plants of water, creating the bizarre pattern. Fairy circles are best known from the Namib Sand Desert but also occur in Namib Rock Desert.

The Namib sand sea is restricted by the Kuiseb River. Although this remarkable river is mostly dry, it flows just often enough to wash any sand that blows into it down to just south of Walvis Bay near the Atlantic Ocean, literally halting the progression of the sand sea in its tracks and delimiting its northern extent with a sharp edge.

CONSERVATION: The Namib is remarkably well protected by large national parks and huge diamond-mining concessions. Nonetheless, mining activities threaten certain local areas, such as around Oranjemund, Namibia, and Alexander Bay, South Africa. Climate change, and especially how it will affect fog-driven moisture on the Namib coast, is of major concern. Some climate change models predict less fog, which not only affects the 48 species of fog-dependent animals but may change the way the entire ecosystem functions.

WILDLIFE: Undoubtedly, the wildlife of this habitat is among the most fascinating of all arid zones on Earth. Namib Sand Desert is less productive than the NAMIB ROCK DESERT and supports far fewer large mammals. However, species composition is largely similar (see pp. 77–81 for details). Two notable mammals are common scavengers, Black-backed Jackal and Brown Hyena (locally called "Strandlopers," meaning beach walkers in Afrikaans). These survive largely on the massive Cape Fur Seal colonies that rest and breed on coastal beaches, where pups die or are opportunistically seized when parents are foraging. Among small mammals, the Namaqua Dune Mole-rat hardly ever comes to the surface, utilizing tubers and bulbs that it stores underground for food and water. Grant's Golden Mole (locally called Dune Sharks) are perfectly adapted to swimming through the loose, dry sand of the dunes and, despite being warm-blooded, behave physiologically like lizards. Because sand (rather than air) is in direct contact with the golden mole's body, it would take too much energy to maintain constant body temperature overnight. As the sand temperature drops, the mole's body temperature follows suit, reaching about 68°F (20°C) and entering torpor, a suspended physiological state, until sand temperature warms up the following morning. Dune Hairy-footed Gerbil is endemic to these dunes, making burrows adjacent to Herero Purslane or grass plants where soils are marginally more stable. Active at night, it is able to concentrate its urine and obtain all water from invertebrate prey, which helps this dune dweller maintain its water balance. It also plugs its entrance hole by day to increase humidity in the burrow.

Very few birds can survive in the sand dune sea, but there are a couple of endemics that specialize in the habitat: Dune Lark (IS) in Namibia and Barlow's Lark (IS) in s. Namibia and nw. South Africa, which despite looking very different might actually be the same species! These ultimate desert survivors can exist almost entirely off metabolic water derived from their seed diet,

Brown Hyena survives in the Namib Sand Desert mainly by scavenging and predating Cape Fur Seal pups along the coast. © SEAN BRAINE

and they can reduce their energy needs by manipulating their metabolic rate. Their extreme physiological adaptation means that Dune Larks have this habitat almost entirely to themselves. In the south, where there is more moisture and vegetation, Barlow's Lark shares its sandy habitat with Cape Lark and Tractrac Chat. At the base of dunes or along larger rivers, Western Riparian Woodland corridors (subhabitat Af2–1) with isolated Camel Thorn trees support Orange River White-eye, Chestnut-vented Warbler, Red-eyed Bulbul, and Red-faced Mousebird.

The reptile assemblage of the Namib Sand Desert is exceptional, comprising dozens of species, many of which are endemic. Several species have adapted to living on shifting dunes, such as Peringuey's Adder (*Bitis peringueyi*), Desert Plated Lizard (*Gerrhosaurus skoogi*), and Shovel-snouted (*Meroles anchietae*) (IS) and Wedge-snouted Lizards (*M. cuneirostris*) (IS). All these species are capable of "swimming" into the sand to hide. The adder uses a side-winding motion to cross soft sand and drinks droplets of fog that condense on its body. It also buries itself in the sand to wait stealthily for passing prey, with eyes unusually perched on top (not sides, like other snakes) of its head, making it look like a sand flounder! The Shovel-snouted Lizard has developed a "thermal dance," alternately lifting their front leg and opposite back leg to get relief from hot sand. They also have two bladders, one for urine and the other like a spare water-bottle that is filled on good days when dew forms on sand dunes. The Namib Sand (*Pachydactylus rangei*) and Namib Dune Geckos (*Pachydactylus vanzyli*) both have stilt-like legs to keep their bodies away from the hot sand. They also have webbed feet, allowing additional resistance to sinking into soft sand when foraging and extra digging capabilities when under it. These geckos will often stand on dune crests in the early morning, and when dew

Dune Lark, Namibia's national bird, is a true dune specialist. © KEN BEHRENS, TROPICAL BIRDING TOURS

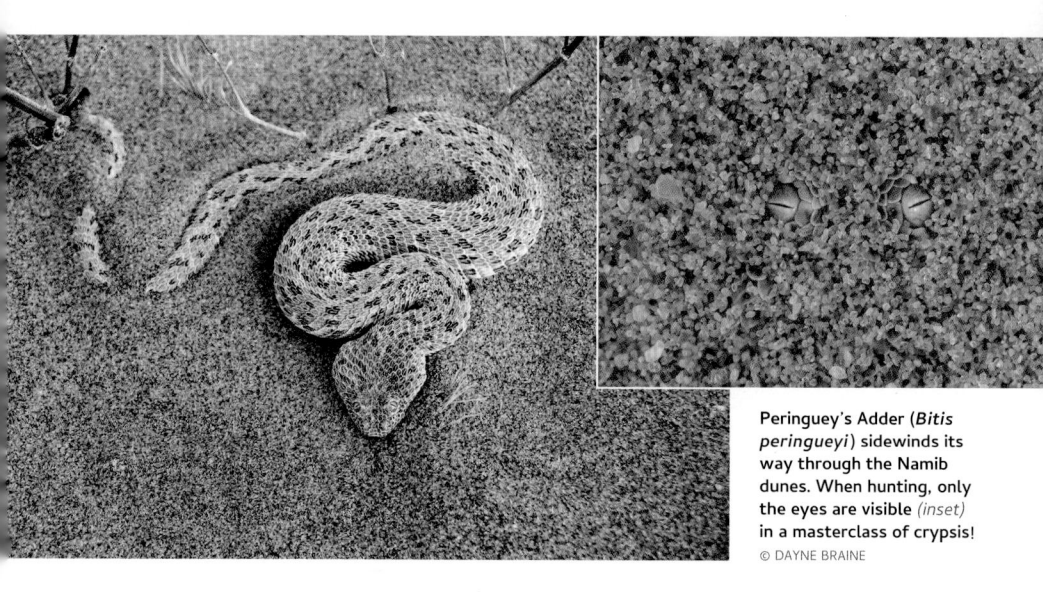

Peringuey's Adder (*Bitis peringueyi*) sidewinds its way through the Namib dunes. When hunting, only the eyes are visible *(inset)* in a masterclass of crypsis!
© DAYNE BRAINE

condenses on them, they lick it off their faces and bodies. Living in open environments seems like a bad idea for a slow mover like the Namaqua Chameleon (*Chamaeleo namaquensis*), but clocking top speeds of 1.8 mi./hr (3 km/hr) actually makes it the fastest chameleon on Earth. Turning dark and angling toward the sun gets the chameleon's metabolism going in the cool early mornings, but when temperatures rise, the chameleon climbs bushes, turns white, and angles its body away from the sun. Other reptiles include legless lizards, desert lizards (*Meroles*), sand lizards (*Pedioplanis*), barking geckos (*Ptenopus*), and Namib day geckos (*Rhoptropus*). Despite this desert being one of the most arid on Earth, it supports frogs, including a few endemics such as Desert Rain Frog (*Breviceps macrops*), a Jabba-the-Hutt-like creature that depends on coastal fog but remains underground by day.

Although invertebrates do not feature much in this book, one cannot ignore the remarkable desert-adapted and diverse flightless darkling beetles (Tenebrionidae) of the Namib. Around 20 species are found in the Namib, most endemic. These beetles have specialized respiration that reduces water loss, and many are covered in wax blooms that help retain water and deflect direct sunlight. A handful of species, especially those of the more arid sand desert, are also adapted to use fog to get water.

Above: **A Namib Sand Gecko (*Pachydactylus rangei*)** engaging in classic gecko behavior and cleaning its eyes with its tongue.
© DAYNE BRAINE

Right: **Namaqua Chameleon (*Chamaeleo namaquensis*)** not only survives in the Namib Sand Desert but is also the world's fastest chameleon! © DAYNE BRAINE

Some dig trenches to enhance dew collection, while others adopt odd poses on dune crests, like a head stand facing into the wind, so water will condense on their body and filter toward their mouths. Others have a physiology that can tolerate body temperatures rising to a phenomenal 123°F (51°C). Even their larvae have the capacity to absorb water vapor directly from the air.

Endemism: The Namib has a rich set of endemic species of plants, birds, and reptiles.

DISTRIBUTION: The epic inland sand sea is confined to the s. Namib, between Luderitz and the Kuiseb River, Namibia. But small isolated patches of this habitat occur along the entire coast of sw. Africa, patchily from s. Angola to nw. South Africa.

WHERE TO SEE: Sossusvlei, Namibia.

SIDEBAR 8 THE EFFECTS OF WORLD OCEAN CURRENTS ON CLIMATE

Continental climates are those that are not really modified by ocean currents. However, many of the world's climates are significantly modified by currents, making them very different from what they would be in the absence of these oceanic influences. This may also be said of the corresponding habitats found within these current-affected areas. Western Europe offers a prime example: Scotland and Norway are kept warmer than places farther east in Europe, like Poland and Belarus, because the North Atlantic Current brings warm water from the tropics. Areas such as the southwest coasts of South America, Africa, and Australia, as well as the Mediterranean Sea coast and the southwest coast of North America, are influenced by cold offshore waters. As a result, these areas all have Mediterranean climates, with hot dry summers and cooler rainy winters.

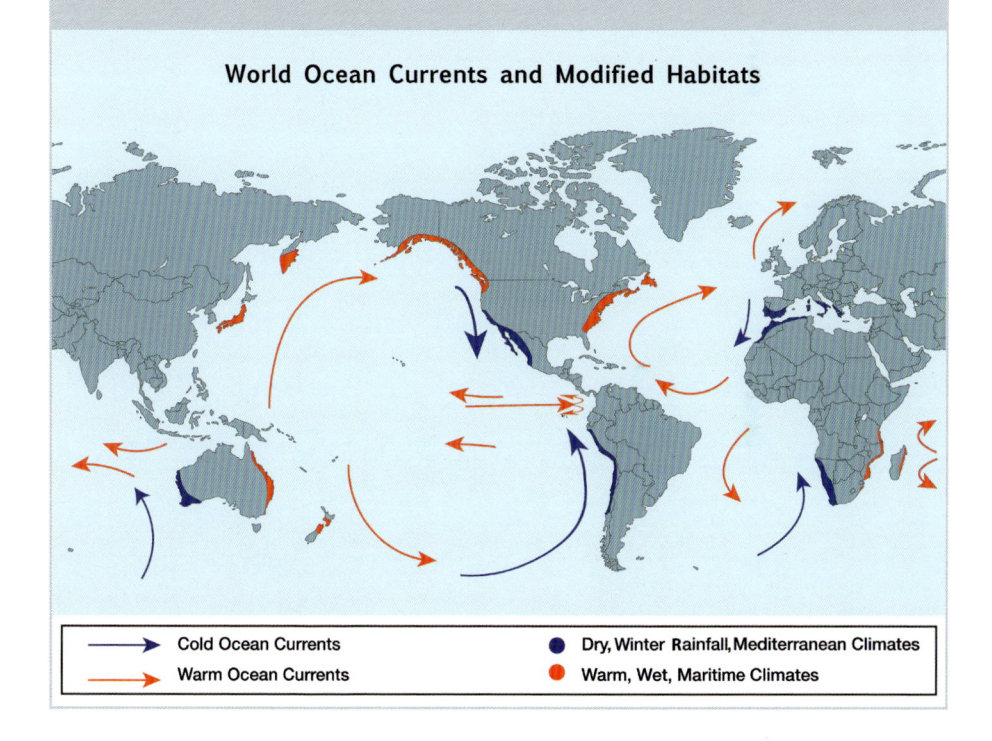

World Ocean Currents and Modified Habitats

- → Cold Ocean Currents
- → Warm Ocean Currents
- ● Dry, Winter Rainfall, Mediterranean Climates
- ● Warm, Wet, Maritime Climates

Af2E ROCKY HAMADA AND MASSIF

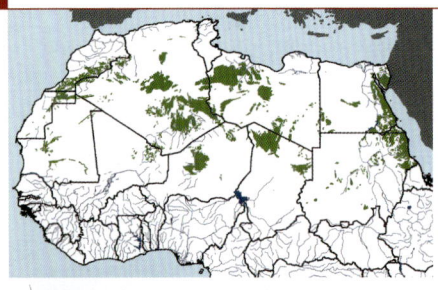

IN A NUTSHELL: Hamadas are desolate residual rock platforms formed through deflation; Massifs are mountains with almost no vegetation that are hot in summer and warm in winter. **Global Habitat Affinities:** WEST ASIAN HAMADA; ROCKY SPINIFEX. **Continental Habitat Affinities:** NAMIB ROCK DESERT; SAHARAN REG DESERT. **Species Overlap:** SAHARAN REG DESERT; SAHARAN ERG DESERT; MAGHREB HOT SHRUB DESERT; CAUCASIAN SHRUB DESERT.

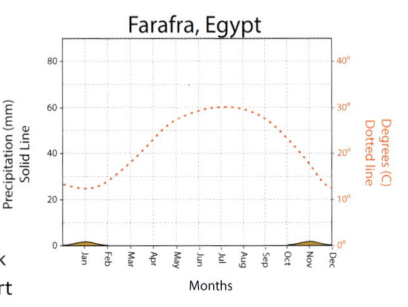

DESCRIPTION: Hamadas are the plain rock surfaces, mountains, and cliff faces of the Sahara. Although on the plains there are polished rocks and desert pavement that can appear very similar to that of SAHARAN REG DESERT, they differ in that Reg develops a carapace on mostly deep colluvial deposits that are winnowed (blown away) but have soil development below the carapace. By contrast, Hamada has cobble-size rocks laying directly on bedrock that can also be abraded and polished. So Hamada desert pavement can progress to form Reg if the wind conditions become much more intense, but the process is not reversible. Once you have cobbles sitting on bare rock (Hamada), the only thing that can happen is that they get covered in a varnish of silica, mixed with iron and manganese oxides such as magnetite and pyrolusite, respectively (becoming Reg).

Hamadas have a similar climate to Reg and Erg desert and are some of the hottest places on Earth (Köppen **Bfh**). Away from the moderating effect of the coast, average summer temperatures reach 113°F (45°C) and maxima top a phenomenal 130°F (50°C). Daily temperature range can be extreme, with fluctuations of 70°F (21°C). Many areas receive less than 1 in. (25 mm) per year and sometimes go years without rain. The average annual rainfall is 2 in. (50 mm), which, when combined with evapotranspiration rates as high as 15 ft. (4.5 m) annually, makes for hyperarid conditions.

Large flat surfaces such as plateau tops and mesas are nearly devoid of vegetation because the small amount of groundwater is typically laden with dissolved salts and compounds. Plant species richness, however, is higher than would be expected given the limited extent of growth. One of the most striking Hamada plants is the Bou Amama (*Anabasis aretioides*), which at first appearance— because they are often dominant in an area—look like a stromatolite field from 3 billion years ago! Also called Desert Cauliflower, they look like cushion plants from South American Paramo, but close examination reveals that they are dwarf microphyllous- and sclerophyllous-leaved shrubs growing in dense ball cushions up to 6 ft. (2 m) across and 2 ft. (60 cm) tall. In other areas, small shrubs and grasses cover less than 1% of surface area, with plant species including the tough Desert Nettle (*Forsskaolea tenacissima*), Fragrant Star (*Asteriscus graveolens*), an attractive yellow daisy, Hammada Scorparia (*Haloxylon scoparium*), Egyptian Sage (*Salvia aegyptiaca*), various small herbs like *Fagonia*, the very colorful Purple Mistress (*Moricandia arvensis*), and grasses such as Camel Grass (*Cymbopogon schoenanthus*).

The climate in the central-southern Sahara is even more brutal than that in the subtropical Mediterranean-influenced northern Sahara, and vegetation persists only in small cracks where tenacious herbs and shrubs, like Storkbill (*Erodium glaucophyllum*), Cairo Sun Rose (*Helianthemum kahiricum*), Melliha (*Reaumuria hirtella*), and Desert Damasa (*Fagonia mollis*) eke out a tenuous existence.

And then there are the great Saharan mountain ranges (called Massifs). Some are huge, but below 6500 ft. (2000 m) in elevation they tend to retain a climate and vegetation similar to the surrounding plains. However, some are much taller, such as the Tibesti Massifs of nw. Chad, with Emi Koussi reaching 11,200 ft. (3415 m), the highest point in the Sahara; Ahaggar in s. Algeria; and Tassili n'Ajjer in se. Algeria. The elevation where these Massifs develop their own orographic rainfall can be as low as 5900 ft. (1800 m) in the w. Sahara but rises to above 6500 ft. (2000 m) in the e. Sahara. These Massifs have much lower temperatures and receive much more precipitation than the surrounding desert. For example, the Tibesti Massif has peaks above 8530 ft. (2600 m) and receives up to 11 in. (280 mm) precipitation a year as rain, mist, and snow.

More rain falls on the Saharan mountains than in the surrounding lowlands, and even misty cloud banks can form, delivering precious water that accumulates in areas called "guelta" that support Hamada Montane Woodland (subhabitat Af2-3). Grasses that can survive only as ephemerals in the lowlands persist in the highlands as perennials, including Habob (*Stipagrostis obtusa*) and *Aristida coerulescens*. Where soil development is more advanced above 6500 ft. (2000 m), shrubs such as *Artemisia*, Bedouin Soapbush (*Anabasis articulata*), and Shabram (*Zilla spinosa*) grow alongside the gymnosperm shrub *Ephedra tilhoana*—a bizarre relative of the Namib's Welwitschia (see p. 78). Together these plants grow up to 3 ft. (1 m) to form a shrubland that resembles the NAMA KAROO from s. Africa. A handful of trees persist, including three of significant interest: Saharan

Classic Saharan Massif landscape in the Aïr Mountains of n. Niger.
© RICHARD JULIA

Basalt Hamada plain at Jebel es Soda, Libya. © GRAY TAPPAN

The capacity of Saharan Massifs to gather water means that wadis are more common in that habitat than in other types of desert. Ennedi Plateau, Chad. © JON HALL

Myrtle (*Myrtus nivellei*) for its sun-protecting chemicals; Laperrine's Olive Tree (*Olea europaea laperrinei*), a drought resistant relict that may help the closely related cultivated trees grow in more arid climates; and perhaps one of Earth's most remarkable conifers, the Saharan Cypress (*Cupressus dupreziana*). With only 233 individuals remaining, this Endangered species is a relic from wetter times. Most individuals are estimated to be over 2000 years old, and they are now barely growing because of increasing desertification and the miniscule amount of annual rainfall (circa 1.2 in./30 mm) they receive. These cypresses are evolved from trees that may be the remnants of a much broader coniferous (nonflowering gymnosperm) woodland that existed during the early Mesozoic. The existence of such a forest is interesting because it implies that the Sahara had a climate that allowed the existence of widespread forest, but it is not surprising considering that angiosperms (flowering plants) only evolved in the Cretaceous (the later part of the Mesozoic), so as long as the climate was conducive to plant growth, conifers had little competition from angiosperms. Riparian strips called Wadis (subhabitat Af2–4) are most common in Hamada but can also occur in Reg and Erg (see sidebar 9, p. 100).

Af2E ROCKY HAMADA AND MASSIF

CONSERVATION: See Conservation section for the SAHARAN REG DESERT account (p. 71). There are some remarkable conservation areas in this region, especially the mountain Massifs, like Tassili n'Ajjer National Park in Algeria. This is a World Heritage Site for the stunning collection of prehistoric rock art, which depicts many of the animals that have vanished from here, including Addax, Dama Gazelle, Scimitar-horned Oryx, and Ostrich.

WILDLIFE: For wildlife widespread in all Saharan habitats, see the Wildlife section of the SAHARAN REG DESERT account (p. 71). Although Rocky Hamada and Massif is a tough habitat, the mountainous sectors have more water and wadis than either Reg or Erg and so, comparatively, it supports more wildlife, especially where there is some vegetation. Many of the most charismatic big mammals are extinct, but Rhim and Dorcas Gazelle persist on plains, often close to mountain bases, where they are hunted by the final 250 or so "Saharan" Cheetah, a unique and Critically Endangered subspecies that is paler and with less distinct markings than its southern cousin. On plains Hamada, the African Wild Ass used to be more widespread, but relentless domestication means that both wild races are Critically Endangered. After dark the Desert Hedgehog emerges, and Striped Hyena lurk in the same areas where small predators, such Rüppell's and Fennec Foxes and African Wild and Sand Cats, hunt several species of hares, gerbils, and jirds. Other nocturnal stunners here include mini-kangaroo-like Greater and Lesser Egyptian Jerboas. Jerboas are phenomenal jumpers, capable of sustained bouncing of 10 ft. (3 m) per leap. The male performs a bizarre mating ritual, rearing up on his hind legs, and when a female approaches, slapping her repeatedly with his stubby forelimbs. Strange, but true.

In the craggy hills and mountains, Barbary Sheep (Aoudad) persist, although the Leopards that predate them are much rarer; Nubian Ibex inhabit similar landscapes in Egypt. Saharan mountains are one of the few places where Palearctic lineages like goat and sheep live adjacent to Afrotropical ones like Rock Hyrax (see sidebar 30, p. 272).

Sand Partridge has a daft name, as it is mostly a rocky Hamada specialist. © KEN BEHRENS, TROPICAL BIRDING TOURS

Moroccan Mastigure (*Uromastyx nigriventris*) is a beastly lizard of rocky outcrops. © DANIELE ARDIZZONE, FLUYENDO PHOTOGRAPHY

and Patas Monkey! But some of the most remarkable Hamada residents are the seldom-seen and poorly known gundis. This entire family (Ctenodactlidae) of rodents comprises five species in four genera endemic to North Africa and the Horn, having diverged from ancestors of Laotian Rock Rat some 44 MYA! Gundis resemble a pika × guineapig hybrid (pika-pig would be a good alternate name!). Three species including Val's, Felou, and Mzab Gundis are Rocky Hamada and Massif specialists. Their short tails are covered in hair to aid balance over uneven terrain, and their ribcages are flexible, allowing them to slip into fissures in rocks. They form sleepover piles of 20+ animals in cold weather, have squeaks to communicate, and thump their hind feet to warn fellow gundis of impending doom. Several bats use fissures and caves in Hamada, including the odd-looking Desert Long-eared Bat, Kuhl's Pipistrelle, and Egyptian Free-tailed Bat.

Birds of this habitat include Barbary Partridge, the poorly named Sand Partridge, Great Gray

Left: **Patas Monkey scale the Massifs in the southern Sahara, a link between the Afrotropics and Palearctic.**
© JON HALL

Below: **Gundis are bizarre pika-like animals that form their own family, which is endemic to North Africa and the Horn. Most species, such as this Mzab Gundi in Algeria, are Rocky Hamada and Massif specialists.**
© VLADIMIR DINETS

Shrike, House Bunting, and Hooded and White-crowned Wheatears. After dark, Egyptian and Nubian Nightjars may be found. The mountains and wadis with steep cliffs hold Rock Martin and raptors such as Sooty, Lanner, and Barbary Falcons; Pharaoh Eagle-Owl; and Desert Owl. Nearby wadis support Pallid Scops-Owl. Other widespread regulars include multiple species of sandgrouse, larks, wheatears, and ravens. The entire region is important for resting migratory Palearctic birds (see sidebar 6, p. 76).

A dwarf form of Nile Crocodile (*Crocodylus niloticus*) persisted in Tassili n'Ajjer until 1940. In rocky crevasses or other hideaways lurk the likes of Algerian Gecko (*Tropiocolotes algericus*), Mali Agama (*Agama boueti*), African Wall Gecko (*Tarentola hoggarensis*), and White-spotted Wall Gecko (*Tarentola annularis*), which are large and aggressive enough to eat gerbils for dinner! Other Hamada denizens are the beastly Sudan (*Uromastyx dispar*) and Moroccan Mastigures (*Uromastyx nigriventris*), while predators include Desert Monitor (*Varanus griseus*), Algerian Whip Snake (*Coluber algirus*), and North African Cat Snake (*Telescopus tripolitanus*). Where water occurs, frogs may be found.

DISTRIBUTION: The Hamadas of the Sahara are sprinkled across North Africa and mostly form a mosaic with Reg and Erg, from the Atlantic coast of Western Sahara and Mauritania to the Red Sea, a distance of 3300 mi. (5300 km). The mountain Massifs run east–west, mostly in a band between 18°N and 28°N.

WHERE TO SEE: Tibesti Massif and Ennedi Plateau, nw. Chad; Tassili n'Ajjer in se. Algeria.

Right: **Rocky desert habitat provides breeding sites for Pharaoh Eagle-Owl.** © CARLOS N. G. BOCOS

Below right: **Wind-scoured (ablation) Massif in Chad.** © JON HALL

| SIDEBAR 9 | WADIS: NOT JUST AN OASIS IN THE DESERT |

Oases are areas where water comes to the surface in a spring, but because they are tiny dots in a sea of sand, they are described in the SAHARAN ERG DESERT account since that is the broader habitat where they tend to be more frequent. Like oases, wadis also have trees and large bushes, and they occur as a microhabitat throughout ROCKY HAMADA AND MASSIF, SAHARAN REG DESERT, and MAGHREB HOT SHRUB DESERT. But they are most common in Hamadas, and their most distinctive and widespread wildlife are described in that account. Both oases and wadis are too limited in distribution to be mapped.

In contrast to oases, wadis are more diverse in nature, being dry riverbeds, colluvial fans (where material is transported through gravity), gullies, and the bases of desert escarpments. Wadis accumulate more water than precipitation alone would suggest through runoff from other areas. So although rainfall may be extremely low on the rock surfaces of Hamada or cobbled plains of Reg, some rainfall avoids immediate evaporation or capture by plants and is accumulated in lower lying areas; hence, the effective water availability is often many times that of the surrounding areas. Think of it like the lusher plant growth on the side of paved highways through arid regions.

Around many Hamadas, the moister areas near the rock surface support Doum Palm (*Hyphaene thebaica*), which fits the image most people have of an oasis. They occur alongside a variety of tamarisk shrubs (*Tamarix*), which can grow as small trees up to 25 ft. (8 m), and Toothbrush Tree (*Salvadora persica*), which can grow in groves that look like a hybrid between a Cork Oak (*Quercus suber*) and Weeping Willow (*Salix babylonica*). These wadis function the same as an oasis for migrating birds and therefore have very similar bird assemblages to oases.

The vegetation of wadis tends to have affinities with the prevailing climatic zone; in the north, they have a Mediterranean or temperate mix of plants, including species like poplars (*Populus*). Throughout most of the Sahara, wadis are dominated by plants of the Sahel, such as the iconic Umbrella Thorn (*Vachellia tortilis*), and plants that have been cultivated, such as the Meru (*Maerua crassifolia*) and Desert Date (*Balanites aegyptiaca*), which both can grow up to 30 ft. (10 m) tall. These stands tend to be widely spaced with very open canopies, little understory, and a ground cover of grasses dominated by the perennial bunchgrass Taman (*Panicum turgidum*). In many areas where the water table sits around 30 ft. (10 m) deep, the tamarisk or "Salt Cedars" (*Tamarix* spp.) dominate to an extent that they form low impenetrable thickets 6 ft. (2 m) tall; in better watered areas, they resemble riparian forest up to 40 ft. (13 m) tall. In these tamarisk areas there is little understory, save for other shrubs like Fire Bush (*Calligonum comosum*) and perennial grasses such as *Stipagrostis pungens*. Tamarisk groves support birds similar to those of Maghreb Hot Shrub Desert.

Hamada plateau with a wadi that supports a palm oasis. Libya. © MICHIEL KUPERS

Af2F NAMA KAROO

IN A NUTSHELL: Arid but highly diverse shrubland of sw. Africa with significant grass cover. **Global Habitat Affinities:** PATAGONIAN STEPPE; CASPIAN WORMWOOD DESERT; SAGEBRUSH SHRUBLAND; GIBBER CHENOPODLANDS; CAUCASIAN SHRUB DESERT. **Continental Habitat Affinities:** SUCCULENT KAROO; MAGHREB HOT SHRUB DESERT; AFROTROPICAL HOT SHRUB DESERT. **Species Overlap:** SUCCULENT KAROO; KALAHARI DRY THORN SAVANNA; NAMIB ROCK DESERT; FYNBOS; STRANDVELD; RENOSTERVELD; MONTANE GRASSLAND.

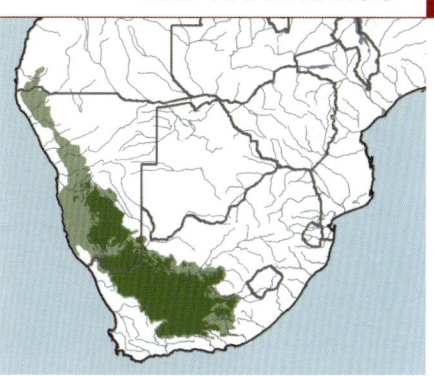

DESCRIPTION: The Nama Karoo is a dry and open grassy shrubland of inland sw. South Africa that extends into Namibia, along the fringes of the Namib Desert. From a distance, the Nama Karoo looks like a typical desert or semi-desert, often reminding visitors of the American Southwest. But a close look reveals an extraordinarily diverse and distinctive plant community unlike any other on Earth, plus an associated

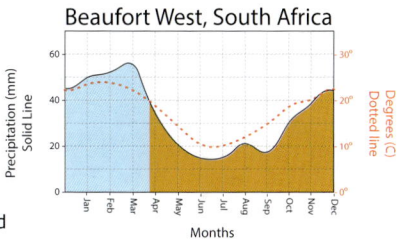

Classic thigh-high Nama Karoo shrubland with few succulents and bulbs, in South Africa's Karoo National Park.
© KEITH BARNES, TROPICAL BIRDING TOURS

Nama Karoo can approach lushness after good rains *(main photo)* **or be incredibly stark and resemble rocky desert in a drought** *(inset).* MAIN PHOTO © KEN BEHRENS; INSET © KEITH BARNES, TROPICAL BIRDING TOURS

community of wildlife. Essentially restricted to the higher reaches of the inland plateau, it mostly falls between 3280 and 4600 ft. (1000–1400 m), with extremes of 1640–6560 ft. (500–2000 m). This is a dry and harsh environment (Köppen **Bwk** and **Bwh**), with huge fluctuations in temperature: summer maxima can exceed 104°F (40°C), while nights below freezing are common in the winter. In the interior, a single day can see temperatures fluctuate by 45°F (25°C). Rainfall is low, generally 4–20 in. (100–500 mm) annually, falling primarily in the austral summer.

The dominant plants of the Nama Karoo are shrubs shorter than 3 ft. (1 m) tall; it is less diverse botanically than the SUCCULENT KAROO with far fewer succulents and bulbs. The most important shrub families are asters (Asteraceae), such as Kopokbush (*Eriocephalus africanus*), Anchor Karoo (*Pentzia incana*), and Bitterthorn (*Chrsocoma ciliata*), as well as Threethorn (*Rhigozium trichotomum*), kareethorn (*Lycium*), and shepherd's trees (*Boscia*), occasionally with scattered euphorbias. There is a significant component of grasses, often dominated by Silky Bushman Grass (*Stipagrostis uniplumis*), Red Oat Grass (*Themeda triandra*), copperwire grasses (*Aristida* and *Elionurus*), or lovegrasses (*Eragrostis*), which are perennial and flush after good summer rains, turning barren plains into something akin to prairie. In places, annuals like *Pentzia* and *Zygophyllum* are common, and trees like guarri (*Euclea*) or Kunibush (*Rhus undulata*) occur but never dominate. Larger bushes and trees are mostly confined to drainage lines that constitute Western Riparian Woodland (subhabitat Af2–1). Some common species include Wild Tamarisk (*Tamarix usneoides*), Karee (*Searsia lancea*), and particularly Sweet Thorn (*Vachellia karoo*), which can create dense groves, and these bands allow species more typical of thorn savanna to survive in the otherwise inhospitable shrubby plains. Despite this habitat's aridity, it does have wetlands, including many artificial farm ponds, which attract a wide variety of waterbirds.

Nama Karoo serves as a transition between the heathlands along the southern coast of Africa and the KALAHARI DRY THORN SAVANNA to the north. Grasses become increasingly common to the east, creating a nebulous transition to the Afrotropical MONTANE GRASSLAND of the South African Highveld.

CONSERVATION: Although the Nama Karoo appears like pristine habitat to the casual observer, it has been profoundly affected by humans. The vast majority of this habitat is on private farms, where sheep and other animals are grazed. Competition from domestic animals and hunting have decimated the region's large mammal populations. Overgrazing also has the insidious effect of slowly replacing the full range of indigenous plants—especially grassy elements, which would have been sustainably grazed by migratory wild animals—with a poor subset of plants that are toxic or unpalatable to domestic animals. Some moister parts of the Nama Karoo, such as the Little Karoo, which borders the RENOSTERVELD and FYNBOS, have been largely converted to agriculture. Another major problem is the introduction of invasive exotic plants, which can further exacerbate the environmental changes wrought by livestock. Nama Karoo remains underrepresented in the formal protected area network.

WILDLIFE: In terms of mammals, the Nama Karoo fauna is a shadow of what it once was. After the rains, with the subsequent flush of grasses, the region used to host massive migrations of antelopes, mainly millions of Springbok, a natural spectacle that matched or exceeded the wildebeest migration of the Serengeti. This habitat also supported nearly a full set of Africa's charismatic big mammals, but many of these species either have been completely eliminated or are now confined to small fenced-in reserves. This area was once the main realm of the now-extinct Quagga, a distinctive and beautiful animal that may have been a subspecies of Plains Zebra. The Nama Karoo was the summer range of migratory Black Wildebeest, which was reduced to close to extinction in the wild before being resurrected by captive breeding. The formerly widespread "Cape" Mountain Zebra and Gemsbok are now largely confined to a handful of conservation areas. Although much has been lost, there are still some big mammals ranging throughout the Nama Karoo, like Kudu. Despite heavy persecution, predators like Caracal and Black-backed Jackal remain. The Nama Karoo is also an excellent habitat for smaller predators, such as Bat-eared and Cape Foxes, Aardwolf (a termite-eating Hyena), Cape Gray and Yellow Mongooses, and Meerkat. Many of these are best sought on night drives, which can count among Africa's

Karoo Bustard in full croaking "song," surrounded by springtime blooms. **Karoo National Park, South Africa.** © KEITH BARNES, TROPICAL BIRDING TOURS

The hematite-stained Koa Dunes with large-seeded *Stipagrostis* and *Brachiaria* tussock grasses are the key habitat for the highly localized endemic Red Lark. During the dry season the habitat is devoid of flowers. Northern Cape, South Africa.
© KEITH BARNES, TROPICAL BIRDING TOURS

most exciting and may also turn up an Aardvark or a Cape Porcupine. One small predator, the Black-footed Cat, is near-endemic to the Nama Karoo and certainly best sought in this habitat; it is a very high-value target for any wildlife enthusiast. Although the vast migratory herds are gone, there are still plenty of Chacma Baboon, Cape and Scrub Hares, Steenbok, and Common Duiker. Springbok and Red Hartebeest have been restocked in various state and private reserves. The Riverine Rabbit is in its own genus and is a Critically Endangered charismatic endemic of the Nama Karoo, where it inhabits rocky gorges where it browses shrubs and grasses. Other special and very localized small mammals include Karoo Rock Sengi, Grant's Rock Rat, and Visagie's Golden Mole.

The Nama Karoo has two endemic birds: the Red Lark (IS) prefers arid dune shrubland with large-seeded tussock grasses, and Sclater's Lark (IS) occupies barren stony plains where it feeds mainly on the seeds of the persistent Eight-day Grass (*Enneapogon desvauxii*). The summer rains, and subsequent flush of grasses, attract rain-tracking granivorous nomads that can be totally absent then suddenly omnipresent a few weeks later (particularly in Bushmanland). Clouds of Gray-backed Sparrowlarks are joined by the local Black-eared Sparrowlark, alongside Lark-like Bunting and Black-headed Canary. The Nama Karoo and SUCCULENT KAROO share a slew of near-endemic bird species, making these habitats an essential destination for birders. Open shrubby plains attract Ludwig's Bustard, Karoo Bustard, Karoo Lark, Rufous-eared Warbler, Yellow-rumped Eremomela, and Sickle-winged and Karoo Chats. Rocky ravines support Layard's Warbler and the local Kopje Warbler. The Nama Karoo also still holds some of Africa's charismatic

Top right: **Namaqua Sandgrouse** is partially nomadic in a variety of dry habitats, including Nama Karoo. It flies long distances each day to access water, and its pleasant calls overhead are a classic sound of arid sw. Africa. © KEN BEHRENS, TROPICAL BIRDING TOURS

Above: **Ground Woodpecker** prefers rocky microhabitats within Nama Karoo and other arid habitats. © KEITH BARNES, TROPICAL BIRDING TOURS

big birds, including Secretarybird and Tawny and Martial Eagles, although these are increasingly rare. More widespread birds that are at home include Jackal Buzzard, Greater Kestrel, Burchell's and Double-banded Coursers, Namaqua Sandgrouse, Bokmakierie, Karoo Scrub-Robin, and Lark-like Bunting. Rivers and streams that penetrate the dry Nama Karoo provide habitat for the endemic Namaqua Warbler and support some species more typical of **KALAHARI DRY THORN SAVANNA**, like Pied Barbet, Pririt Batis, and Cape Crombec.

The Nama Karoo has a diverse set of reptiles, including several charismatic ones that are very localized: Anchieta's Agama (*Agama anchietae*), Karoo Dwarf Chameleon (*Bradypodion karrooicum*), Namaqua Chameleon (*Chamaeleo namaquensis*), and Western Dwarf Girdled Lizard (*Cordylus minor*). The stunning and local Red Adder (*Bitis rubida*) is restricted to rocky outcrops. Another group with several endemics is the tortoises: Boulenger's Padloper (*Chersobius boulengeri*) is restricted to rocky dolerite ridges, Greater Padloper (*Homopus femoralis*) likes grassy hills, and Parrot-beaked Padloper (*Homopus areolatus*) and Tent Tortoise (*Psammobates tentorius*) also occur. The beastly Leopard Tortoise (*Stigmochelys pardalis*) is common but shared with many other habitats. Other groups with high endemism are sandveld lizards (*Nucras*), sand lizards (*Pedioplanis*), girdled lizards (*Cordylus*), crag lizards (*Pseudocordylus*), and thick-toed geckos (*Pachydactylus*). Most of the endemic reptiles are highly localized, and while some are conspicuous, the majority will be found only by those who search carefully. Despite being arid, the Nama Karoo holds several interesting frogs including Southern Pygmy Toad (*Poyntonophrynus vertebralis*), Karoo Toad (*Vandijkophrynus gariepensis*), Karoo Caco (*Cacosternum Karooicum*), and Giant Bullfrog (*Pyxicephalus adspersus*). However, all these are also found in the adjacent Succulent Karoo or Kalahari Dry Thorn Savanna habitats.

DISTRIBUTION: Nama Karoo habitat is landlocked, mostly in central and western South Africa, though it extends well into Namibia, eventually terminating in s. Angola, mostly lying above 3000 ft. (900 m) on the vast Southern African Plateau. Nama Karoo is adjacent to many other habitats and is replaced by SUCCULENT KAROO to the southwest, NAMIB ROCK DESERT to the northwest, FYNBOS and RENOSTERVELD to the south and on mountain ranges, ALBANY THICKET to the southeast, MONTANE GRASSLAND to the east, and KALAHARI DRY THORN SAVANNA to the northeast.

WHERE TO SEE: Karoo National Park, South Africa.

Above: **Horned Adder (*Bitis caudalis*) is a widespread and beautiful snake of sw. Africa's arid habitats, including Nama Karoo.** © KEIR AND ALOUISE LYNCH

Left: **Common Giant Ground Gecko (*Chondrodactylus anguilifer*) is a spectacular reptile of the arid southwest.** © KEIR AND ALOUISE LYNCH

Af2G SUCCULENT KAROO

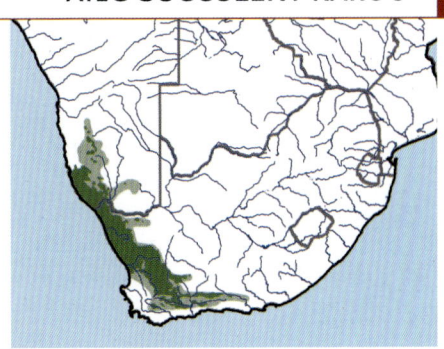

IN A NUTSHELL: Arid but highly diverse shrubland of sw. Africa with a ground cover of diverse succulents and perennial and annual cover. **Global Habitat Affinities:** PLAINS MONTE; CASPIAN WORMWOOD DESERT; SAGEBRUSH SHRUBLAND; GIBBER CHENOPODLANDS; CHENOPOD SHRUBLAND; CAUCASIAN SHRUB DESERT. **Continental Habitat Affinities:** NAMA KAROO; MAGHREB HOT SHRUB DESERT; AFROTROPICAL HOT SHRUB DESERT. **Species Overlap:** NAMA KAROO; KALAHARI DRY THORN SAVANNA; NAMIB ROCK DESERT; FYNBOS; STRANDVELD; RENOSTERVELD; MONTANE GRASSLAND.

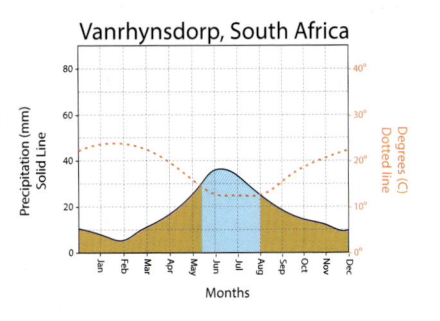

DESCRIPTION: The Succulent Karoo is an open semiarid shrubland found in coastal w. South Africa and s. Namibia, extending onto the western fringes of the great escarpment up to 4920 ft. (1500 m). The vast array and abundance of succulents make this habitat unique. Of the world's 36 biodiversity hotspots designated by Conservation International, the Succulent Karoo is the only one in an arid zone, with plant diversity and endemism rivaling some rainforest hotspots. Despite this, physiognomically it resembles other semiarid shrublands around the world, such as the Sonoran and Atacama deserts of the Western Hemisphere. This is a dry and harsh environment (Köppen **Bwk**), with huge fluctuations in temperature: summer maxima can exceed 104°F (40°C), while nights below freezing are common in the winter. In the interior, a single day can see a temperature fluctuation of 45°F (25°C). However, most of the year is more moderate with a mean annual temperature of 62°F (17°C). Rainfall is very low throughout, ranging annually from 0.8 to 11 in. (20–290 mm) but generally 4–8 in. (100–200 mm), almost all of it falling as intense storms from cyclonic fronts during the austral winter (Jun–Aug). Coastal mists and dew are common and another form of regular precipitation. Snowfall, hail, and frost are rare. Hot desiccating winds (katabatic winds) descend from the escarpment and scour the plains, locally called "berg winds"; these can occur year-round.

The dominant plants of the Succulent Karoo are dwarf succulent shrubs less than 3 ft. (1 m) tall, floristically more closely aligned with FYNBOS than NAMA KAROO. This habitat supports a remarkably diverse plant community of at least 6000 species, including around 17% of the world's succulents (circa 1700 species), of which 40% are endemic or near endemic. Most of the 80 endemic genera are succulents or geophytes. The most important family is the fig marigolds (Aizoaceae), including thorn vygies (*Eberlanzia* and *Ruschia*), various species of dewflowers (*Drosanthemum*), and mesembs

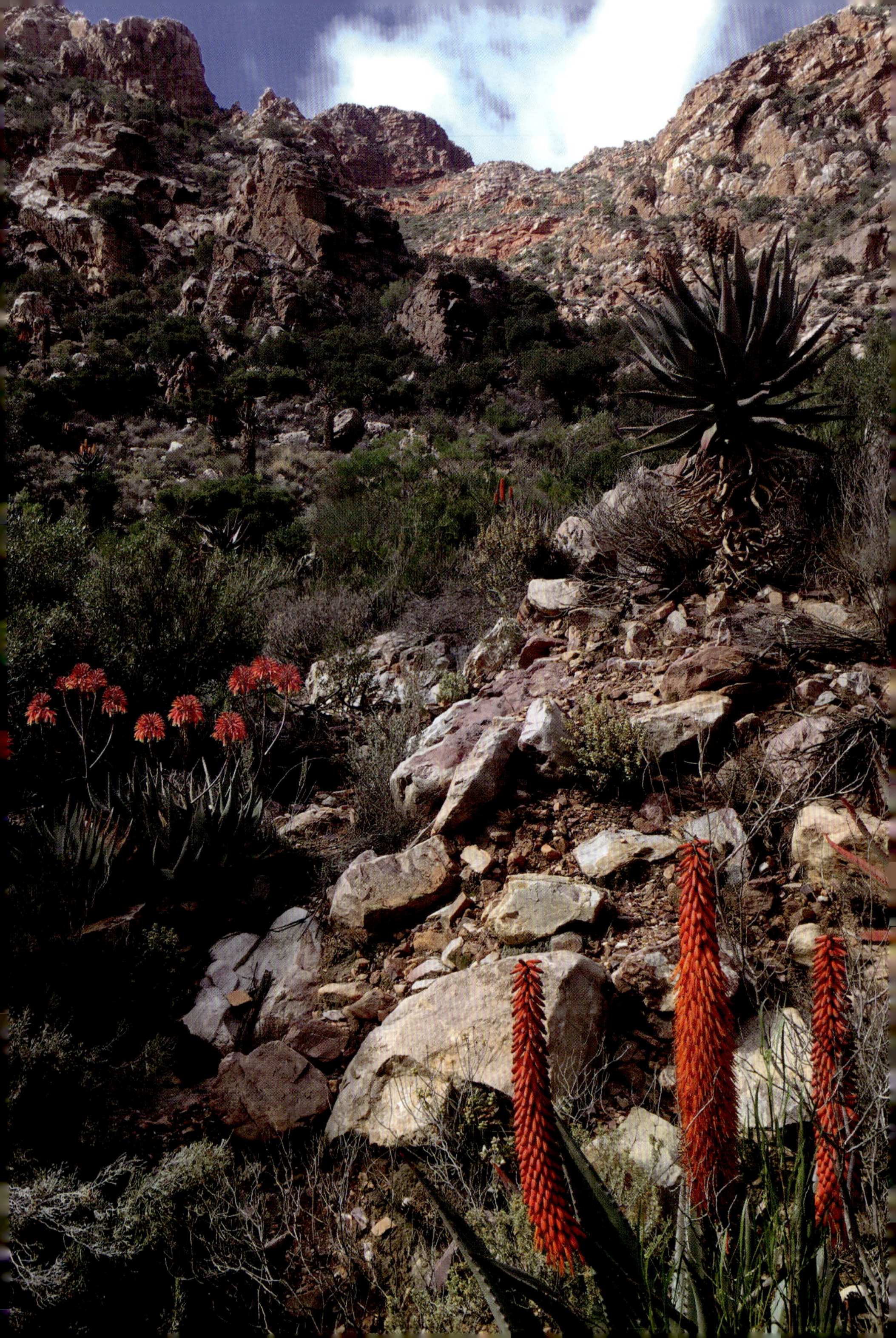

(*Malephora*). Other distinctive succulent groups are the spurges (Euphorbiaceae) and stonecrops (Crassulaceae). Some succulents are able to use unique biochemical pathways to remain drought resistant (see sidebar 10, p. 111). Geophytes and bulbs, constituting 18% of the flora, are diverse, particularly irises (Iridaceae) and squills (Hyacinthaceae). The upper portion of the Namaqualand escarpment near Niewoudtville is particularly famous as a location to view these during the flowering season. Although not particularly diverse, another important floral component is the annuals (primarily daisies in the family Asteraceae), often on degraded or fallow lands that, together with the bulbs, create one of the most spectacular mass flowering events on Earth, the "Namaqualand blooms" in August and September after the winter (Jun–Aug) rains. Grass is scarce in the Succulent Karoo, unlike the more eastern Nama Karoo.

Opposite: **Rugged canyon below Swartberg Pass, with a mix of Succulent Karoo, Nama Karoo, and Albany Thicket elements. Western Cape, South Africa.** © KEN BEHRENS, TROPICAL BIRDING TOURS

Succulent Karoo plants: bulbs from the family Asphodelaceae *(top left)* **are common features, two species of fig marigold** *(top right, bottom left)* **and a small euphorbia** *(bottom right)*.
© KEITH BARNES (TOP LEFT, BOTTOM RIGHT) AND
© KEN BEHRENS (TOP RIGHT, BOTTOM LEFT),
TROPICAL BIRDING TOURS

110 DESERTS AND ARID LANDS

Right: **Heuweltjies are distinct circular patches comprising completely different vegetation to neighboring areas. Their formation is likely linked to ancient or modern termite occurrence, and they feature throughout the Karoo.** © KEITH BARNES, TROPICAL BIRDING TOURS

Springtime Succulent Karoo, a typically disturbed area has fewer succulents and more annual components (usually daisies). Near Port Nolloth, Northern Cape, South Africa. Contrasted with desolate summer drought Succulent Karoo *(inset),* **Western Cape.** © KEITH BARNES, TROPICAL BIRDING TOURS

Heuweltjies (an Afrikaans term meaning "Little Mound") litter the landscape in the Succulent Karoo (also marginally into other habitats). These are slightly raised, 3 ft. (1 m) high, evenly distributed mounds around 65–130 ft. (20–40 m) in diameter and composed of calcium-rich soils, which often support completely different plant communities. Sometimes the soil is richer with more diverse plant communities, and sometimes the soil is more compacted, inhibiting plant growth. Their origin is a mystery, with various theories proposed: some think they are fossil termite mounds 4–40 thousand years old; others believe they are created by a combination of termites and fossorial mammals like Aardvark and mole-rats; and yet others think they are formed through differential erosion patterns. However they came to be, they create visibly distinct circular

SIDEBAR 10 — PHOTOSYNTHESIS—BEATING DROUGHT THROUGH BIOCHEMISTRY

All plants convert sunlight to energy using a process called photosynthesis. The exchange of gasses is conducted via pores in the leaves called stomata that can open and close. Most plants use two very similar processes, called C3 and C4, by which carbon dioxide enters the leaf and water departs (via a process called evapotranspiration) during the daytime. Most of these plants lose over 80% of the water absorbed by the roots during this process. But other groups, including some primarily Southern African plants, have evolved an entirely different way of fixing carbon. Crassulacean acid metabolism (CAM) photosynthesis is an adaptation to drought. In CAM, stomata open only at night to collect carbon dioxide that is then transported into specialized cells where it is converted to malic acid and stored. During the day, the malic acid is reconverted into carbon dioxide, which is then photosynthesized while the stomata remain closed, critically retaining water in blazing hot temperatures. This mechanism was first discovered in stonecrops (Crassulaceae), which have many representatives in Africa's arid zones. But drought-efficient CAM photosynthesis has evolved independently in a wide variety of plants.

A member of the stonecrops, this "pig's ear" (*Cotyledon*) is capable of fixing carbon using a unique biochemical process. © KEITH BARNES, TROPICAL BIRDING TOURS

patches of habitat in vast portions of the Succulent Karoo, enhancing diversity by allowing the growth of different species from the surrounding terrain.

Larger bushes and trees are mostly confined to drainage lines that constitute Western Riparian Woodland (subhabitat Af2-1). Some common species include Wild Tamarisk (*Tamarix usneoides*), Karee (*Searsia lancea*), and particularly Sweet Thorn (*Vachellia karoo*), which can create dense groves; these bands allow species more typical of thorn savanna to survive in the otherwise inhospitable shrubby plains. Copses of distinctive-looking Kokerboom or Quiver Tree (*Aloe dichotoma*) grow on boulder-strewn hills (or "koppies") in n. Namaqualand. Despite the habitat's aridity, it does have wetlands, including many artificial farm ponds, which attract a wide variety of waterbirds.

CONSERVATION: Only a tiny percentage of this crucial biodiversity hotspot has been formally conserved, a conservation issue that is currently being addressed by NGOs, like World Wildlife Fund, and South African National Parks. With limited agricultural potential and not much grass, the land in this biome is not highly sought after for human activity, but there has been significant sheep grazing and subsequent erosion.

WILDLIFE: The NAMA KAROO and Succulent Karoo share a significant community of near-endemic large mammals and birds, making these habitats an essential destination for the wildlife enthusiast (see Nama Karoo account for main discussion). Beyond that the Succulent Karoo is a mecca for small mammals, including Golden Moles, with De Winton's and Van Zyl's Golden Moles very local threatened endemics and Cape and Grant's Golden Moles shared only marginally with neighboring habitats. The Namaqua Dune Mole-rat is another near-endemic fossorial specialty that may be

Rocky areas rich in aloes are the microhabitat for the scarce Kopje, or Cinnamon-breasted Warbler. Skitterykloof, Western Cape, South Africa. © KEN BEHRENS, TROPICAL BIRDING TOURS

Af2G SUCCULENT KAROO   113

Springtime in Succulent Karoo. Ideal habitat for Rufous-eared Warbler. Northern Cape, South Africa.
MAIN PHOTO © KEN BEHRENS; INSET © KEITH BARNES, TROPICAL BIRDING TOURS

Springbok in the mountainous landscape of Tanqua Karoo National Park, Western Cape, South Africa.
© KEN BEHRENS, TROPICAL BIRDING TOURS

Left: **The strange Namaqua Rain Frog (*Breviceps namaquensis*) is endemic to the Succulent Karoo, spending most time underground, surfacing to feed mostly after rains.** © KEIR AND ALOUISE LYNCH

Right: **The Succulent Karoo is a reptile paradise, with marquee species including Armadillo Girdled Lizard (*Ouroborus cataphractus*), which gives birth to live young which the female sometimes even feeds!** © KEITH BARNES, TROPICAL BIRDING TOURS

involved in the creation and maintenance of distinctive habitat patches on Heuweltjies (see Description above). Most birds are shared with the Nama Karoo, but Cape Lark (IS) is near-endemic to the Succulent Karoo's coastal dunes and adjacent soft-sand areas. After winter rains, nomadic granivores like buntings and sparrowlarks may move in from the Nama Karoo or s. Namib, including the very local "Damara" Black-headed Canary, which is near-endemic to the northern mountains of the Succulent Karoo.

The Succulent Karoo has a diverse set of reptiles, many found only here. Charismatic species include tortoises like Tent Tortoise (*Psammobates tentorius*) and Speckled Padloper (*Chersobius signatus*); stunning small snakes like Many-horned (*Bitis cornuta*) and Namaqua Dwarf Adders (*B. schneideri*); and gorgeous lizards like Namaqua (*Chamaeleo namaquensis*) and Western Dwarf (*Bradypodion occidentale*) Chameleons, Namaqua Flat (*Platysaurus capensis*), Armadillo Girdled (*Ouroborus cataphractus*), and Large-scaled Girdled (*Cordylus macropholis*) Lizards, sandveld (*Nucras*), sand (*Pedioplanis*), and crag (*Pseudocordylus*) lizards, and thick-toed geckos (*Pachydactylus*).

Despite its aridity, the Succulent Karoo has a diverse frog fauna, including three endemics: the weird and wonderful Namaqua Rain Frog (*Breviceps namaquensis*), Namaqua Stream Frog (*Strongylopus springbokensis*), and Namaqua Caco (*Cacosternum namaquense*). Slightly more widespread near-endemics are Karoo Toad (*Vandijkophrynus gariepensis*), Karoo Caco (*Cacosternum karooicum*), and Cape Sand Frog (*Tomopterna delalandii*). Most of the herps are highly localized to some small portion of the Succulent Karoo. And while some are conspicuous, the majority will be found only by those who search carefully. Amongst the invertebrates, scorpions, monkey-beetles, bee flies, bees, and wasps all have concentrations of diversity and endemism in the Succulent Karoo.

DISTRIBUTION: Succulent Karoo is restricted to w. South Africa and s. Namibia. Much of the Succulent Karoo is coastal, but part of it lies above the escarpment at 3000 ft. (900 m) on the vast Southern African Plateau. Succulent Karoo is replaced by NAMA KAROO to the east, FYNBOS and STRANDVELD to the south and southwest and on mountain ranges, and NAMIB SAND DESERT and NAMIB ROCK DESERT to the north.

WHERE TO SEE: Goegap Nature Reserve, South Africa; Namaqua National Park, South Africa.

Af2H MAGHREB HOT SHRUB DESERT

IN A NUTSHELL: Low arid shrublands that fringe the northern part of the Sahara Desert. **Global Habitat Affinities:** CHENOPOD SHRUBLAND; CHIHUAHUAN DESERT; SUCCULENT PUNA; CASPIAN WORMWOOD DESERT; CAUCASIAN SHRUB DESERT; SAGEBRUSH SHRUBLAND. **Continental Habitat Affinities:** AFROTROPICAL HOT SHRUB DESERT; NAMA KAROO. **Species Overlap:** SAHARAN REG DESERT; SAHARAN ERG DESERT; ROCKY HAMADA AND MASSIF; AFROTROPICAL HOT SHRUB DESERT; CAUCASIAN SHRUB DESERT.

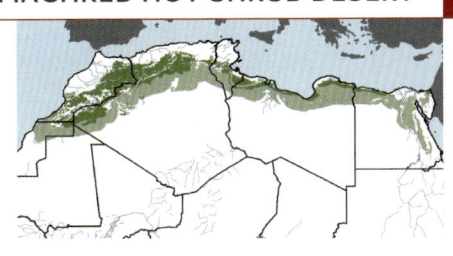

DESCRIPTION: The Maghreb Hot Shrub Desert is similar to some of the CAUCASIAN SHRUB DESERT of the Middle East and other shrub deserts and arid lands in Africa, such as the AFROTROPICAL HOT SHRUB DESERT and NAMA KAROO, in that it does not seem desolate enough to be regarded as desert, seemingly having too many trees and shrubs. But despite plant cover reaching over 50% and being much lusher than the barren Erg, Reg, and Hamada of the central Sahara, it still has an evaporation level of over 10 ft. (3 m) annually, which exceeds the level of precipitation by an order of magnitude. Precipitation is concentrated in the winter months (Nov–Mar), while summer (Jun–Aug) is devoid of rain or, at best, limited to less than half an inch (12 mm) a month; annually this habitat receives between 4 and 10 in. (100–250 mm) of rain. Hot shrub desert forms in the part of the Sahara that has a temperate and subtropical Mediterranean climate, whereas areas of North Africa that receive less than 4 in. of precipitation are generally unable to support this habitat but are rather barren Erg or Reg desert.

This habitat mainly occurs on plains or low rises and over a wide variety of substrates. These substrates include

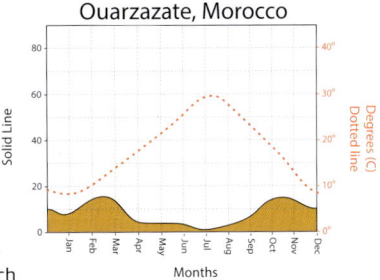

Ouarzazate, Morocco

Shrub Desert growing on the colluvial slope of a Hamada, something of a transitional habitat, which highlights the trickiness of classifying the habitat of any single location. Errachidia, Morocco.
© KEITH BARNES, TROPICAL BIRDING TOURS

116 DESERTS AND ARID LANDS

Maghreb Hot Shrub Desert is knee high and frequently found in the plains immediately below the Atlas Mountains. Tagdilt Track, Morocco. © KEN BEHRENS, TROPICAL BIRDING TOURS

This is the typical aspect of Maghreb Hot Shrub Desert, with widely spaced shrubs, reminiscent of Nama Karoo. Tizgaghine, Morocco. © NICK ATHANAS, TROPICAL BIRDING TOURS

lithosols (thin soil dominated by coarse rock fragments), formed on nutrient-deficient crystalline metamorphic and igneous bedrocks like granites, rhyolites, and quartzites, and also include aridisols (desert soils), developed on sedimentary rocks and unconsolidated tertiary sediments. The reason that small shrubs can establish, rather than woodland, is that all these soils have stable, well-developed soil profiles and more precipitation than Reg and Erg, but less precipitation than juniper, pine, or sclerophyll woodlands of the Atlas Mountains and Mediterranean coastal areas.

The vegetation is a shorter, more open, scrubbier, euphorbia-rich version of **NORTHERN DRY THORN SAVANNA**. Tamarisks are one of the dominant larger plants, mainly being around 5 ft. (1.5 m) tall but in places growing into full trees of 33 ft. (10 m) and spaced around 165 ft. (50 m) apart. Tamarisks are associated with trees such as Gum Acacia (*Senegalia senegal*), Lentisk (*Pistacia lentiscus*), *Ziziphus lotus* (one of the multiple plants called "jujube"), and Arartree (*Tetraclinis articulata*). Cactus-like spurges (*Euphorbia*) form an important component of the ground cover. These have thick succulent stems and are usually around 1 ft. (30 cm) tall but can grow to over 3 ft. (1 m) tall in dense clumps. The dominant species of spurge change geographically. The most obvious shrub in the west is the King Jubia Euphorbia (*Euphorbia regis-jubae*), which is stout-trunked with very thick branches and leaves that sprout like a clump of bananas. They rarely take the form of a 6.5 ft. (2 m) tall tree with a thick trunk, rounded crown, looking very similar to the Tenerife Milk Spurge (*Euphorbia lamarckii*) of the Canary Islands. One common species, *E. beaumierana*, has no common name, and another two, *E. resinifera* and *E. echinus*, are both, confusingly, called Desert Spurge, African Spurge, and Tikiwt. Bushes include numerous smaller tamarisks (*Tamarix*) and Southern Boxhorn (*Lycium intricatum*) but also Coastal Ragwort (*Senecio leucanthemifolius*) and various halophytic saltbushes. Perennial ground cover is often formed by White Wormwood (*Artemisia herba-alba*), which will whittle down in dry years,

After a good rainfall, a variety of both annuals and perennials grow quickly and bloom, making the semi-desert burst into ephemeral lushness. Agdz, Morocco. © KEN BEHRENS, TROPICAL BIRDING TOURS

then regrow rapidly when conditions are favorable. Between these ground shrubs are perennial grasses that remain low, generally around 15 in. (38 cm) tall. After seasonal rains, perennial grasses are augmented by a burst of ephemeral grasses and shrubs, such as False Esparto Grass (*Lygeum spartum*), and sedges like Wild Onion (*Cyperus bulbosus*) and Virgin's Mantle (*Fagonia cretica*).

In all the variations of Maghreb Hot Shrub Desert, seasonal and annual precipitation fluctuations can be marked, and rainfall can result in a burst of grasses between shrubs that can actually grow above them and make the habitats look like a grassland for a short period. Some people may treat this as desert grassland, but for the vast majority of the time, it is a shrubland, and the wildlife is associated with the shrubs and small bushes rather than with the ephemeral grasses.

At a subhabitat level the Maghreb Hot Shrub Desert can be divided into the Atlantic (Af2H-1) west of Egypt and the Red Sea (Af2H-2) in Egypt. Generally, the plants are thornier in the east, and the bird assemblage more closely resembles that on the Arabian Peninsula.

CONSERVATION: Many of these shrublands are anthropogenic in nature with euphorbias invading what was previously Argan and juniper woodland because of overcutting and overgrazing. Although the majority of this habitat is natural, this northern movement toward the coast and mountains at the expense of woodland and MAGHREB GARRIGUE is the first step toward desertification in North Africa.

WILDLIFE: The wildlife of this zone has a lot in common with CAUCASIAN SHRUB DESERT and AFROTROPICAL HOT SHRUB DESERT, along with the barren Saharan habitats. See the Wildlife section of the SAHARAN REG DESERT account (p. 71) for a discussion of the overlapping wildlife. This habitat may have been a key one for the now-extinct Aurochs, a cattle-like beast depicted in Egyptian reliefs and featured in Roman gladiatorial amphitheaters, which suddenly vanished in the early 1600s. Barbary Sheep and Nubian Ibex occasionally descend from their hilly enclaves to forage here. On the plains, one can find Barbary Ground Squirrel, Long-eared Hedgehog, and African Wolf, which was previously thought to be part of the Golden Jackal complex but has proven to be more closely related to Gray Wolf and Coyote. This animal is probably Anubis, the Egyptian god of funerary rites, protector of graves and guide to the underworld! Other predators include the bird-hunting Caracal and the striking black-and-white Saharan Striped Polecat. Hills and rocky enclaves support North African Elephant Shrew, an Afrotropical stowaway (see sidebars 20, p. 234 and 42, p. 383) that since the extinction of the North African Elephant is the only Afrotherian (a massive group including golden moles, tenrecs, aardvarks, and sea cows) remaining in the Palearctic!

Wheatears such as Desert Wheatear are some of the most common and characteristic birds of shrub desert. © DANIELE ARDIZZONE, FLUYENDO PHOTOGRAPHY

Thick-billed Lark, perhaps the most striking of a range of North Africa desert larks, often gathers in small to large groups. © HAMID MEZANE, GAYUIN BIRDING TOURS

In tough conditions these animals will enter torpor, dropping their body temperature to 41°F (5°C), but instead of rolling into a ball they remain in a crouching position with eyes partially open, ready to escape potential predators. Another small mammal odd-job is the Fat Sand Rat, replete with the scientific name *Psammomys obesus*! It specializes on halophytic (salt-tolerant) steppe, eating stems and leaves of chenopod succulents and Amaranthaceae plants almost exclusively. Provided they stick to their natural "greens" they remain lean, but as soon as they veer toward a standard rodent diet of grains, they quickly become obese and can develop type-2 diabetes. Most of the small mammals are gerbils, jirds, mice, and shrews.

The birds of this habitat are mostly shared with AFROTROPICAL HOT SHRUB DESERT, this habitat's twin along the southern side of the central Saharan barren desert. Examples include Houbara Bustard; Egyptian and Nubian Nightjars (often near wadis or water); Black-bellied Sandgrouse; Fulvous Chatterer; Arabian Babbler; Palestine Sunbird; Brown-necked Raven; Dunn's, Crested, and Lesser Short-toed Larks; Blackstart; Mourning, White-crowned, and Hooded Wheatears; and the nasal-voiced Trumpeter Finch. Red-rumped Wheatear and African Desert Warbler like flat saline steppe with chenopods. This habitat is also used by passage and wintering migrants, especially *Curruca* warblers like Cyprus and Tristram's Warblers. Scrub Warbler is in a monotypic genus that is essentially endemic to this type of scrub, but it occurs across the Sahara-Sindian region, reaching Pakistan. Black-crowned Sparrowlark is nomadic and can turn up in numbers after rainfall.

This desert is fairly rich in reptiles and amphibians. Many species are able to tolerate other habitats, sneaking up into the foothills of the Atlas Mountains or out into the drier, more desolate deserts, especially if they follow wadis or use oasis habitat. But most of the species mentioned here are best found in Maghreb Hot Shrub Desert. Böhme's Gecko (*Tarentola boehmei*) and

Scrub Warbler makes up a monotypic genus and is a specialist of shrub desert.
© DANIELE ARDIZZONE, FLUYENDO PHOTOGRAPHY

Morocco Lizard-fingered Gecko (*Saurodactylus mauretanicus*) are hot shrub specialists, while Manuel's (*Chalcides manueli*) and Moorish Skinks (*C. polylepis*), Bosc's Fringe-toed Lizard (*Acanthodactylus boskianus*), North African Mastigure (*Uromastyx acanthinura*), and Small-spotted Lizard (*Mesalina guttulata*) are a little more widespread. The bizarre Helmethead Gecko (*Tarentola chazaliae*) is most closely linked to shrubs on coastal dunes and in saline environments. North African Rock Agama (*Agama bibronii*), Oudri's Fan-footed Gecko (*Ptyodactylus oudrii*), and Ocellated Skink (*Chalcides ocellatus*) like to hide in crevasses and under rocks. Predators of these lizards are snake species like Schokari Sand Racer (*Psammophis schokari*), Egyptian Cobra (*Naja haje*), Saw-scaled Viper (*Echis pyramidum*), the fierce Saharan Horned Viper (*Cerastes cerastes*), and the Maghreb-endemic Moorish Viper (*Daboia mauritanica*). False Cobra (*Rhagerhis moilensis*) is another predator of small vertebrates that mimics a cobra by spreading its neck and hissing!

Frogs in this zone prefer streams and oasis environments and include Brongersma's Toad (*Barbarophryne brongersmai*), North African Green Toad (*Bufotes boulangeri*), Mediterranean Tree Frog (*Hyla meridionalis*), and Sahara Frog (*Pelophylax saharicus*). Reptiles favoring the oasis environment include Saharan Pond Turtle (*Mauremys leprosa saharica*), Viperine Water Snake (*Natrix maura*), Mograbin Diademed Snake (*Spalerosophis dolichospilus*), and the tiny and blind Beaked Thread Snake (*Myriopholis algeriensis*).

DISTRIBUTION: Occurs over a massive region, all the way from the Mauritania coastline to the Red Sea, dividing the Sahara from the Mediterranean Sea and the Atlas Mountains. In the south this habitat merges into Erg and Reg deserts, though pockets of shrub desert can be found deep into the Sahara in slightly moister microclimates. The arid e. Canary Islands are dominated by this habitat.

WHERE TO SEE: To the east and south of Agadir, Morocco; along the coast near Tripoli in Libya.

Af2l AFROTROPICAL HOT SHRUB DESERT

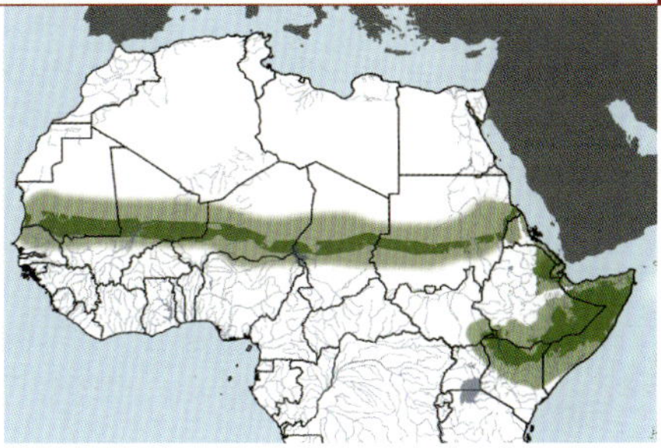

IN A NUTSHELL: Very dry habitat south of the central Sahara with few if any trees, an abundance of bare ground, a scattering of shrubs, and local or seasonal grass cover, especially after sporadic rainfall. **Global Habitat Affinities:** CAUCASIAN SHRUB DESERT; ARID TUSSOCK ACACIA SHRUBLAND; BLUEBUSH AND SALTBUSH; CHENOPOD SHRUBLAND; SUCCULENT PUNA; CASPIAN WORMWOOD DESERT; SAGEBRUSH SHRUBLAND. **Continental Habitat Affinities:** MAGHREB HOT SHRUB DESERT; SUCCULENT KAROO; NAMA KAROO. **Species Overlap:** NORTHERN DRY THORN SAVANNA; MAGHREB HOT SHRUB DESERT; AFROTROPICAL GRASSLAND; CAUCASIAN SHRUB DESERT.

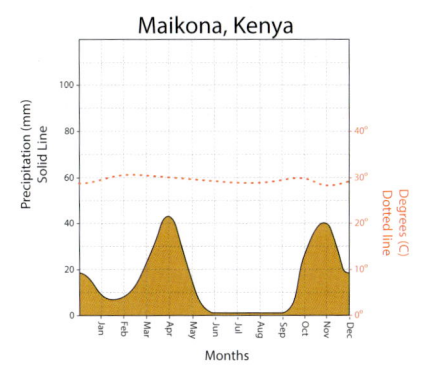

DESCRIPTION: This habitat usually has a great deal of bare ground but is still somewhat vegetated with low plants. So it is drier and generally lacks trees, unlike NORTHERN DRY THORN SAVANNA, but is moister and less barren than SAHARAN ERG or REG DESERT. In general, rocky areas have a scattering of shrubs, whereas sandy areas are steppes dominated by grass. There is a strong division within the Afrotropical Hot Shrub Desert, with the Sahel Desert (subhabitat Af2l-1) running from Mauritania to Sudan and the Horn of Africa Desert (subhabitat Af2l-2) found from Eritrea down through Ethiopia and Somalia and into n. Kenya.

This desert receives between 4 and 10 in. (100–250 mm) of rain annually (Köppen **Bwh**), though rainfall is sporadic. Areas of North Africa that receive less than 4 in. (100 mm) of precipitation are generally unable to support this habitat and are rather barren Saharan Erg or Reg Desert. In North Africa, the winter is cool and dry, though temperatures rarely drop below freezing. The summers are hot, and this is when the Intertropical Convergence Zone brings sparse rainfall. Northeast Africa has two annual periods during which sparse rains can fall. The northeastern deserts are hot year-round, though they cool off at night.

Typical grasses throughout are lovegrasses (*Eragrostis*) and *Panicum turgidum* bunchgrass. In the Horn other common grasses include Ghagras (*Centropodia glauca*) and Egyptian Crowfoot Grass (*Dactyloctenium aegyptium*). In North Africa, *Stipagrostis pungens* is shared with the deserts to the north, whereas Cram-Cram (*Cenchrus biflorus*), *Schoenefeldia gracilis*, and Stalked Bur Grass (*Tragus racemosus*) show a more southern, Sahelian thorn savanna influence. Other n. African desert grasses include *Cympopogon proximus*, needlegrasses (*Aristida*), and *Lasiurus hirsutus*. The sand-loving sedge *Cyperus conglomeratus* can also be part of the ground cover. Herbaceous plants are common, especially after good rains fall. *Indigofera* can be found throughout. Common herbaceous species in the Horn are copperleaf (*Acalypha*), bush violet (*Barleria*), and *Aerva*, while in North Africa typical genera include *Tribulus*, heliotropes (*Heliotropium*), and *Pulicaria*.

Shrubby versions of this desert have a variety of species. Along the coast of Somalia, Kapok Bush (*Aerva javanica*) and *Jatropha pelargoniifolia* are typical. Elsewhere in the Horn, common shrubs include euphorbias, aloes, *Ipomoea sultani*, *Leucas argyrophylla*, *Ochradenus baccatus*, and *Zygophyllum hildebrandtii*. Rocky areas of the Horn often feature a fascinating assemblage of small succulent plants, reminiscent of Southern Africa's SUCCULENT KAROO. Succulent genera include *Dracaena*, euphorbias, *Caralluma*, aloes, *Echidnopsis*, *Kleinia*, and Climbing Milkweeds (*Sarcostemma*). Typical North African shrubs are Small-leaved Cross-Berry (*Grewia tenax*), Hanza (*Boscia senegalensis*), Fire Bush (*Calligonum comosum*), Mustard Tree (*Salvadora persica*), *Combretum glutinosum*, Meru (*Maerua crassifolia*), and Karira (*Capparis decidua*). The spiky and leathery shrub Tahara (*Cornulaca monacantha*) is often used as an indicator species to delineate the southern limit of the Sahara Desert proper.

Some desert areas have shrubby growth forms of species, including acacias and commiphoras, that more typically grow as taller trees in adjacent Northern Dry Thorn Savanna. It is startling to see a shrubby Umbrella Thorn (*Vachellia tortilis*), normally a stately, spreading tree, growing as a 3 ft. (1 m) shrub! The lava fields of n. Kenya support scrubby growth of *Rhigozum somalense* and Blackthorn (*Senegalia mellifera*). Taller trees grow very locally in areas with groundwater, such as valley bottoms, or at the desert fringe, creating an outlier of Northern Dry Thorn Savanna. Typical groups include *Commiphora*, euphorbias, acacias, Desert Dates (*Balanites aegyptiaca*), and golds (*Rhigozum*).

CONSERVATION: The conservation situation throughout this habitat is poor, partially due to long-term political instability. The situation in North Africa is especially grim, and the loss of the vast majority of this region's large mammal populations, described in the Wildlife section below,

Central Kenya holds the southernmost outposts of this barren habitat, home to localized William's and Masked Larks. Dida Galgalu, Kenya. © KEITH BARNES, TROPICAL BIRDING TOURS

Af2I AFROTROPICAL HOT SHRUB DESERT

counts among the continent's greatest conservation tragedies. The North African deserts have been heavily used by humans for a very long time, making it hard to be certain of the natural state of this habitat. Another problem across Africa's deserts is off-road driving, which can have a long-lasting effect on this fragile habitat.

WILDLIFE: Larger wildlife naturally tends to be thin on the ground in the desert due to the scarcity of food. Nonetheless, deserts are home to a fair number of mammals and birds and are excellent habitat for reptiles. Classic desert animals include gazelles, oryxes, foxes, bustards, sandgrouse, and larks. There is a major biogeographic divide between the deserts of the Horn of Africa and those of Northern Africa, with only a narrow connection between these two zones. The Northern African desert is more closely allied with the MAGHREB HOT SHRUB DESERT than are the Horn of Africa deserts, though both share clear biogeographic links with the deserts of the Arabian Peninsula.

The Horn deserts share a few species with the Namib Desert (see NAMIB ROCK DESERT and NAMIB SAND DESERT), including Black-backed Jackal, Common Ostrich, and Double-banded Courser, remnants of times in the geological past when these two arid areas were connected.

North Africa's desert was once one of the continent's great mammal ranges, supporting vast herds of gazelles, Addax, and Scimitar-horned Oryx. An aggregation of some 10,000

Left: **Aardwolf is found in the shrub deserts of Africa that have a plentiful supply of termites for food, including those in ne. Africa.** © KEITH BARNES, TROPICAL BIRDING TOURS

Below: **Scimitar-horned Oryx was declared extinct in the wild, but a small population has been reintroduced to Ouadi Rimé-Ouadi Achim Game Reserve in c. Chad, where they seem to be thriving.** © JON HALL

oryx in Chad, witnessed in 1936, boggles the mind. Typical of desert environments, these mammals migrated around their vast range, following seasonal flushes of abundant food. These large herds of herbivores were stalked by a healthy population of predators, including Cheetah, Leopard, Striped and Spotted Hyenas, and Common Jackal. What remains today is only a dim shadow of this past glory. The larger predators are nearly eliminated, though smaller and/or more nocturnal predators, such as jackals, Pale Fox, and Fennec Fox persist. Scimitar-horned Oryx was declared extinct in the wild in 2000, though a small population has been reintroduced to Chad. Addax, once seen in herds of over 1000, is now restricted to Chad and Mali. Dama and Rhim Gazelles have been reduced to total populations in the hundreds and are considered Critically Endangered and Endangered, respectively. The widespread Red-fronted and Dorcas Gazelles have maintained larger populations, though both still have Vulnerable status.

The northeastern desert is less contiguous than that of North Africa, and many species are found in only a small portion of this region. This area has a strong affinity with the CAUCASIAN SHRUB DESERT of the Arabian Peninsula, and many species are found on both sides of the Red Sea and Gulf of Aden. As in North Africa, the mammal population has been greatly reduced by hunting and overgrazing, and most of the larger species are now regionally extinct. Some fascinating mammals do remain locally, including Hamadryas Baboon; Speke's, Dorcas, and Soemmerring's Gazelles; Beira; Rüppell's Fox; Leopard; and Striped Hyena. Deserts in Ethiopia and Eritrea support the world's last few African Wild Asses, the wild progenitor of the domestic donkey.

The birds of this habitat are mostly shared with NORTHERN DRY THORN SAVANNA, the Maghreb Hot Shrub Desert, this habitat's twin, along the northern side of the central Saharan barren desert, and the deserts of the Arabian Peninsula and Central Asia. Birds found throughout this habitat's extent include Chestnut-bellied, Lichtenstein's, and Spotted Sandgrouse; Lanner; Crested and Desert (IS) Larks; and Black-crowned Sparrowlark. Common Ostrich is found in North Africa and locally in the northeast, and Somali Ostrich is found widely in the deserts of the Horn. Cream-colored Courser is found in desert across North Africa and into the Danakil. Farther east and south it is replaced by Somali Courser. Birds found only in the North African desert, though shared with other adjacent deserts, include Pharaoh Eagle-Owl, Egyptian Nightjar, Greater Hoopoe-Lark, Greater Short-toed Lark (winter visitor), Dunn's Lark, and Brown-necked Raven. Nubian Bustard is a special bird of the North African desert, though it is shared with the adjacent Northern Dry Thorn Savanna. Birds of the Horn deserts include Arabian Bustard, Nubian Nightjar, Somali Bee-eater,

Despite the name, most of the range of Arabian Bustard is in Africa, where it is found in both shrub desert and dry thorn savanna. © KEN BEHRENS, TROPICAL BIRDING TOURS

Dama Gazelle is now on the very edge of extinction. Hopefully the reintroduction programs in Chad succeed, since to lose such a beautiful and distinctive species would be tragic. © JON HALL

Thekla Lark, Chestnut-headed Sparrow-Lark, Somali Crow, and Somali Fiscal. Several species, mostly larks, are virtually endemic to this habitat in the Horn, though some of these will venture into adjacent dry thorn savanna. These are Heuglin's Bustard; William's, Obbia, Ash's, and Masked Larks; Somali Long-billed Lark; and Lesser Hoopoe-Lark.

The northeast is rich in reptiles, including wall lizards (*Mesalina*), Eritrean Rock Agama (*Acanthocercus annectans*), Ocellated Spinytail (*Uromastyx ocellata*), and Saw-scaled Viper (*Echis pyramidum*). Somali endemic reptiles include Haacke-Greer's Skink (*Haackgreerius miopus*) and *Latastia cherchii*.

Endemism: The Horn of Africa Desert has a few endemic bird species, though most of them also use NORTHERN DRY THORN SAVANNA habitat. The Horn deserts have some endemic succulent plants, including several species of euphorbia.

DISTRIBUTION: This desert makes up the southern fringes of the Sahara, south of the barren Reg and Erg deserts of the central Sahara, and is immediately north of the Sahel NORTHERN DRY THORN SAVANNA. Afrotropical Hot Shrub Desert and grassland runs from the Atlantic coast in w. Mauritania, across n. Mali, Chad, and Niger, to the Red Sea coast of ne. Sudan. From here a narrow band follows the coast south, wrapping around the northern outliers of the Abyssinian Highlands in Eritrea. Much of the Danakil Depression of Djibouti and Ethiopia is this habitat, and it is widespread in eastern Ethiopia and most of Somalia. Desert also occurs locally in the driest portions of northern Kenya. The northern Afrotropical Hot Shrub Desert is bordered by Northern Dry Thorn Savanna to the south, and much of the northeastern desert occurs in a fine matrix alongside this habitat. This desert is mainly found between sea level and 2600 ft. (800 m), though there are hills well over 3300 ft. (1000 m).

The south-to-north transition in North Africa from Sahel dry thorn savanna to Afrotropical Hot Shrub Desert to completely barren desert is far from simple and does not simply follow moisture gradients. In sandy areas, the exposed tops of dunes can have a desert character deep into the Northern Dry Thorn Savanna zone. Meanwhile, drainage lines and other moister areas can support trees and create a microhabitat that has the character of Northern Dry Thorn Savanna, penetrating deep into this desert and even locally into the generally barren Erg, Reg, and Hamada desert of the central Sahara. So these habitats form a complex, interlocking mosaic over thousands of square miles.

WHERE TO SEE: Dida Galgalu Desert, n. Kenya; Ouadi Rimé-Ouadi Achim Faunal Reserve, Chad.

Hot shrub desert with a temporary growth of grass. Chad. © JON HALL

SIDEBAR 11 DURICRUSTS AND DESERTIFICATION

Resistant elements such as iron, aluminum, silica, and calcium move around naturally in soils for thousands of years, causing gradual changes in soil and vegetation types. But when vegetation over a soil that is balanced with a hot, shaded environment is cleared or overgrazed, these elements solidify, and the soils turn into desert wasteland, regardless of the natural precursor, be it desert, savanna, dry deciduous forest, or even semi-evergreen forest habitat.

In some environments, such as warm, wet regions, prolonged chemical weathering and seasonal movement of elements in solution over thousands of years result in complex regolith (weathered material overlying fresh rock) profiles, including laterite formed from iron oxides (ferricrete) and aluminum oxides (bauxite or alcrete). In arid terrains, over very large timescales, natural silcretes and calcretes form as hardpans in natural soils. In both humid and arid terrains, the soils on top of these ancient surfaces are very nutrient poor, and the vegetation on hillsides (where soils are younger and contain more nutrients) is usually lusher than on top of plateaus. In some cases, such movements can be much faster and form duricrusts—hard concretions of residual elements in soils. In more arid environments where chemical weathering is less intense, silica and calcium get concentrated as silcrete and calcrete duricrusts at the bottom of slopes. In wetter environments, iron and aluminum moved downslope in groundwater crystallize on slopes and in unconsolidated colluvial regolith in valley bottoms. Duricrusts can be exposed and baked on valley sides but usually they remain as concretions in soil profiles. However, when these materials are exposed to desiccation and heating through forest clearing and topsoil removal, materials that may have lasted for hundreds of years in colloidal form can harden and solidify into an impermeable and durable duricrust cap. Once formed, either as ferricrete in savanna environments or as silcrete and calcrete in more arid environments, these duricrust caps form a wasteland in which forests, savannas, and desert shrublands can no longer grow. Areas such as the small patches on this diagram marked as A, B, C, and D can exist as massive desertified wastelands in areas that were formerly semi-evergreen forests surrounding the Congo, or dry thorn savanna in the Sahel of Africa. In the figure, (A) represents a human-induced hardening of laterite and infusion of ferricrete formed from clearing and topsoil removal in a savanna landscape. (B) is a natural ferricrete formed at a break in slope and where vegetation is stunted. (C) is a human-induced ferricrete barren area formed when forest is cleared and topsoil is removed. (D) is human-induced silcrete/calcrete desertification formed from clearing and overgrazing in an arid environment.

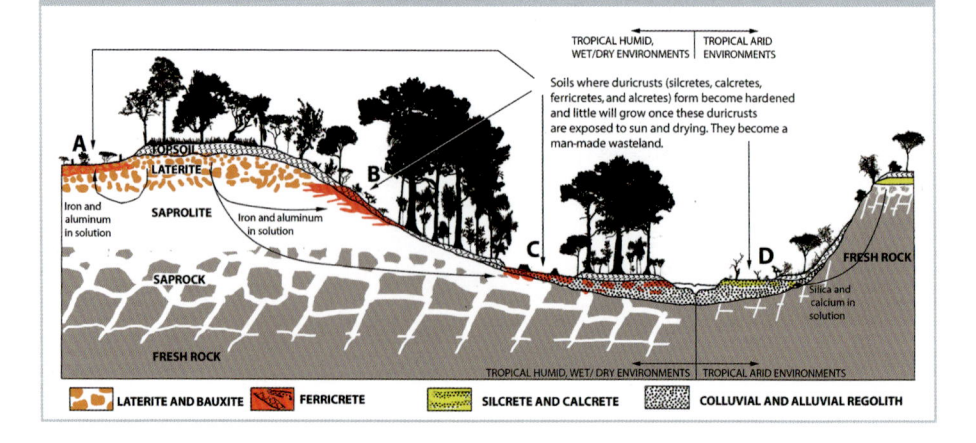

Af2J SPINY FOREST

IN A NUTSHELL: Dry forest or thicket on Madagascar that includes mostly deciduous trees and in which many plants are spiny. **Global Habitat Affinities:** SERTAO CAATINGA; SONORAN DESERT. **Continental Habitat Affinities:** ALBANY THICKET. **Species Overlap:** MALAGASY DECIDUOUS FOREST; INDIAN OCEAN LOWLAND RAINFOREST; INDIAN OCEAN MONTANE RAINFOREST; MALAGASY GRASSLAND AND SAVANNA.

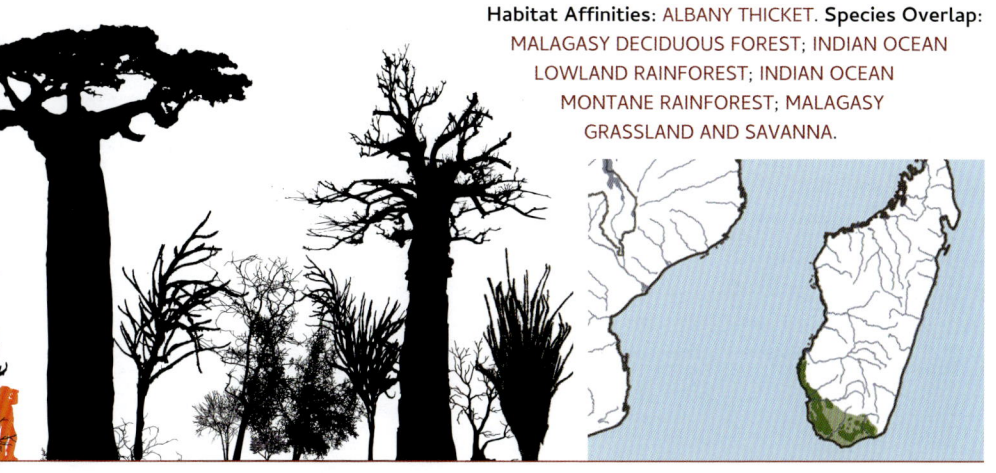

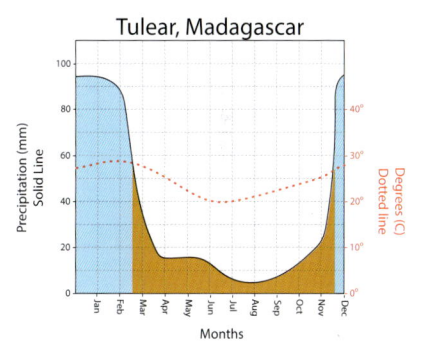

DESCRIPTION: This arid habitat of sw. Madagascar is one of the Afrotropics' most visually distinctive; it is unlike any other place on Earth and rather like something dreamed up for an episode of *Star Trek*. All the plants have adapted to an arid climate. Some are partly or completely deciduous; some have photosynthesizing branches or trunks; some have thick and leathery leaves; others have structures that serve to precipitate fog, a precious source of moisture. Some of the most striking trees in this habitat are fat-trunked, water-absorbing baobabs and pachypodiums, many-armed octopus trees (Didiereaceae), and a variety of succulent euphorbias. Other common trees include commiphoras, flame trees (*Delonix*), Helicopter Tree (*Gyrocarpus americanus*), and moringas. Succulent plants such as aloes and kalanchoes are prominent in the understory.

There are clues suggesting that, prior to the ancient moistening of the climate and development of the rainforest, Madagascar was entirely arid, likely making this biome the island's oldest. The Spiny Forest actually has strong biogeographic links to the Namib Desert, which is probably the world's oldest desert. The antiquity of this habitat is borne out by the Spiny Forest's extraordinary diversity of endemic plants: 95% of the plants are endemic not just to Madagascar but to this habitat. Usually referred to as "spiny forest" (as it is here) and sometimes as "spiny desert," it is actually neither a true forest nor a true desert but rather a semiarid or sub-desert brushland or thicket.

Southwestern Madagascar lies in the rain shadow of the eastern mountains and is too far south to be much influenced by the Intertropical Convergence Zone. As such, its climate (Köppen **Bsh**) is dry, with annual rainfall of 12–20 in. (300–500 mm) and a dry season that can be as long as 10 months. During drought periods, some areas can receive little rainfall for more than a year.

Classic Spiny Forest, growing on red sand and dominated by baobabs and octopus trees *(inset, octopus tree limb shown from the top).* **Beza Mahafaly Reserve, Madagascar.** © KEN BEHRENS, TROPICAL BIRDING TOURS

The exact species composition and the feel of Spiny Forest vary a lot across its range. These variations basically fall out as three distinct subhabitats: White-sand Coastal (Af2J-1), Red-sand Lush (Af2J-2), and Limestone Plateau (Af2J-3). The white-sand coastal version is often remarkably dominated by *Euphorbia stenoclada*, which are well spaced and interspersed with a variety of shorter and more generic bushes. The canopy is 10–13 ft. (3–4 m) tall and very open. The substrate of red-sand forest varies from rich red to pale orange and is the lushest and biologically richest version of Spiny Forest. The canopy can be as tall as 33 ft. (10 m), and varies from moderately open to nearly closed. There is a great diversity of plants, but many-limbed octopus trees are usually prominent. The limestone plateau version of Spiny Forest is generally shorter, 13–20 ft. (4–6 m) tall, and moderately open. Baobabs, euphorbias, smaller octopus trees (such as *Alluaudia procera*), and pachypodiums are especially prominent. This limestone Spiny Forest bears a strong resemblance to the stunted MALAGASY DECIDUOUS FOREST growing on Tsingy limestone rock formations, farther north, though its suite of wildlife is more aligned with Spiny Forest. In some places, Spiny Forest can take the form of fairly generic thornscrub, especially where Spiny Forest transitions to Malagasy Deciduous Forest. Degraded Spiny Forest areas start to resemble an arid savanna but usually lack grassy ground cover.

White-sand Coastal Spiny Forest is usually easy to walk through, but the other versions of this habitat make for difficult walking. Red-sand Lush Spiny Forest is usually quite thick, and although Limestone Plateau Spiny Forest can be quite open, the often-jagged limestone substrate and abundance of spiny plants mean that it is still difficult to walk through.

White-sand Coastal Spiny Forest, a distinct subhabitat, is dominated by *Euphorbia stenoclada*. Exclusive habitat of the Littoral Rock-Thrush *(inset)*. Anakao, Madagascar. © KEN BEHRENS, TROPICAL BIRDING TOURS

CONSERVATION: Although the Spiny Forest is less degraded than the other three Malagasy forest habitats (MALAGASY DECIDUOUS FOREST, INDIAN OCEAN MONTANE, and LOWLAND RAINFOREST), it is under increasing pressure as the local human population increases and people struggle to simply survive in an impoverished region with a harsh climate. Although fires don't seem to have been a major factor historically, fires started by humans now destroy large chunks of Spiny Forest each year, usually in preparation for cultivation.

WILDLIFE: This is the poorest of Madagascar's habitats for wildlife overall, though it holds a large number of local endemics. Verreaux's Sifaka and Ring-tailed Lemur are the classic big lemurs of Spiny Forest, though they have been eliminated from much of their former range. The other larger mammals of the Spiny Forest are nocturnal species, namely Gray-brown (IS) and Gray Mouse Lemurs, White-footed and Petter's Sportive Lemurs, and Lesser Hedgehog Tenrec. Fossa was once widespread but is now seemingly rare, probably due to the elimination of the large lemurs upon which it preys. Two other members of the endemic Malagasy carnivorans family (Eupleridae) are found in highly localized zones within Spiny Forest: Narrow-striped and Grandidier's Vontsiras.

The avifauna of the Spiny Forest consists mainly of widespread forest birds, plus a few species shared with the MALAGASY DECIDUOUS FOREST. But there is a fascinating set of endemic birds, including Subdesert Mesite, Long-tailed Ground-Roller, Running Coua, Littoral Rock-Thrush, Subdesert Brush-Warbler (IS), Thamnornis, Archbold's Vanga, and Lafresnaye's Vanga (IS). Most are widespread in the southwest. The Littoral Rock-Thrush is restricted to White-sand Coastal Spiny Forest, south of the Onilahy River. Red-shouldered Vanga is endemic to the Limestone

130 DESERTS AND ARID LANDS

Limestone Plateau Spiny Forest with a stunted Fony Baobab (*Adansonia rubrostipa*). The inset shows a huge ancient Fony Baobab. Tsimanampetsotsa National Park, Madagascar. © KEN BEHRENS, TROPICAL BIRDING TOURS

Watercourses through the Spiny Forest have a lusher character, resembling deciduous forest. The Radiated Tortoise (*Astrochelys radiata*), once abundant, is now Critically Endangered, mainly due to collection for the pet trade, domestic and international. Beza Mahafaly Reserve, Madagascar. © KEN BEHRENS, TROPICAL BIRDING TOURS

Af2J SPINY FOREST   131

Though the large lemurs have been eliminated from much of the Spiny Forest, Verreaux's Sifaka remains locally. This species has a remarkable ability to jump between thorn-studded octopus trees without injuring its leathery hands!
© KEN BEHRENS TROPICAL BIRDING TOURS

Deep red sand Spiny Forest at sunset. Octopus trees "reaching for the sky" make for arresting silhouettes! Parc Mosa, Ifaty, Madagascar.
© KEN BEHRENS, TROPICAL BIRDING TOURS

Plateau Spiny Forest, and Verreaux's Coua is also virtually restricted to this habitat. Two bird subspecies, which are sometimes considered full species, are Spiny Forest endemics: "Subdesert" Stripe-throated Jery and "Green-capped" Red-capped Coua. Other typical species of Spiny Forest, but which aren't restricted to this habitat, include Madagascar Buzzard; Madagascar Cuckoo; Madagascar Turtle-Dove; Crested Drongo; Madagascar Magpie-Robin; Chabert, Sickle-billed, and White-headed Vangas; and Sakalava Weaver.

The Spiny Forest is replete with reptiles, and the open nature of the habitat makes them easy to see. There are two endemic tortoises, Radiated Tortoise (*Astrochelys radiata*) and Spider Tortoise (*Pyxis arachnoides*), which were formerly common and widespread but are now considered Critically

*Ocelot or Madagascar Ground Gecko (*Paroedura picta*) is mainly seen at night and can be abundant during the rainy season.* © KEN BEHRENS, TROPICAL BIRDING TOURS

132   DESERTS AND ARID LANDS

Left: **The spectacular and roadrunner-like Long-tailed Ground-Roller is restricted to a narrow enclave of Spiny Forest.** © KEN BEHRENS, TROPICAL BIRDING TOURS

Below: **The Lesser Hedgehog Tenrec belongs to the tenrec family, Madagascar's most diverse mammal family, with members ranging from hedgehog-like to shrew-like and even otter-like!** © KEN BEHRENS, TROPICAL BIRDING TOURS

Endangered. Several members of the endemic Malagasy iguanid family (Opluridae) are prominent, including Merrem's Madagascar Swift (*Oplurus cyclurus*) and Three-eyed Lizard (*Chalarodon madagascariensis*). Other typical reptiles include Warty Chameleon (*Furcifer verrucosus*), Three-lined (*Zonosaurus trilineatus*) and Four-lined (*Z. quadrilineatus*) Plated Lizards, Madagascar Keeled Plated Lizard (*Tracheloptychus madagascariensis*), Gold-spotted Skink (*Trachylepis aureopunctata*), Sakalava Madagascar Velvet Gecko (*Blaesodactylus sakalava*), fish-scale geckos (*Geckolepis*), and Ocelot Gecko (*Paroedura picta*). Snake diversity is low, but Dumeril's Boa (*Acrantophis dumerili*) and Mahafaly Sand Snake (*Mimophis mahfalensis*) are both present throughout.

Endemism: Although plant distribution is complex, with several centers of endemism within the Spiny Forest, the wildlife of this habitat is rather uniform throughout. Bizarrely, though, two of the habitat's most interesting birds, Long-tailed Ground-Roller and Subdesert Mesite, are found only in the northwestern corner of the Spiny Forest, between the Mangoky and Fiheranana Rivers.

DISTRIBUTION: Malagasy Spiny Forest is confined to areas below 1300 ft. (400 m) in sw. Madagascar. Watercourses in this dry area support tall gallery forest that remains lush year-round. This riparian forest has a character reminiscent of MALAGASY DECIDUOUS FOREST, though it is far less diverse. To the north, as moisture levels increase, Spiny Forest slowly blends into Malagasy Deciduous Forest. To the east there is a remarkable, stark transition to INDIAN OCEAN MONTANE and LOWLAND RAINFOREST; some mountains have rainforest on their eastern slopes and Spiny Forest on their western slopes.

WHERE TO SEE: Parc Mosa, Reniala, and other small private reserves near Ifaty.

Af2L DRAGON'S BLOOD TREE SEMI-DESERT

IN A NUTSHELL: Botanically unique semi-desert habitat on Socotra and adjacent Indian Ocean islands, dominated in parts of the highlands by the umbrella-shaped Dragon's Blood Tree. **Global Habitat Affinities:** GALÁPAGOS LOWLAND DESERT; SONORAN DESERT. **Continental Habitat Affinities:** SUCCULENT KAROO; SPINY FOREST. **Species Overlap:** NORTHERN DRY THORN SAVANNA; AFROTROPICAL HOT SHRUB DESERT.

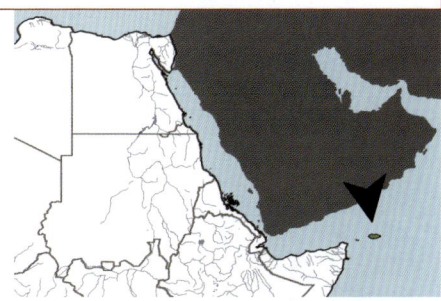

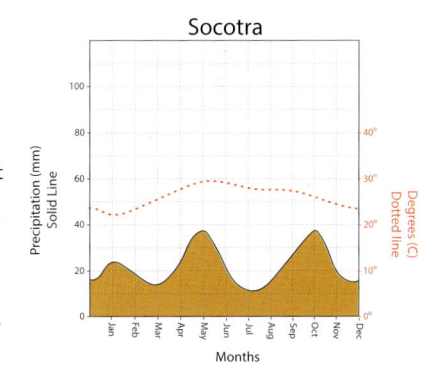

DESCRIPTION: The large island of Socotra and smaller adjacent islands, including Abd al-Kuri, Samhah, and Darsah, lie in the Indian Ocean south of the Arabian Peninsula and 140 mi. (240 km) from the Horn of Africa. The isolation and arid climate here have produced an unusual and distinctive semi-desert flora unlike any other on Earth. Socotra emerged from the sea 40 MYA and has been completely isolated for around 20 million years. Although poor in wildlife, this habitat is given separate coverage here based on its botanical and visual distinctiveness. From a biogeographic perspective, Socotra is firmly aligned with the arid habitats of the Horn of Africa and the s. Arabian Peninsula rather than the Palearctic. Despite its position in the w. Indian Ocean, it has little in common with that region's other islands, probably because the arid climate is unlike that of the other islands, save for the distant southwest of Madagascar.

Much of the moisture in this arid environment (Köppen **Bwh**) comes from mist, though some areas experience a rainy season from October to March. Typical rainfall is 5–8 in. (125–200 mm), though the wettest parts of the highlands receive as much as 40 in. (1000 mm) of precipitation annually.

Dragon's Blood Tree Semi-desert is generally rugged, rocky, and open and is sparsely vegetated with various plants that are adapted to a dry climate. The coastal plains are covered in arid shrubland that strongly resembles AFROTROPICAL HOT SHRUB DESERT. The central plateau has a larger variety of taller bushes, and the highest and moistest areas support an evergreen shrubland. The highest point on Socotra is 4931 ft. (1503 m). Typical plants of this montane microhabitat include

The namesake Dragon's Blood Tree, *Dracaena cinnabari*, has an unmistakable silhouette. Diksam Plateau, Socotra. © JAMES EATON, BIRDTOUR ASIA

Scrubby coastal plain habitat. Qashio, Socotra. © JONATHAN NEWMAN

Endemic Socotran Chameleon (*Chamaeleo monachus*).
© JAMES EATON, BIRDTOUR ASIA

Cephalocroton socotranus, Conkerberry (*Carissa edulis*), *Indigofera sokotrana*, Broadleaf Hopbush (*Dodonaea viscosa*), Socotra Fig (*Ficus socotrana*), and *Euphorbia socotrana*. Wadis can support dense thickets of woody plants, including *Vachellia pennivenia*, (*Dorstenia gigas*), *Searsia thyrsifolia*, and *Ruellia insignis*.

This habitat's namesake and most famous species is the markedly symmetrical, umbrella-shaped Dragon's Blood Tree (*Dracaena cinnabari*), one of the world's most bizarre and photogenic trees. This beautiful tree is the last remaining relic of a lost age, a whole suite of plants called the Tethys flora, which thrived on the shores of the Tethys Sea during the Mio-Pliocene. It is restricted to the higher parts of the limestone plateau and usually grows alongside various frankincense trees (*Boswellia*), aloes, and Impala Lily (*Adenium obesum*). Another unusual plant is the fat-trunked Cucumber Tree (*Dendrosicyos socotranus*), the only member of the cucumber family that grows in tree form. It usually occurs as part of a remarkable community of succulents growing on limestone hillsides, especially on the mountains' northern slopes. Along with Cucumber Trees, this habitat includes euphorbias, Impala Lily, *Kleinia scottii*, *Kalanchoe robusta*, and Socotra Fig. This is also prime habitat for the endemic Socotrine Aloe (*Aloe perryi*), which is famous for its cosmetic and medicinal properties and is sought-after around the world.

Other characteristic trees of the Socotra archipelago include frankincense, Christ's Thorn Jujube (*Ziziphus spina-christi*), commiphoras, African Star-Chestnut (*Sterculia africana*), and *Euphorbia arbuscula*. Overall, there are around 250 species of plants, of which 30% are endemic, including 10 endemic genera.

CONSERVATION: Despite being remote, sparsely populated, and arid, Socotra has been profoundly affected by human activity. It is hard to know the extent of what has been lost, as this area is virtually untouched by paleontology. But it is certain that there have been extinctions; there are tantalizing hints in the historical record of giant lizards, tortoises, and perhaps an endemic crocodile. Socotra once had a much moister climate—and supported a larger human population—which may have been pushed into ecological collapse by natural climate change, human habitat modification, or a combination thereof. Wetlands, rivers, and pastures were once common; the continued existence of three endemic species of freshwater crabs hints at this past wetland richness. Portuguese visitors reported the presence of domestic water buffalo in the early 17th century, whereas only more hardy domestic animals remain today. Regardless of the details, it is certain that overgrazing and tree cutting have contributed to the desertification of these islands. Feral cats also pose a threat, especially to the endemic birds. Despite its losses, Socotra remains a biological wonderland that has been described as "the most alien-looking place on Earth"!

A wadi like this is perfect habitat for many of Socotra's endemic birds. Ayhaft Canyon, Socotra.
© JAMES EATON, BIRDTOUR ASIA

WILDLIFE: These islands have few mammals; the only indigenous species are a few shrews and bats, one of which is endemic. Most of the terrestrial birds and reptiles are endemic. There are 10 endemic bird species: Socotra Buzzard, Scops-Owl, Warbler (IS), Cisticola, Starling, Sunbird (IS), Sparrow, Grosbeak, Bunting, and Abd al Kuri Sparrow. The Socotra Warbler makes up a monotypic endemic genus. Additionally, there are many endemic bird subspecies. Aside from the endemics, most bird species are shared with the AFROTROPICAL HOT SHRUB DESERT and NORTHERN DRY THORN SAVANNA of the adjacent Horn of Africa. Examples include Laughing Dove, Bruce's Green-Pigeon, Lichtenstein's Sandgrouse, White-browed Coucal, Cream-colored Courser, Brown-necked Raven, Black-crowned Sparrow-Lark, Somali Starling, and Cinnamon-breasted Bunting. There are also some dry-country-loving long-distance migrants that pass through and winter, including Red-tailed and Isabelline Shrikes, Greater Short-toed Lark, Isabelline and Desert Wheatears, and Tawny Pipit. Socotra is one of the strongholds of the enigmatic Forbes-Watson's Swift.

Reptiles show low diversity but a high degree of endemism; the majority of the 30-odd species are endemic. These include Socotra Chameleon (*Chamaeleo monachus*), several rock geckos (*Pristurus*) and blind snakes (*Myriopholis*), and many leaf-toed geckos (*Hemidactylus*). Strangely enough, amphibians are entirely lacking. There are around 600 insect species, of which 90% are endemic. Socotra is famous among spider lovers as the home of one of the world's most beautiful tarantulas, the Socotra Island Blue Baboon (*Monocentropus balfouri*).

AF2L DRAGON'S BLOOD TREE SEMI-DESERT 137

DISTRIBUTION: Dragon's Blood Tree Semi-desert habitat is found throughout the archipelago, though biologically Socotra is by far the richest of the islands.

WHERE TO SEE: Rokeb di Firmihin, Socotra.

Left: **The endemic Socotra Grosbeak belongs to a small genus whose other two members are found in the s. Arabian Peninsula and n. Somalia.** © ROB HUTCHINSON, BIRDTOUR ASIA

Below left: **Socotran Giant Gecko (*Haemodracon riebeckii*) on a Dragon's Blood Tree.**
© JAMES EATON, BIRDTOUR ASIA

Below: **Sunset on the Diksam Plateau, in the Dragon's Blood Tree Semi-desert of Socotra.**
© JAMES EATON, BIRDTOUR ASIA

TEMPERATE BROADLEAF FORESTS

Af3A LAUREL FOREST

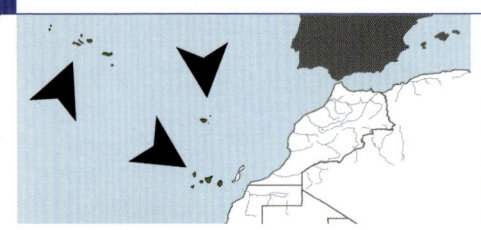

IN A NUTSHELL: Unique, relict subtropical wet evergreen forest with laurel-shaped leaves and open understory, restricted to the cloud belt of the Macaronesian islands. **Global Habitat Affinities:** BEECH FOREST; EUROPEAN DECIDUOUS RAINFOREST. **Continental Habitat Affinities:** MOIST MONTANE FOREST (in temperate South Africa). **Species Overlap:** MAGHREB MAQUIS; MAGHREB PINE FOREST.

DESCRIPTION: Laurisilva (Laurel Forest) is a very tall, subtropical forest, generally with a closed, interlocking canopy of trees with dark green, laurel-shaped leaves (mostly oval leaves approximately 4 in./10 cm long and 2 in./5 cm broad). Trees average around 33 ft. (10 m) high but reach up to 67–100 ft. (20 m to rarely 30 m) in gullies and basins. Little light passes through the canopy, and the understory is very open with occasional shrubs. Laurel Forest extended across much of s. Europe 14–40 MYA but they dwindled over time, mostly being driven extinct by the extreme climates of the ice ages. They remain trapped on Macaronesia where climate has fluctuated little since the Tertiary and are now globally rare. The flora and fauna of this relict forest is quite unique, and many of the plants are endemic within ancient lineages that remain only here, having gone extinct elsewhere. The vegetation is very like parkland, with mature trees and open understory, and is easy to walk through, save for the fact that it often grows on steep slopes. Owing to the moisture content, the ground is often covered in a thick carpet of ferns and mosses.

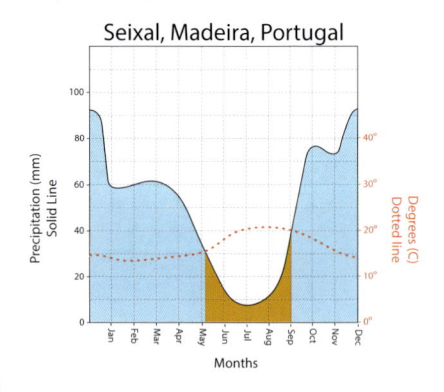

Af3A LAUREL FOREST

Above: **Classic moist, mossy, and cloud-draped Laurel Forest on Madeira.** © CARLOS MANUAL MARTIN JIMENEZ

Right: **Madeira Firecrest is sometimes found in Laurel Forest, though it actually prefers the heath-dominated habitat found at slightly higher elevations.** © NIGEL VOADEN

Despite their temperate latitudinal distribution (28–40°N), these forests have several tropical features. They are evergreen, growing in areas where frost is absent and constant humidity guaranteed. Their biomass amounts to 121 ton/acre, or 300 tonnes/ha, and their canopies can grow very tall. They may contain up to 20 different tree species in a few hectares, quite diverse for temperate forest.

All the islands of this group are volcanic, and Laurel Forest grows on deep soils at elevations of 1000–5000 ft. (300–1500 m) that are influenced by moisture-laden oceanic winds that form a cloud belt, mostly on north-facing slopes and especially in deep gorges or inaccessible inland valleys. The climate is temperate oceanic (Köppen **Cfa**) and is regulated by a branch of the Gulf Stream and a high pressure zone. Cloud, mists, and fog are common here, particularly at the highest elevations. There is a fairly balanced annual average temperature around 55–66°F (13–19°C), and the annual precipitation

Laurel Forest on the Macaronesian Islands supports a handful of localized endemic bird species, such as Madeira Chaffinch. © NIGEL VOADEN

ranges from 16 to 60 in. (400–1500 mm) and even reaches an incredible 149 in. (3800 mm) in the wettest portions of the Azores, where there is a constant fog drip and no climatic stress. The massive moisture gradient means there are several types of Laurisilva: dry Laurisilva (<19 in./480 mm); humid Laurisilva (19–47 in./480–1200 mm]); and hyperhumid Laurisilva (>59 in./1500 mm). In addition to the different types, the species composition differs on the three main archipelagos: Azores, Madeira, and Canaries.

The main characteristic species are trees in the family Lauraceae, several of which are endemic, such as Azores Laurel (*Laurus azorica*), Bay Laurel (*Laurus novocanariensis*), Canary Laurel (*Apollonias barbujana*), Madeira Mahogany (*Persea indica*), and Stinkwood (*Ocotea foetens*). Other species include Firetree (*Myrica faya*), Lily-of-the-valley-tree (*Clethra arborea*), picconia (*Oleaceae*), Madeira Cheesewood (*Pittosporum coriaceum*), Small-leaved Holly (*Ilex canariensis*), Azores Juniper (*Juniperus brevifolia*), and Madeira Elder (*Sambucus lanceolata*). Most of these are ancient endemic representatives of flora that was formerly more widespread across North Africa and Europe. Some species that occur in Macaronesia survived in Europe, like Portuguese Laurel Cherry (*Prunus lusitanica*), and herbs include foxgloves and *Ixanthus*. The shrub layer has elements including *Erica* species like Tree Heather (*E. arborea*) but also endemic *E. maderensis*; where *Erica* spp. dominate at higher elevations the habitat becomes MONTANE HEATH. Ferns and mosses abound in shadowy valleys and smother the ground, banks, rocks, and tree trunks. These grow alongside abundant liverworts and lichens.

CONSERVATION: Macaronesian Laurisilva forests have been exploited since people first arrived in these island groups around 2500 YBP. More intensive use since European colonization in the 15th century has significantly reduced their surface area. The original forest cover was largely razed to create farmland and degraded by forest exploitation. Significantly reduced nowadays, in some areas habitat is still being degraded by exploitation and livestock. Elsewhere severe fragmentation is threatening habitat diversity and leading to species extinction. Other current threats are the spread of exotic species, such as Maritime Pine (*Pinus pinaster*) and Southern Blue Gum (*Eucalyptus globulus*), introductions for pine plantations, especially in the Azores and Madeira, and forest fires, especially serious in the Canary Islands. Given the steep topography of Madeira, erosion has become a serious problem in the south, where logging and overgrazing have removed vegetation cover. Introduced herbivores (goats, sheep, and rabbits) constitute a major threat to the natural vegetation, while cats and rats present a major threat to the avifauna. In the Azores, less than 2% of the original forest remains, but in Madeira around 15–20% remains. What does remain is currently well protected in conservation areas. Climate change models suggest that the fog belt that Macaronesian Laurisilva depends on may shift to lower elevations, which are currently completely covered in arable crops. The endemic rodents of the Canaries have gone extinct, recently including the Canary Islands Giant Rat.

WILDLIFE: Vertebrates include a limited number of species with high rates of endemism, including shrews and bats. Those that use Laurel Forest include Tenerife Long-eared Bat and Leisler's Bat.

Several birds are almost restricted to these forests and adjacent heath, including three endemic pigeons: Laurel and Bolle's Pigeons (both Canary Islands) and Trocaz Pigeon (Madiera). Fruits are available to birds year-round for various reasons: the influence of elevation on the ripening process—fruits lower down are ripe while those higher up are still unripe; different trees produce fruits at different times of the year; and the most important tree species (*Laurus*, *Ilex*, *Myrica*, *Picconia*, *Persea*) present an almost continuously, nonseasonally controlled fruit production pattern, with peaks and troughs that usually vary from one year to another. Azores Bullfinch (Azores) is also near-endemic to this habitat, Island Canary is an island generalist that uses Laurel Forest, and although Madeira Firecrest (Madeira) is also found in this habitat, it prefers the MONTANE HEATH slightly higher up. Some other endemic birds use Laurel Forest but only marginally, including Tenerife and Gran Canaria Blue Chaffinches (both endemic to Canaries), which prefer Canary Pine (*Pinus canariensis*) forests. Zino's Petrel breeds only on Madeira and Boyd's Shearwater only on the Canaries in the Laurisilva zone, but mostly on sheer cliffs inaccessible to goats. Other more widespread bird species using Laurel Forest include common western Palearctic generalists such as Eurasian Blackbird, European Robin, Common Woodpigeon, Eurasian Blackcap, Common Chaffinch, and European Goldfinch (the latter two each have four to five endemic subspecies on different Macaronesian islands). Other species include African Blue Tit, endemic Canary Islands Chiffchaff, Eurasian Woodcock, and Long-eared Owl.

Endemism: There are huge numbers of endemic plants on each of the Macaronesian Laurisilva archipelagos. Land snails (41% in Azores, 84% in Madeira, and 94% in Canaries) and arthropods are also rich, with nearly 500 species of endemics, but this does not translate to vertebrate endemism, which is relatively poor.

DISTRIBUTION: Restricted to the three main Macaronesian archipelagos: Azores and Madeira (both Portugal) and the Canaries (Spain). Macaronesian Laurel Forest is patchy but limited to elevations of 1000–5000 ft. (300–1500 m).

Above: **Trocaz Pigeon is a specialist of Laurel Forest on the island of Madeira. Detail of the leaves and flowers** *(inset)* **of Tilo or Stinkwood (*Ocotea foetens*), a common member of the Lauraceae on Madeira and the Canary Islands.** MAIN PHOTO © NIGEL VOADEN; INSET © KEITH BARNES, TROPICAL BIRDING TOURS

Left: **Laurel Forest on Tenerife, Canary Islands.**
© VINCENT LEGRAND

African Humid Forest Dendrogram

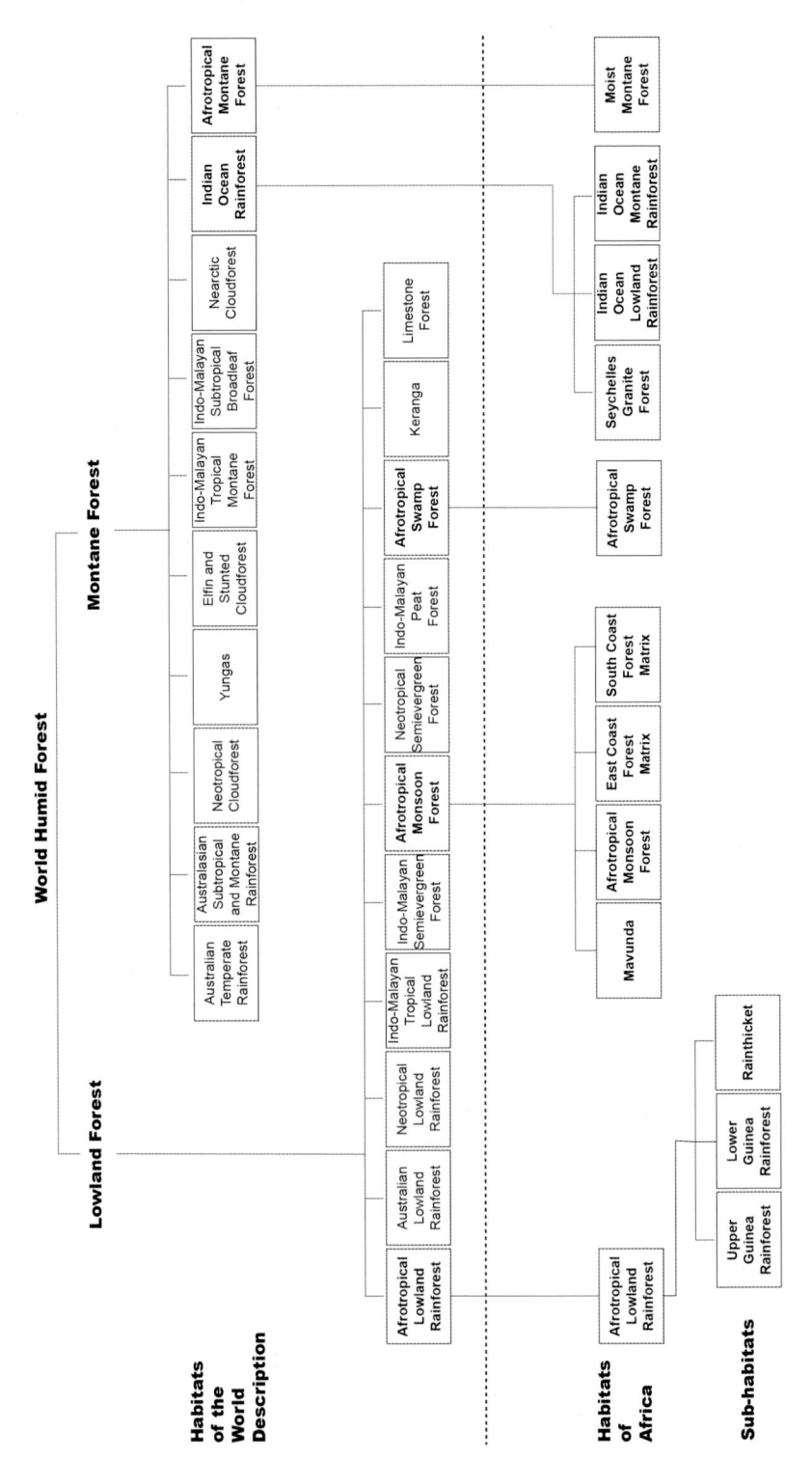

World Humid Forest

Lowland Forest — **Montane Forest**

Habitats of the World Description

- Afrotropical Lowland Rainforest
- Australian Lowland Rainforest
- Neotropical Lowland Rainforest
- Indo-Malayan Tropical Lowland Rainforest
- Indo-Malayan Semievergreen Forest
- Australian Temperate Rainforest
- Australasian Subtropical and Montane Rainforest
- Neotropical Cloudforest
- Yungas
- Elfin and Stunted Cloudforest
- Indo-Malayan Tropical Montane Forest
- Indo-Malayan Subtropical Broadleaf Forest
- Nearctic Cloudforest
- Indian Ocean Rainforest
- Afrotropical Montane Forest

- Afrotropical Monsoon Forest
- Neotropical Semievergreen Forest
- Indo-Malayan Peat Forest
- Afrotropical Swamp Forest
- Keranga
- Limestone Forest

Habitats of Africa

- Afrotropical Lowland Rainforest
- Mavunda
- Afrotropical Monsoon Forest
- East Coast Forest Matrix
- South Coast Forest Matrix
- Afrotropical Swamp Forest
- Seychelles Granite Forest
- Indian Ocean Lowland Rainforest
- Indian Ocean Montane Rainforest
- Moist Montane Forest

Sub-habitats

- Upper Guinea Rainforest
- Lower Guinea Rainforest
- Rainthicket

WARM HUMID BROADLEAF FORESTS

Af4A AFROTROPICAL LOWLAND RAINFOREST

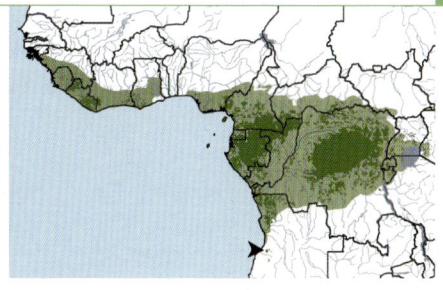

IN A NUTSHELL: Lush forest with a tall canopy and complex structure that remains moist and humid throughout most or all of the year. **Global Habitat Affinities:** AMAZON TERRAFIRMA; AUSTRALASIAN TROPICAL LOWLAND RAINFOREST; MALABAR LOWLAND RAINFOREST; MESOAMERICAN LOWLAND RAINFOREST. **Continental Habitat Affinities:** INDIAN OCEAN LOWLAND RAINFOREST; SEYCHELLES GRANITE FOREST. **Species Overlap:** SWAMP FOREST; MONSOON FOREST; AFROTROPICAL MANGROVE; MOIST MONTANE FOREST; GUINEA SAVANNA.

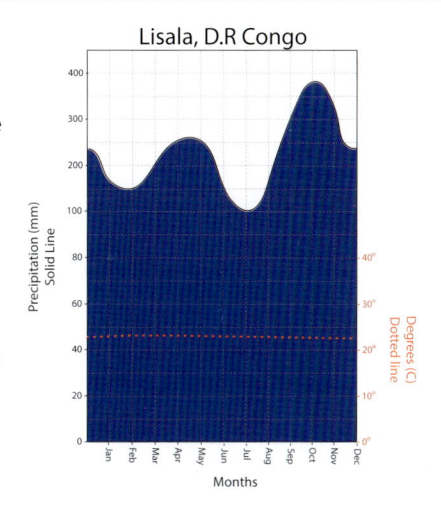

DESCRIPTION: Rainforest is renowned as a habitat of incredibly high biodiversity, and African rainforest is no exception. The continent holds one of the world's great tropical lowland rainforests, comparable to South America's Amazon (Amazon Terrafirma) and the forests of se. Asia and Australasia. This habitat grows only in areas with high rainfall for most of the year (Köppen **Afa**), although this African rainforest zone is drier than the world's other rainforests, and rainfall is less reliable throughout the year. For first-time visitors, especially during one of the short dry seasons, its relatively dry nature might come as a surprise. Typical annual rainfall is 63–80 in. (1600–2000 mm), though locally can be as low as 40 in. (1000 mm) or as high as 200 in. (5000 mm) in areas around the Gulf of Guinea or places that precipitate orographic (mountain-associated) rainfall. Some areas remain drenched year-round, but most of the region's lowland rainforest experiences a short dry season (Dec–Feb). Other areas, such as much of the Congo Basin, have two short dry periods: December–January and June–July. Even the dry periods have some rainfall. Unsurprisingly, humidity and temperature remain high year-round. This complex habitat creates a variety of microclimates. The canopy is subject to daily temperature fluctuations between lows of around 70°F (20°C) and highs of 85–90°F (30–33°C), whereas the sheltered and largely windless forest understory sees only minor temperature fluctuations of one to two degrees each day.

This habitat is found from sea level to around 5000 ft. (1500 m) elevation in Uganda and w. Kenya and on the mountains of Upper Guinea. It is mostly found in very flat terrain, though there are some hilly areas, in particular the foothills of the Albertine Rift and West African mountains. In other areas, such as the East Usambaras of Tanzania, the transition from lowland rainforest to **MOIST MONTANE FOREST** occurs around 2600 ft. (800 m) elevation.

The Afrotropical Lowland Rainforest has a high canopy that averages 100–130 ft. (30–40 m), with some towering emergent trees reaching 200 ft. (60 m). Although tall, it is not nearly as tall on average as the rainforests of se. Asia; it is also less diverse botanically. Nonetheless, it is still impressive that botanists have identified up to 200 plant species in a tiny 0.14 acre (0.06 ha) plot! While dominant rainforest trees in Asia are usually dipterocarps, in Africa they tend to be members of the Caesalpinioideae subfamily of the enormous legume family, and dipterocarps are virtually absent. Most trees have smooth bark on a tall narrow trunk that supports a correspondingly narrow crown, a typical growth form where sunlight is at a premium.

The canopy walkway at Kakum National Park, Ghana, gives a rare opportunity to be in the upper levels of the rainforest. A good place for Blue-throated Roller.
MAIN PHOTO © KEN BEHRENS; INSET © KEITH BARNES, TROPICAL BIRDING TOURS

Lowland rainforest has a complex midstory that includes smaller trees and many lianas; climbers can make up one-third of the plant assemblage in some places. These climbers include giant *Combretum*, *Agelaea*, and *Strychnos* lianas. Epiphytes,

The forest understory along the Bigodi Trail near Kibale National Park, Uganda. © KEITH BARNES, TROPICAL BIRDING TOURS

especially orchids and ferns, are common. Undisturbed areas, where little light reaches the forest floor, can have a remarkably open understory lacking much herbaceous growth and are easy to walk through. An example of this is in Uganda's Kibale National Park, where trackers of Chimpanzees often lead visitors off trails without much trouble. Disturbed areas quickly develop a virtually impenetrable understory. Tree-fall gaps are an important habitat in mature lowland rainforest. When a single, massive tree—connected by lianas to dozens of other trees—falls, it can create a considerable gap where sunshine suddenly penetrates to the floor. Fast-growing plants like members of the ginger family spring up and provide an important food source for mammals such as African Forest Elephant and Eastern and Western Gorillas. There are patches of seemingly natural grassland, mostly edaphic (caused by nutrient-poor soil), throughout the lowland rainforest zone (see AFROTROPICAL GRASSLAND). In Togo and Benin, in an area called the Dahomey Gap, the GUINEA SAVANNA reaches the coast, dividing the Afrotropical Lowland Rainforest belt into two zones. Although the habitat looks and feels almost identical on either side, there is a major floral and faunal division between the blocks worthy of subhabitat status: Upper Guinea (Af4A-1) and Lower Guinea (Af4A-2).

It is difficult to generalize about tree species, as there is considerable diversity within small areas and common trees change across the vast range of this habitat. Limbali (*Gilbertiodendron dewevrei*), an exception, can sometimes dominate large stretches of forest. Other typical lowland rainforest trees are cola (*Cola*), Newtonia, stinkwood (*Celtis*), Bilinga (*Nauclea diderrichii*), tropical chestnuts (*Sterculia*), Canarium, Kapok (*Ceiba pentandra*), ebony (*Diosporys*), miombo (*Brachystegia*), *Cynometra*, *Berlinia*, African mahogany (*Khaya*), *Albizia*, Gabon Nut (*Coula edulis*), and African Teak (*Pericopsis elata*). Guinea Plum (*Parinari excelsa*) is fascinating, as it occurs throughout most African rainforests as well as in South America. Although often uncommon, it is the predominant tree above 3000 ft. (1000 m) in the Upper Guinea highlands of Guinea, Sierra Leone, and Liberia. This forest occurs up to 5000 ft. (1500 m) and has the shorter canopy and abundant epiphytes more typical of Moist Montane Forest but remains fundamentally

Lowland rainforest is the habitat of Green-breasted Pitta, one of Africa's two pitta species.
© IAIN CAMPBELL, TROPICAL BIRDING TOURS

Rainforest with caves or rocks holds two species of Rockfowl, or Picathartes. This is the more westerly species, the White-necked Picathartes. Bonkro, Ghana. © BEN KNOOT, TROPICAL BIRDING TOURS

Forest streams like this one in Ankasa National Park, Ghana, are rich in river specialist birds like kingfishers as well as an abundance of frogs.
© KEN BEHRENS, TROPICAL BIRDING TOURS

lowland rainforest in terms of both its wildlife and plant assemblages.

Much of this habitat functions more like **MONSOON FOREST** than classic rainforest, meaning that it has a prolonged dry season, and many large trees lose some or all their leaves during the dry season. This drier forest generally occurs in a broad "buffer" zone surrounding the moister core of the Congo Basin and in a narrow band between the Upper Guinea moist rainforest and adjacent **GUINEA SAVANNA**, often along watercourses in that zone. The reason why this book has chosen to retain these forests in the lowland rainforest category is that their wildlife is fundamentally similar to the adjacent unequivocal rainforests. The situation is most confusing in the south, such as in ne. Angola, s. DRC, and nw. Zambia, where Monsoon Forest comes into contact with the southern monsoon-forest-like rainforests. There is no obvious dividing line between these habitats, and they merge in a nebulous transition. It is fascinating to note that although this habitat maintains cohesion in terms of its wildlife assemblage, there is a nearly complete turnover in the tree species between the driest and moistest rainforests. Many trees that are predominant canopy species in the semi-evergreen peripheral forest zones occur only in secondary forest in the moister core rainforests. By the same token, a sizeable minority of the plants of the moistest rainforest zones around the Gulf of Guinea are restricted to them and are not widespread across the Congo Basin.

Rainforest that has been heavily degraded often forms a surprisingly stable thicket-like formation (see sidebar 37, p. 335) that supports a distinctive mix of species. The stability of this habitat is maintained by human timber cutting as soon as trees reach a suitable diameter. Grassy glades are common and begin to attract grassland and savanna species. This Anthropogenic Rainthicket (subhabitat Af4A-3) is now actually the default habitat in the forest zone of West Africa as well as in parts of the Congo Basin. Even this secondary rainthicket habitat eventually gives way under human pressure, as it is increasingly being completely cleared for agriculture. As elsewhere in the world, rainforest habitat degrades quickly once its canopy is breached (see sidebar 11, p. 126).

CONSERVATION: Africa's lowland rainforest zone shows stark contrast in terms of the degree of human impact. The forest from Nigeria west has been drastically affected by the large human population, which can exceed hundreds of people per square mile. Larger mammals, as well as squirrels and bats, have been wiped out of most of the region. This extirpation is driven by a booming market for bushmeat; in Nigeria, a large Grasscutter (also known as Greater Cane Rat) can sell for the equivalent of two weeks' salary for an average person. Logging, forest clearing for agriculture and firewood, and even mining have all taken a toll. Meanwhile, the Congo Basin has

Af4A AFROTROPICAL LOWLAND RAINFOREST

vast stretches of nearly pristine forest, where the human population density can be 10 or fewer per square mile (4/km^2). But even the Congo Basin has been affected by hunting for bushmeat and ivory, and larger mammals are now scarce along navigable waterways. Poverty and political instability have plagued much of the region and hamstrung most attempts at forest protection.

WILDLIFE: Very few mammals or birds are found across the entirety of the range of Afrotropical Lowland Rainforest, but there are many groups of animals that are found primarily or exclusively in this habitat. In conjunction with SWAMP FOREST, this habitat has the continent's most distinctive assemblage of wildlife. There is little overlap between the wildlife of lowland rainforest/Swamp Forest and that of other habitats, save for some species shared with MOIST MONTANE FOREST, MONSOON FOREST, and/or EAST COAST FOREST MATRIX.

Lowland rainforest holds a bounty of remarkable mammals. Unfortunately, owing to their shy nature, the prevalence of hunting, and the density of the habitat itself, most are very difficult to see. A successful rainforest walk might include a good look at one or two squirrels and fleeting glimpses of a troop of fleeing guenon monkeys. The major rivers, especially the Sassandra, Volta, Niger, Cross, and Congo, are serious barriers to mammals and profoundly affect speciation and their distributions, tightly delineating the ranges of many species. Classic lowland rainforest mammals are African Forest Elephant; African Brush-tailed Porcupine; pangolins, including the hefty Giant Pangolin; Red River Hog; Giant Forest Hog; "Forest" African Buffalo; and Bongo. The large predators are Leopard and African Golden Cat, while smaller predators include African Civet; African linsangs; many genets, including the Giant Genet; and kusimanses (rainforest mongooses). The "holy grail" mammal of this habitat is Okapi, a beautiful, dark brown and white giraffe relative that seems virtually impossible to glimpse in the wild.

This habitat is one of the world's most important areas for primate diversity. This includes four great apes, humans' closest relatives: Eastern Gorilla, Western Gorilla, Chimpanzee, and Bonobo. Chimpanzee and Bonobo show classic distributions separated by the Congo River: Chimpanzee to the north and Bonobo to the south. The great apes are joined by a wide range of monkeys: Mandrill and Drill; a diverse set of colobuses; mangabeys; a plethora of guenons, such as the Red-tailed

Rainforest is the main habitat for Africa's great apes. Among these, perhaps the least frequently seen is Bonobo, which is also the species that is genetically closest to humans. © VLADIMIR DINETS

The Endangered Uganda Red Colobus is unusual among primates in that it is a folivore, eating leaves as its main diet. © KEITH BARNES, TROPICAL BIRDING TOURS

Mandrill, an unusual and colorful primate, found exclusively in lowland rainforest in Lower Guinea, occasionally gathers in "supertroops" 100+ strong.
© VLADIMIR DINETS

Monkey (IS south of Congo River); Potto; angwantibos; and galagos, such as the widespread Thomas's Galago (IS). Squirrels are also prominent. These include the otter-size Forest Giant Squirrel; sun squirrels, such as Red-legged Sun Squirrel (IS north of Congo River); African Pygmy Squirrel; Slender-tailed Squirrel; palm squirrels; at least nine species of rope squirrels; flying mice; and anomalures, such as Cameroon Scaly-Tail. The forest floor shelters a remarkable diversity of small antelopes, namely a diverse range of duikers and Royal and Dwarf Antelopes, plus a tiny deer biogeographically linked to Asia, the Water Chevrotain.

Lowland rainforest is a very rich habitat for birds, with many centers of endemism. Birders wanting to come to grips with the majority of rainforest birds will have to make multiple trips, and most of the countries that host this habitat are challenging to access. While time spent in this forest will yield far more bird sightings than mammal sightings, birding here is hard work. Many species are scarce, and most are shy and unapproachable, hiding in the canopy or in the foliage away from trails. Some of the groups that reach peak diversity in rainforest are fishing-owls, spinetail swifts, trogons, kingfishers, hornbills, barbets, honeyguides, woodpeckers, broadbills, cuckooshrikes, illadopsises, greenbuls, flycatcher-thrushes, forest robins, longbills, flycatchers, wattle-eyes, malimbes, and nigritas. Of all these groups, the most diverse is the greenbuls, a cryptic horde that gives headaches to birders trying to identify them.

Common and widespread species include Red-tailed Greenbul (IS) and Green Hylia (IS), the latter one of two members of the newly recognized Hyliidae family, which is virtually endemic to this habitat, along with Swamp Forest. Just a few more of the fascinating birds of the lowland rainforest include Long-tailed Hawk; Congo Serpent-Eagle; Nkulengu Rail; Black-collared Lovebird; Great Blue Turaco; Chocolate-backed Kingfisher; Black, Blue-headed, and Black-headed Bee-eaters; African Piculet; Blue Cuckooshrike; two wattled-cuckooshrikes; and Tit-hylia. This habitat has wonderful hornbills, ranging from the tiny dwarf hornbills to the weird, long-tailed hornbills and the massive Black-casqued and Yellow-casqued Hornbills. There are two *Picathartes* rockfowl, odd passerines with bare heads and long legs that nest in forest caves. The avian equivalents of the near-mythical Okapi are Congo Peacock and Shelley's Eagle-Owl.

São Tomé and Príncipe are like the Galápagos or Madagascar in serving as a laboratory of evolution. They support a rich set of forest endemic birds, including weird examples of island gigantism (Giant Sunbird) and dwarfism (Dwarf Ibis). Other notable birds include the São Tomé Shorttail, a bizarre forest-dwelling wagtail, São Tomé Grosbeak, a finch, and Newton's Fiscal, the only fiscal whose habitat is forest interior.

SIDEBAR 12 — OWLNIGMAS

Africa has its fair share of enigmatic birds: species that have not been seen for decades or that are so rare that they are hardly ever seen by birders and ornithologists. Perhaps the most desirable of all is the Congo Bay (or Itombwe) Owl. It is in the barn owl genus *Tyto*, though was previously in *Phodilus* alongside the two bay-owls that are restricted to Asia. This species was last seen in 1996, and before that it was known from only a single specimen collected in 1951! But it is not Africa's only mystery owl. Maned Owl, in the monotypic genus *Jubula*, and the beastly Shelley's Eagle-Owl, are almost as rare, with few verified records in the last 20 years. Last but not least, Rufous Fishing-Owl and Albertine Owlet are seen more frequently but remain high-value quarries for anyone with an interest in birds. So, do you want to buy a thermal scope?!

A mist-netted Congo Bay Owl. © TOM BUTINSKI

Anthropogenic Rainthicket holds a distinctive mix of birds: a subset of savanna species mixed with a subset of rainforest species. To give a feeling for this mix, here are some frequently sighted species in this habitat: Blue-spotted Wood-Dove, African Pied Hornbill, Woodland Kingfisher, Red-rumped Tinkerbird, Double-toothed Barbet, Red-faced Cisticola, Little Greenbul, Dusky-blue Flycatcher, Splendid Starling, Collared Sunbirds, Village and Compact Weavers, Bronze and Black-and-white Mannikins, and Orange-cheeked Waxbill. Most larger mammals, raptors, and hornbills are lacking because of the lack of big trees and/or the heavy hunting pressure.

Reptiles are not easily seen, but that doesn't mean that lowland rainforest lacks reptile diversity. Typical reptiles include forest geckos (*Cnemaspis*), tree snakes (*Thrasops*), Forest Vine Snake, snake-eaters (*Polemon*), Forest Cobra, Jameson's Mamba, night adders (*Causus*), Gaboon Viper (*Bitis gabonica*), Rhinoceros Viper (*Bitis nasicornis*), and chameleons (*Chamaeleo*). Afrotropical rainforest is extremely rich in frogs, and this habitat holds a high proportion of the continent's 800-some amphibian species. It is also Africa's most diverse habitat for butterflies, which can be easier to spot than its mammals, birds, or reptiles.

Lowland rainforest birding is usually tough, but the rewards, like this Yellow-bellied Wattle-eye, are great.
© DUBI SHAPIRO

Endemism: The Lower Guinea (Af4A-2) and Upper Guinea (Af4A-1) forests are quite distinctive, and there are many sister species on either side. Within Upper Guinea are two minor centers of endemism, which for mammals are divided by the Sassandra River. In the Congo Basin, many

Left: **The long-eared and long-faced Red River Hog is found both in rainforest and Swamp Forest.**
© VLADIMIR DINETS

Below: **Rainforest is the continent's primate stronghold. Monkeys rule the day, but the night belongs to prosimians like this West African Potto.**
© JANUS OLAH, BIRDQUEST

species are found only on the river's east or west sides, either in the w. Cameroon–Gabon region or the parts of e. DRC and Uganda that lie below the Albertine Rift mountains. The ranges of many local mammals are delineated by major rivers, especially the largest, the Congo.

The volcanic islands of São Tomé, Príncipe, and Annobón, which lie in the Gulf of Guinea, are rich in endemics, including more than 20 endemic birds. The southern outliers of rainforest along the Angolan escarpment are another minor center of endemism, holding a few endemic birds. The African rainforest has around 10,000 plants, of which 80% are endemic. There are several endemic plant families, and around one-quarter of the plant genera are endemic.

DISTRIBUTION: The heart of the African rainforest is the Congo Basin. From there it extends in a relatively narrow band into the Upper Guinea forests of West Africa; these are isolated by the dry Dahomey Gap, which covers much of e. Ghana, Togo, and Benin. The gap is dominated by the more open GUINEA SAVANNA habitat, with lowland rainforest restricted to rivers and other wet spots. To the north, lowland rainforest quickly gives way to Guinea Savanna and again is found locally only around rivers. To the south of the Congo Basin, the transition to MIOMBO is more gradual, and rainforest, AFROTROPICAL GRASSLAND, Miombo, and other savanna habitats form a complex matrix. The southernmost rainforest outposts are found along the Angolan escarpment. East Africa holds the easternmost lowland rainforest, terminating at Kakamega Forest in w. Kenya.

WHERE TO SEE: Kibale National Park, Uganda; Kakum National Park, Ghana; Ivindo National Park, Gabon; Korup National Park, Cameroon.

Af4C MONSOON FOREST

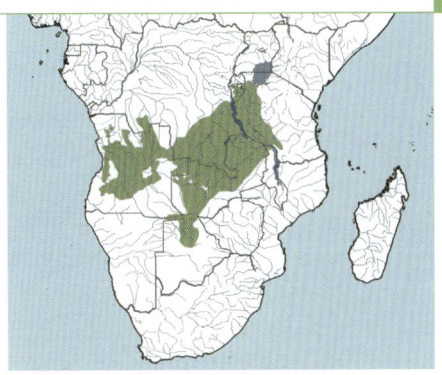

IN A NUTSHELL: Lush, closed-canopy forest characterized by leaf litter on the forest floor and abundant lianas and palms in the midstory; drier and more seasonal than AFROTROPICAL LOWLAND RAINFOREST and occurs mainly along watercourses. **Global Habitat Affinities:** NEOTROPICAL SEMI-EVERGREEN FOREST; AUSTRALASIAN MONSOON VINEFOREST; SOUTHEAST ASIAN SEMI-EVERGREEN FOREST; MESOAMERICAN SEMI-EVERGREEN FOREST. **Continental Habitat Affinities:** EAST COAST FOREST MATRIX; SOUTH COAST FOREST MATRIX; ANGOLAN DECIDUOUS FOREST. **Species Overlap:** AFROTROPICAL LOWLAND RAINFOREST; EAST COAST FOREST MATRIX; MIOMBO; MOIST MIXED SAVANNA; MOIST MONTANE FOREST.

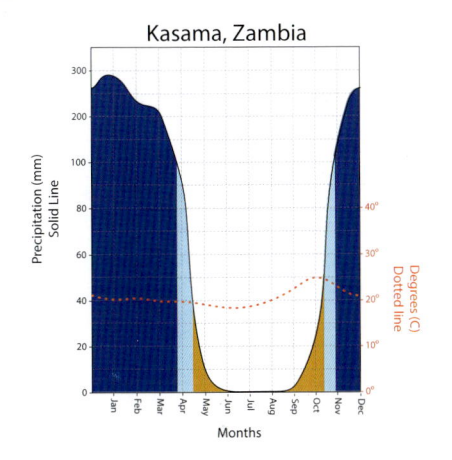

DESCRIPTION: To the east and south of the vast Congo Basin rainforest is another type of humid forest called Monsoon Forest. Also lush, and with a closed canopy, this habitat has several significant differences from AFROTROPICAL LOWLAND RAINFOREST. Most significantly, for several months each year, little or no rain falls in Monsoon Forest, resulting in a level of stress to plants that is rare in true rainforest. This is manifested by the presence of many deciduous and semi-deciduous trees and shrubs, which produce an accumulation of leaf litter on the forest floor, giving Monsoon Forest during the dry season an atmosphere closer to that of a Northern Hemisphere deciduous broadleaf forest in autumn than a rainforest. Another general difference is that Monsoon Forest has more lianas (hanging vines) and palms in its midstory. The understory is often quite open, allowing relatively easy exploration on foot. Leaves tend to be leathery to resist desiccation during the dry season, and few trees have the leaves with the "drip tips" that are common in rainforest. The canopy is lower than in true rainforest, rarely taller than 65 ft. (20 m).

Monsoon Forest rarely occurs in vast blocks, unlike lowland rainforest. It is found patchily, primarily along watercourses. In this climate (Köppen **Ama**), most rain falls during the austral summer, approximately November–April. Typical rainfall is in the range of 30–60 in. (800–1500 mm), depending on the year and the location. In general, the seasons are somewhat unpredictable and variable, and

Above: **Classic Monsoon Forest, with abundant lianas and dry leaf litter. Kinjila, Angola.**
© NONY DATTNER, ECOPLANET FILMS

Right: **"Brown-faced" Black-backed Barbet from Angola. This species can be found both in Monsoon Forest and Moist Montane Forest.** © DUBI SHAPIRO

dry spells are common. The plant species of this forest are not fire resistant, unlike those of adjacent savannas like MIOMBO. The exact character of Monsoon Forest varies a lot throughout its range. Plant diversity is high, with many localized endemics. Most Monsoon Forest includes at least a few of the following, often as some of the dominant tree species: Natal Guarri (*Euclea natalensis*), *Berlinia giorgii*, Guinea Plum (*Parinari excelsa*), figs, *Daniellia alsteeniana*, *Marquesia*, Pod Mahogany (*Afzelia quanzensis*), *Entandrophragma delevoyi*, and Waterberry (*Syzygium guineense*). Small patches of SWAMP FOREST can be found within Monsoon Forest. Though these don't support a notably different set of wildlife, they do contain some trees shared with the more extensive Swamp Forest of the Congo Basin, including *Mitragyna stipulosa*, *Uapaca*, *Syzygium owariense*, and *Xylopia*.

In drier Miombo, Monsoon Forest occurs along watercourses in thin bands that may be only a few trees wide and can be walked through in mere moments. Farther north in the Miombo zone are wider forests called *Mushitu*, which can be 1500 ft. (450 m) or more across. In w. Zambia and just across the border into Angola are extensive stands of dry evergreen forests known as MAVUNDA, which are very similar to Monsoon Forest but distinctive enough to split as their own habitat. Even the lushest Monsoon Forest generally has a lower canopy than lowland rainforest, around 80 ft. (25 m) maximum.

One odd type of Monsoon Forest, called Groundwater Forest, grows in parts of inland East Africa where rainfall is insufficient to support forest but where oddly out-of-place patches of lush forest are nurtured by abundant groundwater. Woodcutting and frequent fire degrade Monsoon Forest into a secondary habitat known in Zambia as "Chipya." Although it is structurally similar to Miombo, it lacks the characteristic tree genera of that habitat. Instead, it has a mix of very tall grass, which often burns fiercely, and a set of widespread trees that are fire resistant. These include Wild Syringa (*Burkea africana*), African Blackwood (*Erythrophleum africanum*), Mobola Plum (*Parinari curatellifolia*), and African Teak (*Pterocarpus angolensis*). Itigi Thicket (see sidebar below) from ne. Zambia and Tanzania is yet another subtype of Monsoon Forest (subhabitat Af4C-1).

SIDEBAR 13 — ITIGI THICKET

While many types of forest and savanna sometimes take on the form of a thicket (see sidebar 37, p. 335), MONSOON FOREST seems especially prone to grow into a thicket formation. One type of thicket is especially distinctive and well studied—the "Itigi Thicket" (subhabitat Af4C-1), found disjunctly in ne. Zambia and c. Tanzania. This thicket is so botanically distinct that it almost merits coverage as a full habitat, like the structurally very similar ALBANY THICKET of South Africa, but it doesn't support a distinctive set of wildlife and is quite small in extent. There is dense growth of trees and shrubs up to 10–16 ft. (3–5 m). The canopy is largely deciduous, but the midstory contains many evergreen or semi-deciduous shrubs and small trees. Little sunlight reaches the thicket floor, so grass and herbaceous growth is largely lacking. Itigi Thicket seems to grow in response to some discreet soil parameters: well-aerated soil without many rocks that receives heavy rainfall during part of the year but dries out quickly during the dry season. This enigmatic habitat may well be a relict of a different climate in the past, a sort of "missing link" from a moister corridor that once connected East and Southern Africa.

Itigi Thicket in ne. Zambia. © ELISE ZYTKOW

Monsoon Forest (Mushitu) and ephemeral shallow freshwater wetland (Dambo) that are often found together in the Miombo Zone of Central Africa during the wet season. © FRANK WILLEMS

CONSERVATION: Hunting has eliminated most of the big mammals, even where human population densities are low. Few formally protected areas include Monsoon Forest habitat, and most of those that exist are poorly policed.

WILDLIFE: Most common species of Monsoon Forest are shared with AFROTROPICAL LOWLAND RAINFOREST. Typical mammals include Blue Monkey, Yellow Baboon, Malbrouck (formerly Vervet Monkey), Gambian Sun Squirrel, Smith's Bush Squirrel, several species of galagos, Bushy-tailed Mongoose, Bushpig, Bushbuck, and Blue Duiker. Many mammals that mainly live in MIOMBO venture into Monsoon Forest to search for food during the dry season. African Savanna Elephant can still be found locally, in remote areas and large national parks. The secretive ambush-hunting Leopard does well in this habitat.

Monsoon Forest supports a distinctive and exciting mix of birds, though virtually none of them is restricted to this habitat. Some widespread and typical birds include Blue-spotted Wood-Dove, Black Cuckoo, Crowned and Trumpeter Hornbills, Schalow's and Ross's Turacos, Black-backed Barbet, Pallid Honeyguide, Common Square-tailed Drongo, Ashy Flycatcher, White-tailed Blue Flycatcher, Cabanis's and Sombre Greenbuls, Black-throated Wattle-eye, Olive Sunbird, and Forest Weaver. Red-capped Robin-Chat and Bocage's Akalat skulk in the understory. African Barred Owlets hunt smaller birds and often attract a crowd of antagonistic neighbors. This habitat is quite

Af4C MONSOON FOREST

Left: **Gallery forest is one of the main habitats of Pel's Fishing-Owl, a huge ginger owl that is one of Africa's most sought-after birds.**
© CHRISTIAN BOIX, AFRICA GEOGRAPHIC

Below: **Rusty-spotted or Large-spotted Genet is a habitat generalist that readily uses Monsoon Forest.**
© KEITH BARNES, TROPICAL BIRDING TOURS

Left: **Several species of turacos, including Schalow's Turaco, are partial to Monsoon Forest. This family of birds is endemic to sub-Saharan Africa.** © KEN BEHRENS, TROPICAL BIRDING TOURS

rich in bushshrikes, including Black-backed Puffback; Black-fronted, Sulphur-breasted, and Four-colored Bushshrikes; and Tropical Boubou. The Monsoon Forest edges are good for seed-eating birds like Grosbeak Weaver, Black-faced Canary, Peters's Twinspot, and Black-tailed Waxbill. Some species that are closely associated with Miombo, such as Shelley's Sunbird and Bar-winged Weaver, will also forage along the edges of Monsoon Forest. A handful of bird species are virtually restricted to this habitat, namely the localized Bannerman's Sunbird and Laura's Woodland-Warbler and the more widespread Gray-olive Greenbul (IS). Many species are shared with the Congo Basin lowland rainforest; just a few exciting examples are Palm-nut Vulture, Western Crested Guineafowl, Pel's Fishing-Owl, Narina Trogon, and African Broadbill. Monsoon Forest is an important habitat for the enigmatic and migratory African Pitta.

Although reptiles are not especially conspicuous, this habitat does hold many species, including dozens of endemics. Typical reptiles include dwarf geckos (*Lygodactylus*), Kalahari Plated Lizard (*Gerrhosaurus multilineatus*), Flap-necked Chameleon (*Chamaeleo dilepis*), Black-necked Spitting Cobra (*Naja nigricollis*), blind snakes (*Typhlops*), Southeastern Green Snake (*Philothamnus hoplogaster*), and Savanna Vine Snake (*Thelotornis capensis*). Monsoon Forest shares several venomous snakes, such as Jameson's Mamba (*Dendroaspis jamesoni*), Forest Cobra (*Naja melanoleuca*), Forest Vine Snake (*Thelotornis kirtlandii*), and Gaboon Viper (*Bitis gabonica*), with the central and West African lowland rainforests.

Endemism: Some of the localized endemic birds that are associated with the Angolan escarpment will use Monsoon Forest habitats in w. Angola. The Itigi Thicket is poorly known but likely contains some endemic plants.

DISTRIBUTION: Found in the moister portions of the MIOMBO zone, where Monsoon Forest acts as a transitional habitat between Miombo broadleaf savanna and AFROTROPICAL LOWLAND RAINFOREST. Monsoon Forest is found mostly along watercourses and in other moister, lower-lying areas and is tenuously connected with EAST COAST FOREST MATRIX by the rivers in Tanzania and Mozambique. These connections are of great biogeographic importance, as they periodically would have connected the forests along the Indian Ocean with the greater Congo Basin, allowing for species exchange. The borders between Monsoon Forest and adjacent MOIST MONTANE FOREST or Afrotropical Lowland Rainforest are not well-defined. The forest changes character slowly with the transition to less seasonality and more rainfall (in the case of lowland rainforest) or higher elevation and a different plant community (in the case of montane forest). This habitat is found primarily on the Central African Plateau, mostly at elevations between 3300 and 5000 ft. (1000–1500 m). Some of the forests in n. Tanzania, Kenya, and Ethiopia could be classified as Monsoon Forest but in this book are considered montane forest because they have wildlife more typical of that habitat. Another area that could be considered Monsoon Forest is the most northerly portions of the Congo Basin and Upper Guinea rainforest. These forests have a drier character than typical rainforest but don't have a set of wildlife that is distinctly different from that of lowland rainforest.

WHERE TO SEE: Lake Manyara National Park, Tanzania; Okavango Delta, Botswana.

Groundwater Forest presents an oddly lush aspect in the midst of dry, savanna-dominated areas. Lake Manyara National Park, Tanzania. © KEN BEHRENS, TROPICAL BIRDING TOURS

Af4D EAST COAST FOREST MATRIX

IN A NUTSHELL: A complex mix of forest, thicket, and lush woodland along the Indian Ocean coast of tropical Africa. **Global Habitat Affinities:** AUSTRALASIAN LITTORAL RAINFOREST; NEOTROPICAL SEMI-EVERGREEN FOREST; AUSTRALASIAN MONSOON VINEFOREST; MALABAR SEMI-EVERGREEN FOREST. **Continental Habitat Affinities:** SOUTH COAST FOREST MATRIX; MONSOON FOREST. **Species Overlap:** SOUTH COAST FOREST MATRIX; MONSOON FOREST; MOIST MIXED SAVANNA; MOIST MONTANE FOREST; AFROTROPICAL LOWLAND RAINFOREST; MIOMBO.

DESCRIPTION: This is one of the most internally complex habitats covered in this book. While structures vary from tall forest to dense thicket to lush woodland, which all seem strikingly different and in other cases might be split as separate habitats, they occur together in a fine matrix holding a fairly cohesive set of wildlife, justifying the consideration of this habitat as a single unit, albeit one rich in microhabitats. East Coast Forest Matrix occurs patchily and locally. Long ago, there may have been a nearly solid band of coastal forest, but it has been fragmented, first by a gradual drying trend and more recently and significantly by human activity. There is very little forest left along the northern part of the coast. Most of Africa's Indian Ocean coastline is covered in a matrix of grassland and savanna with scattered remnant forest patches.

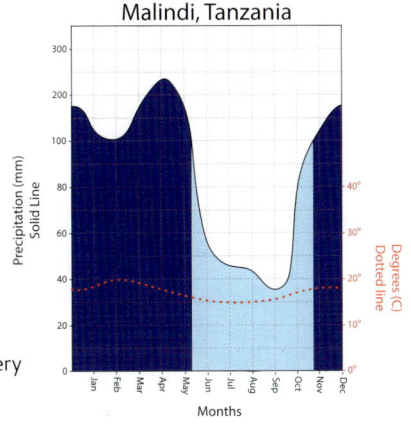

The timing of the distinct wet and dry seasons (Köppen **Ama**) in this habitat varies regionally. Along the n. Indian Ocean coast, near the equator, there are two rainy seasons: a long one (Apr–Jun) and a short one (Nov–Dec). Elsewhere, most rain falls during the austral summer, approximately November–April. Typical rainfall is in the range of 30–60 in. (800–1500 mm), depending on the year and the location. In general, the seasons are somewhat unpredictable and variable, and drought is common.

The most typical forest formation within the East Coast matrix is semi-evergreen forest that has strong affinities with the Monsoon Forest found inland. It is lush and has a closed canopy, but for several months of the year little or no rain falls, resulting in a level of stress to the plants that is rare in a true rainforest. This stress is manifested in the accumulation of leaf litter on the forest floor. The canopy is typically shorter than true rainforest, 65–80 ft. (20–25 m), with emergents reaching 130 ft. (40 m) tall. Lianas are abundant, but epiphytes are less common than in rainforest. The midstory is complex and well developed, and the understory varies from dense and thicket-like to quite open and easy to walk through. The moistest versions of this habitat, such as those that occur in the East

Right: **Areas of East Coast Forest Matrix that transition from forest structures to savanna structures often have a dense thicket understory. Save, Mozambique.** © KEN BEHRENS, TROPICAL BIRDING TOURS

Below: **Even in generally forested areas there are frequent more open stretches that allow enough sunlight to support the growth of grass. Coutada 12, Zambezi Valley, Mozambique.** © KEN BEHRENS, TROPICAL BIRDING TOURS

Usambara Mountains of Tanzania, resemble full rainforest, especially at elevations around 2300 ft. (700 m), just below the transition to MOIST MONTANE FOREST. Small pockets of SWAMP FOREST exist within the forest matrix and have a distinctive plant community but not a different set of wildlife.

When walking a transect through this habitat, it is common to transition from closed-canopy forest, where the forest floor is littered with leaf litter, into a more open woodland-type formation with around 50% canopy coverage and abundant grass in the understory, then back into closed forest again. In situations like this, the wildlife assemblages change at different rates. The woodland formations have some savanna-type birds in the understory, whereas their canopy retains a bird community that is similar to immediately adjacent forests.

Given this habitat's complexity, it is hard to generalize about its constituent trees. But many forests include *Hymenaea verrucosa*, *Diospyros*, Panga Panga (*Millettia stuhlmannii*), *Newtonia buchananii*, Dune Myrtle (*Eugenia capensis*), *Drypetes*, Silver-leaf Milkplum (*Englerophytum natalense*), *Terminalia sambesiaca*, Pod Mahogany (*Afzelia quanzensis*), Coastal Red Milkwood (*Mimusops caffra*), Natal Guarri (*Euclea natalensis*), Flat-crown (*Albizia adianthifolia*), African Teak (*Milicia excelsa*), *Brexia madagascariensis*, *Olax*, and figs. Some drier forests are dominated by legumes, with common genera including *Scorodophloeus*, *Erythrophleum*, *Berlinia*, *Julbernardia*, *Guibourtia*, and *Hymenaea*. Some of the woodland growing on infertile sands along the coast of Kenya is open and Miombo-like and mainly comprises *Brachystegia*, Miombo's signature tree genus; it is retained in this habitat category because its wildlife assemblage is closely aligned with that of coastal forest. Elsewhere on the Kenyan coast, but in close proximity, occurs a very different sort of formation: dense evergreen thickets of *Cynometra webberi*. From Bazaruto south, along the s. Mozambique coast, there are stunted forests growing on tall sand dunes. These are closely aligned with similar habitat within the SOUTH COAST FOREST MATRIX and are addressed there.

CONSERVATION: Much of the e. African coast is heavily populated and has been affected by woodcutting and conversion to agriculture for thousands of years. Hunting has eliminated most of the big mammals, even in areas where human population densities are low. Few formally protected areas include East Coast Forest Matrix habitat, and most of those are poorly policed. The dominant habitats along the coast today are towns and villages, small-scale mixed agriculture, in which coconut palms, cashew trees, and mango trees are prominent, and a form of MOIST MIXED SAVANNA that may have widely replaced forest because of anthropogenic pressures.

WILDLIFE: East Coast Forest Matrix is a complex and exciting habitat for wildlife. Very few species are found

South African Porcupine is a habitat generalist that can be seen at night in East Coast Forest Matrix. © KEITH BARNES, TROPICAL BIRDING TOURS

Dune matrix, including grassland, savanna, and forest, with a very low canopy in Pomene National Reserve, Mozambique. © KEN BEHRENS, TROPICAL BIRDING TOURS

throughout the complete range of this habitat and its many distinctive variations. Most of the species that are found throughout are also shared with **AFROTROPICAL LOWLAND RAINFOREST**. Typical mammals include Blue Monkey, Red-bellied Coast Squirrel, Suni, Natal Red Duiker, several species of galagos and sun squirrels, Bushy-tailed Mongoose, Bushpig, and Bushbuck. A few genera have different representatives in the north and south, such as Yellow Baboon and Ochre Bush Squirrel in the north and Chacma Baboon and Smith's Bush Squirrel in the south. The region along the Zambezi River is a stronghold for the scarce Nyala, one of Africa's more beautiful antelope. This is the habitat for three of the continent's five awesome species of giant sengi, or elephant shrew, as well as the Sokoke Dog Mongoose. Coastal forest in Tanzania and s. Kenya has the fur-cape-wearing black-and-white Angola Colobus. Two beautiful red colobuses are localized endemics in the north: Tana River Red Colobus along its

Left: **East Coast Forest Matrix is a major stronghold of the scarce and handsome Nyala.** © KEN BEHRENS, TROPICAL BIRDING TOURS

Below: **This habitat holds three of Africa's five giant sengis, or elephant shrews, including Golden-rumped Elephant Shrew.** © JANOS OLAH

Black-backed Puffback male in fully puffed display.
© KEITH BARNES, TROPICAL BIRDING TOURS

namesake river and Zanzibar Red Colobus on its namesake island. Pemba and Mafia Islands host Africa's only flying foxes: an endemic species on Pemba and a species shared with the Comoros and Seychelles on Mafia.

East Coast Forest Matrix is one of Africa's more fascinating birding habitats. This is not only because of its own richness but because it tends to occur alongside MOIST MIXED SAVANNA, MIOMBO, and freshwater wetlands and/or AFROTROPICAL MANGROVE, attracting a huge variety of birds.

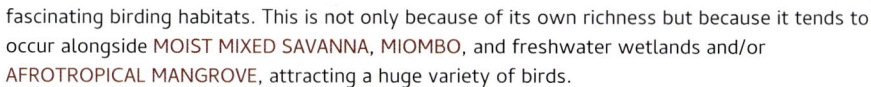

Very few birds are found throughout the entirety of the East Coast Forest Matrix discussed here, but some widespread and typical birds include Fasciated Snake-Eagle, Brown-headed Parrot, Green Malkoha, Trumpeter Hornbill, African Broadbill, Livingstone's Turaco, White-eared Barbet, Black Cuckooshrike, Square-tailed Drongo, Sombre Greenbul, Eastern Nicator, Four-colored Bushshrike, Chestnut-fronted and Retz's Helmetshrikes, Black-and-white Shrike-flycatcher, African Crested Flycatcher, Livingstone's Flycatcher, Black-bellied Starling, Olive Sunbird, Forest Weaver, and Plain-backed Sunbird. Understory species, mostly skulkers, include Southern Crested Guineafowl, African Pygmy-Kingfisher, Terrestrial Brownbul, Red-tailed Ant-Thrush, Bearded Scrub-Robin, Green-backed Camaroptera, and Peters's Twinspot. All these are shared with SOUTH COAST FOREST MATRIX and/or MONSOON FOREST. A few species are virtually restricted to this habitat, namely Green Tinkerbird, Mombasa Woodpecker, Fischer's Greenbul, and East Coast Akalat. Forests along the Zambezi River are amongst the best places in Africa to see the enigmatic African Pitta, which jumps into the air and gives its frog-like call just after its arrival to breed during the rainy season. For visiting birders, some of the most appealing birds are localized endemics. The northern coastal zone is especially rich, boasting the likes of Fischer's Turaco, Sokoke Scops-Owl, Sokoke Pipit, Yellow Flycatcher, Zanzibar Boubou, and Clarke's Weaver. Several species are confined to the southern portions of this habitat, along with the adjacent South Coast Forest Matrix. These include Rudd's Apalis, Livingstone's Flycatcher, and Woodward's Batis. Usambara Hyliota is endemic to the foothills of the East Usambara mountains.

Although reptiles are not especially conspicuous, this habitat does hold many species, including dozens of endemics. Typical

East Coast Forest Matrix supports some of Africa's most impressive large snakes like Mozambique Spitting Cobra (*Naja mossambica*). © MARIUS BURGER

reptiles include dwarf geckos (*Lygodactylus*), Tropical Girdled Lizard (*Cordylus tropidosternum*), Flap-necked Chameleon (*Chamaeleo dilepis*), several pygmy chameleons (*Rhampholeon*), blind snakes (*Typhlops*), Southeastern Green Snake (*Philothamnus hoplogaster*), Cross-barred Tree Snake (*Dipsadoboa flavida*), Mozambique Spitting Cobra (*Naja mossambica*), and Savanna Vine Snake (*Thelotornis capensis*). East Coast Forest Matrix shares some of Africa's large and venomous snakes, such as Forest Cobra (*Naja melanoleuca*) and Gaboon Viper (*Bitis gabonica*), with the Central and West African lowland rainforest and Monsoon Forest.

Endemism: Within this habitat, there are many localized endemic birds, and some endemic smaller mammals. South Kenya and n. Tanzania are especially rich in localized endemics, and there are also a couple along the Tana River in c. Kenya. The islands of Pemba, Zanzibar, and Mafia, off the coast of Tanzania, have a few endemics each.

DISTRIBUTION: Found along the Indian Ocean coast from s. Somalia south to s. Mozambique. Mostly within 30 mi. (50 km) of the coast but occurs inland 100 mi. (160 km) or more along major rivers like the Zambezi. Some extreme outlier forests, which largely have the character of coastal forest, occur as far inland as Malawi. The boundary line with the SOUTH COAST FOREST MATRIX is poorly defined but occurs in the vicinity of Inhambane, Mozambique, at the transition point from the subtropics to the tropics. Stunted forest growing on tall dunes, a subhabitat that is closely associated with the South Coast Forest Matrix, continues slightly farther north, to the Bazaruto

Some of the lushest east coast forest is found in the lower elevations of the East Usambara Mountains, Tanzania. Dragon trees (*Dracaena*) *(inset)* are common. Despite the similarity to rainforest there is still abundant leaf litter *(inset)*. © KEN BEHRENS, TROPICAL BIRDING TOURS

Af4D EAST COAST FOREST MATRIX

area. In terms of elevation, this habitat is mostly low lying, below 500 ft. (150 m), but it occurs up to 1300 ft. (400 m) in Kenya's Shimba Hills and 2600 ft. (800 m) on the lower slopes of Tanzania's East Usambara Mountains.

WHERE TO SEE: Arabuko Sokoke National Park, Kenya; East Usambara Mountains, Tanzania; Coutada 12, Mozambique.

Old termite mounds, with a dense thicket-like growth, are a common sight. Coutada 12, Zambezi Valley, Mozambique. © KEN BEHRENS, TROPICAL BIRDING TOURS

SIDEBAR 14 — WEAVERS, A MEGADIVERSE AFROTROPICAL FAMILY

The weaver family is one of the most diverse in sub-Saharan Africa. It is also one of the most African of bird families: of 123 species in this expansive family, only 5 are found outside Africa, in the Asian tropics. And all 16 genera within the family are found in Africa, with one confined to the Indian Ocean islands. Virtually any day spent in the field in the Afrotropics, whether in rainforest or semi-desert, will be spent in the company of at least a couple of weavers. In size, weavers range from the bulky buffalo-weavers down to the diminutive queleas. Weavers' namesake behavior is their nest construction, for which they are justly famous. Within this family different species have a huge variety of nest structures and breeding strategies. These range from cooperative breeders that construct enormous stick or straw communal nests to solitary breeders that weave a fine grass basket.

Vitelline Masked-Weaver visiting its nest. Can you imagine weaving something this fine and complicated with only a pair of tweezers in your mouth?!
© KEN BEHRENS, TROPICAL BIRDING TOURS

Af4E SOUTH COAST FOREST MATRIX

IN A NUTSHELL: Different types of tall closed-canopy forest and dense scrub thicket of subtropical and temperate se. Africa. **Global Habitat Affinities:** AUSTRALASIAN LITTORAL RAINFOREST; NEOTROPICAL SEMI-EVERGREEN FOREST; AUSTRALASIAN MONSOON VINEFOREST; MALABAR SEMI-EVERGREEN FOREST. **Continental Habitat Affinities:** EAST COAST FOREST MATRIX; MONSOON FOREST. **Species Overlap:** EAST COAST FOREST MATRIX; ALBANY THICKET; MOIST MONTANE FOREST; MONSOON FOREST.

DESCRIPTION: The differing nature of this habitat's microhabitats, and the way they meld into one another, make it difficult to define, but South Coast Forest Matrix is essentially a mosaic of closed-canopy dune forest, sand forest, riverine forest, thickets, and lush woodland less than 30 mi. (50 km) from the coast and below 1310 ft. (400 m) in elevation. Forest patches are often small and very patchy in nature, rapidly transitioning into other habitats, such as STRANDVELD, MOIST MIXED SAVANNA, and AFROTROPICAL GRASSLAND (often flooded). They also are adjacent to a matrix of various freshwater habitats and agriculture.

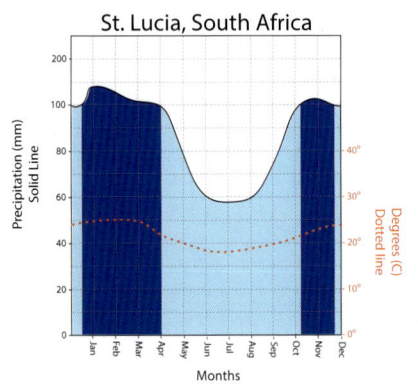

Generally, the main habitat is Dune Forest (subhabitat Af4E-1), which comprises well-developed species-rich 80–100 ft. (25–30 m) tall subtropical and temperate forest on coastal plains and stabilized dunes. There is a continuous canopy with a multilayered mid- and understory 16–30 ft. (5–10 m) tall where lianas are common. The ground layer is open and mostly covered in leaf litter, with some grasses present. Typically, this forest is taller than MOIST MONTANE FOREST and with fewer epiphytes but more lianas and tangles. In places, edaphic factors (poor soil) lead to forest becoming dense and stunted, with a canopy as low as 16–22 ft. (5–7 m). While most trees are evergreen, there are also semi-deciduous and deciduous trees. These stunted forests generally grow on well-developed sandy–loamy soils on a young sand-dune system still in the process of sedimentation. Sand Forest (subhabitat Af4E-2) is very distinct and found mostly in n. KwaZulu-Natal, South Africa, and s. Mozambique on red-brown

Above: **As it approaches a wind-swept beach, coastal Dune Forest shrinks down to a dense ground-hugging thicket. Alexandria Dunefield, Eastern Cape, South Africa.** © KEITH BARNES, TROPICAL BIRDING TOURS

acidic soils on ancient dune systems. Sand Forest comprises 16–19 ft. (5–6 m) tall super-dense thickets with lianas and emergents forming a 50 ft. (15 m) tall disjointed canopy of primarily deciduous and semi-deciduous elements. It has a longer dry season and more leaf litter than Dune Forest, growing in conditions shared with **MONSOON FOREST.**

Dune Forest at Lagoa Inhampavala, Mozambique. This formation is very dense but with a low canopy. © KEN BEHRENS, TROPICAL BIRDING TOURS

South Coast Forest Matrix stretches through a wide range of latitudes, and although the climate (Köppen **Cma**) is moderate, with largely consistent temperatures, temperature differs considerably between subtropical and temperate locations. Subtropical annual highs range from 90 to 77°F (32–25°C), while lows range from 75 to 61°F (24–16°C). In more temperate areas, lows may reach 46°F (8°C) in winter. South Coast Forest Matrix is a fairly moist system with more rain falling in the

South coast forest at Lagoa Inhampavala, Mozambique: great habitat for White-throated Robin-Chat. © KEN BEHRENS, TROPICAL BIRDING TOURS

subtropics than temperate zones, with 39 in. (1000 mm) in Durban and 25 in. (625 mm) in Port Elizabeth. Rain falls year-round but more in the summer (Oct–Mar). Frost and snow are basically unheard of. Fire is rare in this constantly wet, often fire-protected habitat, although adjacent grassland and savanna burn frequently, and fire may slowly eat into this habitat, reducing it incrementally.

South Coast Forest Matrix is notable for its absence of bastard yellowwood (*Afrocarpus*) and true yellowwood (*Podocarpus*) species typical of Moist Montane Forest. Instead the most important and dominant plants include White Milkwood (*Sideroxylon inerme*), Coast Red Milkwood (*Mimusops caffra*), White Stinkwood (*Celtis africana*), Dune Sourberry (*Dovyalis rotundifolia*), Flat-crown (*Albizia adianthifolia*), False Forest Ironplum (*Drypetes reticulata*), Natal Guarri (*Euclea natalensis*), Cape Plane (*Ochna arborea*), and Kooboo Berry (*Mystroxylon aethiopicum*). Shrubs and scramblers include Cat-thorn (*Scutia myrtina*), Wild Caperbush (*Capparis sepiaria*), Coast Silver Oak (*Brachylaena discolor*), Cape Teak (*Strychnos decussata*), and num-nums (*Carissa*). Sand Forest is shorter and denser with different species composition and smaller trees. Bastard Tamboti (*Cleistanthus schlechteri*), Torchwood (*Balanites maughamii*), Sand Canaryberry (*Suregada zanzibarensis*), Hairy Cola (*Cola greenwayi*), Water Ironplum (*Drypetes arguta*), and Green Apple (*Monodora junodii*) are characteristic. Sand Lemon-rope (*Salacia leptoclada*) forms lianas, and palms and bird-of-paradise plants (*Strelitzia*) are abundant. The grass component of South Coast Forest Matrix is poorly developed, though forest patches may be completely surrounded by open grassland and savanna.

Af4E SOUTH COAST FOREST MATRIX

CONSERVATION: Around 20% of this habitat is found in various protected areas in South Africa and Mozambique, including iSimangaliso Wetland Park. Outside the protected area network threats include habitat destruction and degradation, often for coastal development. Introduced plants and plantations have replaced much former habitat. Mining of dunes for titanium also occurs locally.

WILDLIFE: South Coast Forest Matrix has much in common with MOIST MONTANE FOREST and ALBANY THICKET. Owing to the patchy nature of this habitat and its intermingling with more open grasslands and savanna habitats, most large mammals that do use it also use adjacent habitats. Good examples include African Savanna Elephant, African Buffalo, and Giraffe.

More typical ungulates are smaller, stealthier ones capable of moving quickly and silently through thick vegetation, such as Suni, Natal Red Duiker, and diminutive Blue Duiker, but larger Bushbuck and elegant Nyala are also typical of this habitat. Common Warthog and Bushpig join the antelope, regularly rooting around on the forest floor. All these are hunted by Leopard, especially in dense tangles. The canopy holds Vervet and Sykes' Blue Monkeys and Red Bush and Smith's Bush Squirrels.

This is one of Africa's more fascinating birding habitats, not only because of its own richness but because it tends to occur in a complex matrix alongside Albany Thicket, MOIST MIXED SAVANNA, AFROTROPICAL GRASSLAND, freshwater wetlands, and AFROTROPICAL MANGROVE, attracting a huge variety of birds. Crowned Eagle is a key predator, a threat to any canopy monkey or squirrel or any small ungulate on the forest floor. Fasciated Snake-Eagle specializes in perch hunting reptiles, often from forest edges. Fruits are eaten and dispersed by Livingstone's Turaco, Crowned and Trumpeter Hornbills, White-eared Barbet, Red-fronted and Yellow-rumped Tinkerbirds, African Olive

Southern Crested Guineafowl are one of the few gamebirds that penetrate the dense coastal forest floor. © KEITH BARNES, TROPICAL BIRDING TOURS

Mozambique Forest Tree Frog (*Leptopelis mossambicus*).
© MARIUS BURGER

Wahlberg's Velvet Gecko (*Homopholis wahlbergi*) is a typical reptile species of South Coast Forest Matrix. © MARIUS BURGER

Pigeon, Sombre Greenbul, and Black-bellied Starling. Canopy flocks led by Square-tailed Drongo often also contain Green Malkoha, African Crested Flycatcher, Black-backed Puffback, and Forest Weaver. Insects are hunted by Eastern Nicator and delightfully vocal Four-colored Bushshrike, while flowering trees attract Purple-banded, Olive, Mouse-colored, and Plain-backed Sunbirds. Southern Crested Guineafowl runs along the forest floor. The forest edge, especially where there are seeding grasses, attracts Green Twinspot, Red-backed Mannikin, and Black-tailed Waxbill. Many species are shared with EAST COAST FOREST MATRIX. For visiting birders, some of the most appealing birds are localized endemics and near-endemics, including Dune Forest specialties like Brown Scrub-Robin (IS) and Woodward's Batis (IS), and Sand Forest specialties like Rudd's Apalis, Pink-throated Twinspot (IS), and Neergaard's Sunbird (IS). Some species that breed on nearby scarp Moist Montane Forest spend the nonbreeding season here, including the globally Endangered Spotted Ground-Thrush.

At night, African Wood-Owls are common. The uber-secretive but common Buff-spotted Flufftail mostly calls its eerie fog-horn-like hoot song on spring and summer nights. A specialist of Lala-Palm savanna (a subtype of Moist Mixed Savanna), often adjacent to the forest, the Lemon-breasted Canary is another local but interesting species associated with this landscape.

Although reptiles are not especially conspicuous, this habitat does hold many species, including several endemics. Typical reptiles include dwarf geckos (*Lygodactylus*), Tropical Girdled Lizard (*Cordylus tropidosternum*), Flap-necked Chameleon (*Chamaeleo dilepis*), Pondo (*Bradypodion caffer*) and Setaro's (*B. setaroi*) Dwarf Chameleons, blind snakes, and Savanna Vine Snake (*Thelotornis capensis*). South Coast Forest Matrix supports some of Africa's largest and most-spectacular snakes, such as Forest Cobra (*Naja melanoleuca*), Eastern Green Mamba (*Dendroaspis angusticeps*), and Gaboon Viper (*Bitis gabonica*). This forest is moist, and with adjacent wetland habitats, it is a haven for frogs, including several that are endemic and near-endemic. These include Bush Squeaker (*Arthroleptis wahlbergii*), Natal Tree Frog (*Leptopelis natalensis*), Delicate Leaf-folding Frog (*Afrixalus delicatus*), Power's Long Reed Frog (*Hyperolius poweri*), Pickersgill's Reed Frog (*H. pickersgilli*), Yellow-striped Reed Frog (*H. semidiscus*), Tinker Reed Frog (*H. tuberilinguis*), and Red-legged Kassina (*Phlyctimantis maculatus*).

DISTRIBUTION: All forest patches within approximately 30 mi. (50 km) of the coast and located between the mouth of the Limpopo River in s. Mozambique to, and including, Alexandria Forest, Eastern Cape, South Africa, are classified as coastal forests; however, there are outliers farther inland and farther west along the coast.

WHERE TO SEE: iSimangaliso Wetland Park, South Africa.

Af4F SWAMP FOREST

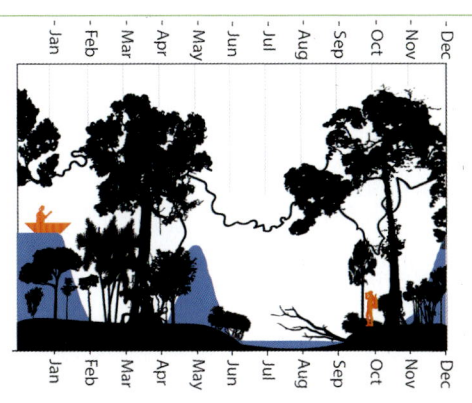

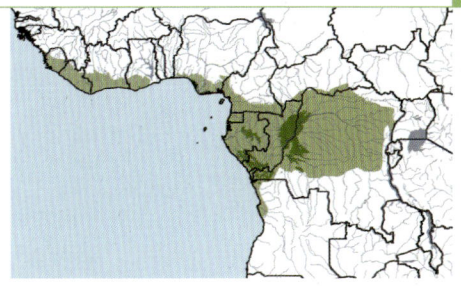

IN A NUTSHELL: Lowland rainforest of Central and West Africa that is completely flooded for at least part of the year, supporting a similar though subtly different set of wildlife than nonflooded forest and a very different set of plants. **Global Habitat Affinities:** IGAPÓ; VÁRZEA; AUSTRALASIAN SWAMP FOREST; PALM SWAMP; ASIAN PEAT SWAMP FOREST. **Continental Habitat Affinities:** INDIAN OCEAN LOWLAND RAINFOREST (very locally contains Swamp Forest). **Species Overlap:** AFROTROPICAL LOWLAND RAINFOREST; AFROTROPICAL MANGROVE; MOIST MONTANE FOREST.

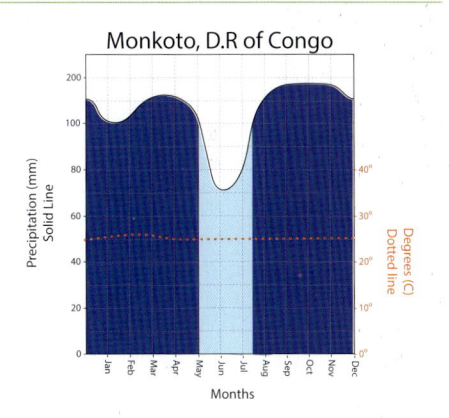

DESCRIPTION: The Afrotropical equivalent of the Igapó and Várzea flooded forests of South America, this habitat consists of forest that is completely flooded for at least part of the year. Although it looks very different when flooded and has a profoundly different botanical composition from normal AFROTROPICAL LOWLAND RAINFOREST, it is not drastically different in terms of its general structure and wildlife. The canopy of Swamp Forest is lower than that of rainforest, averaging 65–80 ft. (20–25 m), with emergent trees up to 150 ft. (45 m) tall. Many trees have aerial roots, and some have stilt roots, reminiscent of mangroves. The climate here is the same as Afrotropical Lowland Rainforest (**Afa**), and the different habitat is a result of geomorphology impeding drainage, producing flooding, and promoting the development of gleysols (saturated gray-colored soils).

The diverse set of trees includes *Spondianthus preussii*, Boarwood (*Symphonia globulifera*), *Mitragyna*, pandanus, *Nauclea pobeguinii*, African Crabwood (*Carapa procera*), Wild Frangipani (*Voacanga thouarsii*), *Uapaca*, Wild Date Palm (*Phoenix reclinata*), and *Pseudospondias*, to name just a few. Interestingly, some tree genera, especially *Carapa* and *Symphonia*, are found only in Swamp Forest in the lowlands but are common in well-drained MOIST MONTANE FOREST. The understory is usually dense and often includes raffia palms (*Raphia*) and a great abundance of lianas, especially spiny rattans. Another common understory component is African Oil Palm (*Elaeis guineensis*), here in its natural range. In some places, large areas of Swamp Forest are dominated by raffia palms, often mixed with *Baphia laurifolia*, *Xylopia*, and *Cryptosepalum congolanum*. Another

Degraded Swamp Forest on the verge of Afrotropical Mangrove in s. Ghana. © KEITH BARNES, TROPICAL BIRDING TOURS

Classic slow-moving backwaters and inundated Swamp Forest in Loango National Park, Gabon. © CHRISTIAN BOIX, AFRICA GEOGRAPHIC

distinctive subtype of Swamp Forest is that dominated by *Alstonia congensis*; this subtype is prevalent in the Upper Guinea region, especially the Niger Delta.

Accessing this habitat is even tougher than accessing typical lowland rainforest and is usually possible only by boating along the open waterways. The defining characteristic of this habitat is flooding, which happens on a seasonal basis, inundating the forest to a depth of up to 3 ft. (1 m), then slowly drying out. The main flooding in the Niger Delta is August–December, while in the vast Congo Basin there are two major flood periods: December–January and May.

CONSERVATION: This is one of the most poorly known and least frequently visited of Africa's habitats. Despite its inaccessibility, this habitat has many conservation challenges. The Niger Delta is surrounded by heavily populated areas and is under great pressure for timber, fish, and bushmeat. The valuable Abura (*Mitragyna*) tree has already been virtually wiped out of the delta. The Congo Basin is much larger and has a much lower population, but African Forest Elephant and "Forest" African Buffalo have still been wiped out of most areas adjacent to navigable waterways. Swamp Forest was traditionally viewed as unsuitable for cultivation, but in many areas has recently been converted into rice fields.

WILDLIFE: This habitat supports a similar set of wildlife to AFROTROPICAL LOWLAND RAINFOREST, though it is preferred over lowland rainforest by some species, notably primates. Unfortunately, wildlife watching here is difficult. Swamp Forest occurs mostly in countries and areas that are difficult to access, and the mammals and larger birds in this part of the world are heavily hunted and tend to be very shy. The Critically Endangered Niger Delta Red Colobus is endemic to its

Central African Slender-snouted Crocodile (*Mecistops leptorhynchus*) has a narrow snout like the Asian Gharial, equipping it to hunt for fish, amphibians, and crustaceans. © CHRISTIAN BOIX, AFRICA GEOGRAPHIC

namesake delta. Pygmy Hippopotamus is associated with Swamp Forest, though sadly it has been eliminated from the eastern portion of its range, including the Niger Delta. Widespread West African species that are fond of this habitat include African Forest Elephant, Allen's Swamp Monkey, De Brazza's and Mona Monkeys, Forest Giant Squirrel, Thomas's Rope Squirrel, Alexander's Kusimanse, Long-tailed Pangolin, Red River Hog, and Water Chevrotain. The Congo River is a major biogeographic barrier for mammals; among species that favor Swamp Forest, Gray-cheeked Mangabey occurs exclusively on the north side, while Golden-bellied and Black-crested Mangabeys and Dryas Monkey live exclusively on the south side.

The Swamp Forest bird community has not been well studied; in general, it is a subset of the birds of Afrotropical Lowland Rainforest. No species are restricted to Swamp Forest, though it is prime habitat for White-crested Bittern, White-backed Night-Heron, Spot-breasted Ibis, Hartlaub's Duck, African Finfoot, White-spotted Flufftail, Vermiculated and Rufous Fishing-Owls, Swamp Palm Bulbul, Cassin's Flycatcher, Congo Sunbird, and Orange Weaver.

Vermiculated Fishing-Owl is a Swamp Forest specialist.
© CHRISTIAN BOIX, AFRICA GEOGRAPHIC

Left: **Swamp Forest is relatively poor in mammals but is a preferred habitat for many primates, such as De Brazza's Monkey.** © MARTIN VAN ROOYEN

Below: **The scarce African Finfoot is the African representative of a tiny family that has only three members worldwide.** © KEN BEHRENS, TROPICAL BIRDING TOURS

Af4F SWAMP FOREST

DISTRIBUTION: This is the main habitat in the central part of the Congo River basin and on the Niger River delta. Away from those vast areas, Swamp Forest is present only locally in a mosaic within AFROTROPICAL LOWLAND RAINFOREST. It also can be found very locally in the rainforest buffer zone and even within MONSOON FOREST and EAST COAST FOREST MATRIX habitats (not mapped).

WHERE TO SEE: Semuliki National Park, Uganda; Loango National Park, Gabon.

White-backed Night-Heron is a nocturnal heron that spends its days sleeping in dense waterside vegetation.
© KEITH BARNES, TROPICAL BIRDING TOURS

SIDEBAR 15 — GALAGOS—THE PRIMATE NIGHTSHIFT

In the Southern African language Afrikaans, galagos are called "nagapies," which means little night monkeys. In their own family, they are an ancient lineage of primates that originated 40–50 MYA and dominated their niche prior to the evolution of the smarter and more aggressive monkeys, which quickly outcompeted them during the day. But galagos had features that made them better nocturnal denizens than monkeys. Hiding in hollows by day, they are able to avoid predation. Large globular eyes are packed with rod cells that make them red-green colorblind but enables excellent vision in the pitch dark. Their night vision is enhanced further by a mirror-like layer behind the retina (the tapetum lucidum), which is why their eyes shine so brightly at night. The ability to move their ears independently helps them locate insects making the slightest movement, and their spring-like hind legs and long tails make them excellent jumpers with a cantilever counterbalance.

Southern Lesser Galago, also known as the Mohol Bushbaby.
© KEITH BARNES, TROPICAL BIRDING TOURS

Af4G MOIST MONTANE FOREST

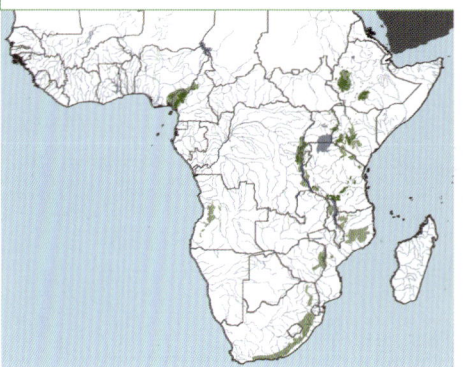

IN A NUTSHELL: Wet forest growing at moderate and high elevations and hosting a distinctive assemblage of wildlife. **Global Habitat Affinities:** AUSTRALASIAN TROPICAL MONTANE RAINFOREST; ANDEAN CLOUDFOREST; MESOAMERICAN CLOUDFOREST; SUNDA MONTANE RAINFOREST. **Continental Habitat Affinities:** AFROTROPICAL MONTANE DRY MIXED WOODLAND; INDIAN OCEAN MONTANE RAINFOREST. **Species Overlap:** AFROTROPICAL MONTANE DRY MIXED WOODLAND; MONSOON FOREST; AFROTROPICAL LOWLAND RAINFOREST; EAST COAST FOREST MATRIX; SOUTH COAST FOREST MATRIX.

DESCRIPTION: Mountains often give rise to moist microclimates in which forest grows because of the orographic effect: mountains force air upward, cooling it and precipitating its moisture. Even within very arid climates, such as that of n. Kenya, isolated mountains still produce enough moisture to foster forest. Although Moist Montane Forest can look superficially similar to AFROTROPICAL LOWLAND RAINFOREST, the cooler, high-elevation climate produces a subtly different forest with a distinctive set of wildlife. One key difference is that in montane forest, much precipitation is mist, which gives rise to abundant mosses and epiphytes (and gives similar habitat in South America the name "cloudforest"). Compared with superficially homogeneous lowland rainforest, montane forest is quite varied. Its classic form is similar to that of lowland rainforest: a lush and complex closed-canopy forest composed of broadleaf trees. Subtle differences include a shorter canopy; more mosses, epiphytes, and lianas; and a different mix of plants. The canopy height is 80–150 ft. (25–45 m), and the midstory and understory are complex and well vegetated. But this classic form tends to grow only in the wettest areas, usually on mountain slopes that receive the most moisture. Drier slopes have more open, semi-deciduous forest at lower elevations and coniferous *Podocarpus*-dominated forest higher up. Although the extremes are obviously different, the dividing line between moist and dry montane forest is often unclear. Montane forest often grows

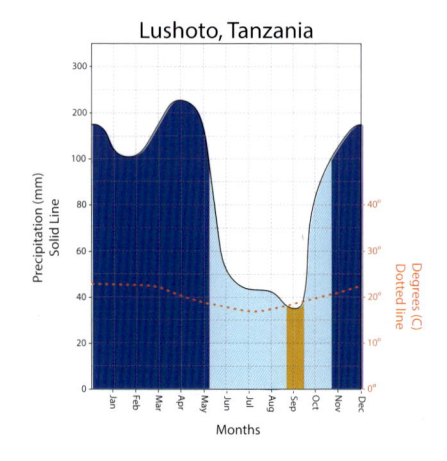

Above: **The Moist Montane Forest of Africa is the globe's second major chameleon hotspot, second only to Madagascar. This is the hefty Usambara Three-horned Chameleon (*Trioceros deremensis*).** © ADAM SCOTT KENNEDY

Left: **Nyungwe National Park, Rwanda, showing the misty precipitation that keeps these forests permanently wet.** © KEITH BARNES, TROPICAL BIRDING TOURS

patchily within a natural matrix of lush scrub and Afrotropical MONTANE GRASSLAND. One common aspect of most of its permutations is a thick understory that makes it difficult to walk through except on trails. There is often a band of African Alpine Bamboo (*Oldeania alpina*) just above the forest zone and below the heath zone, and bamboo is commonly mixed into the forest too.

The climate that supports montane forest is generally warm during the day and cool at night (Köppen **Afb**). Temperatures below freezing sometimes occur in the temperate southern part of the continent. Rainfall varies widely, and much of the moisture arrives in the form of mist rather than rain. Mists form year-round, providing moisture even during the dry season, explaining why montane forest has fewer semi-evergreen and deciduous species than lowland rainforest types that receive a similar amount of annual precipitation but which are more stressed by longer dry periods with little mist.

The driest montane forest receives around 40 in. (1000 mm) of precipitation annually; this seems to be the approximate lower limit for supporting this habitat. The wettest forests on Mt. Cameroon and Mt. Bioko can receive 400 in., or 33 ft. (10,000 mm), of rain in a year, making them among the wettest places on Earth. Montane forests tend to receive some moisture throughout the year, but most have at least one major rainy season, and some mountains in East Africa have two distinct rainy seasons each year.

It is also hard to generalize about the elevation at which montane forest is found. In most places with lowland forest, the transition to montane forest occurs at around 2600 ft. (800 m), though it can sometimes be as low as 1600 ft. (500 m). In Uganda, the transition occurs at around 5000 ft. (1500 m), while in the cool climate of s. South Africa, montane forest is somewhat paradoxically found all the way to sea level. The highest montane forests are found at around 11,500 ft. (3500 m), though

Fire can transform montane environments into grasslands, leaving narrow strands of Moist Montane Forest in drainage lines. Morro de Moco, Angola.
© KEN BEHRENS, PROMISED LAND VENTURES

these are exceptional, and the transition from forest to MONTANE HEATH or AFROTROPICAL MONTANE DRY MIXED WOODLAND is much lower on most mountains.

Owing to the vast and widely scattered range and complexity of Moist Montane Forest, its tree species vary. On one level, the forest seems uniform, as most of the trees are superficially similar, aside from the coniferous yellowwoods (*Podocarpus* and *Afrocarpus*). This stands in contrast to more arid environments, especially MOIST MIXED SAVANNA, which has an astounding diversity of very obviously different-looking trees. But the superficial uniformity of montane forest hides great diversity, and a remarkable range of plants can be found in a small area, especially in the moister and more extensive montane forests near the center of the continent. The Eastern Arc, Albertine Rift, and Cameroon mountains each have around 1000 endemic plants. With all that said, there are some classic Moist Montane Forest trees with wide ranges. These include olives (Oleacea), Cape Beech (*Rapanea melanophloeos*), Forest Ironplum (*Drypetes gerrardii*), crotons (*Croton*), Cape Ash (*Ekebergia capensis*), wild elders (*Nuxia*), hollies (*Ilex*), African Cherry (*Prunus africana*), Guinea Plum (*Parinari excelsa*), *Ocotea*, Common Forest Ochna (*Ochna holstii*), African Elm (*Celtis africana*), Lemonwood (*Xymalos monospora*), East African Mahogany (*Khaya anthotheca*), *Tabernaemontana*, and waterberry (*Syzygium*). The frequent presence of conifers (*Podocarpus* and *Afrocarpus*) and tree ferns distinguish montane forest from lowland forest. The ground cover includes some grasses, abundant ferns, and other herbaceous plants. Epiphytes, including stranglers, are common.

The mix of trees changes slightly in temperate South Africa, though many of the species and genera remain the same, and there is no major change of the wildlife. Some tree species only found in the temperate southern forests include Candlewood (*Pterocelastrus tricuspidatus*), Mountain Hardpear (*Olinia emarginata*), False Olive (*Buddleja saligna*), Pock-Ironwood (*Chionanthus foveolatus*), Forest Bushwillow (*Combretum kraussii*), and Common Wild Quince (*Cryptocarya transvaalensis*).

Af4G MOIST MONTANE FOREST

> **SIDEBAR 16   MOIST MONTANE FOREST SUBHABITATS**
>
> Moist Montane Forests are like isolated sky islands, and although the habitat looks and feels similar on most mountain blocks, there are minor differences and major floral and faunal divisions worthy of subhabitat status: Cameroon-Nigeria Highlands (Af4G-1); Gulf of Guinea Islands (Af4G-2); Angolan Scarp (Af4G-3); Albertine Rift (Af4G-4); Abyssinian Highlands (Af4G-5); North Tanzania–Kenya Volcanoes (Af4G-6); Eastern Arc (Af4G-7); Southeastern Highlands (Af4G-8); and South African Temperate (Af4G-9).

The Witels tree (*Platylophus trifoliatus*) is remarkable as the closest relative of a tree endemic to the temperate rainforest of Tasmania! In a similar vein, the Butterspoon Tree (*Cunonia capensis*) belongs to a genus that, aside from this one South African representative, is endemic to New Caledonia, in the South Pacific! The Assegai Tree (*Curtisia dentata*) is a typical montane forest tree from Mozambique south and remarkably makes up its own plant family.

CONSERVATION: The already scattered and isolated nature of this habitat makes it especially susceptible to human pressures. Foremost among these is the conversion of forest into farmland, as it grows in prime places for agriculture. Conflict between humans and animals can become a major issue where plantations and villages run right up to the edge of a national park. Another major threat, especially in areas with abundant natural grassland, is perennial human-started fires, which gradually shrink the forest, eventually leaving it confined to narrow gorges and valleys.

WILDLIFE: While montane forest is not as diverse as lowland forest or savanna, it does have fairly high diversity along with many localized endemics. It is a challenging habitat for mammal viewing—though not as tough as AFROTROPICAL LOWLAND RAINFOREST. The many localized endemic mammals of Africa's montane forest are mostly small and inconspicuous: mice, rats, and shrews. The patches of forest apparently weren't large or stable enough to allow for the evolution of many large endemic mammals. Some of the smaller mammals that specialize in this habitat are Eastern Tree Hyrax, L'Hoest's and Preuss's Monkeys, Carruther's Mountain Squirrel, the four subspecies of Tanganyika Mountain Squirrel, and Jackson's Mongoose. The Eastern Arc Mountains are home to some rare monkeys, of which the strangest is the Kipunji. Described only in 2005, this arboreal brown monkey has similarities to both baboons and colobuses. Most of the large mammals of montane forest are widespread species that use a broad range of habitats. These include Angola and Guereza Colobuses, African Savanna Elephant, Leopard, Lion, Slender Mongoose, African Buffalo, Blue Duiker, and Bushbuck. Quite a few species are shared with either the MONSOON FOREST, the Afrotropical Lowland Rainforest, or both. Examples of these are African Forest Elephant, Giant Forest Hog, and Harvey's and Weyns's

Remnant Moist Montane Forest in a tea estate in the East Usambaras, Tanzania. Conservation of crucial blocks of forest are in the hands of private commercial owners. © KEN BEHRENS, TROPICAL BIRDING TOURS

WARM HUMID BROADLEAF FORESTS

Tracking "Mountain" Eastern Gorillas in Volcanoes National Park, Rwanda. © KEN BEHRENS (RIGHT); © KEITH BARNES (LEFT), TROPICAL BIRDING TOURS

Duikers. Montane forest (especially its edges) is the most important component in the mix of habitats used by "Mountain" Eastern Gorilla.

For birders, montane forest is one of the continent's most exciting habitats due to its many localized endemics. For keen birders, while the sweeping savannas might inspire their first couple of trips to Africa, it is the forests that bring them back over and over. Some of the diverse and widespread groups of birds in montane forest include pigeons, turacos, greenbuls, robins, apalises, and sunbirds. Classic species of this habitat are Crowned Eagle, Rameron Pigeon, Bar-tailed Trogon, Silvery-cheeked and Black-and-white-casqued Hornbills, Western and Moustached Tinkerbirds, Gray Cuckooshrike (IS), White-starred Robin, Pink-footed Puffback, Sharpe's Starling, and Brown-capped Weaver. The entire small bird family Modulatricidae, comprising Spot-throat, Dapple-throat, and Gray-chested Babbler, is endemic to this habitat.

Endemism: The Albertine Rift holds the most diverse montane forest overall, and among the most interesting of its dozens of endemics are African Bay-Owl, Grauer's Broadbill, Yellow-crested Helmetshrike, Red-collared Mountain-Babbler, and Neumann's and Grauer's Warblers.

The next most endemic-rich forests are those in the ancient, non-volcanic Eastern Arc Mountains, which run from se. Kenya all the way through Tanzania into n. Malawi and n. Mozambique. The most remarkable bird here is the Udzungwa

The massive Crowned Eagle hunts monkeys and small antelope in Moist Montane Forest.
© KEN BEHRENS, TROPICAL BIRDING TOURS

Partridge, which is genetically similar to Asian partridges and was described only in the 1990s when discovered in a soup pot! Other Eastern Arc endemic birds include Usambara Eagle-Owl, Dapple-throat, Spot-throat, African and Long-billed Tailorbirds, and Uluguru Bushshrike.

The mountains of n. Tanzania and Kenya are much newer, mostly of volcanic origin, and have a few endemics plus a mix of the species of adjacent montane blocks. Such a mix of species is also found in the mountains of far s. Tanzania, Malawi, and Mozambique. The Ethiopian Highlands have many endemic species, though most key into that region's more common **AFROTROPICAL MONTANE DRY MIXED WOODLAND** rather than to Moist Montane Forest.

In the southern part of the continent, there are two further minor centers of avian endemism: the highlands of e. Zimbabwe, w. Mozambique, and s. Malawi and the scattered montane forests of South Africa. The Angolan escarpment has some forest endemics such as Swierstra's Francolin and Red-crested Turaco. West Africa has one major area of mountains, along the west side of Cameroon, as well as on the islands of Bioko and São Tomé. These hold the richest montane forests outside East Africa, with endemic birds that include Mount Cameroon Spurfowl, Bannerman's Turaco, White-throated Mountain-Babbler, Green-breasted and Mount Kupe Bushshrikes, Green Longtail, White-tailed Warbler, São Tomé Sunbird, and São Tomé Grosbeak. In far w. Africa, in the Upper Guinea region, are some low mountains with relict moist forests, but these mainly support species typical of Afrotropical Lowland Rainforest.

Montane forest holds many endemic reptiles. Some of the prominent groups include forest geckos (*Cnemaspis*), dwarf geckos (*Lygodactylus*), forest lizards (*Adolfus*), and bush vipers (*Atheris*). Around half of the world's chameleons are endemic to Madagascar, but most of the rest are found in African Moist Montane Forest. The Eastern Arc Mountains are the richest, with 10 endemic chameleons, ranging from the hefty Usambara Two-horned Chameleon (*Kinyongia fischeri*) to the diminutive Uluguru Pygmy Chameleon (*Rhampholeon uluguruensis*). Nearly all the major mountain regions have at least one endemic chameleon. The abundant moisture in montane forest makes this a good habitat for amphibians. The Cameroon mountains have about 40 endemics; the Albertine Rift, 32; and the Eastern Arc, 25. Some of the diverse groups include reed frogs (*Hyperolius*), long-fingered frogs (*Cardioglossa*), puddle frogs (*Phrynobatrachus*), screeching frogs (*Arthroleptis*), and tree frogs (*Leptopelis*).

DISTRIBUTION: "Sky islands" of Moist Montane Forest dot the east side of Africa from s. Ethiopia all the way to s. South Africa. Outliers exist in Cameroon, Upper Guinea, and Angola. The majority of Africa's mountains are associated with the Great Rift Valley, whose formation forced up blocks of mountains (sidebar 39, p. 349).

WHERE TO SEE: Nyungwe National Park, Rwanda; Ngorongoro Conservancy, Tanzania; Wilderness National Park, South Africa.

Tree ferns are highly characteristic of Moist Montane Forest. Nyungwe National Park, Rwanda. © KEN BEHRENS, TROPICAL BIRDING TOURS

Af4J INDIAN OCEAN LOWLAND RAINFOREST

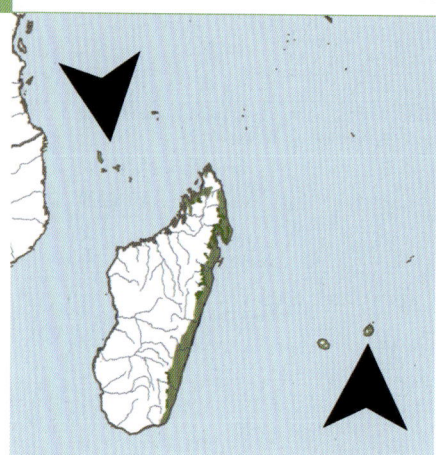

IN A NUTSHELL: Lush evergreen broadleaf forest with a complex structure and a tall canopy that remains moist and humid year-round. Found below 2600 ft. (800 m) on Madagascar, with small remnants in the Mascarene and Comoros archipelagoes. **Global Habitat Affinities:** ISLAND ARC LOWLAND RAINFOREST; AMAZON TERRAFIRMA; AUSTRALASIAN TROPICAL LOWLAND RAINFOREST; MALABAR LOWLAND RAINFOREST; MESOAMERICAN LOWLAND RAINFOREST. **Continental Habitat Affinities:** AFROTROPICAL LOWLAND RAINFOREST; SWAMP FOREST; SEYCHELLES GRANITE FOREST. **Species Overlap:** INDIAN OCEAN MONTANE RAINFOREST; MALAGASY DECIDUOUS FOREST; SPINY FOREST.

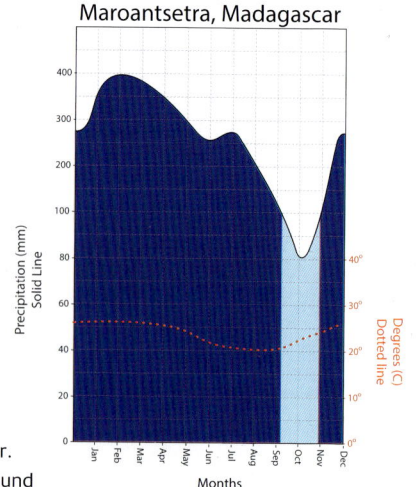

DESCRIPTION: The overall character of Indian Ocean Lowland Rainforest is superficially similar to that of other tropical rainforests, but a close look reveals a high diversity of plants, the vast majority of which are endemic, along with an incredibly rich set of endemic wildlife. Within the region, Madagascar's rainforests are by far the most diverse, while the other Indian Ocean islands have vastly reduced sets of plants and animals. Because of its rainforests, Madagascar has almost continental-level diversity in some groups and fully deserves the moniker "the eighth continent." Much of this account will address Malagasy rainforest as a whole, comprising both lowland and montane components. The subsequent account, covering montane rainforest, will detail how rainforest is different at higher elevations, especially on Madagascar.

Lowland rainforest is found from sea level up to around 2600 ft. (800 m), where it rapidly transitions to INDIAN OCEAN MONTANE RAINFOREST. While transitions between closely related habitats are often

Af4J INDIAN OCEAN LOWLAND RAINFOREST

gradual and difficult to pinpoint, this transition in the Malagasy region is often quite striking, though the exact elevation of the transition varies somewhat depending on slope, aspect, and latitude. For example, exposed ridgelines with poor soil have montane rainforest far below the typical cutoff elevation of 2600 ft.

This forest is tall and lush, up to 130 ft. (40 m), though more typically 80–100 ft. (25–30 m). Compared with AFROTROPICAL LOWLAND RAINFOREST, the canopy is slightly lower and lacks towering emergent trees. The driest lowland areas receive as little as 30 in. (800 mm) of rain annually, though 80 in. (2000 mm) is typical, and the wettest areas can receive well over 120 in. (3000 mm) of rain in a year (Köppen **Awa**). Some rain falls in every month of the year, but there is a cooler, drier season from April to October and a warmer, wetter season from November to March. This pattern is created by the Intertropical Convergence Zone, which moves south during the austral summer. Some drier semi-deciduous forest was formerly found in the lowlands on some of the islands, especially Mauritius and Rodrigues in the Mascarenes, but this forest has been virtually obliterated by humans. Some of this forest may have been palm dominated and must have been a fascinating habitat.

Madagascar's rainforest holds thousands of plants, some 80% of which are endemic. The other islands have less diverse plant assemblages, though endemism remains high. Unlike some other

Left: **Like other rainforests around the world, Indian Ocean rainforest has a complex, multitiered structure. Amber Mountain National Park, Madagascar.**
© KEN BEHRENS, TROPICAL BIRDING TOURS

Above: **The blue pigeons of the Indian Ocean islands, like this Madagascar Blue-Pigeon, are a biogeographic enigma; their closest living relative appears to be the Cloven-feathered Dove of distant New Caledonia!**
© KEN BEHRENS, TROPICAL BIRDING TOURS

rainforests around the world, the rainforest on Madagascar is never dominated by one species or a small set of species; it is always diverse and complex.

Some widespread tree genera on Madagascar are *Symphonia, Ficus, Canarium, Tambourissa, Albizia, Diospyros, Ocotea, Dracaena, Dombeya,* and *Olea*. Typical families include coffee (Rubiaceae), euphorbia (Euphorbiaceae), ebony (Ebenaceae), myrtle (Myrtaceae), and laurel (Lauraceae). Members of the legume family are far less common and important than in the African lowland rainforest. There are many endemic palms, and this group is both more diverse and more conspicuous than in African rainforest. Large lianas are surprisingly scarce, but pandanus is abundant, and epiphytes and tree ferns both occur, though at much lower densities than in montane rainforest.

Scrubby secondary forest, known locally as *savoka*, is common on Madagascar and is the default semi-natural habitat on all the other Indian Ocean islands. It typically contains widespread exotic plants like Strawberry Guava (*Psidium cattleyanum*), Common Lantana (*Lantana camara*), gingers (*Zingiber*), and brambles (*Rubus*). Although widespread on Madagascar, secondary forest is actually less common there than on the African mainland. This difference seems to be due to Madagascar's much poorer and less vigorous set of secondary species, which struggle to outcompete aggressive exotics. On Madagascar, vast areas of *savoka* are now dominated by the indigenous Traveler's Palm (*Ravenala madagascariensis*), or Ravenala (its Malagasy name). But even this secondary formation tends to give way after persistent burning and is converted into depauperate grassland. Ravenala is actually not a palm but rather a member of the bird-of-paradise family, with links to South Africa and ne. South America that date back to Gondwanaland. It has palm-like leaves in a graceful fan along a single plane. Ravenala is one of the most distinctive plants on Earth, in the same league as the Norfolk Pine, and has become a pantropical ornamental tree, though sadly its Malagasy origins are known by very few.

CONSERVATION: Eastern Madagascar has been subject to vast deforestation. Lower-elevation forests away from the west side of the Masoala Peninsula, and other parts of the northeast, have been decimated; most of the remaining rainforest is INDIAN OCEAN MONTANE RAINFOREST, higher up the slopes. This loss is of grave concern for the future integrity of Madagascar's biodiversity. There are national parks that protect large tracts of lowland rainforest, though their integrity is poorly enforced.

As concerning as the conservation situation is on Madagascar, it is much worse on the other Indian Ocean islands. Owing to hunting, the introduction of exotic mammals, and the destruction of forest, the Mascarene Islands have lost much of their wildlife, including the charismatic Dodo of Mauritius. What other

Streams within lowland rainforest are a rich microhabitat for frogs. Masoala National Park, Madagascar. © KEN BEHRENS, TROPICAL BIRDING TOURS

Af4J INDIAN OCEAN LOWLAND RAINFOREST

Rainforest bamboo is a key microhabitat for some wildlife, including bamboo lemurs. The main photo shows Northern Bamboo Lemur in Marojejy National Park, Madagascar, and the inset Gray Bamboo Lemur.
© KEN BEHRENS, TROPICAL BIRDING TOURS

place on Earth is more famous for an extinct animal it once hosted? There have been fewer extinctions on the Comoros, but many of the endemic species there hang on by a thread. Well-organized conservation efforts are afoot in the Mascarenes, but the same cannot be said for the poor and heavily populated Comoros.

WILDLIFE: Any discussion of Indian Ocean rainforest habitat will revolve around Madagascar, as it is far more diverse than the other islands and has been the main engine of evolution in the region. The other island groups have a much-reduced subset of Malagasy wildlife. Rainforest is by far Madagascar's richest habitat, and the majority of this diversity is shared by lowland and montane rainforest, though some differences in their wildlife assemblages will be highlighted.

The marquee mammals are the lemurs, a whole endemic radiation of the primate order. The largest is the Indri, which sings a shockingly loud song that is simultaneously reminiscent of both whales and wolves. There are also three beautiful species of lowland rainforest sifakas and Black-and-white Ruffed and Red Ruffed Lemurs. These big lemurs have a diverse supporting cast of smaller species: mouse lemurs, dwarf lemurs, bamboo lemurs, brown lemurs, and woolly lemurs. Like all Malagasy forest, lowland rainforest supports the rarely seen Aye-aye, a bizarre lemur with huge eyes, bushy fur, and an elongate, skeletal middle finger, which makes up its own family.

Much of Madagascar's mammal diversity is within the endemic tenrec family. Its members include the large Tailless or Common Tenrec; several hedgehog-like tenrecs; the stream-dwelling, otter-like Web-footed Tenrec; and a diverse range of smaller shrew tenrecs. Unfortunately, Tailless Tenrec has been introduced to most of the other Indian Ocean islands and has become an invasive species there. Madagascar has an endemic subfamily of rodents that includes the odd tree-dwelling Madagascar rats and a range of tuft-tailed rats. The Malagasy carnivorans are another endemic family, most of whose members are rainforest dwelling. The largest is the lemur-hunting Fossa, which looks like a long and lean arboreal Puma. Smaller and more mongoose-like carnivorans are

Black-and-white Ruffed Lemur is one of the biggest and most showy of a rich set of rainforest lemurs. © KEN BEHRENS, TROPICAL BIRDING TOURS

Aye-aye. So many things are bizarre about this lemur, including its skeletal middle finger that helps it to occupy the woodpecker niche *(inset)*. MAIN PHOTO © KEITH BARNES; INSET © KEN BEHRENS, TROPICAL BIRDING TOURS

the Fanaloka, Ring-tailed Vontsira, and Eastern Falanouc. Madagascar has a rich assemblage of bats. The other Indian Ocean islands are virtually devoid of indigenous mammals save for bats, the most notable of which are several large species of flying foxes.

Rainforest supports more Malagasy endemic birds than any other habitat on the island. These include the majority of the members of the island's endemic families. There are four spectacular ground-rollers, two of which are found exclusively in lowland rainforest: Scaly and Short-legged. Lowland rainforest also supports one of the island's three mesites, three of four asities, and many members of the endemic Malagasy warbler family. The enigmatic Dusky Tetraka seems to be entirely restricted to lowland rainforest. Other typical lowland rainforest birds include Madagascar Ibis; Madagascar Blue-Pigeon; Greater and Lesser Vasa Parrots; Blue, Red-breasted (IS), and Red-fronted Couas; Madagascar Spinetail; Madagascar Wood-Rail; Madagascar Pygmy-Kingfisher; Cuckoo-roller; Madagascar Paradise-Flycatcher; and Nelicourvi Weaver. Rainforest supports a bounty of vangas, which run the gamut from the small and chickadee-like Red-tailed Vanga to the hefty and huge-billed Helmet Vanga. This is the main habitat for two near-mythical raptors: Madagascar Serpent-Eagle and Red Owl.

Surprisingly, the avifauna throughout the rainforest remains largely uniform, although diversity peaks at around 2600 ft. (800 m), around the transition from lowland to montane rainforest. A handful of species, including Helmet and Bernier's Vangas, seem to be restricted to the northern half of Madagascar's lowland rainforest belt, though for unknown reasons.

Each of the other groups of Indian Ocean islands has its own assemblage of rainforest endemic birds. The Comoros have the richest set, more than 20 endemics, and the Mascarenes have around 20 endemics. Most of these are closely related sister species to Malagasy birds, though there are exceptions, such as Echo Parakeet on Mauritius,

Above: **Madagascar Ibis, a shy but spectacular rainforest denizen.** © ROB HUTCHINSON, BIRDTOUR ASIA

Right: **Lowland rainforest specialist ground-rollers: Short-legged** *(left)* **and Scaly** *(right)*. **These birds belong to a Malagasy endemic family, all of whose members are delightfully attractive.** © KEN BEHRENS, TROPICAL BIRDING TOURS

186   WARM HUMID BROADLEAF FORESTS

Above: **Helmet Vanga is seemingly restricted to the northern lowland rainforests of Madagascar, for unknown reasons.** © KEN BEHRENS, TROPICAL BIRDING TOURS

Below: **Collared Nightjar seems to be an ancient member of the nightjar family, a sort of "dinosaur nightjar"!** © KEN BEHRENS, TROPICAL BIRDING TOURS

Comoro Pigeon on the Comoros, and the odd Grand Comoro Flycatcher on Grande Comore. Sadly, the most charismatic and distinctive birds of the Mascarene Islands, the likes of Dodo, Rodrigues Solitaire, Mascarene Parrot, and Réunion Starling, are long extinct.

Madagascar is fabulously rich in reptiles, boasting over 400 species, compared with around 800 for the whole continent of Africa. More than half of these are endemic to the rainforest. Madagascar's most famous reptiles are chameleons, and around half of the world's chameleons are endemic to the island, mostly to the rainforest. These include the world's two largest species, Parson's (*Calumma parsonii*) and Oustalet's (*Furcifer oustaleti*) Chameleons; a range of midsize species; and the dwarf chameleons (*Brookesia* and

Palleon), some of which are among the world's smallest vertebrates. Another rich group in the rainforest is the colorful, diurnal day geckos (*Phelsuma*). The leaf-tailed geckos (*Uroplatus*) are wonderfully cryptic and include the huge Common Leaf-tailed Gecko (*U. fimbriatus*), the bamboo-mimicking Lined Leaf-tailed Gecko (*U. lineatus*), and the red-eyed Satanic Leaf-tailed Gecko (*U. phantasticus*). Other Malagasy rainforest lizards include plated lizards (*Zonosaurus*), Afro-Malagasy skinks (*Trachylepis*), and ground geckos (*Paroedura*). The diverse range of snakes includes the bizarre leaf-nosed snakes (*Langaha*) and a small group of biogeographically mysterious boas (*Sanzinia* and *Acrantophis*).

Rainforest on the other islands is far poorer in reptiles, which number no more than a few dozen species on each archipelago, mostly members of groups that are far more diverse in Madagascar, such as day geckos and skinks.

Starry-night Reed Frog (*Heterixalus alboguttatus*) is one of the showiest of hundreds of species of frogs in Malagasy lowland rainforest. © KEN BEHRENS, TROPICAL BIRDING TOURS

Over 300 frogs have been described in Madagascar, and DNA barcoding suggests there may be around 500 species. The vast majority of these are endemic to the rainforest. Just a few of the noteworthy lowland rainforest frogs are the Starry-night Reed Frog (*Heterixalus alboguttatus*), Marbled Rain Frog (*Scaphiophryne marmorata*), Madagascar fringed frogs (*Spinomantis*), a large array of bright-eyed frogs (*Boophis*), and the mantellas (*Mantella*), which resemble the poison dart frogs of the Neotropics. Along with the smallest dwarf chameleons, the tiny stump-toed frogs (*Stumpffia*) count among the world's smallest vertebrates. Diverse genera in Malagasy lowland rainforest include pointy frogs (*Blommersia*), bridge frogs (*Gephyromantis*), pandanus frogs (*Guibemantis*), and Madagascar frogs (*Mantidactylus*). The other Indian Ocean islands are very poor in amphibians.

Madagascar is also rich in fascinating invertebrates, including a diverse array of butterflies, the huge Comet Moth, and the bizarre Giraffe-necked Weevil.

Endemism: On Madagascar, there are many mammals, reptiles, and amphibians that are restricted to a small portion of the rainforest. Most birds are found throughout the Madagascar rainforest, both lowland and montane, though a few species are restricted to the northern half of the rainforest belt. The flora and fauna of the Comoros and Mascarenes are quite distinct, with many endemics on each island group. These islands do not show a strong difference between lowland and montane avifaunas.

Tomato Frog (*Dyscophus antongilii*), a big juicy amphibian of lowland rainforest and adjacent habitats. © KEN BEHRENS, TROPICAL BIRDING TOURS

DISTRIBUTION: Lowland rainforest formerly extended contiguously along the length of e. Madagascar, from the lower portions of Tsaratanana in the north to near Fort Dauphin in the south, and also on Amber Mountain in the far north. The vast majority of the lowland forest has been destroyed, and the only large tracts remaining are in the northeast, including Masoala National Park.

Lowland rainforest formerly blanketed much of the other Indian Ocean islands but is now restricted to small, inaccessible, or formally protected areas on the Comoros and Mauritius. Even Réunion, which has abundant forest in its rugged highlands, has lost nearly all its lowland forest.

WHERE TO SEE: Masoala National Park, Madagascar; Black River Gorges National Park, Mauritius.

SIDEBAR 17 — HABITAT HELPERS—DISPERSERS

Bruce's Green-pigeon feasts on fruit of a fig in Lalibela, Ethiopia. © KEITH BARNES, TROPICAL BIRDING TOURS

One of the epauletted fruit bats—they are nearly impossible to tell apart—a key fruit disperser in the African savannas.
© KEITH BARNES, TROPICAL BIRDING TOURS

Plants and wildlife coexist in a complex dynamic in which one is constantly affecting the other. Many plants produce edible fruits that contain seeds. Animals eat the fruit and move away before defecating, acting as seed dispersal agents while simultaneously providing the seeds with a nice little pile of organic matter to help the saplings germinate! A fruiting fig tree anywhere in Africa is worth watching. The odd cackles of fruit-eating green-pigeons are a sure sign that there is some action, and parrots and glossy-starlings are likely to join in the melee. Overripe fruit attract fruit flies, which bring in paradise-flycatchers. *Chlorocebus* (Vervet) monkeys, baboons, and squirrels will all enjoy the bounty of fruit while even Bushbuck and Nyala antelope will feed on fallen fruits on the ground. In closed forests, small antelopes like duikers are crucial for dispersing larger fruits, while small fruits are distributed by the greenbuls—a massive guild of fruit-eating birds. After dark, palm-civets and other small mammals take over. But above all groups, the least frequently encountered yet perhaps most crucial are the fruit-eating bats. In one species, the Straw-colored Fruit Bat, a single large colony was calculated to disperse 300,000 small seeds per night, enough to kickstart the regrowth of 2000 acres (800 ha) of forest! The potential for these animals to help us reforest areas is crucial, and their conservation as keystone species (a species that crucially supports an ecosystem) is essential. Wildlife also acts as pollinators of flowers (see sidebar 34, p. 306).

Af4K INDIAN OCEAN MONTANE RAINFOREST

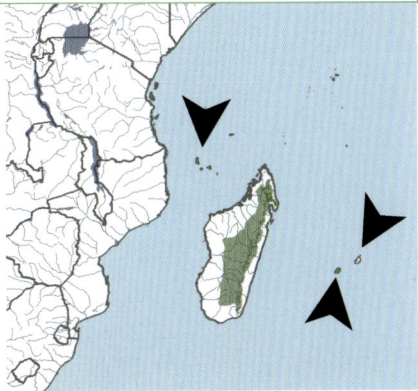

IN A NUTSHELL: Lush evergreen or semi-evergreen broadleaf forest with many epiphytes, tree ferns, and mosses and a low to moderate canopy, that remains moist and humid throughout the year. Found above 2600 ft. (800 m) on Madagascar and locally in the Mascarene and Comoros archipelagoes. **Global Habitat Affinities:** PHILIPPINE MONTANE RAINFOREST; AUSTRALASIAN TROPICAL MONTANE RAINFOREST; ANDEAN CLOUDFOREST; MESOAMERICAN CLOUDFOREST; SUNDA MONTANE RAINFOREST. **Continental Habitat Affinities:** MOIST MONTANE FOREST; SEYCHELLES GRANITE FOREST. **Species Overlap:** INDIAN OCEAN LOWLAND RAINFOREST; MALAGASY DECIDUOUS FOREST; SPINY FOREST.

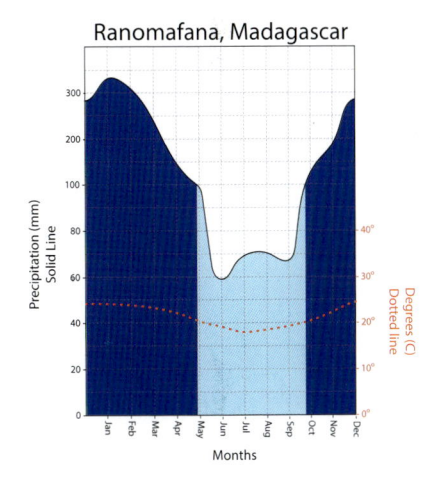

DESCRIPTION: Indian Ocean Montane Rainforest shares its general attributes, and most of its plants and wildlife, with INDIAN OCEAN LOWLAND RAINFOREST (previous account). In this respect, Madagascar shows less of a continental pattern of speciation and more of the uniform flora and fauna typical of islands. Nonetheless, there is a significant and perceptible divide between lowland and montane forests, occurring typically around 2600 ft. (800 m). Most general aspects of Malagasy rainforest were addressed in the previous account; this account will focus on the distinctive attributes of higher-elevation humid forest.

Montane rainforest is found from around 2600 ft. (800 m) up to tree line, which is 6500–9000 ft. (2000–2750 m) on the Indian Ocean islands. Compared with lowland rainforest, this habitat has a lower canopy. The lower-lying and lushest version of montane forest can be as tall as 65–80 ft. (20–25 m), while 30–40 ft. (10–12 m) is more typical at higher elevations; extremely stunted versions can be as short as 10 ft. (3 m). Montane rainforest tends to have trees that branch quickly after emerging from the ground, as opposed to the long, straight bare trunks (boles) that characterize canopy trees in the lowland rainforest. Higher-elevation forest has the characteristics of montane or cloudforest all around the world: trees covered with abundant tree ferns and epiphytes; tree trunks and branches enveloped in moss and lichens; and the ground thickly blanketed at the highest elevations.

Above: **Marojejy National Park, Madagascar.**
© KEN BEHRENS, TROPICAL BIRDING TOURS

Left: **Indian Ocean Montane Forest is misty and moss-draped, classic "cloudforest." Mount Karthala, Grande Comore.**
© KEN BEHRENS, TROPICAL BIRDING TOURS

Mistletoe cactus (*Rhipsalis baccifera*) is a common epiphyte, especially at lower elevations.

Though this habitat is generally drier and more water-stressed than lowland rainforest, it also shows a higher internal climatic variation, ranging from quite seasonal and dry to some of the wettest forest in the region, which can receive up to 400 in. (10,000 mm) of rain in a year (Köppen **Amb**). Some rain falls in every month of the year, but there is a pronounced cool, drier season from April to October and a warmer, wetter season from November to March.

The majority of the plants of montane rainforest are shared with lowland rainforest. Some characteristic trees of higher-elevation forest are species of *Schefflera* and *Weinmannia*. Two genera of the Monimiaceae family are endemic to montane rainforest. Lianas are abundant in the midstory, as are herbaceous

Right: **Rivers and streams are frequent and a crucial microhabitat for many frog species. Ranomafana National Park, Madagascar.** © KEN BEHRENS, TROPICAL BIRDING TOURS

plants in the understory. Bamboo is a common element, especially small bamboo species that grow low to the ground. Palms and pandanus are frequent, though less common than in lowland rainforest. The cooler climates of this habitat support a set of plants that are more typical of temperate regions. These include bugles (*Ajuga*), *Sanicula*, lady's mantles (*Alchemilla*), buttercups (*Ranunculus*), *Vaccinium* (the genus to which blueberry and cranberry belong), and violets (*Viola*). Orchids occur at high density and diversity. Montane forest is the habitat of *Ascarinopsis coursii*, one of Madagascar's many biogeographic enigmas. This woody plant belongs to Chloranthaceae, a family that is mostly found in Asia, the Pacific, and the Neotropics, but which is absent from continental Africa. On Mt. Karthala, on Grand Comoro, prominent trees include olives (*Olea*), *Ocotea comoriensis*, African Cherry (*Prunus africana*), *Gambeya boiviniana*, and *Khaya comorensis*.

As in African Moist Montane Forest, tree ferns are common. Analamazaotra Reserve, Madagascar. © KEN BEHRENS, TROPICAL BIRDING TOURS

CONSERVATION: Madagascar's montane rainforests have fared far better than their lowland counterpart but have still been subject to rampant deforestation. Until recently there was a cohesive belt of montane rainforest running along the escarpment, but a look at satellite imagery shows that the continuity of this rainforest belt is being lost: in several places, there is no longer any intact rainforest remaining between the top and bottom of the east slope. The montane forest on Grande Comore and Réunion is still extensive, but only small patches remain on the other Comoros and Mauritius.

WILDLIFE: The community of larger mammals is not drastically different between lowland and montane rainforest. However, the best places to see most rainforest lemur species are now in montane rainforest because of the destruction of most lowland forest. Montane rainforest is the stronghold of the Indri, the largest lemur, which gives a shockingly loud song that is simultaneously reminiscent of both whales and wolves.

Montane rainforest in Madagascar is slightly more distinctive in terms of birds, though again the majority of species is shared with lowland rainforest. There are two members of the ground-roller family that are found in this habitat: Rufous-headed and Pitta-like Ground-Rollers. Yellow-bellied Sunbird-Asity is a specialist of stunted ridgeline montane rainforest, where it replaces the common and widespread Common Sunbird-Asity. Cryptic Warbler is also restricted to this ridgeline habitat, where it stridently sings from the canopy. Brown Emutail (IS) and Yellow-browed Oxylabes are two incredibly skulking understory birds that are found exclusively in the mossy understory of montane rainforest. Other typical montane rainforest birds include Madagascar Flufftail, Madagascar Blue-Pigeon (IS), Greater and Lesser Vasa Parrots, Blue and Red-fronted Couas, Cuckoo-roller, Madagascar White-eye, a wide range of vangas, and Forest Fody. Some of the last strongholds of Madagascar Serpent-Eagle and Red Owl are montane rainforests.

Most of the endemic birds of the other Indian Ocean islands are found throughout those islands' rainforests, but a few species are restricted to montane rainforest. These include Comoro Scops-Owl and Grand Comoro Flycatcher on the island of Grand Comoro and Réunion Cuckooshrike and Réunion White-eye on Réunion.

Malagasy montane rainforest is as diverse for amphibians, if not slightly more diverse than lowland rainforest. Common, widespread, and frequently seen species include Madagascar Jumping Frog (*Aglyptodactylus madagascariensis*) and Madagascar Bright-eyed Frog (*Boophis madagascariensis*).

Af4K INDIAN OCEAN MONTANE RAINFOREST

Indri, which has a black morph *(inset)*, makes one of the world's most remarkable natural sounds, reminiscent of both whales and wolves. © KEN BEHRENS, TROPICAL BIRDING TOURS

| SIDEBAR 18 | SAMBIRANO RAINFOREST |

Most of Madagascar's rainforest is in a narrow belt, but the biome broadens in the north of the island. This westward extension of the rainforest zone, known as the Sambirano rainforest, is produced by the moist winds of the Intertropical Convergence Zone, which lack a significant mountain obstacle north of the Tsaratanana Massif and wrap around that massif during the wet season, bringing moisture all the way to the west coast and the interior. This complex Sambirano area is rich in localized reptiles and amphibians and has a fascinating mixture of eastern rainforest and western dry forest bird species.

Pitta-like Ground-Roller is a classic species of Malagasy montane forest. It is highly vocal and relatively conspicuous in the austral spring, then goes incognito for the rest of the year.
© KEN BEHRENS, TROPICAL BIRDING TOURS

Af4K INDIAN OCEAN MONTANE RAINFOREST

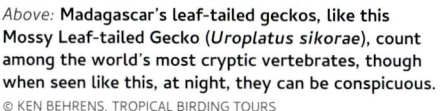

Above: **Madagascar's leaf-tailed geckos, like this Mossy Leaf-tailed Gecko (*Uroplatus sikorae*), count among the world's most cryptic vertebrates, though when seen like this, at night, they can be conspicuous.**
© KEN BEHRENS, TROPICAL BIRDING TOURS

Top right: **Elongate Leaf Chameleon (*Palleon nasus*) belongs to the smallest of Madagascar's four chameleon genera.** © KEN BEHRENS, TROPICAL BIRDING TOURS

Middle right: **Male Giraffe-necked Weevils (*Trachelophorus giraffa*) use their long neck for fighting, while females use their shorter necks to roll leaves into rolls in which eggs are laid.**
© KEN BEHRENS, TROPICAL BIRDING TOURS

Below right: **Spectacular Baron's Mantella (*Mantella baroni*) is the Malagasy equivalent of a poison dart frog. They secrete a toxic alkaloid that they get from eating mites and warn off potential predators using their technicolor patterns!** © KEN BEHRENS, TROPICAL BIRDING TOURS

Diverse genera include *Cophyla*, *Platypelis*, bright-eyed frogs (*Boophis*), bridge frogs (*Gephyromantis*), Madagascar frogs (*Mantidactylus*), and *Spinomantis*. A few species, such as *Platypelis olgae* and *Rhombophryne tany* are restricted to montane forest and other habitats at the highest elevations of the Tsaratanana Massif, at 6500 ft. (2000 m) or higher.

Ring-tailed Vontsira, a member of the endemic Malagasy carnivorans family. © KEN BEHRENS, TROPICAL BIRDING TOURS

Endemism: The most significant endemic areas for reptiles and amphibians are the n. Tsaratanana Massif and the Sambirano rainforest (see sidebar 18, p. 193).

DISTRIBUTION: Montane rainforest is found along most of the length of the eastern Madagascar escarpment, from Tsaratanana in the north to near Fort Dauphin in the south and in isolated fragments in the central highlands. Though generally restricted to a narrow zone, the montane rainforest belt broadens in the north, nearly reaching the west coast near the island of Nosy Be. Extensive tracts of montane rainforest remain on Grande Comore and Réunion, which has managed to hold on to around 40% of its forest cover. The very highest portions of the other Comorian islands and Mauritius have forest remnants with the character of montane rainforest. At the highest elevations in Madagascar, montane rainforest gives way to **MONTANE HEATH**, and even before the forest gives out, heath species and trees with sclerophyllous leaves start to be an important component of these high forests.

WHERE TO SEE: Andasibe-Mantadia National Park, Madagascar; Ranomafana National Park, Madagascar; Réserve Naturelle de la Roche Écrite, Réunion.

Fascinating high-elevation rainforest on Réunion that is dominated by tree ferns. Piton de la Fournaise, Réunion. © KEN BEHRENS, TROPICAL BIRDING TOURS

Af4L SEYCHELLES GRANITE FOREST

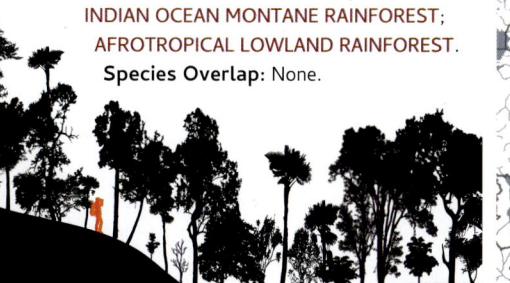

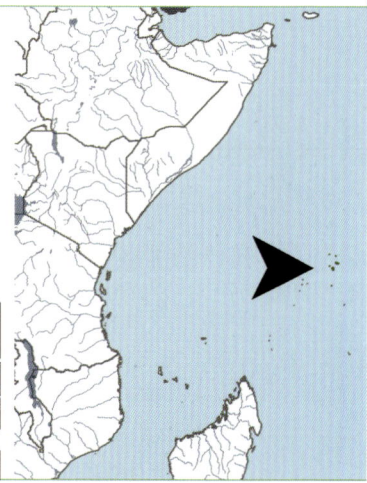

IN A NUTSHELL: Moist broadleaf evergreen rainforest found on a granitic substrate on the Inner Islands of the Seychelles archipelago. **Global Habitat Affinities:** INDO-MALAYAN LIMESTONE FOREST. **Continental Habitat Affinities:** INDIAN OCEAN LOWLAND RAINFOREST; INDIAN OCEAN MONTANE RAINFOREST; AFROTROPICAL LOWLAND RAINFOREST. **Species Overlap:** None.

DESCRIPTION: The Seychelles is a far-flung archipelago of over 100 islands in the western Indian Ocean. The main population centers and the largest islands are found in the Central Seychelles (or the "Inner Islands"), including Mahé, Silhouette, Praslin, and La Digue. While most such remote islands on Earth are volcanic or coralline, the Central Seychelles are remarkable for their continental origin. Essentially, these islands are tiny bits of ancient shrapnel from the breakup of the vast southern supercontinent of Gondwana, which rifted apart from Madagascar and India around 75 MYA. Most of the Central Seychelles are almost purely composed of ancient granite rock, which is around 750 million years old. These islands' isolation, in combination with this rocky substrate, have produced a highly distinctive forest type.

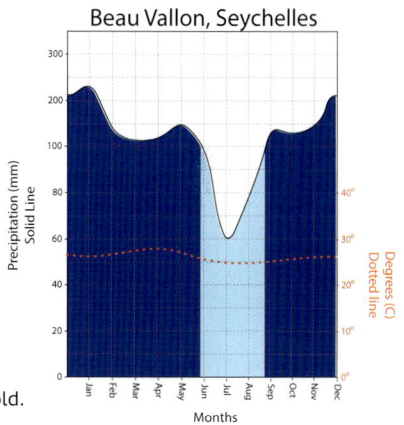

The Central Seychelles are extremely moist (Köppen **Ama**), receiving 90–200 in. (2300–5000 mm) of rain annually. The climate remains humid year-round, but there are two seasons: the monsoon season of heavy rains, from November to April, and a slightly drier and cooler season from May to October, when the trade winds blow strongly.

The rainforest of the Central Seychelles is fundamentally quite similar to other rainforests around the world, with complete canopy coverage and a complex multilayered structure. The substrate is uniformly rocky and almost always includes striking granite boulders and expansive areas of open rock on the steeper slopes. These islands have been heavily affected by human activities, including logging, clearing for agriculture, and the introduction of exotic species. Even the better-preserved tracts of granite forest are likely quite different from what they were before the arrival of people. Nonetheless, several distinct subtypes of forest can still be observed. In general, the canopy gets shorter as the elevation increases. Lowland forest has a canopy around 100 ft. (30 m),

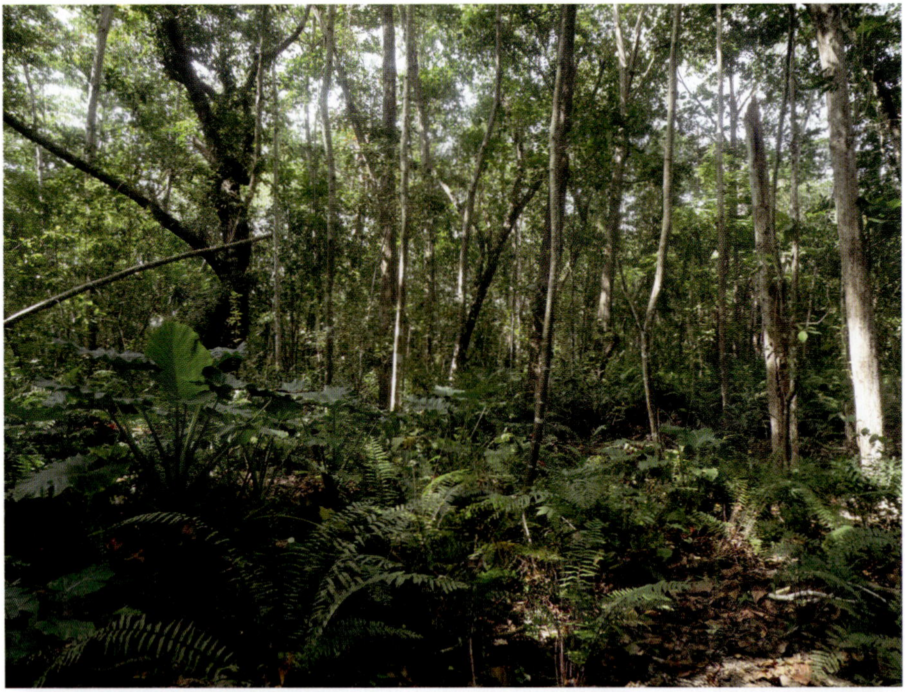

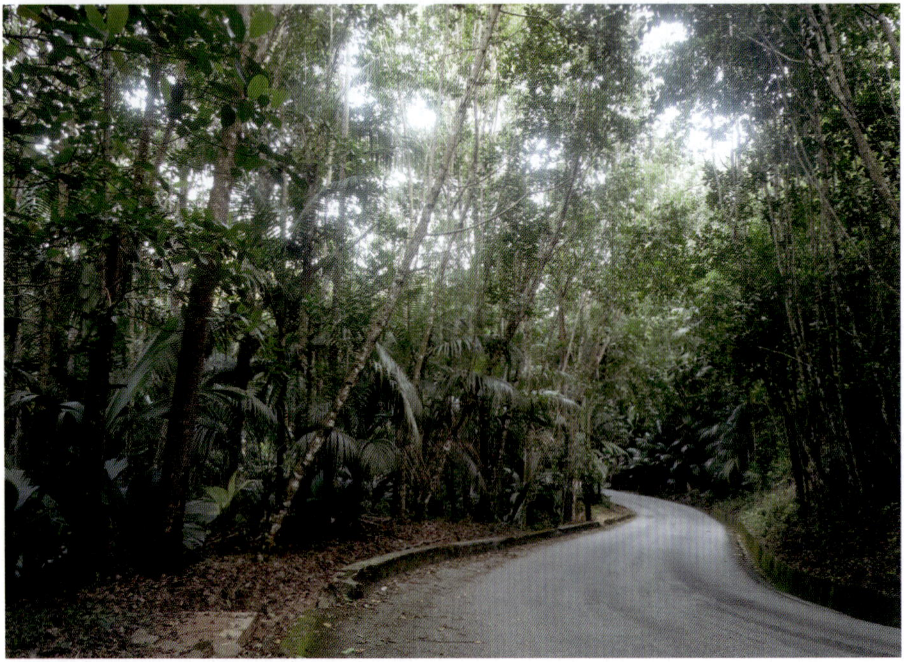

Af4L SEYCHELLES GRANITE FOREST

OPPOSITE PAGE:
Top: **Flat lowland granite forest dominated by Fish Poison Tree (*Barringtonia asiatica*).** © KEN BEHRENS, TROPICAL BIRDING TOURS

Bottom: **Palm forest on Praslin, Seychelles. This is the natural habitat of the extraordinary Sea Coconut (*Lodoicea maldivica*), which boasts the largest seed in the plant kingdom.** © KEN BEHRENS, TROPICAL BIRDING TOURS

THIS PAGE:
This is why it is called "Granite Forest"! Mahé, Seychelles.
© KEN BEHRENS, TROPICAL BIRDING TOURS

but the canopy decreases to around 40 ft. (12 m) on the highest mountains of Mahé, above 1800 ft. (550 m). The highest-elevation forest has abundant moss, much like other montane rainforest around the world, including INDIAN OCEAN MONTANE RAINFOREST. The most remarkable subtype of granite forest is a slightly drier lowland forest in which palms are dominant. This habitat is the refuge of the spectacular endemic Sea Coconut (*Lodoicea maldivica*), a towering tree that boasts the largest seed in the plant kingdom. Palms are common throughout the granite forest, and this habitat holds several monotypic endemic palm genera. Some of the most common trees, especially in the lowlands along the coast, have been introduced by humans. These include Coconut Palm (*Cocos nucifera*), Tamanu (*Calophyllum inophyllum*), Sea Trumpet (*Cordia subcordata*), Noni (*Morinda citrifolia*), and Moluccan Albizia (*Falcataria falcata*). The albizia is especially widespread and prominent as an emergent from the forest canopy. It has a tall, spreading structure with an umbrella canopy reminiscent of a savanna acacia. Common trees that are likely or definitely indigenous include *Northia seychellana*, Pacific Teak (*Intsia bijuga*), *Dillenia ferruginea*, and *Mimusops sechellarum*. The most remarkable plant of the Seychelles is the Jellyfish Tree (*Medusagyne oppositifolia*), which is named for the jellyfish-like appearance of its dehisced fruit (fruit that have opened to release seeds). It is sometimes placed in its own plant family but is now usually included within the Ochnaceae family. This ancient tree is incredibly rare—there are less than 30 plants on three Mahé mountaintops. Despite the sodden climate, the Jellyfish Tree shows adaptations to dry climates, suggesting that it originates all the way back to before the breakup of

WARM HUMID BROADLEAF FORESTS

Introduced Moluccan Albizia (*Falcataria falcata*) is prominent in the canopy. © KEN BEHRENS, TROPICAL BIRDING TOURS

Gondwanaland. Young trees have not been observed in the wild; this may be a species that was outcompeted by more modern plants and is on the way to natural extinction.

CONSERVATION: As on most island groups, humans have had a cataclysmic effect on the Seychelles. Most of the natural habitat has been logged or cleared for agriculture, and introduced plants and animals, including cats, dogs, goats, Common Myna, and Tailless Tenrec from Madagascar, run amok in the remaining natural areas. At least two bird species went extinct (Seychelles Parakeet and Marianne White-eye), several other species came close to extinction, and Aldabra Giant Tortoises of the locally endemic subspecies were wiped out. Fortunately, though, the Seychelles avoided the mass extinctions that occurred on islands like Hawaii and Mauritius. And further good news is that the Seychelles have been serious about conservation in recent decades. More than one-third of the islands' territory is now set aside for conservation. There have been aggressive habitat restoration and exotic species elimination programs on several offshore islands, which has allowed the rebound of the populations of several rare bird species.

WILDLIFE: As would be expected on such remote islands, the granite forest of the Seychelles is very low in diversity but high in endemism. The only indigenous mammals are bats, including a couple of widespread species and a couple of endemics. The impressively large Seychelles Flying Fox resembles a pterodactyl in flight! Despite the name, this species is shared with the Comoros and Tanzania's Mafia Island. The chunky and spiky-haired Tailless Tenrec is now more easily seen in the Seychelles than in Madagascar, from where it originates.

The most common and conspicuous birds are introduced species: Common Myna and Red Fody, and these

The flying foxes, including the Seychelles Flying Fox, are a classic group of the Indian Ocean islands.
© KEN BEHRENS, TROPICAL BIRDING TOURS

penetrate even into closed-canopy granite forest. The introduced Madagascar subspecies of Malagasy Turtle-Dove has largely outcompeted and replaced the endemic subspecies, which may actually be a full species. A few species of endemic birds are widespread in granite forests on the larger islands: Seychelles Blue-Pigeon, Seychelles Swiftlet, Seychelles Kestrel, Seychelles Bulbul, and Seychelles Sunbird. Other species are restricted to only one of the main islands. Mahé holds the Critically Endangered Seychelles Scops-Owl and Vulnerable Seychelles White-eye (which is also on the smaller adjacent island of Conception). Praslin has Seychelles Black Parrot, another odd "dinosaur parrot" alongside the vasa parrots of Madagascar and Comoros. La Digue holds the gorgeous Seychelles Paradise-Flycatcher, in which males are all black with purplish gloss and females are rufous-and-white with a black head. Three bird species flirted with extinction but have been saved by intense conservation efforts, especially the elimination of invasive species, on several offshore islands. These are Seychelles Warbler, Seychelles Magpie-Robin, and Seychelles Fody. Among these, Seychelles Magpie-Robin came the closest to extinction, with a population of only around a dozen individuals in the 1960s. All the Seychelles Granite Forest endemic birds were formerly more widespread; the island patchwork distribution of endemic species described above is mainly a function of human pressure.

Arguably the most spectacular creature of the Seychelles is a reptile, the Aldabra Giant Tortoise (*Aldabrachelys gigantea*). This is the only other place in the world, besides the Galápagos, where you

Aldabra Giant Tortoise (*Aldabrachelys gigantea***) on the small offshore island of Cousin, Seychelles. The Seychelles are one of the last places on Earth where you can see a member of the giant tortoises, formerly a widespread group.** © KEN BEHRENS, TROPICAL BIRDING TOURS

One of several species endemic to granite forest, the Seychelles Blue-Pigeon. © KEN BEHRENS, TROPICAL BIRDING TOURS

can see one of the formerly widespread giant tortoises. The average weight of a male tortoise is 550 lb. (250 kg)! This tortoise was extirpated from the central granite islands, but thankfully a large population (now estimated at 100,000!) remained on the corraline island of Aldabra, providing a source for reintroduction to the Central Seychelles. Overall, the granite forests are poor in herps. There are six endemic caecillians, 20 lizards, and three snakes—two endemic and one introduced. Remarkably, despite the formidable barrier of surrounding saltwater, there are six species of frog—five endemic and one introduced. The small Seychelles Tiger Chameleon (*Archaius tigris*) is an endemic beauty with a greenish base color, usually overlaid with black spotting. It belongs to a monotypic genus and is quite distinct from the chameleons of Madagascar. The Seychelles hold a rich suite of terrestrial crabs and hermit crabs. Among these is the widespread but remarkable Coconut Crab (*Birgus latro*), the largest terrestrial invertebrate on Earth.

Endemism: Most indigenous vertebrates are endemic. Of around 200 indigenous flowering plants, only one-third are endemic, a surprisingly low proportion.

DISTRIBUTION: Granitic Central Seychelles islands, approximately 700 mi. (1000 km) northeast of Madagascar, in the w. Indian Ocean.

WHERE TO SEE: Vallée de Mai Nature Reserve, Praslin (palm forest); Morne Seychellois National Park, Mahé (middle and high elevation forest); Aride Island (coastal/small island forest).

Seychelles Paradise-Flycatcher, in which males are completely glossy black, is arguably the most spectacular of the Seychelles endemic birds.
© KEN BEHRENS, TROPICAL BIRDING TOURS

Af4M MAVUNDA

IN A NUTSHELL: Open, dry broadleaf evergreen forest, with a dense impenetrable understory, that grows on soft Kalahari sand. **Global Habitat Affinities:** AUSTRALIAN DRY VINEFOREST; MALABAR SEMI-EVERGREEN FOREST; CERRADÃO. **Continental Habitat Affinities:** MONSOON FOREST; EAST COAST FOREST MATRIX; SOUTH COAST FOREST MATRIX. **Species Overlap:** MONSOON FOREST; MIOMBO; ANGOLAN DECIDUOUS FOREST; AFROTROPICAL LOWLAND RAINFOREST.

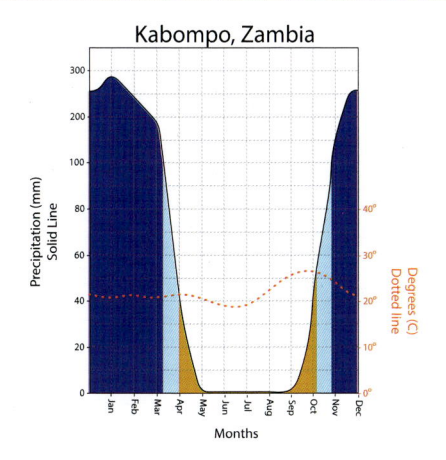

DESCRIPTION: Like the geographically adjacent MONSOON FOREST, Mavunda is somewhat transitional between MIOMBO savanna and AFROTROPICAL LOWLAND RAINFOREST. Mavunda is tall, open evergreen forest with an impassable understory of thicket and lianas. The name seems to be derived from a local word for "problem," perhaps referring to the difficulty of accessing it. It occurs patchily, often surrounded by a sea of Miombo or AFROTROPICAL GRASSLAND, and structurally most resembles dry deciduous and monsoon forests. It differs in having a much higher proportion of the canopy trees evergreen and an understory that is always impenetrable. Mavunda occurs almost exclusively on infertile Kalahari sands in flat areas at middle elevations of around 3600 ft. (1100 m). The upper sandy layers drain well and fast, lending it that dry forest feel; however, aquifers may collect underground, and Mavunda is often adjacent to grassy plains that are seasonally saturated. Monsoon Forest, on the other hand, grows almost exclusively along watercourses. This habitat grows in a moderate tropical climate (Köppen **Ama**), with temperatures fluctuating from 44 to 86°F (7–30°C). Rainfall ranges from 32 to 47 in. (800–1200 mm) in three distinct seasons: a hot dry season (Aug–Oct), a hot wet season (Nov–Apr), and a cool dry season (May–Jul). Despite considerable rain, the nature of the fast-draining soil means surface water is rare, and even though the seasons suggest a deciduous nature, most trees do not lose their leaves. Natural fires do not tend to penetrate this forest, but human clearing followed by deliberate burning has degraded many Mavunda forest patches into Miombo-like woodlands, which are then known as "Chipya."

Mavunda is a three-story forest with a 50–80 ft. (15–24 m) evergreen or semi-evergreen canopy, usually dominated by *Cryptosepalum exfoliatum* trees (called Mukwe locally). It also frequently contains species of *Brachystegia*, which are more typical of Miombo, Large False Mopane (*Guibourtia coleosperma*) with its stunning rosewood timber, and *Marquesia*. A discontinuous evergreen midcanopy, 10–50 ft. (3–15 m) tall, is made up of the same tree species. The understory, 10–13 ft. (3–4 m) tall, is a dense tangle of evergreen shrubs, scramblers, and climbers. Shrubs and small trees include Kalahari White Bauhinia (*Bauhinia petersiana*), Stamvrug (*Englerophytum magalismontanum*), bushwillows (*Combretum*), and *Canthium*. The ground is mostly covered with leaf litter, but in gaps there can be mosses and typical "rainforest" grasses. In places, a stunted version of Mavunda occurs on gneiss-derived soils. There are a few endemic plant species, though most are shared with adjacent habitats. In the south, Mavunda transforms into GUSU where rainfall drops below 24 in. (600 mm), and elsewhere it is replaced by Monsoon Forest or Miombo where soils change from well-drained Kalahari sands.

CONSERVATION: Mavunda is naturally localized due to its very specific soil and climatic constraints, and outside West Lunga National Park this habitat is disappearing rapidly for slash-and-burn agriculture. Unfortunately, owing to the nutrients being stored in the trees, once the habitat is burned and used for crops for 1 or 2 years, the land becomes an ecological wasteland, and the forest does not return. This habitat is in slightly better shape in remote e. Angola, but even there degradation is starting. Logging of Large False Mopane is intense throughout. Although a significant amount of habitat is protected in West Lunga National Park and adjoining conservation areas, this is a drier sector and not representative of all types of Mavunda. Outside the national park, degradation, cultivation, and deliberate fires lead to Mavunda transforming into Chipya, a

Well-preserved Mavunda in West Lunga National Park, Zambia. © FRANK WILLEMS

MIOMBO-like woodland. In places, firewood and timber are extracted and poaching is common. Unfortunately, much of the wildlife in Mavunda has been hunted unsustainably and many large (and even small) mammals that it used to support are now gone.

WILDLIFE: Species richness of mammals is low with no known endemics to the habitat, which shares most of its fauna with AFROTROPICAL LOWLAND RAINFOREST, MONSOON FOREST, or MIOMBO. The only near-endemic mammal is Rosevear's Striped Grass Mouse, known from two localities in Zambia. The forest does provide habitat for three special and secretive forest animals: Yellow-backed and Blue Duikers and White-bellied Tree Pangolin. All are seldom seen, and the pangolin is

Right: **Mavunda Canopy. Zambia.**
© FRANK WILLEMS

Below: **Succulent epiphytes and a termite mound in Mavunda. This is the main habitat of Margaret's Batis** *(inset)*. **Zambia.** © FRANK WILLEMS

heavily trafficked for its scales for the East Asian market. Bushpig, South African Porcupine, and Blue Monkeys also use Mavunda. In places like West Lunga National Park, animals such as African Buffalo, Sitatunga, and African Savanna Elephant use this habitat extensively.

One of the most famous birds of this habitat is likely more myth than reality. The White-chested Tinkerbird, collected in 1964 in Mavunda, has never been seen again. It is likely an aberrant Yellow-rumped Tinkerbird or hybrid, and many authorities disregard its specific validity. The near-endemic is Margaret's Batis, a canopy sprite that follows flocks. The other birdlife is a combination of forest species (often in the understory) and woodland species (mostly in the canopy). The understory is alive with the ringing calls of the local "Perrin's" form of Four-colored Bushshrike. African Broadbill is common in the dense tangles, while Western Crested Guineafowl charge around with their punk-like hairstyles. The understory is enlivened by vibrant species like Red-capped Robin-Chat, Blue-mantled Flycatcher, and Tropical Boubou, while roads with grassy edges attract shy Red-throated Twinspot and Black-tailed Waxbill. Cabanis's and Yellow-bellied Greenbul betray their presence with their nagging chattering. In the canopy, Many-colored Bushshrike and Purple-throated Cuckooshrike join small flocks, often led by Square-tailed Drongo. These flocks are almost always joined by species more typical

Above: **Fire degrades Mavunda into a Miombo-like woodland locally called "Chipya." Zambia.** © FRANK WILLEMS

Left: **Blue Monkey is a widespread forest-dwelling primate that readily uses Mavunda.** © KEN BEHRENS, TROPICAL BIRDING TOURS

Above: **Mavunda has an incredibly dense understory, paradise for shy species like Red-capped Robin-Chat.**
© KEITH BARNES, TROPICAL BIRDING TOURS

Left: **Black-throated Wattle-eye delivers its piping vocalizations from the forest midstory.**
© KEN BEHRENS, PROMISED LAND VENTURES.

of Miombo, including Rufous-bellied Tit, Red-capped Crombec, Southern Hyliota, and Bar-winged Weaver. All the while, Schalow's and Ross's Turaco cackle loudly and bounce around the canopy, which is also inhabited by the ever-present Yellow-rumped Tinkerbird. Mavunda locally supports more secretive lowland forest species like Olive Long-tailed Cuckoo, Least Honeyguide, Lemon Dove, Green Twinspot, and, in the wettest areas, Red-tailed Bristlebill.

DISTRIBUTION: Highly localized, with many patches being small, mostly in w. Zambia north and south of the Kabompo River, north to Mwinilunga town, extending marginally into e. Angola.

WHERE TO SEE: West Lunga National Park, Zambia.

African Dry Deciduous Forests Associations

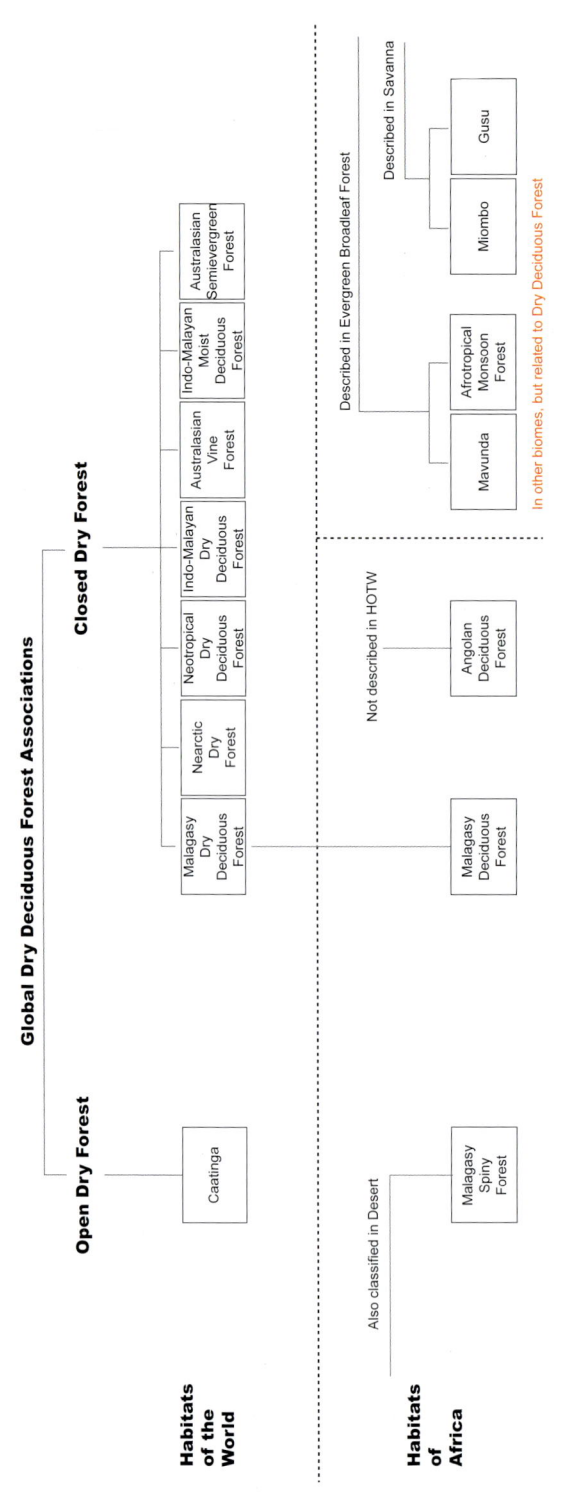

Global Dry Deciduous Forest Associations

Habitats of the World

- Open Dry Forest
  - Caatinga
- Closed Dry Forest
  - Malagasy Dry Deciduous Forest
  - Nearctic Dry Forest
  - Neotropical Dry Deciduous Forest
  - Indo-Malayan Dry Deciduous Forest
  - Australasian Vine Forest
  - Indo-Malayan Moist Deciduous Forest
  - Australasian Semievergreen Forest

Habitats of Africa

- Malagasy Spiny Forest — Also classified in Desert
- Malagasy Deciduous Forest
- Angolan Deciduous Forest — Not described in HOTW
- Mavunda, Afrotropical Monsoon Forest — Described in Evergreen Broadleaf Forest
- Miombo, Gusu — Described in Savanna

In other biomes, but related to Dry Deciduous Forest

TROPICAL DECIDUOUS FORESTS

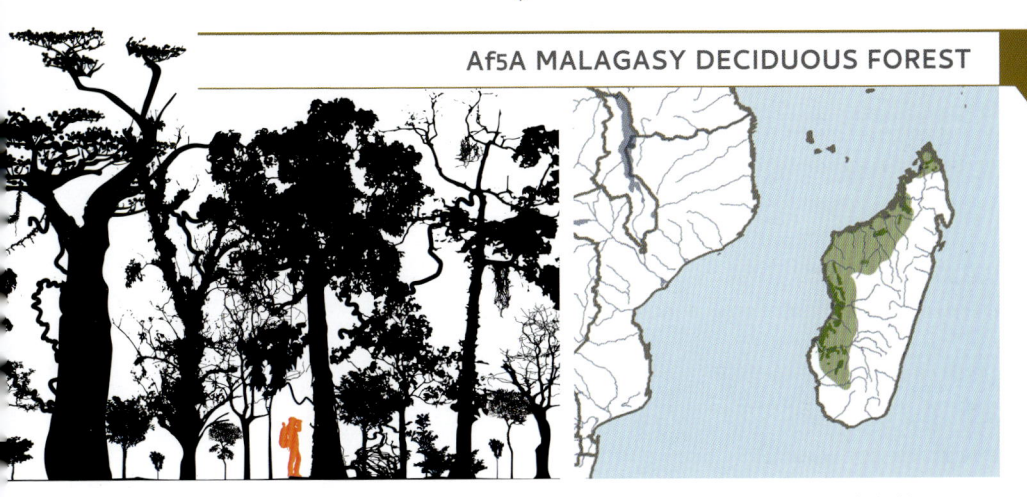

Af5A MALAGASY DECIDUOUS FOREST

IN A NUTSHELL: Tall forest with a closed canopy that is dominated by deciduous and semi-deciduous trees and has a long dry season. **Global Habitat Affinities:** DRY DECIDUOUS YUNGAS; SOUTHEAST ASIAN DRY DECIDUOUS FOREST; INDIAN DRY DECIDUOUS FOREST; MESOAMERICAN PACIFIC DRY DECIDUOUS FOREST. **Continental Habitat Affinities:** ANGOLAN DECIDUOUS FOREST. **Species Overlap:** SPINY FOREST; INDIAN OCEAN LOWLAND RAINFOREST; MALAGASY GRASSLAND AND SAVANNA.

DESCRIPTION: This is one of the world's richest dry forest habitats and one of only two such habitats in the Afrotropics. Like all the natural habitats of Madagascar, it is incredibly rich in fascinating, distinctive, and mostly approachable wildlife. Malagasy Deciduous Forest is fairly tall, usually reaching 30–50 ft. (10–15 m), and has a closed canopy. In some areas, on rich soil, it can be taller, reaching 80 ft. (25 m). Away from watercourses, most of the trees are deciduous, losing their leaves during the dry season. The abundant leaf litter and open character of this habitat in the dry season almost make it resemble a temperate broadleaf deciduous forest in autumn. But at the peak of the wet season, it feels rainforest-like, dense and green. As with most forested habitats, there is little or no grass in the understory.

This habitat lies in the rain shadow of Madagascar's eastern escarpment. Its climate (Köppen **Awa**) is much drier than the rainforest but wetter than the SPINY FOREST, as more moisture manages to curl around the northern part of the island and reach the northwest. The annual rainfall is generally 40–60 in.

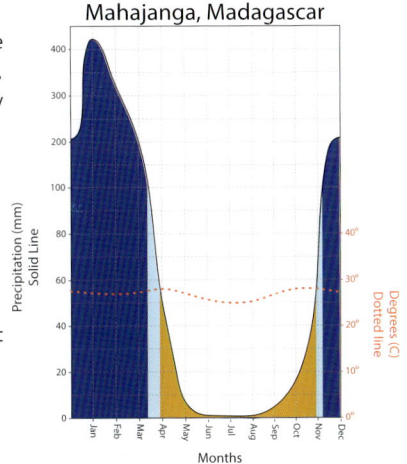

Deciduous forest bursting into green lushness after the first rains at the start of the austral summer. Ankarana National Park, Madagascar. Madagascar Blue Vanga *(inset)*. © KEN BEHRENS, TROPICAL BIRDING TOURS

(1000–1500 mm), though it can be as low as 20 in. (500 mm) in the area of transition to Spiny Forest and as high as 80 in. (2000 mm) in the Sambirano region of the north, where this habitat blends into Sambirano rainforest (sidebar 18, p. 193). Most of the year is dry, and the vast majority of the rain falls in a well-defined rainy season between November and March.

The plants of this habitat show a high degree of endemism, even greater than that of Madagascar's rainforest. There are multiple species of baobabs, including the towering, columnar Grandidier's Baobab (*Adansonia grandidieri*). Other typical trees include ebonies (*Diospyros*), rosewoods (*Dalbergia*), *Dupuya madagascariensis*, *Cedrelopsis*, pandanus, flame trees (*Delonix*), and figs (*Ficus*). Lianas are abundant, but epiphytes, moss, and lichens are scarce, and ferns are absent.

There are three distinct types of deciduous forest, which grow on three different substrates. The tallest and lushest version of this habitat grows on lateritic clays, though only small remnants of this type remain. Characteristic trees include Farafatsy (*Givotia madagascariensis*), *Stereospermum euphorioides*, and Ravintsara (*Cryptocarya agathophylla*), the source of an essential oil that is prized within Madagascar and by the perfume industry abroad.

A second type grows on sandy soils and has not been as thoroughly destroyed as the clay forest. The canopy height is variable but generally lower than on clay soils. Typical trees are Hazomalany (*Hernandia voyronii*), *Securinega*, *Abrahamia deflexa*, and Governor's Plum (*Flacourtia indica*). In some areas, usually those degraded by humans, or in areas of transition to Spiny Forest, this habitat takes the form of thick scrub or thicket. Such places are likely to hold *Albizia*, commiphoras, acacias, and donkeyberry (*Grewia*). Along rivers, riparian forest of Tamarind (*Tamarindus indica*), figs, and Mango (*Mangifera indica*) grows.

Af5A MALAGASY DECIDUOUS FOREST

The third type of Malagasy Deciduous Forest is that growing on porous limestone, such as the limestone karst known as *Tsingy*. This is a distinctive and beautiful community of plants that is nearly distinct enough to split as its own full habitat. This limestone forest and thicket has even harsher conditions than the other types of deciduous forest and is closely allied to the Spiny Forest. Like Spiny Forest, it includes many succulent plants, including leaf succulents like aloe and kalanchoe; bottle-trees like baobabs, commiphora, flame trees, and pachypodium; and stem succulents such as *Adenia* and euphorbias.

CONSERVATION: This habitat has been decimated by human activity. About 97% of the original forest has already been destroyed, making it the Afrotropics' most severely threatened forest habitat. Owing mainly to fires but also to clearing for agriculture and wood, the vast majority of w. Madagascar is now MALAGASY GRASSLAND AND SAVANNA, in which Mango and fire-resistant palms survive. It is well on its way to following the central highlands down the path of near-complete deforestation.

Deciduous Forest has an open character and abundant leaf litter. Ankarana National Park, Madagascar. © KEN BEHRENS, TROPICAL BIRDING TOURS

WILDLIFE: Deciduous forest holds nearly as many mammals as the eastern rainforest (INDIAN OCEAN LOWLAND and MONTANE RAINFOREST), and they are much easier to see in this more open habitat, which is a lemur-watching paradise. The mix varies locally, as most of the mouse lemurs, sportive lemurs, and sifakas are restricted to small portions of this habitat. The standout mammals are the big, diurnal sifakas. The six that use this habitat are, from south to north, Verreaux's, Von der Decken's, Crowned, Coquerel's, Perrier's, and Golden-crowned. Other diurnal lemurs include Western Lesser Bamboo Lemur; Common, Sanford's, and Red-fronted Brown Lemurs; and Crowned and Mongoose Lemurs. There are

In the dry season, most canopy trees lose their leaves, though many midstory and understory species retain leaves. Baobabs are prominent. Kirindy Forest, Madagascar. © KEITH BARNES, TROPICAL BIRDING TOURS

Above: **Red-fronted Brown Lemur in Kirindy Forest, which it shares with Giant Coua** *(inset).* **Note the abundance of dry leaf litter, typical of the end of the dry season.** © KEN BEHRENS, TROPICAL BIRDING TOURS

Below: **Like the Malagasy rainforests, deciduous forest is remarkably rich in nocturnal lemurs, such as this Tavaratra Mouse Lemur.** © KEN BEHRENS, TROPICAL BIRDING TOURS

Above: **Fossa, Madagascar's largest predator, primarily an arboreal lemur hunter, is a bit like a mini-Puma.** © KEN BEHRENS, TROPICAL BIRDING TOURS

even more nocturnal lemurs: a bunch of species of mouse lemurs, including the Critically Endangered Madame Berthe's Mouse Lemur, the world's smallest primate; Fat-tailed Dwarf Lemur; Coquerel's Giant Mouse Lemur; Pale Fork-marked Lemur; a range of localized sportive lemurs; and Western and Cleese's Woolly Lemurs. This forest is also home to the Malagasy Giant Jumping Rat and is the best habitat for seeing Fossa, a mongoose-cum-Puma and the island's largest carnivore.

There is not a large number of highly localized bird species in western dry forest, unlike mammals and reptiles. The majority of birds here are widespread forest species that are shared with the rainforest and/or SPINY FOREST, such as Madagascar Green-Pigeon; Giant Coua; Long-billed Bernieria; Blue, White-headed, Hook-billed, Sickle-billed, and Rufous Vangas; and Sakalava Weaver. Among the few birds found almost exclusively in western dry forest are White-breasted Mesite, Tsingy Wood-Rail, Coquerel's Coua (IS), Schlegel's Asity, Appert's Tetraka, and Van Dam's Vanga. Most of these are quite local, even within the biome; this pattern may have been exacerbated by severe forest fragmentation, though forest destruction doesn't fully account for their scarcity and localized distribution.

Dry deciduous forest makes excellent habitat for reptiles, which are especially conspicuous during the rainy season. There are many chameleons, including a bunch of dwarf chameleons (*Brookesia*), Rhinoceros Chameleon (*Furcifer rhinoceratus*), and Labord's Chameleon (*F. labordi*). The latter is notable as the world's only "annual" reptile: its entire population dies off during the dry season, to be replaced by the hatching of eggs during the next rainy season. Lizards of western dry forest include Cuvier's Madagascar Swift (*Oplurus cuvieri*), Western Plated Lizard (*Zonosaurus laticaudatus*), Western Skink (*Trachylepis tandrefana*), two Madagascar velvet geckos (*Blaesodactylus*), several ground geckos (*Paroedura*), and three leaf-tailed geckos (*Uroplatus*). Snakes are common and frequently

White-breasted Mesite is a rare understory skulker known from only a handful of sites, most of which are deciduous forest.
© KEN BEHRENS, TROPICAL BIRDING TOURS

Crowned Lemur is found both in deciduous forest and rainforest in Madagascar's far north.
© KEN BEHRENS, TROPICAL BIRDING TOURS

Above: **Madagascar Leafnose Snake (*Langaha madagascariensis*)** has a bizarre nose appendage, whose function is unknown, though it may enhance their camouflage. © KEN BEHRENS, TROPICAL BIRDING TOURS

Below: **Schlegel's Asity** is the only member of the Malagasy-endemic asity family that is found in deciduous forest habitat. © KEN BEHRENS, TROPICAL BIRDING TOURS

sighted. Some of the most widespread include Western Madagascar Tree Boa (*Sanzinia volontany*), Madagascar Ground Boa (*Acrantophis madagascariensis*), Madagascar (*Leioheterodon madagascariensis*) and Blond (*L. modestus*) Hognose Snakes, and Four-lined Snake (*Dromicodryas quadrilineatus*).

The western dry forest has high frog diversity. A few widespread species and groups are the reed frogs (*Heterixalus*), Madagascar Bullfrog (*Laliostoma labrosum*), *Scaphiophryne*, Western Bright-eyed Frog (*Boophis doulioti*), bridge frogs (*Gephyromantis*), Madagascar frogs (*Mantidactylus*), and Betsileo Mantella (*Mantella betsileo*).

Endemism: Major limestone massifs, especially the Tsingy de Bemaraha, the Tsingy of Ankarana, and the Montagne des Français, support some highly localized endemics, mostly reptiles and amphibians but a couple of mammals and birds as well. Examples are the Tsingy Wood-Rail and Western Nesomys of Bemaraha, and the Tsingy Plated Lizard of Ankarana and Montagne des Français. Quite a few species are restricted to the southern portion of the western dry forest, as around Kirindy Forest. These include the Malagasy Giant Rat and Narrow-striped Vontsira, shared with the Spiny Forest.

DISTRIBUTION: Dry deciduous forest is the original habitat in the northern two-thirds of w. Madagascar and well into the southern interior, up to around 3000 ft. (900 m). It grades into rainforest in the Sambirano region of the north, around the Tsaratanana Massif. North of Tsaratanana, there is additional western dry forest, which locally reaches the east coast in the dry northern microclimate. Some of this northern forest is so dry that it actually resembles southwestern SPINY FOREST, though without any of that biome's unique wildlife. Wetlands in this zone have a character similar to AFROTROPICAL DEEP FRESHWATER MARSH and are Madagascar's richest. They support several endemics, such as Madagascar Jacana and Madagascar Fish-Eagle. The Mascarene Islands were historically dominated by humid forest but also had dry forest with a character similar to Malagasy Deciduous Forest. Sadly, this habitat is now virtually extinct.

WHERE TO SEE: Ankarafantsika National Park, Madagascar; Zombitse-Vohibasia National Park, Madagascar.

Af5B ANGOLAN DECIDUOUS FOREST

IN A NUTSHELL: Moderately tall forest with a closed canopy that is dominated by deciduous trees and has a long dry season. **Global Habitat Affinities:** DRY DECIDUOUS YUNGAS; SOUTHEAST ASIAN DRY DECIDUOUS FOREST; MESOAMERICAN PACIFIC DRY DECIDUOUS FOREST. **Continental Habitat Affinities:** MALAGASY DECIDUOUS FOREST; MONSOON FOREST; GUSU. **Species Overlap:** MOIST MIXED SAVANNA; MONSOON FOREST; AFROTROPICAL LOWLAND RAINFOREST; MOPANE; MAVUNDA.

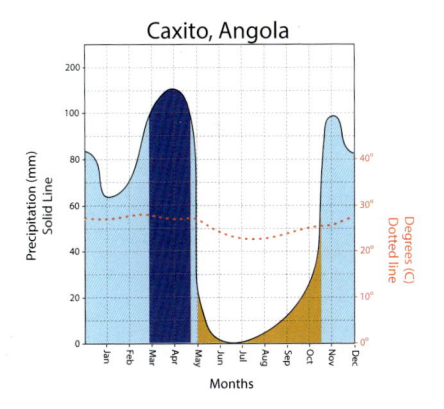

DESCRIPTION: The narrow strip that extends between the Atlantic coast and the top of the Angolan escarpment is one of Africa's most biogeographically complex. A remarkably high proportion of the continent's habitats are found here, at slightly different elevations and precipitation levels. This aggregation is most dense and compacted in s. Angola and becomes broader and less diverse in the north. One component habitat of this mega-diverse escarpment is a deciduous forest, the only one on continental Africa. This deciduous forest has some grass in the understory of more open areas but lacks the uniformly grassy understory of true savanna. Its overall structure is similar to MONSOON FOREST, but a much higher proportion of the canopy trees are fully deciduous. It is fascinating that this distinctive habitat evolved exclusively in this narrow zone of Africa.

As in other deciduous forests around the world, Angolan Deciduous Forest has highly seasonal rainfall (Köppen **Awa**). May to September is very dry, and slightly cooler, while heavy rains begin in October and continue until April. Compared with other deciduous forests, this habitat is unusual in that it receives less rainfall, typically 24–32 in. (600–800 mm), much lower than a typical deciduous forest rainfall of around 50 in. (1200 mm). Its niche at the escarpment base probably has abundant groundwater, augmenting the water received via rainfall. Temperatures remain warm year-round, though with a slight dip during the austral winter (Jun–Aug): lows can drop to 60°F (15°C), especially in s. Angola, with summer highs up to 95°F (35°C).

TROPICAL DECIDUOUS FORESTS

Rainy season aspect, resembling rainforest. Mangueiras, Angola. © ARNON DATTNER, ECOPLANET FILMS

Dry-season deciduous forest, showing the dense semi-evergreen canopy below, and deciduous emergents above—a strange and distinctive type of forest. Santa Ambuleia, Angola. © KEITH BARNES, TROPICAL BIRDING TOURS

Af5B ANGOLAN DECIDUOUS FOREST

The structure of Angolan Deciduous Forest is unusual. There is a complete or near-complete canopy formed by a dense and almost thicket-like growth of trees and shrubs that is 16–33 ft. (5–10 m) tall. Emerging from this dense growth are tall trees, most common and conspicuous of which is African Baobab (*Adansonia digitata*). Another common emergent, which has a conspicuous smooth pale trunk, is *Sterculia setigera*, a member of the same genus as the widespread African Star-Chestnut (*S. africana*). Also common and distinctive is the Angolan endemic *Euphorbia candelabrum*. These emergents can be 80 ft. (25 m) tall. The lower canopy is formed by a diverse mix of woody species. This habitat is remarkably poorly known biologically, and especially botanically, but some of its components can include monkey-orange (*Strychnos*), *Bauhinia*, Bird Plum (*Phyllogeiton discolor*), *Garcinia*, Angolan Green-thorn (*Balanites angolensis*), donkeyberry (*Grewia*), caperbush (*Capparis*), *Gymnosporia senegalensis*, and ebony (*Diospyros*).

During the dry winter, the emergent trees and many of the canopy trees lose their leaves, creating an abundance of leaf litter. Many smaller trees and shrubs in the thicket-like lower portion of the forest retain some or all of their leaves year-round. With the onset of monsoonal summer rain, this habitat transforms into something lush, verdant, and rainforest-like. Some more open areas have a grassy understory, and the transition from this habitat into MOIST MIXED SAVANNA is a gradual and sometimes hazy mélange. In general, lower-lying areas, especially along watercourses, are moister and have a more complete canopy while at slightly higher elevation and in drier areas, this habitat starts to give way to savanna.

Lush deciduous forest along a drainage line. Note the abundant leaf litter. Santa Ambuleia, Angola.
© KEN BEHRENS, PROMISED LAND VENTURES

CONSERVATION: Although this habitat is under increasing human pressure, large tracts remain intact, especially in n. Angola. As elsewhere in Angola, there is rampant hunting, and big mammals are scarce or absent. As humans increasingly create openings within the forest, fire is likely to become a major threat, since the plants of deciduous forest are not fire-resistant. Increasing frequency of fires will start to convert deciduous forest into a habitat that functions as MOIST MIXED SAVANNA.

WILDLIFE: Historically, this habitat likely held a mix of mammals similar to adjacent habitats such as MOIST MIXED SAVANNA and MOPANE. Unfortunately, mammal populations have been vastly reduced by hunting. Smaller species such as Malbrouck (sometimes lumped as Vervet Monkey), Angolan Talapoin, Chacma and Olive Baboons, and Common Duiker persist even in populated areas. Large species, including African Buffalo and African Savanna Elephant, remain in more remote regions.

Euphorbia candelabrum is a common and spectacular emergent from the dense and thicket-like lower canopy layer. Santa Ambuleia, Angola. © ARNON DATTNER, ECOPLANET FILMS

Deciduous forest is quite lively from a birding perspective, as it has a high density and diversity of birds, including several localized Angolan endemic and near-endemic species. Birds in the latter category include White-fronted Wattle-eye, Angola Batis, Angola Helmetshrike, Red-backed Mousebird, Hartert's Camaroptera, Rufous-tailed Palm-Thrush, and Monteiro's Bushshrike. There are also some species that are more typically found in Congo Basin lowland rainforest, including

The Endangered Angola Helmetshrike is endemic to the northern portion of the Angolan Deciduous Forest. © DUBI SHAPIRO

Af5B ANGOLAN DECIDUOUS FOREST

Angolan Deciduous Forest is one of the habitats used by the localized endemic Red-crested Turaco. © DUBI SHAPIRO

Petit's Cuckooshrike, Green Crombec, and Forest Scrub-Robin. During the rainy season, the enigmatic migratory African Pitta can be found in some wet gullies. Overall, deciduous forest has an eclectic mix of species shared with adjacent habitats. Good examples include African Green-Pigeon, Black Scimitarbill, Crowned Hornbill, African Black-headed Oriole, Gabon Boubou, Sulphur-breasted and Four-colored Bushshrikes, White-browed and Red-capped Robin-Chats, Purple-banded and Collared Sunbirds, Forest Weaver, Black-tailed Waxbill, and Black-faced Canary. Greenbul diversity is high, typical of forest rather than savanna habitats, with a mix that includes Yellow-bellied, Yellow-necked, and Yellow-throated Greenbuls and the near-endemic Pale-olive Greenbul. Mottled and Bat-like Spinetails nest in the baobabs, which are a dominant feature of the landscape.

The conspicuous reptiles of Angolan Deciduous Forest are generally shared with adjacent savanna and forest habitats. Examples include the widespread Brown Forest Cobra (*Naja subfulva*), Gaboon Adder (*Bitis gabonica*), Black Mamba (*Dendroaspis polylepis*), Boomslang (*Dispholidus typus*), and Kalahari Plated Lizard (*Gerrhosaurus auritus*). One special reptile is the Angolan endemic Mucoso Agama (*Agama mucosoensis*).

Endemism: Within the relatively narrow range of this habitat, several of the endemic and near-endemic bird species are found in only the northern portions of dry deciduous forest.

DISTRIBUTION: Found in a band along the base of the Angolan escarpment, from just north of Luanda in the north to near Lubango in the south. Owing to the steep slope of the south escarpment, this habitat is found in a slender band there, at approximately 1600–3300 ft. (500–1000 m). In its northern reaches, where the slopes are more gradual, this habitat is found in a broader band, from just above sea level up to at least 1600 ft. (500 m).

Deciduous forest may have once been found on the Cape Verde islands, especially at higher elevations and other areas with higher moisture. Unfortunately the natural habitat there has been almost completely destroyed, leaving only hints about what used to be there. Some forest formations in East Africa strongly resemble deciduous forest. These tend to occur in zones where MOIST MIXED SAVANNA or MIOMBO transition to MONSOON FOREST or EAST COAST FOREST MATRIX. These formations in East Africa are not mapped here, since they do not have a distinctive assemblage of wildlife and are considered as a dry form of Monsoon Forest or East Coast Forest Matrix.

WHERE TO SEE: Santa Amboleia, Angola; Mangueiras, Angola.

African Savannas Dendrogram

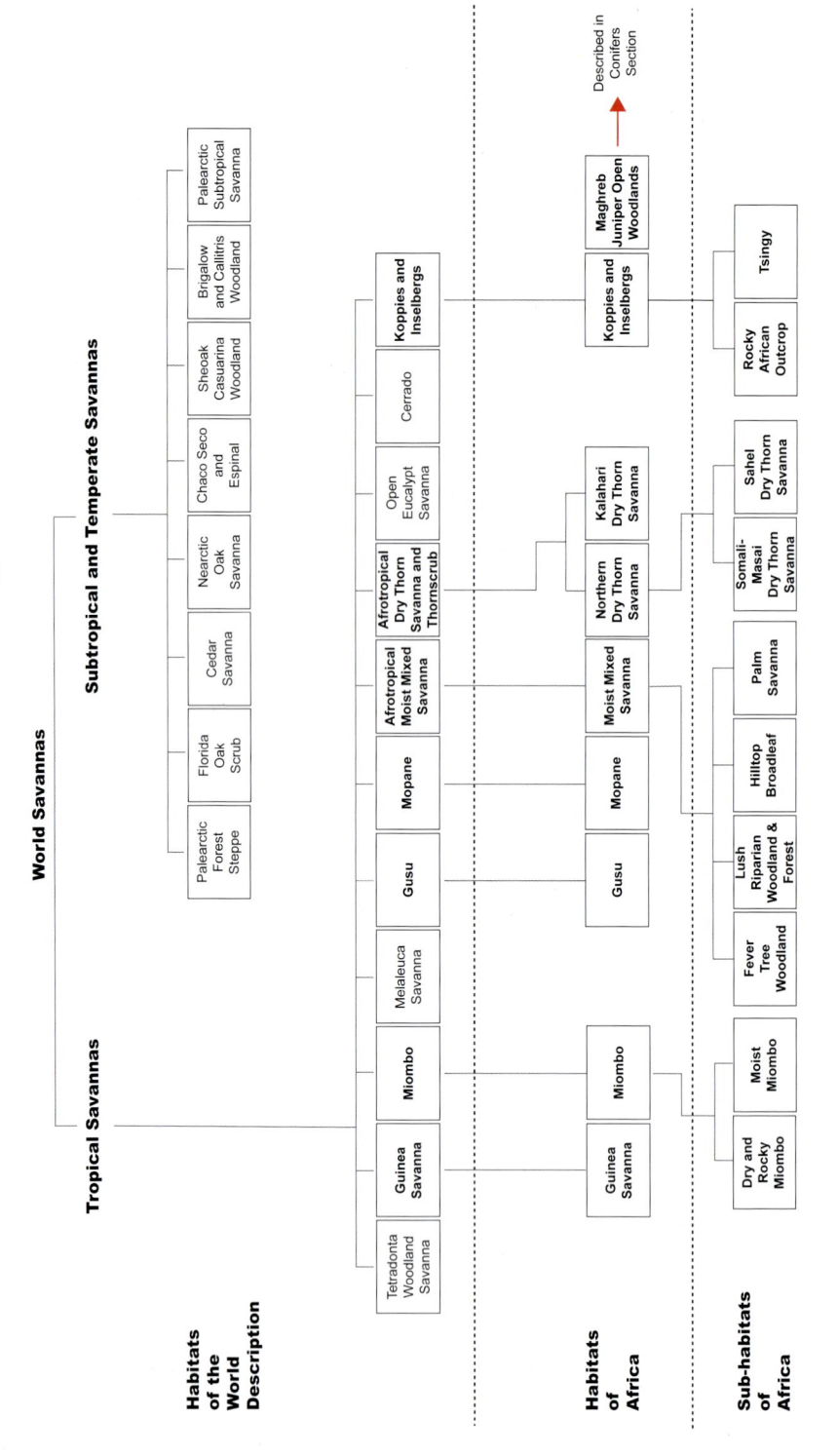

World Savannas

Tropical Savannas | Subtropical and Temperate Savannas

Habitats of the World Description

Tropical Savannas:
- Tetradonta Woodland Savanna
- Guinea Savanna
- Miombo
- Melaleuca Savanna
- Gusu
- Mopane
- Afrotropical Moist Mixed Savanna
- Afrotropical Dry Thorn Savanna and Thornscrub
- Koppies and Inselbergs

Subtropical and Temperate Savannas:
- Palearctic Forest Steppe
- Florida Oak Scrub
- Cedar Savanna
- Nearctic Oak Savanna
- Chaco Seco and Espinal
- Sheoak Casuarina Woodland
- Brigalow and Callitris Woodland
- Palearctic Subtropical Savanna
- Open Eucalypt Savanna
- Cerrado

Habitats of Africa
- Guinea Savanna
- Miombo
- Gusu
- Mopane
- Moist Mixed Savanna
- Northern Dry Thorn Savanna
- Kalahari Dry Thorn Savanna
- Koppies and Inselbergs
- Maghreb Juniper Open Woodlands → Described in Conifers Section

Sub-habitats of Africa
- Dry and Rocky Miombo
- Moist Miombo
- Fever Tree Woodland
- Lush Riparian Woodland & Forest
- Hilltop Broadleaf
- Palm Savanna
- Somali-Masai Dry Thorn Savanna
- Sahel Dry Thorn Savanna
- Rocky African Outcrop
- Tsingy

SAVANNAS

Af6A MOPANE

IN A NUTSHELL: Variable savanna habitat that is dominated by Mopane, a semi-deciduous broadleaf tree with a diverse range of growth forms. **Global Habitat Affinities:** SHRUBBY SAVANNA; NORTHERN ACACIA SHRUBLAND; CAMPO CERRADO; DRY CHACO. **Continental Habitat Affinities:** MIOMBO; GUSU. **Species Overlap:** MOIST MIXED SAVANNA; KALAHARI DRY THORN SAVANNA; GUSU; MIOMBO.

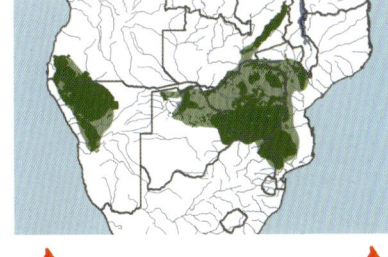

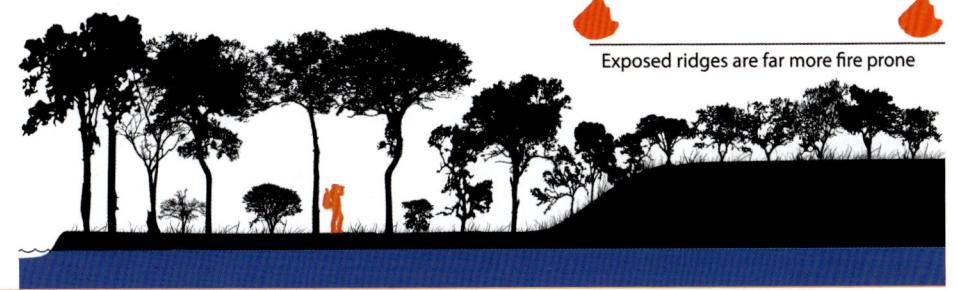

Exposed ridges are far more fire prone

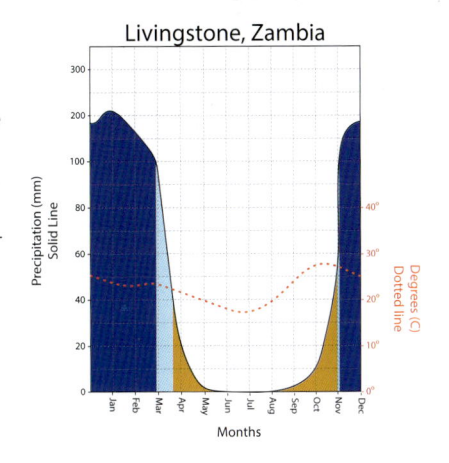

DESCRIPTION: Any discussion of Mopane should begin with its namesake dominant tree, easily recognized by its butterfly-shaped leaves, which smell like turpentine when crushed. The Mopane tree (*Colophospermum mopane*) has several odd characteristics: As a rule, broadleaf woodland grows on nutrient-poor, well-drained soil, but the Mopane is a broadleaf tree that frequently grows on nutrient-rich, poorly-drained soil, though it is also capable of growing on poor and well-drained soils. It is also unusual in its hugely variable growth forms, which are a response to local growing conditions and disturbance, especially from elephants. It most typically forms a broadleaf woodland similar to MIOMBO or GUSU, though with a shorter canopy, at 24–33 ft. (7–10 m). But in the eastern part of its range, it can also grow in forest-like stands known as "cathedral Mopane," with trees as tall as 80 ft. (25 m) and only sparse grass in the understory. On the other end of the size spectrum, in the arid west, it can grow in scrubby stands that are 3–10 ft. (1–3 m) tall and, exceptionally, even as low as 1 ft. (0.3 m) tall in thickets interspersed with grass. Regardless of its height, Mopane has a distinctive structure. There is one main trunk, with the narrow canopy formed by side branches that ascend at sharp angles. This creates a tree with an unusually tight "V" shape, quite unlike

Livingstone, Zambia

Lush Mopane at the base of the Angolan escarpment. Renato Grande, Angola. © KEITH BARNES, TROPICAL BIRDING TOURS

African Savanna Elephants drinking from a waterhole in Mopane habitat in central Etosha National Park, Namibia. Note the austral springtime flush of orange leaves, similar to that seen in Miombo.
© KEN BEHRENS, TROPICAL BIRDING TOURS

"Cathedral" Mopane in the dry season, unburned *(left)* and recently burned *(right)*. Liwonde National Park, Malawi. © KEN BEHRENS, TROPICAL BIRDING TOURS

either the classic "lollipop" tree or the iconic savanna "umbrella" tree. Mopane is variably deciduous, depending on local conditions. In the wettest sites, it is nearly evergreen, while in the driest it loses all leaves for several months. When Mopane trees begin to grow new leaves at or just before the onset of the summer rains, these leaves flush yellow or orange, giving this habitat a beautiful yet strangely autumnal springtime aspect.

Like other woodland and savanna habitats, Mopane savanna grows in areas with highly seasonal rainfall; here the rain falls during the austral summer (Nov–Mar). It is found in a wide range of conditions, from extremely dry areas at the edges of the Namib (Köppen **Bma**) to areas that receive up to 30 in. (800 mm) of rain annually (Köppen **Ama**). At the Namib edge, Mopane often grows alongside Welwitschia (*Welwitschia mirabilis*), an enigmatic plant that is considered a sort of "living fossil" (see p. 78 for more information).

Mopane is fire-resistant, like many savanna trees, allowing it to persist after fires sweep through and consume the understory grass. Although Mopane tends to dominate, other trees are often mixed into this habitat, such as acacias, donkeyberry (*Grewia*), African Baobab (*Adansonia digitata*), African Blackwood (*Dalbergia melanoxylon*), Trumpet Thorn (*Catophractes alexandri*), Marula (*Sclerocarya birrea*), Tamboti (*Spirostachys africana*), Apple Leaf (*Philenoptera violacea*), Weeping Wattle (*Peltophorum africanum*), Angolan Green-thorn (*Balanites angolensis*), bushwillows (*Combretum*), shepherd's trees (*Boscia*), sourplum (*Ximenia*), *Terminalia*, and commiphoras. These other tree species rarely do much to break up the oddly uniform and distinctive aspect of Mopane habitat when seen at a distance. Another peculiarity of the Mopane tree is that it tends to either dominate a habitat or to be absent, with little middle ground. The mix of other tree species associated with Mopane is quite different in the moister east, compared with the arid west, but across its range Mopane habitat shares few plant species with Miombo.

CONSERVATION: Vast regions of Mopane have been set aside in well-protected reserves, but there are still problems, including poaching, especially of Black Rhinos. Most large mammals in the Angolan and n. Namibian Mopane were killed during the Angolan civil war. On private land, farmers habitually kill predators like jackals, Cheetah, and Caracal.

WILDLIFE: Mopane is generally not the most interesting habitat for wildlife; it does not support a distinctive or cohesive faunal assemblage but rather a much-reduced selection of the wildlife of

224 SAVANNAS

Right: **Mopane makes excellent hunting (and eating!) ground for Leopards.** © KEITH BARNES, TROPICAL BIRDING TOURS

Below: **The localized small antelope Sharpe's Grysbok is quite fond of Mopane. The butterfly-shaped Mopane tree (*Colophospermum mopane*) leaf is unmistakable** *(inset).* © KEITH BARNES, TROPICAL BIRDING TOURS

adjacent savannas. On safari, you may find yourself quickly transiting through Mopane to get to a more productive habitat. Nonetheless, there is one mega-mammal that is very fond of Mopane: the African Savanna Elephant, which eats nearly every part of the Mopane tree and commonly knocks over full-size trees to reach the leaves at the top (see sidebar 40, p. 373). Elephants make openings in the woodland, which fill with grass, providing fuel for bushfires, which further serve to keep the habitat open. There are concerns about the impact of elephants on areas in which their populations are out of balance because of various human interventions. Mopane is one of the trees that have the ability to load tannins into their leaves to deter heavy browsing by herbivores. Not only that, but Mopane can also release pheromones to communicate the presence of herbivores to nearby trees, allowing them to preemptively increase their tannin levels. Impala also thrive in the matrix of dense woodland and open glades that is typical of Mopane. In Namibia's Etosha National Park, Mopane is an important habitat for the localized Black-faced subspecies of Impala. Many other mammals will use Mopane, though it is not a particularly preferred habitat, including Black-backed Jackal, Black Rhinoceros, Plains Zebra, Blue Wildebeest, Giraffe, Sharpe's Grysbok, and Greater Kudu. The western tracts of Mopane are used by Springbok, Gemsbok, "Hartmann's" Mountain Zebra, and Brown Hyena. Many large reserves include swaths of Mopane and support large cats, Spotted Hyena, and African Wild Dog.

The birds of Mopane are heavily influenced by surrounding habitats. The western Mopane habitat is used by some of the endemic birds associated with the Namib Escarpment. In particular, Mopane along watercourses is a crucial habitat for the localized Violet Woodhoopoe and Bare-cheeked Babbler. Upland western Mopane can have White-tailed Shrike and Carp's Tit. In the south-central part of its range, on the moderate elevations of the Central African Plateau, Mopane has a reduced subset of the birds of **KALAHARI DRY THORN SAVANNA**, such as Crimson-breasted Gonolek and Mariqua Flycatcher. Mopane to the north and east is more influenced by **MOIST MIXED SAVANNA** and, to a lesser degree, **MIOMBO**. Widespread savanna species occurring in Mopane habitat include Pearl-spotted Owlet, Green Woodhoopoe, Greater Blue-eared and Meves's Starlings, and White Helmetshrike. Miombo influence is shown by the local presence of Arnot's Chat, White-breasted Cuckooshrike, and Miombo Wren-Warbler. Mopane in s. Malawi holds the localized Reichenow's Woodpecker.

Mopane habitat has a subset of typical savanna reptiles, including conspicuous lizards like Black-lined Plated Lizard (*Gerrhosaurus*

Violet Woodhoopoe is one of a handful of bird species that prefer Mopane habitat. This localized species of Namibia and Angola often interbreeds with the widespread Green Woodhoopoe, and the birds in this photo show some evidence of hybridization.
© KEN BEHRENS, TROPICAL BIRDING TOURS

nigrolineatus) and Tropical Spiny Agama (*Agama armata*). In the west, the diurnal Ovambo Tree Skink (*Trachylepis binotata*) is often found on Mopane trees.

Mopane is the main food plant for the Mopane worm, the caterpillar form of a beautiful species of emperor moth (*Gonimbrasia belina*). These caterpillars emerge in huge numbers in the summer, providing an important food source for both wildlife and humans.

Because Mopane is strongly influenced by more diverse surrounding habitats, there are major differences between the wildlife found in the western Mopane of Angola and Namibia, the higher-elevation Mopane associated with the Kalahari Basin, and the lower-lying eastern and northern Mopane of e. South Africa, Mozambique, Zimbabwe, and Zambia.

DISTRIBUTION: Mopane is widespread though patchy from n. Botswana to s. Malawi and south to e. South Africa. This is an important habitat in the Zambezi, Limpopo, and Luangwa River valleys and locally around the Okavango Delta. Mopane also occurs in a discrete area of n. Namibia and s. Angola. Mopane can be found at both low elevations (such as the lowveld and river valleys of e. South Africa and Mozambique) and moderate ones (as on the Central African Plateau). Mopane and GUSU together provide a transition from the moister MIOMBO to their north and the drier and thornier savanna types to their south and west. Mopane tends to occur in close conjunction with KALAHARI DRY THORN SAVANNA and MOIST MIXED SAVANNA, as it grows on a soil type that is also well suited to those habitats. It is fascinating as a habitat that bridges a broad range of moisture regimens, occurring all the way from the fringes of the Namib Desert to the edges of semi-evergreen MONSOON FOREST.

WHERE TO SEE: Etosha National Park, Namibia (central and western portions); South Luangwa National Park, Zambia; Kruger National Park, South Africa (northern portions).

SIDEBAR 19   DUNG BEETLES, LIKE A ROLLING STONE?

Dung beetles are found all over the globe, on all the continents save Antarctica. There are thousands of species in multiple families, though most belong to the scarab beetle family. Despite their wide distribution, dung beetles are strongly identified with Africa, as its savanna landscapes with abundant large mammal dung lying around are perfect for dung beetles, which can often be seen at work. The "classic" dung beetles are those that roll balls of dung to a place where they bury them underground, then lay eggs on them. There are also tunneling dung beetles that bury dung where they find it and other species that simply inhabit dung. Remarkably, some species have been shown to navigate by the Milky Way and another by the polarization patterns of moonlight!

Dung beetle with its prize, on the move. Note the small flies hitching a ride on the dung beetle's back!
© KEITH BARNES, TROPICAL BIRDING TOURS

Af6B NORTHERN DRY THORN SAVANNA

IN A NUTSHELL: Variably dry, open savanna of North and East Africa that is dominated by thorny acacia and commiphora trees and has grass cover during the rainy season. **Global Habitat Affinities:** GRASSY MULGA; DECCAN THORNSCRUB; CAMPO SUJO. **Continental Habitat Affinities:** KALAHARI DRY THORN SAVANNA. **Species Overlap:** MOIST MIXED SAVANNA; KALAHARI DRY THORN SAVANNA; GUINEA SAVANNA; MIOMBO.

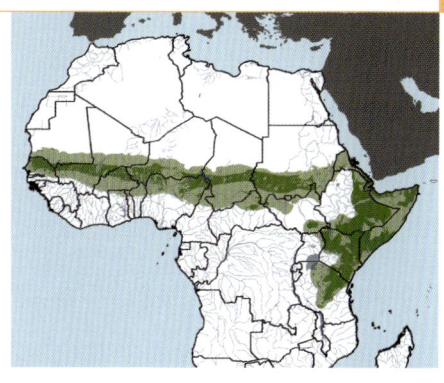

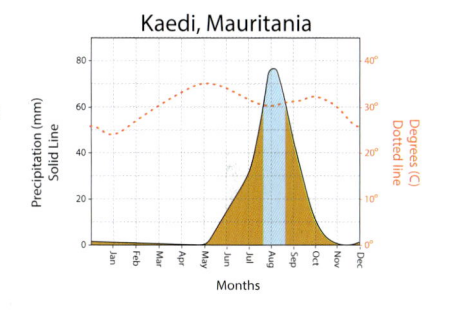

DESCRIPTION: This is a classic savanna environment in which there is variable cover from trees and bushes, and much or all the ground is seasonally covered in grass. What makes this habitat distinctive is its generally dry and highly seasonal climate (Köppen **Bma**) and the predominance of thorny and fine-leaved trees. During the long dry season, this habitat can seem superficially almost barren and desertlike, but during the rains, it bursts into fantastic productivity, an ephemerally lush and green world that bears little resemblance to the dry and dusty landscape of just a few weeks earlier. The grassland and woodland components of this habitat have fairly discrete sets of wildlife, and this account mainly treats the woodland aspect of this habitat; for more detail about the grassland aspect, see the AFROTROPICAL GRASSLAND account. These two habitats often form a mélange, and separating them can seldom be done cleanly.

Northern Dry Thorn Savanna is found in two distinct zones that form discrete subhabitats—in the Sahel (Af6B-1) of northern sub-Saharan Africa and in the Somali-Masai zone (Af6B-2) in the northeast. It is also closely allied with the KALAHARI DRY THORN SAVANNA of the Kalahari and Namib regions of the southwest. Evidence of these ancient cross-continental links is the fact that 5% of the plants of the Sahel are shared with the Kalahari Dry Thorn Savanna, despite their remoteness.

The existence of these three similar, yet discreet zones has driven a fascinating pattern of colonization and speciation (discussed further in the Distribution section). The exact structure of Northern Dry Thorn Savanna is variable. It can be an open landscape covered mainly in grassland with scattered bushes. Or it can be woodland in which trees and bushes predominate, though not to the extent that they prohibit the growth of grass. In general, this habitat is open enough to make it easy to walk through, despite its thorny character. The height of the tallest trees is highly variable. In drier areas, where this habitat undergoes a gradual transition into AFROTROPICAL HOT SHRUB DESERT, this habitat shrinks down to a semi-desert shrubland in which the tallest woody plants are 3 ft. (1 m) or less. In moister areas, and along watercourses or other areas with groundwater, there can be isolated tall trees up to 65 ft. (20 m) tall. Occasionally this habitat will have a nearly complete canopy, but more typically, several canopy diameters of open space separate tall trees.

This is a generally dry and hot habitat in which the vast majority of the rain falls during short rainy seasons, often in violent thunderstorms. The rains are produced by the Intertropical Convergence Zone, which shifts throughout the year. Rain falls from May to October in the north and in a more complicated pattern in parts of East Africa, with "long rains" between March and May and "short rains" sometime between September and December. The rainfall is variable, typically 8–24 in. (200–600 mm) annually, but it can fail entirely in some years. As such, this is an exacting habitat in which the animals and plants have had to adapt to harsh conditions. One such adaptation among animals is local movements, even among species that are considered residents. For example, many big mammals in the Serengeti feed on the open grasslands when they are green after the rains, then retreat to the dry thorn savanna when the grassy plains dry up.

Unlike moister savanna habitats, dry thorn savanna can lose most or all its grass during the dry season because of grazing, fire, and desiccation. Fires have always been frequent, especially at the onset of the rainy season, but have been made more frequent by humans. When present, the grass is a maximum of 2 ft. (60 cm) tall. In terms of its woody vegetation, this habitat is characterized by the dominance of spiny acacias (*Vachellia* and *Senegalia*) and commiphoras. Acacias with the stereotypically African umbrella form can be prominent, such as the widespread, conspicuous, and often dominant Umbrella Thorn (*Vachellia tortilis*). Another distinctive acacia, found in the Somali-Masai zone, is Whistling Thorn (*V. drepanolobium*), which can dominate large stretches of the landscape. Whistling Thorn has a mutualistic relationship with ants, which make nests in its swollen thorns, swarming out to protect the tree from herbivores. Other trees of this habitat

Areas along watercourses support narrow bands of lush woodland, which resembles Moist Mixed Savanna. Sof Omar, Ethiopia. © KEN BEHRENS, TROPICAL BIRDING TOURS

Escarpments like this one hold an interesting set of specialist wildlife. Baringo, Kenya.
© KEN BEHRENS, TROPICAL BIRDING TOURS

include shepherd's trees (*Boscia*), donkeyberry (*Grewia*), Desert Date (*Balanites aegyptiaca*), jujubes (*Ziziphus*), and mustard trees (*Salvadora*). Meru (*Maerua crassifolia*) is a widespread and common tree species of the Sahel. Rattlepods (*Crotalaria*) are common trees and shrubs in the Horn of Africa. Succulent euphorbias and aloes are often common. Bushwillows (*Combretum*) and/or terminalias (*Terminalia*) are prominent in some areas, especially where this habitat transitions to moister savanna and woodland. Tamarisk (*Tamarix*) grows in areas with saline soils and occupies the same niche as casuarina in the dry parts of Australia. Most trees lose at least some of their leaves during the dry season, though the harshness of the environment and the small and fine nature of the leaves mean that there is rarely a significant accumulation of leaf litter.

Grass cover in the Sahel is provided mostly by annual grasses, including Indian Sandbur (*Cenchrus biflorus*), *Schoenefeldia gracilis*, Needlegrass (*Aristida stipoides*), and European Bur-Grass (*Tragus racemosus*). Perennial grasses are less common, though Gamba Grass (*Andropogon gayanus*) may formerly have been common before it was largely replaced by cultivation, in the higher-rainfall portions of the Sahel. Common grasses in the Somali-Masai region include panicgrasses (*Panicum*), Red Oat Grass (*Themeda triandra*), bluestems (*Andropogon*), Vlei Bristlegrass (*Setaria incrassata*), Egyptian Crowfoot Grass (*Dactyloctenium aegyptium*), Sixweeks Threeawn (*Aristida adscensionis*), and lovegrasses (*Eragrostis*).

In the Sahel, most of the ancient rocks have been covered by more recent sediment, so the soils have developed in mostly eolian sands that were deposited by winds through the Miocene (23–2.6 MYA) or the ice ages of the Pleistocene (2.6 MYA–12,000 YBP). There are some clay soils from lakes of the Pleistocene and early Holocene (11,000–6000 YBP) and some ancient rocks on higher ground.

Aside from the Kalahari Dry Thorn Savanna, the most similar habitat covered in this book is MOIST MIXED SAVANNA. These two come together in a complicated mosaic across East Africa. Moist Mixed Savanna is characterized by a higher overall diversity of trees, a more complex and varied structure, and by the local prominence of broadleaf species. In most dry areas that are

SAVANNAS

Right: **Classic Dry Thorn Savanna in n. Tanzania. It seems dry and uninviting but is often pumping with birds and, in protected areas, large mammals as well.** © KEN BEHRENS, TROPICAL BIRDING TOURS

Below: **Whistling Thorn (*Vachellia drepanolobium*), with the ants that nest in its convenient galls.** MAIN PHOTO © KEN BEHRENS; INSET © KEITH BARNES, TROPICAL BIRDING TOURS

dominated by Northern Dry Thorn Savanna, Moist Mixed Savanna or GUINEA SAVANNA (n. Africa) can still be found locally along watercourses.

CONSERVATION: Although the human population is generally low density, their impact upon this habitat is profound as they try to scrape out a living in a tough environment. Cutting trees for firewood and charcoal, grazing, and hunting are widespread. The northeastern zone is in decent condition, with some well-protected areas and good numbers of big mammals. The Sahel, on the other hand, has been more heavily affected by humans for much longer, and the vast majority of its big mammals have been wiped out. Much of the Sahel zone and Somalia experience political instability, meaning that people are struggling to survive, guns are widespread, and law enforcement is virtually nonexistent—a bad mixture when it comes to big mammal conservation. Another challenge facing this habitat, especially in the Sahel, is desertification, which is exacerbated by human activities, such as attempts to cultivate very thin and dry soils during the brief rainy season.

WILDLIFE: In general, this is an excellent habitat for wildlife, harboring a high overall density and diversity of animals that are easy to see because of its open nature. Prominent groups of mammals include oryxes, gazelles, and Giraffe. Northern Dry Thorn Savanna is an important habitat in some of Africa's greatest safari destinations, such as Tanzania's Serengeti National Park.

The birding in this habitat can be among the world's most electrifying, particularly in the Somali-Masai zone and especially after good rains. It can be absolutely pumping with birds and is delightful to walk through in places like s. Ethiopia, where there are no dangerous big mammals. Prominent groups of birds include bustards, sandgrouse, doves, scrub-robins, starlings, shrikes, warblers, sunbirds, and weavers. Many Palearctic migrants use this habitat, probably because of its structural similarity to the Mediterranean woodlands where they breed, as well as its tremendous productivity.

This habitat is also replete with conspicuous reptiles, including sand (multiple genera) and long-tailed (*Latastia*) lizards, agamas, skinks, geckos, and snakes, which you have a greater chance of bumping into here than in most other African habitats. Male agamas, sporting bold colors such as jet black and construction orange, can be especially arresting.

The only animals that occur throughout Africa's dry thorn savanna, both northern and southern, are widespread savanna species that aren't restricted to this habitat, such as Honey Badger, Lion, Caracal, Spotted Hyena, Black Rhinoceros, Giraffe, Tawny Eagle, and Namaqua Dove.

This habitat's two discrete subhabitats are central to any discussion of its wildlife. Much of the most interesting wildlife is restricted to only one of these zones. The Somali-Masai zone is the richest in terms of wildlife, as it lies between the Sahel and Kalahari, both of which historically acted as reservoirs of potential colonists. Among the species shared with the Sahel are Striped Hyena, African Collared-Dove (IS), Blue-naped Mousebird, and Northern Crombec. Those also found in the Kalahari include Southern Springhare, Black-backed Jackal, Bat-eared Fox, and Kori Bustard.

Gerenuk often stand on their hind legs and stretch their necks to reach higher foliage. Yabello, Ethiopia. © KEN BEHRENS, TROPICAL BIRDING TOURS

232 SAVANNAS

Right: **A Northern Carmine Bee-eater hitching a ride on the back of a Kori Bustard; any insects flushed by the bustard are quickly snapped up by the opportunistic bee-eater. Awash National Park, Ethiopia.** © KEN BEHRENS, TROPICAL BIRDING TOURS

Below: **Grévy's Zebra is a Somali-Masai dry thorn savanna endemic. Samburu National Park, Kenya.** © KEN BEHRENS, TROPICAL BIRDING TOURS

The European Starling gives this group a bad name as an often-introduced species, but starlings like the Golden-breasted Starling are among Africa's most spectacularly beautiful birds. © NICK ATHANAS, TROPICAL BIRDING TOURS

The Somali-Masai zone has some wonderful mammals that are virtually endemic, including Lesser Kudu (IS), Gerenuk (IS), Beisa Oryx, Dibatag, and the bizarre eusocial and virtually cold-blooded Naked Mole-rat. Just a few of the many wonderful birds that are endemic or near-endemic are Buff-crested Bustard (IS), Golden-breasted Starling, Eastern Violet-backed Sunbird, Rosy-patched Bushshrike, and Somali Bunting. This zone has fantastic reptile diversity, and there are dozens of endemics.

The big mammals in the Sahel used to rival those in eastern and southern Africa but have been decimated by hunting and other human pressures. This was a major habitat of the Scimitar-horned Oryx, which went extinct in the wild in the year 2000 but has subsequently been reintroduced to Chad. Other Sahel mammals include Patas Monkey, Pale Fox, and Red-fronted (IS), Dama, and Dorcas Gazelles. Classic Sahel birds include Scissor-tailed Kite, Golden Nightjar, Quail-plover, Yellow-breasted Barbet, and Black Scrub-Robin.

Endemism: In general, endemism is fairly low in the Sahel and quite high in the Somali-Masai region. For example, around 3% of the Sahel's plants are endemic, whereas around 50% of the Somali-Masai plants, including dozens of genera, are endemic. Likewise with birds and mammals: a small set of each is endemic to the Sahel, whereas large suites are restricted to the Somali-Masai dry thorn savanna.

DISTRIBUTION: In general, dry thorn savanna lies between true desert environments and broadleaf woodland or moist savanna environments. This habitat comprises two discrete sectors, similar to those of **AFROTROPICAL HOT SHRUB DESERT**: the Sahel of n. Africa, and the Somali-Masai biome of the northeast (extending into the Arabian Peninsula). The Sahel is a relatively narrow band of 125–250 mi. (200–400 km), which runs from Senegal and s. Mauritania east to Eritrea. A very narrow band of this habitat wraps around the northern ramparts of the Abyssinian highlands, then extends south through Ethiopia's Great Rift Valley. To the east, a vast swath of dry thorn savanna runs from eastern Ethiopia and Somalia, south through much of Kenya, and well into Tanzania. The vast majority of this habitat lies below 2600 ft. (800 m), though it is found in much higher areas in the Great Rift Valley of Ethiopia, and in north-central Tanzania.

Over geological time, as Africa alternated between cool wet and warm dry periods, these Sahel and Somali-Masai zones and the dry thorn savanna of the Kalahari and Namib edge, would have been repeatedly isolated and then reconnected, resulting both in many species being shared, especially between the Sahel and Somali-Masai, and many species being isolated in one block.

Along the southern fringe of the Sahel and much of Tanzania, dry thorn savanna mixes into moister habitats and may be restricted to higher, drier areas. Along the northern verge of the Sahel, and in parts of the Horn of Africa, dry thorn savanna extends only along drainage lines and elsewhere gives way to **AFROTROPICAL HOT SHRUB DESERT** habitat in which trees are lacking. These fringe areas generally lack grassy ground cover, though it can develop temporarily after good rains.

WHERE TO SEE: Awash National Park, Ethiopia; Tsavo East and West National Parks, Kenya; Waza National Park, Cameroon.

Golden Nightjar is a Sahel endemic and one of the world's most beautiful nightjars.
© DANI LOPEZ VELASCO, ORNIS BIRDING EXPEDITIONS

| SIDEBAR 20 | **PALEOCLIMATE, HABITAT FLUCTUATIONS, BIOGEOGRAPHY, AND REFUGIA** |

Through the eons, Africa has undergone climatic fluctuations that have dramatically affected the distribution of its habitats and their wildlife. Over evolutionary time this fluctuating habitat distribution has shaped zones of endemism, like the Albertine Rift. But some of these fluctuations have been surprisingly recent; Africa has approached both desert maxima and forest maxima scenarios (see figure) as recently as 18,000 and 5800 YBP, respectively. During dry periods, the forests fragment and form refugia in a sea of savanna and grassland. Even when the climate becomes wetter and the forests expand and reconnect, these refugia remain the most species-rich forest patches, harboring most endemics. Dry conditions permitted the thorn savannas to form an arid corridor linking the Kalahari to the Masai Steppe and across to the Sahel. Connections like this allow widespread dispersal, explaining why birds like African Pygmy Falcon are found in arid thorn savannas on both sides of the continent, separated by moist woodland savanna today. It also explains the presence of closely related sister species like Pale and Eastern Chanting-Goshawks on either side of the (currently closed) arid corridor.

It is notable that the Southern African desert zone (Namib-Karoo) has always remained isolated, even during the driest spells, which is why it harbors so many regional endemics. But the Sahara and Horn of Africa deserts do connect and share elements. During the wettest spells, the great rainforest belt expands, creeping south into Zambia and north and east into Sudan. The coastal forests also expand and are tenuously connected to the Congo Basin forests via fingers of monsoon forest and riparian strips that cross the wet savanna belt, allowing dispersal and speciation between Africa's two great forest systems. Wetter times also allow the moist broadleaf savannas to expand, severing the arid corridor between the Kalahari and Somali-Masai thorn savannas. The true desert elements of the Sahara shrink to become pockets of desert within thorn savanna. These moist conditions also permit the colonization of the mountains in the Sahara by junipers and other Palearctic conifers.

These frequent and dramatic habitat fluctuations with repeated patterns of connection and separation drive speciation in forest and desert refugia through the processes of dispersal followed by isolation and differentiation. But the savannas are not isolated nearly as frequently, which may explain why so many of their species are more widespread generalists, adaptable and capable of living in a variety of savanna habitats.

The maximum extent of desert *(left)*, the maximum extent of rainforest *(middle)*, and "current" pre-industrial 1800s *(right)*.

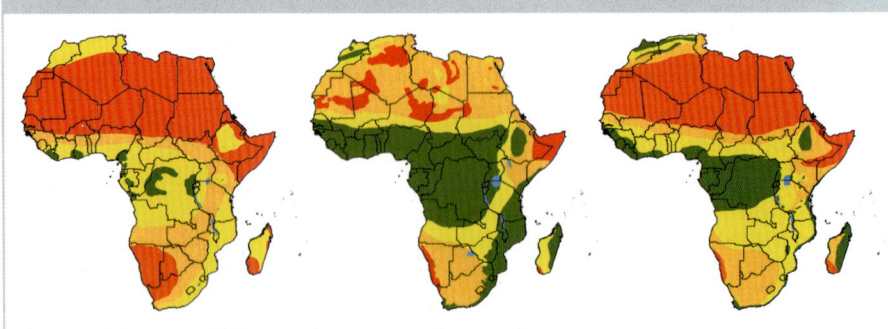

Green = rainforest zone, Yellow = moist savannas and monsoon forests, Orange = thorn savannas, and Red = hot shrub and true deserts.

Af6C KALAHARI DRY THORN SAVANNA

IN A NUTSHELL: Variably dry, open savanna of Southern Africa that is usually dominated by thorn trees and has grass cover during the rainy season. **Global Habitat Affinities:** NORTHERN ACACIA SHRUBLAND; CAMPO SUJO. **Continental Habitat Affinities:** NORTHERN DRY THORN SAVANNA. **Species Overlap:** MOIST MIXED SAVANNA; MOPANE; GUSU; MIOMBO; NAMA KAROO; NORTHERN DRY THORN SAVANNA.

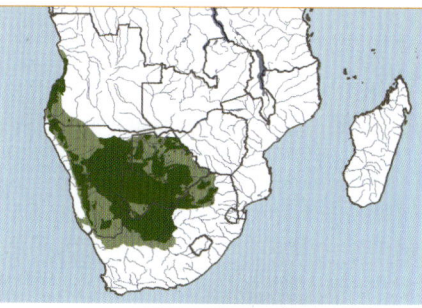

DESCRIPTION: This is a classic savanna environment in which there is variable cover from trees and bushes, and much or all of the ground is seasonally covered in grass. The core portion of this habitat is often referred to as the "Kalahari Desert." Kalahari comes from the Tswana language and is translated as "waterless place" or "great thirst." Although Kalahari Dry Thorn Savanna does grow predominately on the sands of the Kalahari basin, it is not a desert but rather a semiarid savanna (Köppen **Bwa**). Some portions of this habitat grow on non-Kalahari substrates, as along the Namib Escarpment and in n. South Africa.

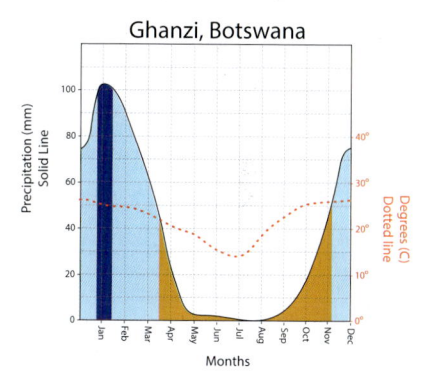

Ghanzi, Botswana

What makes this habitat distinctive is its generally dry and highly seasonal climate and the predominance of thorny and fine-leaved trees. During the long dry season, this habitat can indeed seem superficially almost barren and desertlike, but during the rains, it bursts into fantastic productivity. The grassland and woodland components of this habitat have fairly discrete sets of wildlife, and this account mainly treats the woodland aspect of this habitat; for more detail about the grassland aspect, see the AFROTROPICAL GRASSLAND account.

In Africa, dry thorn savanna is found in three distinct zones—in the Sahel of n. Africa and the Somali-Masai zone in the northeast (see NORTHERN DRY THORN SAVANNA) and in the Kalahari region of the southwest—which has driven a fascinating pattern of colonization and endemic speciation among the zones (discussed further in the Distribution section).

The exact structure of dry thorn savanna is variable. On sandy soil, it is usually an open landscape covered mainly in grassland with scattered trees and bushes. Subtle differences in the depth, water retention capacity, and clay content of the sand produce marked differences in the local vegetation. There are vast parallel dunefields, 10–25 ft. (3–8 m) tall, that run from se. Namibia, through the Kgalagadi, and well into nw. South Africa. The typical structure of vegetation here is taller trees in

Koronaberg, Tswalu, South Africa. Water sources in arid country like this attract small seedeaters like these Violet-eared Waxbills and Scaly Weavers. MAIN PHOTO © KEITH BARNES; INSET © KEN BEHRENS, TROPICAL BIRDING TOURS

the stabilized areas at the bases of the dunes and a complex mix of shrubs, grass, and bare sand on the dunes themselves, determined by aspect and sand stability. In rocky areas, bushes predominate, sometimes forming a dense, thicket-like structure. In general, this habitat is open enough to allow easy visibility and to make it easy to walk through, despite its thorny character. The height of the tallest trees is variable, though typically not taller than 23 ft. (7 m). In drier areas on the fringes of the Namib and Karoo, this habitat shrinks to a semi-desert shrubland in which the tallest woody plants are 3 ft. (1 m) or less. In moister areas, and along watercourses or other areas with groundwater, there can be isolated tall trees up to 65 ft. (20 m) tall.

During austral winter (June–September), this habitat has moderate daytime temperatures and cold nights, often below freezing. Summers (December–March) are hot, though still considerably cooler than in the lower-lying Northern Dry Thorn Savanna. The vast majority of the rain falls during the summer in a short rainy season of 1–4 months, often in intense thunderstorms. The

SIDEBAR 21   WATER IN DRY COUNTRY

Anyone spending time watching wildlife in the dry swath of sw. Africa that includes the Namib, Karoo, and Kalahari, will quickly find themselves at an oasis, borehole, or pond. Because in dry country, any water draws wildlife like a magnet. The quintessential example of this type of wildlife-watching is a late dry season safari in Namibia's Etosha National Park. The way to find big mammals and birds is essentially to drive from waterhole to waterhole. There can be thousands of big mammals and huge flocks of birds at the best spots, a remarkable wildlife spectacle by any measure. Water features, even tiny ones, can also be a great way to photograph birds, especially if you use your vehicle as a blind, or have a hide.

Af6C KALAHARI DRY THORN SAVANNA

Right: **Gemsbok in Mokala National Park, South Africa.** This photo shows typical dry thorn savanna structure: widely spaced thorn trees with grassy ground cover.
© KEITH BARNES, TROPICAL BIRDING TOURS

Below: **Ground Pangolin** is a nocturnal mammal that is sought after but very rarely seen by safari aficionados. The Kalahari remains the most likely habitat for an encounter.
© KEITH BARNES, TROPICAL BIRDING TOURS

rainfall is variable, typically totaling 6–20 in. (150–500 mm) annually, though it can fail entirely in some years. Fires have always occurred, especially at the onset of the rainy season, but have been made more frequent by humans.

Unlike moister savanna habitats, dry thorn savanna can lose most or all its grass during the dry season because of grazing, fire, and desiccation. Even at its wet season height, grasses don't exceed 3 ft. (1 m). Typical grasses include Kalahari Sour Grass (*Schmidtia kalahariensis*), Kalahari Sand Quick (*S. pappophoroides*), Digitgrass (*Digitaria eriantha*), Silver Wool Grass (*Anthephora argentea*), Wool Grass (*A. pubescens*), Sickle Grass (*Pogonarthria squarrosa*), Needlegrass (*Aristida meridionalis*), panicgrasses (*Panicum kalaharense* and *P. lanipes*), and several species of lovegrass (*Eragrostis*) and *Stipagrostis*. Red Oat Grass (*Themeda trianda*) formerly covered vast stretches, but has largely been eliminated by overgrazing.

The dominant tall tree is Camel Thorn (*Vachellia erioloba*), which usually grows in a classic umbrella form, a stately icon of the African savanna. Other trees tend to be shorter, often taking bushy form. Characteristic species include commiphoras, Shepherd's Tree (*Boscia albitrunca*), donkeyberry (*Grewia*), Buffalo Thorn (*Ziziphus mucronata*), Sicklebush (*Dichrostachys cinerea*), Bluebush (*Diospyros lycioides*), Lekkerbreek (*Ochna pulchra*), and a variety of acacias, including *Senegalia flekii*, *S. caffra*, and *S. mellifera* and *Vachellia luederitzii*, *V. hebeclada*, and *V. tortilis* (Umbrella Thorn, also a classic tree of Northern Dry Thorn Savanna).

Although this habitat grows mainly on deep sandy soil, it is also found locally in stony areas. This community is dominated by trees and shrubs, sometimes forming thickets. It is something of a

Giraffe against red dunes in Kgalagadi Transfrontier Park, South Africa. *(inset)* A close-up look at a typical flowering "acacia" in the genus *Vachellia*. © KEN BEHRENS, TROPICAL BIRDING TOURS

transitional microhabitat to MOIST MIXED SAVANNA, sharing that habitat's higher diversity of woody species. Typical species of stony dry thorn savanna are various acacias, Shepherd's Tree, golds (*Rhigozum*), Lavender Fever-Berry (*Croton gratissimus*), Leleshwa (*Tarchonanthus camphoratus*), Puzzle Bush (*Ehretia rigida*), gwarries (*Euclea*), False Olive (*Buddleja saligna*), donkeyberry, euphorbia, *Searsia* (formerly *Rhus*), Common Spike-thorn (*Gymnosporia heterophylla*), and Buffalo Thorn.

A classic Kalahari ground plant is the African Horned Cucumber (*Cucumis metuliferus*). The spiky fruits of this vine are a traditional source of both food and water. Succulents are locally common. The most frequent are euphorbias and aloes, but some areas also have *Crassula* and *Kalanchoe*. Bushwillows (*Combretum*) and/or terminalias are prominent in some areas, especially where this habitat transitions to moister savanna and woodland. Tamarisk grows in areas with saline soils. Most trees lose at least some of their leaves during the dry season, though the harshness of the environment and the small and fine nature of the leaves mean that there is rarely a significant accumulation of leaf litter.

This habitat is often quite similar to Moist Mixed Savanna, and these two mix with each other in South Africa, Botswana, and Zimbabwe. Moist Mixed Savanna is characterized by a higher overall diversity of trees, a more complex and varied structure, and by the local prominence of broadleaf species. In many dry areas that are dominated by dry thorn savanna, Moist Mixed Savanna can still be found locally along watercourses.

CONSERVATION: Although some portions of this habitat have been heavily grazed or had their larger trees cut, vast swaths are in good condition, with many protected areas and lots of big mammals. Botswana's Central Kalahari Game Reserve includes around 10% of the country's land area and is one of the largest protected areas on the continent.

WILDLIFE: In general, this is an excellent habitat for wildlife, harboring a high overall density and diversity of animals, which are easy to see owing to its open nature. It is an important habitat in

---

SIDEBAR 22 — DO GIRAFFES HAVE TOXIC NECK SYNDROME?

Scientists have come up with a range of intriguing explanations for the strangeness of the Giraffe. Some have hypothesized that as Africa's savannas expanded, new plants, including acacia thorn trees, with strong physical and chemical defenses, exposed these animals to a spike in toxins, which then increased their mutation rates, accelerated their evolution, and brought about this weird and wonderful enigma. This scenario does sound rather like science fiction though! Others suggest that their long necks evolved as a result of sexual selection, with longer-necked males winning more of their neck-bashing contests.

A Giraffe mother looks awkward stretching her bizarre neck down to tend to her youngsters. An odd evolutionary development indeed.
© KEITH BARNES, TROPICAL BIRDING TOURS

some of Africa's top safari destinations, such as Namibia's Etosha National Park and South Africa and Botswana's Kgalagadi Transfrontier Park. The wildlife of this habitat has strong affinities with the MOIST MIXED SAVANNA to its east and to a lesser degree to adjacent broadleaf woodlands (especially MOPANE).

The Kalahari Dry Thorn Savanna holds many classic African big mammals, such as Lion, Leopard, Spotted Hyena, Black Rhinoceros, Aardvark, African Savanna Elephant, Ground Pangolin, Plains Zebra, Common Warthog, and Giraffe. Bovid diversity and abundance are excellent, with species including Blue Wildebeest, Hartebeest, Springbok, Steenbok, African Buffalo, Common Eland, Greater Kudu, Gemsbok, and Impala. The attendant diversity of mammalian predators is remarkable, from widespread species like Cheetah, Honey Badger, and Caracal to more localized ones like Brown Hyena, Black-footed Cat, Cape Fox, Aardwolf, and Meerkat. A few mammals are shared with the Somali-Masai NORTHERN DRY THORN SAVANNA, evidence of the past link between these zones. These include Southern Springhare, Black-backed Jackal, and Bat-eared Fox.

The birding can be excellent, with a generally high bird density. Prominent groups include bustards, sandgrouse, doves, scrub-robins, starlings, shrikes, warblers, and sunbirds. A fairly

The Kalahari is famous for big mammals but also harbors many smaller species, such as the diminutive "Damara" Kirk's Dik-Dik.
© KEN BEHRENS, TROPICAL BIRDING TOURS

Although Aardvark is widespread in Africa's more open habitats, it is most frequently sighted in dry thorn savanna. © CHARLEY HESSE, TROPICAL BIRDING TOURS

Left: **The range of Crimson-breasted Gonolek is virtually the same as that of Kalahari Dry Thorn Savanna. This brilliantly colored bird livens up the drab dry season aspect of this habitat.** © KEN BEHRENS, TROPICAL BIRDING TOURS

Below: **Southern White-faced Owl, a fierce small nocturnal predator.** © KEN BEHRENS, TROPICAL BIRDING TOURS

large set of birds is endemic to the Kalahari zone, though none of these are exclusively found in dry thorn savanna but are shared with adjacent habitats. Classic Kalahari Dry Thorn Savanna birds include Burchell's (IS) and Namaqua Sandgrouse, Pale Chanting-Goshawk, Crimson-breasted Gonolek, Southern Pied-Babbler (IS), and Kalahari Scrub-Robin (IS). Fast-moving mixed flocks of midsize birds are a common sight and frequently include Cardinal Woodpecker, Common Scimitarbill, Brubru, Ashy Tit, Southern Penduline-Tit, Barred Wren-Warbler, Yellow-bellied Eremomela, Cape Crombec, Black-chested Prinia, Pririt Batis, and Golden-breasted Bunting. In the dry season, large flocks of seed-eating birds form and gather around waterholes. Typical constituents of such a flock include Red-billed Quelea (often in vast numbers), Southern Masked-Weaver, Violet-eared Waxbill, Southern Cordonbleu, Black-throated Canary, and Shaft-tailed Whydah. Raptors such as Lanner, Gabar Goshawk, and Tawny Eagle, hunt the abundant wildlife that is attracted to waterholes. A few birds, like Pygmy Falcon, Black-faced Waxbill, and Mariqua Sunbird, are shared with the Somali-Masai Northern Dry Thorn Savanna.

This habitat is also replete with conspicuous reptiles, including Flap-necked Chameleon (*Chamaeleo dilepis*), Rock Monitor (*Varanus albigularis*), sand (*Meroles, Nucras, Pedioplanis*) and long-tailed (*Latastia*) lizards, worm-lizards, Common Rough-scaled Lizard (*Meroles squamulosa*), plated lizards (*Gerrhosaurus*), agamas, skinks (*Trachylepis*), and geckos. Stilletto snakes (*Atractapsis*) like to burrow in soft sand. You have a greater chance of bumping into poisonous snakes in this habitat than in most other African habitats. Species include Shield Cobra (*Aspidelaps scutatus*), Snouted Cobra (*Naja annulifera*), Black Mamba (*Dendroaspis polylepis*), Puff Adder (*Bitis arietans*), and Horned Adder (*Bitis caudalis*).

There are not many frogs in this dry environment, but a few species can be found, mostly during the rainy season or in areas with water. These include Bushveld Rain Frog (*Breviceps adspersus*), Western Olive Toad (*Amietophrynus poweri*), Damaraland Pygmy Toad (*Poyntonophrynus damarensis*), Bubbling Kassina (*Kassina senegalensis*), Common Platanna (*Xenopus laevis*), Giant Bullfrog (*Pyxicephalus adspersus*), and Knocking Sand Frog (*Tomopterna krugerensis*).

Dry thorn savanna is a key component of the mix of habitats along the Namib-Angolan Escarpment, which has a set of endemics including Black Mongoose, Hartlaub's Francolin, Rockrunner, White-tailed Shrike, and Namibian Rock Agama (*Agama planiceps*).

DISTRIBUTION: The southwestern African Kalahari Dry Thorn Savanna is completely disjunct from the other two blocks of similar habitat (see NORTHERN DRY THORN SAVANNA). Over geological time, as Africa alternated between cool wet and warm dry periods, these zones would have been repeatedly isolated and then reconnected, as an arid corridor through Central Africa opened and closed, fueling colonization and subsequent speciation across the corridor (see sidebar 20, p. 234).

Ephemeral stream in Kalahari Dry Thorn Savanna, Caraculo, Angola. © KEN BEHRENS, TROPICAL BIRDING TOURS

In general, dry thorn savanna lies between true desert environments and broadleaf woodland or moist savanna environments. Kalahari Dry Thorn Savanna is mostly found from 2800 to 5000 ft. (850–1200 m), though in some places down to 2000 ft. (600 m), and up to 8200 ft. (2500 m) in the mountains of Namibia. In areas like ne. South Africa and Botswana, dry thorn savanna mixes into moister habitats and may be restricted to higher, drier areas. Along the verge of the Namib Desert and the Karoo plateau, dry thorn savanna extends only along drainage lines, forming Western Riparian Woodlands (subhabitat Af2-1), and elsewhere gives way to drier habitats in which trees are lacking.

WHERE TO SEE: Kgalagadi Transfrontier Park, South Africa and Botswana.

The formerly widespread African Wild Dog has been reduced to a few major strongholds, many of which are located in Kalahari Dry Thorn Savanna. © KEITH BARNES, TROPICAL BIRDING TOURS

Af6D MOIST MIXED SAVANNA

IN A NUTSHELL: Fairly moist and very diverse savanna that has a variety of both thorny, fine-leaved trees and broadleaf trees. **Global Habitat Affinities:** CERRADO SENSU STRICTO; DRY CHACO; OPEN EUCALYPT SAVANNA. **Continental Habitat Affinities:** NORTHERN DRY THORN SAVANNA; KALAHARI DRY THORN SAVANNA; EAST COAST FOREST MATRIX; SOUTH COAST FOREST MATRIX. **Species Overlap:** ALBANY THICKET; KALAHARI DRY THORN SAVANNA; NORTHERN DRY THORN SAVANNA; MOPANE; GUSU; GUINEA SAVANNA; MIOMBO.

DESCRIPTION: This habitat, which offers some of Africa's most exciting wildlife watching, may be the most complex of all those covered in this book. It is also one of the trickiest to define and is something of a catchall for several microhabitats that could be split out, except for the fact that these microhabitats occur in such a fine and complex tapestry that they are almost impossible to tease apart. The name "moist mixed savanna" itself falls short, given that this habitat can be quite dry at times. It merges broadleaf woodland/savanna habitats such as GUINEA SAVANNA and MIOMBO with the much drier thorn savanna.

This is generally a warm to hot and seasonally dry habitat (Köppen **Awa**), in which the vast majority of the rain falls during intense summer monsoonal rainy seasons produced by the Intertropical Convergence Zone, which moves north and south throughout the year. In the south, the rains fall during the austral summer (November–March), while in much of East Africa there are two rainy periods: "long rains" March–June and "short rains" September–December, and in the north during the boreal summer (June–September). Typical annual rainfall is 16–40 in.

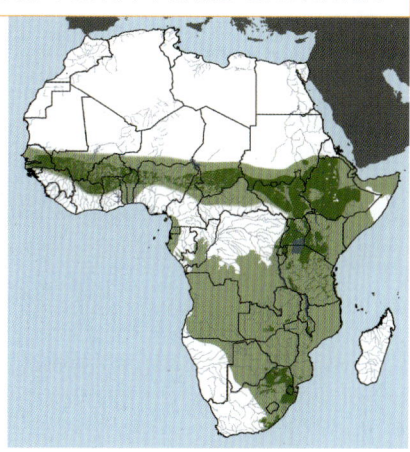

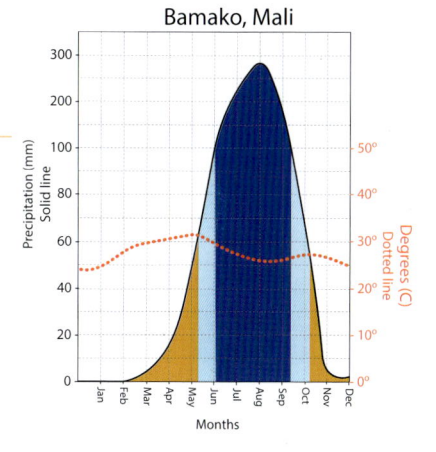

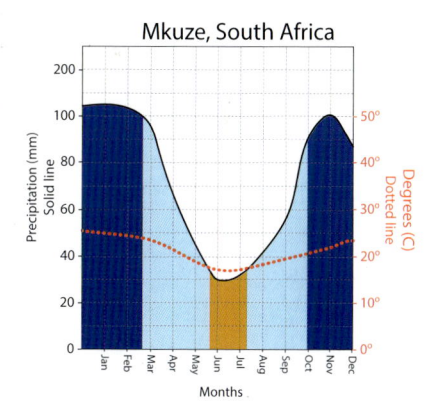

Moist Mixed Savannas often contain rich wetlands. South Luangwa National Park, Zambia.
© KEITH BARNES, TROPICAL BIRDING TOURS

(400–1000 mm). Fires have always been frequent, especially at the onset of the rainy season, but have been made more frequent by humans.

Two characteristics of Moist Mixed Savanna are complexity and diversity. Although it includes broadleaf components, broadleaf trees never dominate the way they do in Miombo, Guinea Savanna, MOPANE, or GUSU. Likewise, although acacia and commiphora are important trees here, they don't completely dominate the way they do in typical dry thorn savannas.

The astounding array of different sorts of very distinctive trees is part of what makes this habitat a delight for a naturalist to explore, and the great diversity here is exposed and easy to enjoy (not concealed, as in a tropical forest). This is a major reason visitors flock to places like South Africa's Kruger National Park, where the abundant wildlife is relatively easy to see and the landscapes invite exploration. There is great structural complexity to this habitat. Like all savannas, it has a grassy ground cover for at least part of the year. And like dry thorn savanna, it can be very open and grassy locally, merging into pure grassland in places, such as the Serengeti. In some areas, this habitat occurs in thicket form (sidebar 37, p. 335), which is not technically savanna as it can be so thick that there is no grassy ground cover. These thickets are included in this habitat because they have much the same bird and mammal species as the more open savanna. In areas with nutrient-poor soil, there can locally be a broadleaf-dominated Moist Mixed Savanna (subhabitat Af6D-1) that is reminiscent of Miombo or Guinea Savanna. Meanwhile, along lakes, rivers, and other watercourses, this habitat forms a lush riparian band with trees to 65 ft. (20 m) or taller (subhabitat Af6D-4). Here it approaches or blends into MONSOON FOREST, SOUTH COAST FOREST MATRIX, and EAST COAST FOREST MATRIX, or even AFROTROPICAL LOWLAND RAINFOREST on the northwest side of Lake Victoria. Along watercourses, narrow bands of moist savanna penetrate deep into arid climates where the dominant habitat is dry thorn savanna. Owing to its well-watered nature, both AFROTROPICAL DEEP FRESHWATER MARSH and more seasonal AFROTROPICAL SHALLOW FRESHWATER MARSH occur within this habitat, adding yet another dimension of diversity.

Af6D MOIST MIXED SAVANNA

Above: **Wet versus dry season aspect – drastically different! Serengeti NP, Tanzania** *(left)*; **Kruger NP, South Africa** *(right).* KEN BEHRENS, TROPICAL BIRDING

Below: **Extremely "mixed" Moist Mixed Savanna that even includes some Mopane trees. Another common component is Bushwillows (***Combretum***;** *inset***). Muar River, Mozambique.** MAIN PHOTO © KEN BEHRENS; *INSET* © KEITH BARNES, TROPICAL BIRDING TOURS

246 SAVANNAS

In terms of trees, various acacias (*Vachellia* and *Senegalia*) are always prominent, especially in lower-lying areas with rich clay soils, as along rivers and other watercourses. These and other spiny trees and bushes such as commiphoras, jujubes (*Ziziphus*), and Sicklebush (*Dichrostachys cinerea*) are much more common than in broadleaf habitats. One distinctive acacia is the yellow-barked Fever Tree

SIDEBAR 23 — MOIST MIXED SAVANNA MICROHABITATS

Tall Whistling Thorn savanna. Serengeti National Park, Tanzania. © KEN BEHRENS, TROPICAL BIRDING TOURS

Euphorbia savanna. Queen Elizabeth National Park, Uganda. © KEN BEHRENS, TROPICAL BIRDING TOURS

Fever Tree Woodland (subhabitat Af6D-3). Nakuru National Park, Kenya. © KEN BEHRENS, TROPICAL BIRDING TOURS

Hilltop Broadleaf Savanna (subhabitat Af6D-1) on a rocky hilltop. Tarangire National Park, Tanzania. © KEN BEHRENS, TROPICAL BIRDING TOURS

Palm Savanna (subhabitat Af6D-2). Lower Cuanza River, Angola. © ARNON DATTNER, ECOPLANÉT FILMS

(*Vachellia xanthophloea*), which often predominates in seasonally flooded areas, forming one of Africa's most distinctive and beautiful microhabitats (Af6D-3). In areas with poor and well-drained sandy soil, as on the tops of hills, the savanna supports mainly broadleaf trees, including terminalias, bushwillows (*Combretum*), and Marula (*Sclerocarya birrea*). Moist Mixed Savanna is one of the key habitats for Africa's most famous tree, the mighty Baobab (*Adansonia digitata*). Other typical trees include *Albizia*, figs (*Ficus*), Jackalberry (*Diospyros mespiliformis*), Sausage Tree (*Kigelia africana*), and Nyala Tree (*Xanthocercis zambesiaca*). Palms, including Wild Date Palm (*Phoenix reclinata*) and Hyphaene, are more abundant and prominent in this habitat than in any other. Palms sometimes dominate a large stretch of country, forming yet another striking microhabitat under the Moist Mixed Savanna umbrella (Af6D-2). Common and characteristic trees in West Africa include Kosso (*Pterocarpus erinaceus*), African Birch (*Anogeissus leiocarpus*), Lannea, Red Kapok Tree (*Bombax costatum*), *Sterculia setigera*, African Mesquite (*Anonychium africanum*), and African Mahogany (*Afzelia africana*). In mountains, where the transition occurs from Moist Mixed Savanna to montane forest, there is sometimes a distinctive form of woodland made up of Flat-topped Acacia (*Vachellia abyssinica*), Lahai Acacia (*V. lahai*), or both (see sidebar below). These stately trees can form a nearly interlocking canopy in the thickest versions of this habitat, quite a striking and beautiful landscape.

Forest tends to turn into savanna-like habitat when it has been degraded by humans or natural climate change. Much of the Moist Mixed Savanna habitat in Africa may have been forest originally,

SIDEBAR 24 EDGE CASE STUDY: HIGHLAND ACACIA WOODLAND

A beautiful and distinctive habitat is found in mountains all the way from Ethiopia to Malawi: high-elevation Acacia woodland, dominated by a couple species, mainly *Vachellia lahai* and *V. abyssinica*. The question is, which of our "Habitats of the World" categories does it fit into, or is it a full habitat in its own right?! This is one example of countless cases where the authors had to make tricky decisions about what to cover and how to cover it. Although this highland acacia has one virtually endemic bird (Brown Parisoma) and shares some species with both MOIST MONTANE FOREST and AFROTROPICAL MONTANE DRY MIXED WOODLAND, we decided to keep it within MOIST MIXED SAVANNA, since the majority of its wildlife is shared with that habitat and its structure and acacia-dominated nature fit within our understanding of Moist Mixed Savanna.

Vachellia in the Ndundulu Mountains of Tanzania is perfect for Brown Parisoma.
© KEITH BARNES, TROPICAL BIRDING TOURS

especially on the Ethiopian Highlands, along the Indian Ocean coast, and around Lake Victoria. It is likely that this habitat has never before been so common on the continent. But unlike human-created habitats in forest-dominated areas, which are remarkably devoid of wildlife, this anthropogenic savanna is often incredibly rich in wildlife, likely due to rapid colonization from adjacent natural savannas.

CONSERVATION: Many well-protected areas include Moist Mixed Savanna, but this habitat certainly has its conservation challenges. Highly organized poaching for rhinos and elephants has reached catastrophic levels in recent years. In more populated areas, cultivation, grazing, hunting, and cutting for firewood and charcoal have eliminated most of the big mammals. The situation is especially grave in West Africa, where most big mammals are only a distant memory.

WILDLIFE: This locally very diverse habitat maintains a surprising cohesiveness across its wide range. Most of the species in e. South Africa are the same as those in c. Tanzania. There is a significant difference between the north and the east, and within the eastern Moist Mixed Savanna there is also a weak north–south division, with a few more restricted-range species found on either side. Diversity and complexity also characterize the animals that use this habitat. Very few are confined to Moist Mixed Savanna; rather, species are shared with adjacent habitats such as MIOMBO, GUINEA SAVANNA, MOPANE, GUSU, KALAHARI DRY THORN SAVANNA, NORTHERN DRY THORN SAVANNA, MONSOON FOREST, EAST COAST FOREST MATRIX, SOUTH COAST FOREST

African Savanna Elephants in a typically diverse and complex Moist Mixed Savanna landscape, with a mix of wetland, grassland, broadleaf, and thorny components. Tarangire National Park, Tanzania.
© KEN BEHRENS, TROPICAL BIRDING TOURS

MATRIX, ALBANY THICKET, and even montane forest. This is a crossover habitat that serves to connect most of Africa's other habitats. The majority of Africa's birds and larger mammals are found in at least two habitats, and Moist Mixed Savanna is situated perfectly, near the middle of the moisture-driven continuum of habitats, allowing it to provide refuge for a remarkably high proportion of the continent's species.

One of the best places on Earth for big mammals, Moist Mixed Savanna hosts widespread and charismatic beasts like African Savanna Elephant, African Wild Dog, Spotted Hyena, Lion, Leopard, White and Black Rhinoceroses, Giraffe, and African Buffalo. A few mammals that are especially typical of this habitat, though not endemic to it, are Side-striped Jackal, Plains Zebra, Greater Kudu, Nyala, Waterbuck, and Impala. The northern band of Moist Mixed Savanna is far poorer in mammals due to the depravations of political instability and widespread hunting.

The bounty of different microhabitats makes Moist Mixed Savanna a birding paradise, often harboring huge bird diversity within a small area. Indeed, this is the main habitat in two national parks reputed to have some of the longest bird lists in Africa: Uganda's Queen Elizabeth National Park and South Africa's Kruger National Park. Moist Mixed Savanna is prime habitat for some of Africa's most beguiling birds, species even mammal-obsessed tourists cannot ignore, like Common Ostrich, Helmeted Guineafowl, Bateleur, ground-hornbills, oxpeckers, and the extremely well-photographed Lilac-breasted Roller. Other classic birds include Dark Chanting-Goshawk, Purple-crested Turaco, Crested Francolin (IS), Brown-headed Parrot, a variety of cuckoos, Green Woodhoopoe, Brown-hooded Kingfisher, Golden-tailed and Bearded Woodpeckers, Black-headed Oriole, Bearded Scrub-Robin (IS, except in Ethiopia and n. Africa), Chinspot Batis, Black-backed Puffback, Southern

Silverbird is a classic species of Moist Mixed Savanna, favoring this habitat over adjacent broadleaf and drier thorn savannas. © KEITH BARNES, TROPICAL BIRDING TOURS

A "tower" of Giraffe in the Masai Mara, Kenya. Many of the trees in this photo are thorny, and this habitat verges on being dry thorn savanna, but there are some more diverse broadleaf components present, especially on the hillside. © KEN BEHRENS, TROPICAL BIRDING TOURS

and Tropical Boubous, Sulphur-breasted and Gray-headed Bushshrikes, Greater Blue-eared Starling, Scarlet-chested Sunbird, various weavers, and Green-winged Pytilia.

Southern areas hold Crested Barbet, White-throated Robin-Chat, and White-breasted Sunbird. The northeast has Silverbird and Variable Sunbird. There is a divide between the eastern and northern Moist Mixed Savanna. The eastern savanna is more complex and diverse overall, with higher bird diversity, whereas the northern savanna has virtually no birds that aren't also found in adjacent dry thorn and/or Guinea savanna. The Moist Mixed Savannas in the Ethiopian Highlands are the primary habitat of several localized endemics, including Black-winged Lovebird, Banded Barbet, and one of Africa's most enigmatic and sought-after birds, Prince Ruspoli's Turaco.

DISTRIBUTION: Moist Mixed Savanna is widespread, from Senegal across to the Ethiopian Highlands and s. Arabian Peninsula, south to e. South Africa, and southwest to Angola. Owing to its somewhat amorphous nature, the limits of Moist Mixed Savanna are tricky to define, especially at the edge of the **KALAHARI DRY THORN SAVANNA** in s. Africa and the edge of the **GUINEA SAVANNA** in n. Africa. It tends to be found in areas with more moisture than dry thorn savannas but less moisture than **MONSOON FOREST**, coastal

Mwanza Flat-headed Agama (*Agama mwanzae*), one of the gaudiest and most conspicuous reptiles of the Serengeti region. © KEN BEHRENS, TROPICAL BIRDING TOURS

Af6D MOIST MIXED SAVANNA 251

forest matrices, or **AFROTROPICAL LOWLAND RAINFOREST** and in more mixed soils and complicated topography than **MIOMBO** and **GUSU** broadleaf woodland. There is also a broad zone of this habitat all the way from West Africa to the Horn of Africa, between the dry thorn savanna of the Sahel and the broadleaf Guinea Savanna, though the farther west one goes in West Africa, the narrower this band becomes. This sector of Moist Mixed Savanna is sometimes called "Sundaic savanna." The highlands of Ethiopia rise like an island of moister habitats in a sea of aridity. Although **AFROTROPICAL MONTANE DRY MIXED WOODLAND** and **MONTANE GRASSLAND** are the classic Ethiopian habitats, this area has also given rise to a montane form of Moist Mixed Savanna. Humans probably played a role in converting highland forest into savanna.

WHERE TO SEE: Kruger National Park, South Africa; Queen Elizabeth National Park, Uganda; Tarangire National Park, Tanzania; Niokolo-Koba National Park, Senegal.

Right: Fire is a central force in this habitat, and some of its wildlife is quick to take advantage of insects fleeing the danger. Fork-tailed Drongo, Mkuze Game Reserve, South Africa. © KEN BEHRENS, TROPICAL BIRDING TOURS

Below: **Black-chested Snake-Eagle and Blue Wildebeest in the Masai Mara, Kenya. Martial Eagle** *(inset)* **is the most fearsome avian predator of this habitat.** MAIN PHOTO © KEITH BARNES; INSET © KEN BEHRENS, TROPICAL BIRDING TOURS

Af6E MIOMBO

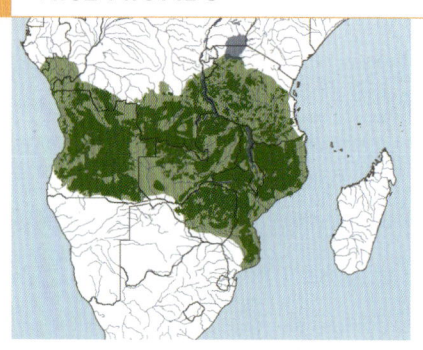

IN A NUTSHELL: Central African savanna habitat that is dominated by broadleaf trees, especially Miombo (*Brachystegia*). **Global Habitat Affinities:** TETRADONTA WOODLAND SAVANNA; BRIGALOW; HUMID CHACO; INDIAN DRY DECIDUOUS FOREST. **Continental Habitat Affinities:** GUINEA SAVANNA; GUSU; MOPANE; EAST COAST FOREST MATRIX. **Species Overlap:** GUSU; MOPANE; MOIST MIXED SAVANNA; GUINEA SAVANNA; EAST COAST FOREST MATRIX; NORTHERN DRY THORN SAVANNA; KALAHARI DRY THORN SAVANNA.

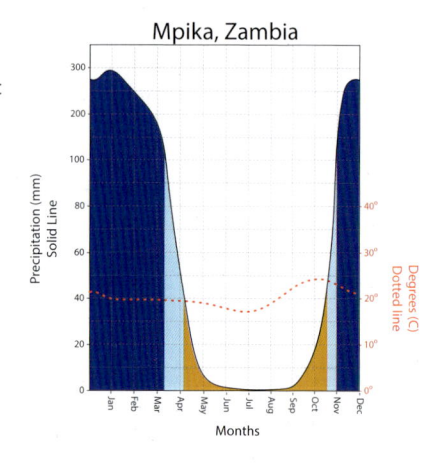

DESCRIPTION: This is a fairly tall broadleaf woodland with abundant grass and distinct wet and dry seasons (Köppen **Ama**). The canopy is typically at around 40 ft. (12 m), though Miombo has a couple of different subhabitats; on stony ground or outcrops Dry and Rocky Miombo (subhabitat Af6E-1) can be stunted and as short as 12 ft. (4 m), while on better developed soils and in higher rainfall regions Moist and Rich Miombo (subhabitat Af6E-2) can be lush, with patches as tall as 60 ft. (18 m). The soils on which Miombo grows are generally poor, usually acidic, heavily leached, and well drained, resulting in a low availability of nutrients compared with other savanna types. Miombo is generally open, with less than 50% canopy cover, allowing abundant light to reach the ground and permitting the growth of thick grass, along with broadleaf shrubs and herbaceous plants. It is park-like and easy to walk through during the dry season, and there is abundant leaf litter. During the wet season, the lush grass (up to 6 ft./2 m tall) and flooded drainage lines make it harder to move through. From April to September, Miombo sees little rainfall, while heavy rains fall during the austral summer. The typical annual precipitation range is 24–60 in. (600–1500 mm).

SIDEBAR 25 — MIOMBO ETYMOLOGY

"Miombo" is a word from the Bemba language, a Bantu tongue that is most prevalent in ne. Zambia but which is used widely in Zambia among speakers of closely related languages. The word Miombo refers to all the trees from the genus *Brachystegia*.

The "springtime flush" in Miombo feels oddly autumnal to those visiting from the Northern Hemisphere. Dzalanyama Forest Reserve, Malawi. © KEITH BARNES, TROPICAL BIRDING TOURS

The vast majority of trees are broadleaf species, which lose some or all of their leaves during the dry season. Many species lose their leaves toward the end of the dry season, then replace them with reddish new foliage, creating the "Miombo flush," which gives this woodland a strangely autumnal feeling, albeit during the austral spring. In the lowest-lying, driest, and hottest areas, Miombo can be completely deciduous, losing all leaves for up to two months.

The most important canopy trees are the namesake Miombo trees (*Brachystegia*), as well as Munondo (*Julbernardia globiflora*), Muchesa (*J. paniculata*), and Doka (*Isoberlinia doka*). These trees dominate Miombo habitat, but their precise mix varies locally, both by microhabitat and between the four major Miombo blocks (see Endemism below). All tend to have a classic lollipop shape, with a rounded leafy ball emerging from a branchless trunk. They are all members of the legume family and share a characteristic leaf type: pinnately compound leaves lacking a terminal leaflet. The leaves vary by species, bearing a few large oval leaflets to many small, fine leaflets. The prevalence of this leaf type gives this forest a distinctive look, recognizable even at a distance. Trees with pinnately compound leaves occur in other habitat types, especially MOIST MIXED SAVANNA, but except in GUSU and GUINEA SAVANNA they rarely predominate to the degree they do in Miombo. Miombo is the southern equivalent of the northern Guinea Savanna. Both are open woodlands with grassy ground cover. Although Doka is important in both habitats, Miombo is characterized by the additional presence of Miombo, Munondo, and Muchesa trees. Miombo is also generally taller, more complex in structure and has a more diverse plant composition. In terms of wildlife, there is only a modest degree of overlap between them.

Although Miombo is dominated by the few tree genera mentioned above, it does hold many other tree species, such as African Teak (*Pterocarpus angolensis*), Pod Mahogany (*Afzelia quanzensis*), *Pericopsis*, *Faurea saligna*, African Blackwood (*Erythrophleum africanum*), African Medlar (*Vangueria*

254 SAVANNAS

Dry season fires dramatically transform Miombo and trigger the growth of many flowering plants. Dzalanyama Forest Reserve, Malawi. © KEN BEHRENS, TROPICAL BIRDING TOURS

Stunted Miombo on rocky hillsides is strikingly different from taller "lollipop" Miombo growing on richer, deeper, better-developed soils. Morro de Moco, Angola. © KEN BEHRENS, PROMISED LAND VENTURES

infausta), *Anisophyllea pomifera*, *Diospyros batocana*, bushwillows (*Combretum*), Muriranyenze (*Albizia antunesiana*), and monkey-orange (*Strychnos*). Stunted Miombo, as along wetland edges and on rocky areas, often supports proteas. Ferns, lianas, and mosses are lacking, while the Spanish Moss-like old man's beard lichen (*Usnea*) is often abundant. The smaller drainage lines in Miombo support open, seasonally flooded grasslands, along with a core of permanent AFROTROPICAL SHALLOW FRESHWATER MARSH. This type of microhabitat within the Miombo zone is called "Dambo." Larger drainage lines and rivers give rise to deeper perennial wetlands, thickets, and MONSOON FOREST.

Fire has always played an important role in Miombo, burning the grass and stimulating new growth, especially at the beginning of the wet season. Today, more frequent fires are started by humans, gradually opening up the habitat. Although Miombo is clearly a natural habitat in many areas, in higher-rainfall areas with frequent human disturbance, it has likely replaced moister and more closed-canopy habitats such as Monsoon Forest. Another profound influence on this habitat is termites, which occur in vast numbers, their towering mounds dotting the landscape.

CONSERVATION: Poor soil, prevalence of the tsetse fly, and recent civil wars have resulted in low human populations in the Miombo realm across much of Angola, Zambia, Tanzania, and Mozambique. Other areas, such as Malawi and Zimbabwe, are heavily populated, and here theoretically protected areas such as Dzalanyama Forest Reserve are being rapidly and heavily logged by charcoal gatherers, despite being a critical watershed for Malawi's capital city, Lilongwe. Hunting is widespread, even in remote areas. Other major threats to Miombo habitat include tree clearing, frequent burning, and overgrazing.

Miombo is one of the favorite habitats of the scarce Sable Antelope. © KEN BEHRENS, TROPICAL BIRDING TOURS

WILDLIFE: Owing to its poor soil and the relatively low availability of nutrients, Miombo has low diversity and density of wildlife compared with other African savanna and woodland habitats. Conditions are most favorable for grazing herbivores with a large body size, which are able to process large amounts of nutrient-poor feed during the dry season. Many mammals and birds also move around during the year, using seasonally flooded AFROTROPICAL GRASSLAND, MONSOON FOREST, and other habitats to get through the extended dry period. Despite its open nature, which is reminiscent of other wildlife-filled savanna environments, Miombo is very different in that you can walk for long periods without seeing much. It feels uncannily quiet until you bump into a feeding flock of birds or enter a glade full of browsing Puku antelope. This is especially true away from the busy austral spring breeding season.

In general, Miombo has a reduced subset of the savanna and grassland mammals typical of southern and eastern Africa. It is the stronghold of the beautiful Sable Antelope, lanky Yellow Baboon, and shy Miombo Genet. African Savanna Elephant is still widespread. Southern Tanzania's Selous Game Reserve has more elephants than anywhere else on the continent, though even here the population has been reduced by poaching from over 100,000 individuals to around 15,000. Selous is also the global stronghold of the endangered African Wild Dog. Other widespread mammals of Miombo include Malbrouck, Vervet Monkey, Side-striped Jackal, Greater Kudu, Bushbuck, Common Duiker, Oribi, and Puku (mainly during the rainy season). Large predators, including Lion, Leopard, and Spotted Hyena, are still fairly common in some areas. Owing to the abundance of ants and termites, Aardvark and Ground Pangolin can be common in Miombo woodland, though they are still hard to see.

Miombo has excellent bird diversity, though it tends to be less diverse than adjacent savannas. A number of birds are restricted to this habitat, but few are found throughout, and many are generally uncommon. In terms of endemic birds, the richest area lies near the center of the biome, in n. and c. Zambia. Miombo endemic birds include Anchieta's, Whyte's, and Miombo Barbets; Pale-billed Hornbill (IS); Miombo Rock-Thrush (IS); Eastern and Western Miombo Sunbirds (IS); Miombo Tit; Miombo Scrub-Robin; Red-capped Crombec; Black-necked Eremomela; Böhm's Flycatcher; Anchieta's Sunbird; Lesser Blue-eared Starling (IS); Chestnut-backed Sparrow-Weaver; and Olive-headed and Bar-winged Weavers. Many bird species are shared by Miombo and other lush savanna and woodland habitats in the region, and some of these are more common in Miombo. Typical nonendemic Miombo birds include Bronze-winged Courser, Pennant-winged Nightjar, Racket-tailed Roller, White-breasted Cuckooshrike, African Golden Oriole, Rufous-bellied Tit, African Spotted Creeper, Arnot's Chat, Greencap Eremomela, Miombo and Stierling's Wren-Warblers, Yellow-bellied Hyliota, Retz's Helmetshrike, Western Violet-backed Sunbird, Orange-winged Pytilia, Broad-tailed Paradise-Whydah, and Cabanis's Bunting. The areas of Monsoon Forest that occur within Miombo are species poor but do support a small set of special birds.

Miombo does not have high reptile diversity, though it does have dozens of species of endemic snakes and lizards, most of which are inconspicuous. One more localized reptile that favors Miombo, and which can be seen during the day, is Angola Dwarf Gecko (*Lygodactylus angolensis*). Widespread and charismatic species of Miombo include Graceful (*Chamaeleo gracilis*) and Flap-necked Chameleons (*C. dilepis*), Cape Rough-scaled Lizard (*Ichnotropis capensis*), Rock Monitor (*Varanus albigularis*), rock pythons (*Python sebae* and *P. natalensis*), Black Mamba (*Dendroaspis polylepis*), Black-necked Spitting Cobra (*Naja nigricollis*), Puff Adder (*Bitis arietans*), Kenyan Bark Snake (*Hemirhagerrhis hildebrandtii*), and Black-necked Agama (*Acanthocercus atricollis*).

Miombo is often draped in *Usnea* lichen. African Spotted Creeper is part of an odd bird genus with only two members, the other occurring in India. MAIN PHOTO © KEN BEHRENS; INSET © KEITH BARNES, TROPICAL BIRDING TOURS

Classic Miombo, with well-spaced lollipop-shaped broadleaf trees and a lush ground cover of grass. Mount Namba, Angola. © KEITH BARNES, TROPICAL BIRDING TOURS

Endemism: Although this habitat is fairly contiguous over a vast area, it is divided into four major blocks: (1) Angola, (2) Northern Zambia, s. DRC, w. Tanzania, and Malawi, (3) s. Zambia and Zimbabwe, and (4) s. Tanzania and Mozambique. There are significant botanical differences between each block. Most endemic birds and reptiles are not found throughout the Miombo but rather in as few as one or as many as three of the major blocks. Another distinction that is often drawn within Miombo is between drier (less than 3.3 ft./1 m precipitation annually) and wetter (more than 3.3 ft. of annual precipitation) Miombo. The wetter Miombo is taller (more than 50 ft./15 m) and far richer botanically.

DISTRIBUTION: This is the prevalent habitat in a vast swath of sc. Africa, from Angola east to nw. Tanzania and south to s. Mozambique. Most of this area is on the Central African Plateau, at elevations of 3300–5000 ft. (1000–1500 m), though portions in Tanzania and Mozambique are much lower, down to 650 ft. (200 m). Most of the terrain is mildly undulating, with isolated rocky outcrops in some areas. Miombo is closely allied with two broadleaf woodland habitats that occur to its south and serve as the transition to the KALAHARI DRY THORN SAVANNA: MOPANE and GUSU. It also shares many species with the MOIST MIXED SAVANNA to the south and east, though it has lower diversity. Throughout the Miombo, there are patches of MONSOON FOREST along drainage lines and rivers; these forests support a mix of species from EAST COAST FOREST MATRIX and the AFROTROPICAL LOWLAND RAINFOREST of the Congo Basin and provide an important link between these regions.

WHERE TO SEE: Kasanka National Park, Zambia; Dzalanyama Forest Reserve, Malawi; Selous Game Reserve, Tanzania.

Pennant-winged Nightjar roosting on a *Brachystegia* branch. If you imagine that this bird is spectacular in flight, you imagine correctly! © KEITH BARNES, TROPICAL BIRDING TOURS

SIDEBAR 26 — SAVANNA ICEBERGS

Termite mounds are an iconic part of the African savanna landscape but occur in most of the continent's habitats. They provide refuge for a variety of wildlife, such as mongooses and reptiles, and are excellent observation posts for mammals. What many people don't realize is that termite mounds are like an iceberg: two-thirds of their structure is below the ground! So, while what is visible can be massive, the subterranean structure is even more expansive.

Vervet Monkey using an elevated termite mound as a lookout point in Tarangire National Park, Tanzania. © KEN BEHRENS, TROPICAL BIRDING TOURS

Af6F GUINEA SAVANNA

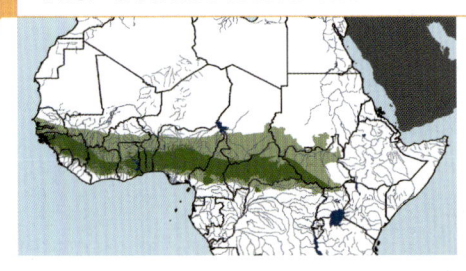

IN A NUTSHELL: Northern African savanna habitat that is dominated by broadleaf trees. **Global Habitat Affinities:** BRIGALOW; GRASSY MULGA; HUMID CHACO; SOUTHEAST ASIAN DRY DECIDUOUS FOREST. **Continental Habitat Affinities:** MIOMBO; GUSU; MOPANE. **Species Overlap:** MOIST MIXED SAVANNA; NORTHERN DRY THORN SAVANNA; AFROTROPICAL LOWLAND RAINFOREST; MIOMBO.

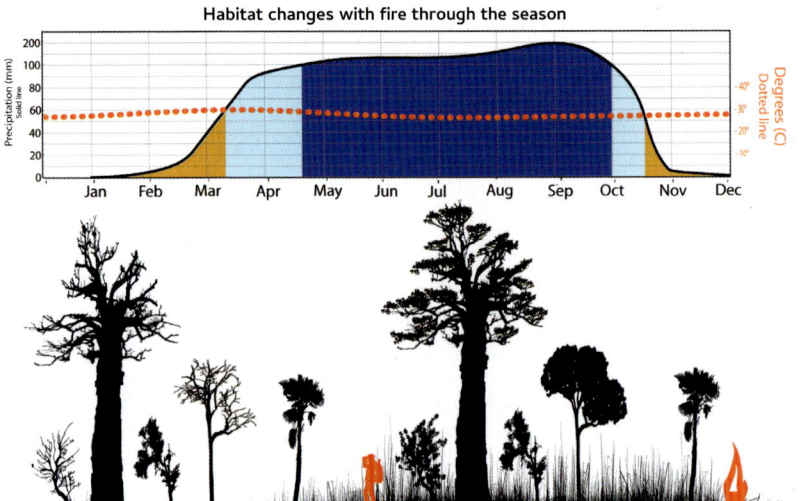

Habitat changes with fire through the season

Larabanga, Ghana

DESCRIPTION: Guinea Savanna is moderately tall broadleaf woodland with abundant grass on the ground and distinct wet and dry seasons (Köppen **Awa**). The canopy ranges from 16–60 ft. (5–18 m) and is generally open, with less than 50% tree cover. This allows abundant light to reach the ground, permitting the growth of a thick cover of mostly perennial grasses, which persists year-round where not burned or grazed. This contrasts with the NORTHERN DRY THORN SAVANNA of the Sahel to the north, where most grasses are annual and grass cover can largely disappear during the dry season. Guinea Savanna is park-like and easy to walk through during the dry season, when much of the ground cover has dried up and burned or been eaten by herbivores. During the wet season, the abundance of tall grass makes it hard to penetrate. For half of the year, Guinea Savanna sees little rainfall, while heavy rains fall during June–October. The typical annual precipitation range is 47–67 in. (600–1700 mm). Temperatures remain warm year-round, typically

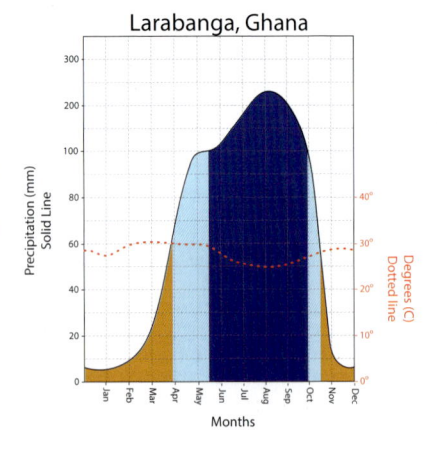

Dry season Guinea Savanna, when many trees are completely or largely leafless and the ground is covered with leaf litter. Mole National Park, Ghana. © KEITH BARNES, TROPICAL BIRDING TOURS

staying between 70–95°F (20–35°C). Fire plays an important role, burning grasses and stimulating new growth, especially at the beginning of the wet season when the understory is still dry and thunderstorms bring lightning strikes. Today most fires are started by humans, resulting in a much higher frequency of burning, and gradually opening up the habitat.

The vast majority of trees are broadleaf species that lose some or all of their leaves during the dry season. The most important of these is doka (*Isoberlinia*), which dominates vast stretches of this habitat, and the Guinea Savanna is sometimes called "Doka" locally. In some areas, doka is absent or less important than other tree species, which can include African Blackwood (*Erythrophleum africanum*), Khaya (*Khaya senegalensis*), Senegal Mahogany (*Khaya senegalensis*), Grenadilla (*Dalbergia melanoxylon*), African Mahogany (*Afzelia africana*), West African Copal Tree (*Daniellia oliveri*), Wild Seringa (*Burkea africana*), African Locust Bean (*Parkia biglobosa*), and Shea (*Vitellaria paradoxa*). Bushwillows (*Combretum*), terminalias, or both dominate some areas, especially in the northeastern and southeastern extensions of Guinea Savanna and in the north where it grades into MOIST MIXED SAVANNA. The most common grass is thatching grass (*Hyparrhenia*), which forms a tall and lush ground cover during the rainy season. In many areas, smaller drainage lines support open AFROTROPICAL GRASSLAND that is seasonally flooded. Larger drainage lines and rivers give rise to deeper perennial wetlands, thickets, and riparian AFROTROPICAL LOWLAND RAINFOREST. These moister corridors allow some rainforest species to penetrate deep into this savanna zone.

Guinea Savanna is the northern equivalent of the MIOMBO of eastern and southern Africa. Both are open woodlands with grassy ground cover. Although Miombo also has doka trees in some areas, its most dominant and important trees are *Brachystegia* species, commonly known as "miombo." Miombo habitat is also generally taller, with a more complex structure and more diverse plant composition. In terms of wildlife, there is only a modest degree of overlap between Guinea Savanna and Miombo.

Left: **Wet season Guinea Savanna is lush and green, with abundant grass. Mole National Park, Ghana.** © KEN BEHRENS, TROPICAL BIRDING TOURS

Below: **Guinea Savanna growing among isolated rocky hills in Shai Hills Reserve, Ghana; perfect habitat for Stone Partridge.** MAIN PHOTO © KEN BEHRENS; INSET © KEITH BARNES, TROPICAL BIRDING TOURS

Af6F GUINEA SAVANNA

SIDEBAR 27 — **WEST AFRICAN DRY VERSUS WET SEASON SAVANNA TRANSFORMATIONS**

Northern Dry Thorn Savanna, top two rows; Moist Mixed Savanna, third and fourth rows; Guinea Savanna, fifth row. Senegal. © GRAY TAPPAN

The horse-like Roan Antelope is a classic Guinea Savanna mammal, though these two individuals were photographed in Kalahari Dry Thorn Savanna. © KEN BEHRENS, TROPICAL BIRDING TOURS

CONSERVATION: Portions of this vast belt are remote and thinly populated by humans, such as ne. Guinea and sw. Mali. Others, such as central and northern Nigeria, are heavily populated. Unfortunately, hunting is widespread, even in remote areas, and has wiped out most of the big mammals. Other major threats include tree clearing, frequent burning, and overgrazing. There are few national parks, and most of those that exist are poorly protected.

WILDLIFE: This habitat historically supported a wide variety of large mammals, though it always had lower diversity than do the varied savannas of East Africa. Unfortunately, the populations of many big mammals have been decimated by hunting and habitat modification. This is the stronghold of Kob (IS) and Giant Eland, though the latter is now restricted mostly to tiny pockets in Central African Republic, Cameroon, and Senegal. It is also one of the strongholds of Roan Antelope, which remains in good numbers, perhaps because of its wary nature. Widespread mammals include Olive Baboon; Patas, Green, and Tantalus Monkeys; Gambian Sun Squirrel; Common Warthog; Bushbuck; Common Duiker; Oribi; and Waterbuck. Black Rhinoceros and White Rhinoceros were once common, but Black Rhino hangs on the brink of local extinction, and the northern subspecies (*cottoni*) of White Rhino is now extinct in the wild. African Buffalo, Lion, Leopard, Cheetah, and African Wild Dog persist, though in ever-shrinking enclaves. This is excellent habitat for African Savanna Elephant, though it has also been widely extirpated.

The wings of the Standard-winged Nightjar are anything but "standard"! © KEITH BARNES, TROPICAL BIRDING TOURS

SIDEBAR 28 — AFRICA'S VANISHING MAMMALS

Africa has retained more of its original big mammal assemblage than any other continent. But that doesn't mean there haven't been significant losses. Both White and Black Rhinos have been eliminated from the vast majority of their historical ranges because of demand for their horns, and elephants have also been greatly reduced in numbers and lost over half of their range, in part because of ivory hunting. Larger predators have also suffered. While the secretive, solitary, and largely nocturnal Leopard retains most of its historical range, diurnal and social predators like African Wild Dog and Lion have been decimated. Smaller herbivores have generally done better, though virtually all the large mammals have been eliminated from vast swaths of n. Africa. An ongoing problem for herbivores is fences, some of which are associated with protected areas, which restrict the movement of animals that have evolved to migrate seasonally. While Africa has so far been spared the mass extinctions of big mammals that have taken place on every other continent, that may not remain the case for much longer.

Left: **In the 20th century, Black Rhino numbers plummeted from hundreds of thousands down to around 2000 individuals.**
© KEN BEHRENS, TROPICAL BIRDING TOURS

Below: **The range of African Lions has contracted by more than 80%.**
© KEITH BARNES, TROPICAL BIRDING TOURS

The Guinea Savanna has excellent bird diversity, though it is inferior to that of the more topographically and climatically complex savannas of East Africa. There are a few species virtually restricted to this habitat, though most birds are shared with the more arid MOIST MIXED SAVANNA and NORTHERN DRY THORN SAVANNA to the north. Typical Guinea Savanna birds include Blue-bellied Roller (IS), Standard-winged Nightjar, Red-throated Bee-eater (IS), Yellow-fronted Tinkerbird, Vieillot's Barbet, Fine-spotted Woodpecker, White-breasted Cuckooshrike, Piapiac, African Spotted Creeper, Yellow-billed Shrike, Western Violet-backed Sunbird, and Purple, Bronze-tailed (IS), and Lesser Blue-eared Starlings. The gallery forests that penetrate the Guinea Savanna generally have an impoverished subset of the birds of true AFROTROPICAL LOWLAND RAINFOREST. But there are a few birds that prefer this habitat, including Violet Turaco, Double-toothed Barbet, Thick-billed Cuckoo, Oriole Warbler, Red-shouldered Cuckooshrike, and White-crowned Robin-Chat. The Guinea Savanna is an important wintering and migratory habitat for large numbers of Palearctic-breeding birds, including raptors, shrikes, and warblers.

Conspicuous reptiles include Common Agama (*Agama agama*), which remains abundant even in farmland and villages.

Above: **Yellow-winged Bat** is one of Africa's most striking bat species, and Guinea Savanna is one of its favored habitats.
© KEITH BARNES, TROPICAL BIRDING TOURS

Right: Rainy season Guinea Savanna in Senegal, dominated by broadleaf trees, especially by doka (*Isoberlinia*) trees.
© GRAY TAPPAN

DISTRIBUTION: Guinea Savanna runs in a broad belt across n. Africa from Senegal to w. Ethiopia and n. Uganda. This broadleaf-dominated habitat occurs between the Congo Basin and Upper Guinea AFROTROPICAL LOWLAND RAINFOREST to the south and the MOIST MIXED SAVANNA and NORTHERN DRY THORN SAVANNA of the Sudanian and Sahel zones to the north. The northern boundary of Guinea Savanna is not well-defined: broadleaf trees gradually give way to more open, mixed, and thorny savanna, and perennial grasses give way to annual ones. Guinea Savanna–like vegetation penetrates deep into the Sudanian zone along watercourses. Guinea Savanna, or something resembling it, is also spreading south into the former rainforest zone as the original rainforest is cut and burned by humans. Heavily degraded rainforest, despite having largely different plant species, quickly takes on the character and wildlife of Guinea Savanna. Guinea Savanna comes to within about 250 mi. (400 km) of the nearest MIOMBO in the area west of Lake Victoria, but despite their structural similarity, these habitats share relatively few species.

WHERE TO SEE: Mole National Park, Ghana; Murchison Falls National Park, Uganda.

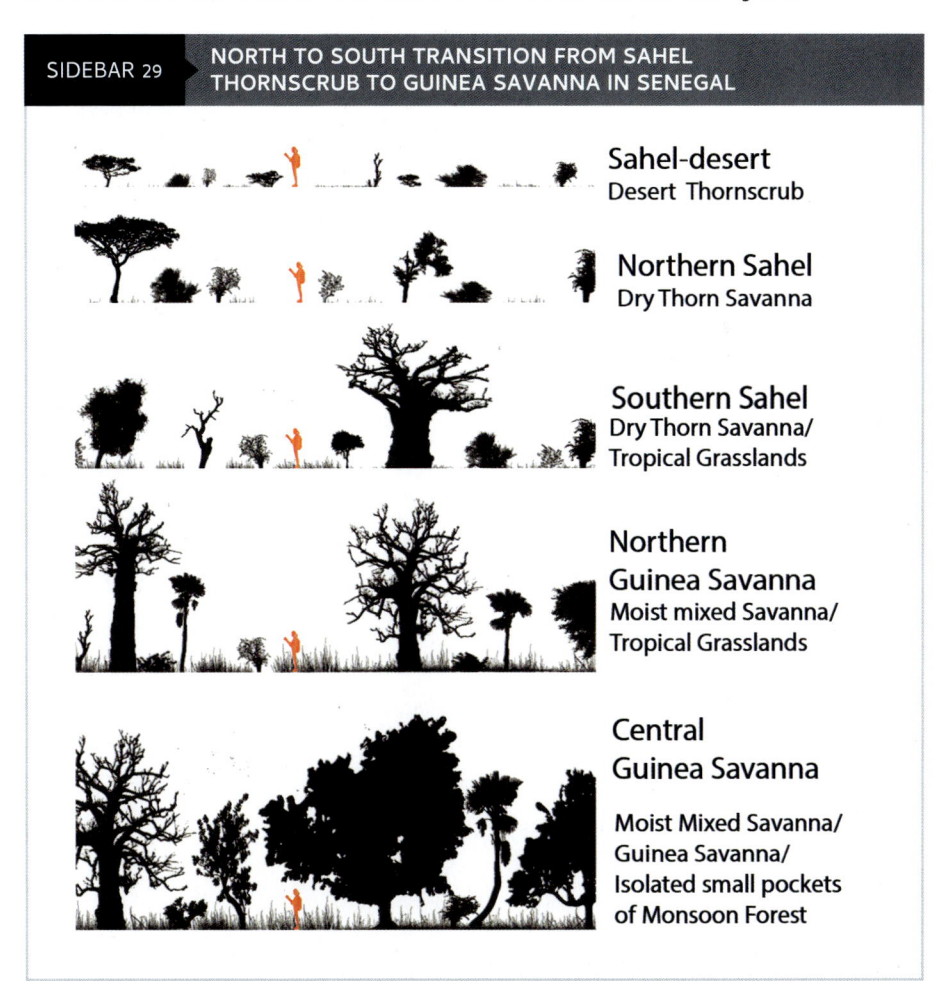

SIDEBAR 29 — NORTH TO SOUTH TRANSITION FROM SAHEL THORNSCRUB TO GUINEA SAVANNA IN SENEGAL

Sahel-desert
Desert Thornscrub

Northern Sahel
Dry Thorn Savanna

Southern Sahel
Dry Thorn Savanna/
Tropical Grasslands

Northern Guinea Savanna
Moist mixed Savanna/
Tropical Grasslands

Central Guinea Savanna
Moist Mixed Savanna/
Guinea Savanna/
Isolated small pockets of Monsoon Forest

268　SAVANNAS

Af6G INSELBERGS AND KOPPIES

IN A NUTSHELL: Exposed rocks that occur within other habitats and support a diverse and distinctive set of plants and wildlife. **Global Habitat Affinities:** SPINIFEX EUCALYPT SAVANNA; ROCKY CERRADO; INDO-MALAYAN LIMESTONE FOREST. **Species Overlap:** Shares species with the surrounding habitats.

A prototypical inselberg: Spitzkoppe, Namibia. Home to mammals including Western Rock Sengi *(left)* and Rock Hyrax *(right)*. MAIN PHOTO AND HYRAX INSET © KEN BEHRENS, TROPICAL BIRDING; SENGI INSET © ANDREY GILJOV

DESCRIPTION: Inselbergs are exposed rocks, usually isolated mountains that stand out of the surrounding terrain. *Koppie*, an Afrikaans word meaning "hill," is often used in the more specific sense of "small rocky hill," usually a pile of boulders. These are just two catchily named examples of rock-dominated habitats, which can also include rocky canyons, eroded hillsides, limestone karst, cliffs, and lava fields. This sort of habitat can be found across Africa, especially in arid zones but locally even in rainforest. While visually striking, such rocks would normally qualify only as a microhabitat within other habitats if not for their possession of a rich and distinctive set of wildlife. Rocky habitat is almost analogous to wetlands: it usually occurs on a small scale within other habitats but must be sought out in order to find its suite of wildlife.

A wide variety of geological phenomena can produce exposed rock. In Africa, one of the most common is ancient granite sticking out of a more recently eroded plain. Sometimes these granite inselbergs are the compressed cores of ancient volcanoes. In some areas, sandstone or limestone massifs have been eroded down into complex series of canyons. The most remarkable example of this sort is the Tsingy of Madagascar (subhabitat Af6G-1): huge blocks of limestone that have eroded into razor-sharp pinnacles. A less common type of exposed rock is the lava fields spewed by recently erupted volcanoes. Inselbergs in savanna environments often support a band of lusher savanna around their base, which is watered by the rain running off the rock faces. Such microhabitats are important for some species, such as the Namib Escarpment endemic Herero Chat. Many rock-loving birds have adapted to human structures and are now found even in the concrete jungles of big cities. Examples are White-collared Pigeon in Ethiopia and various red-winged starlings across the continent, such as Neumann's, Red-winged, and Bristle-crowned Starlings.

The rupicolous (rock-loving) vegetation of this habitat varies a lot depending on local moisture conditions, position on the continent, and surrounding habitats. In deserts, rocks can be virtually bare, while in lowland forest environments, their crevices can be colonized by a scrubby form of rainforest. Classic inselbergs tend to be found in savanna and grassland habitats, and in these

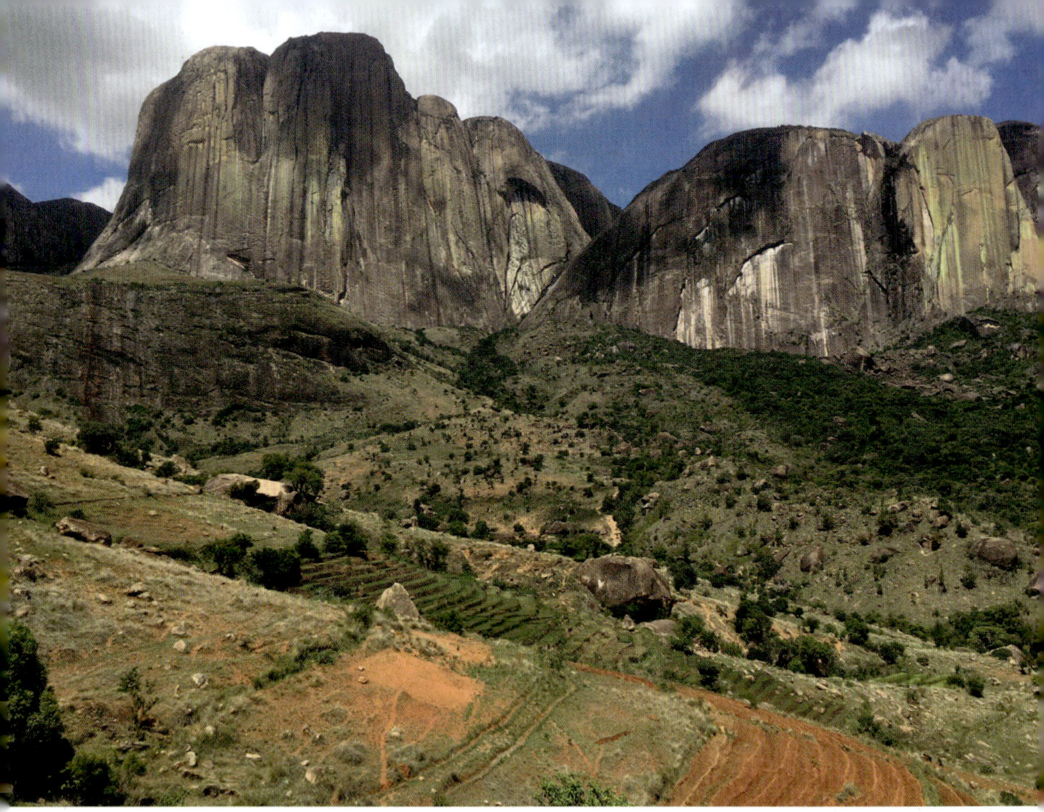

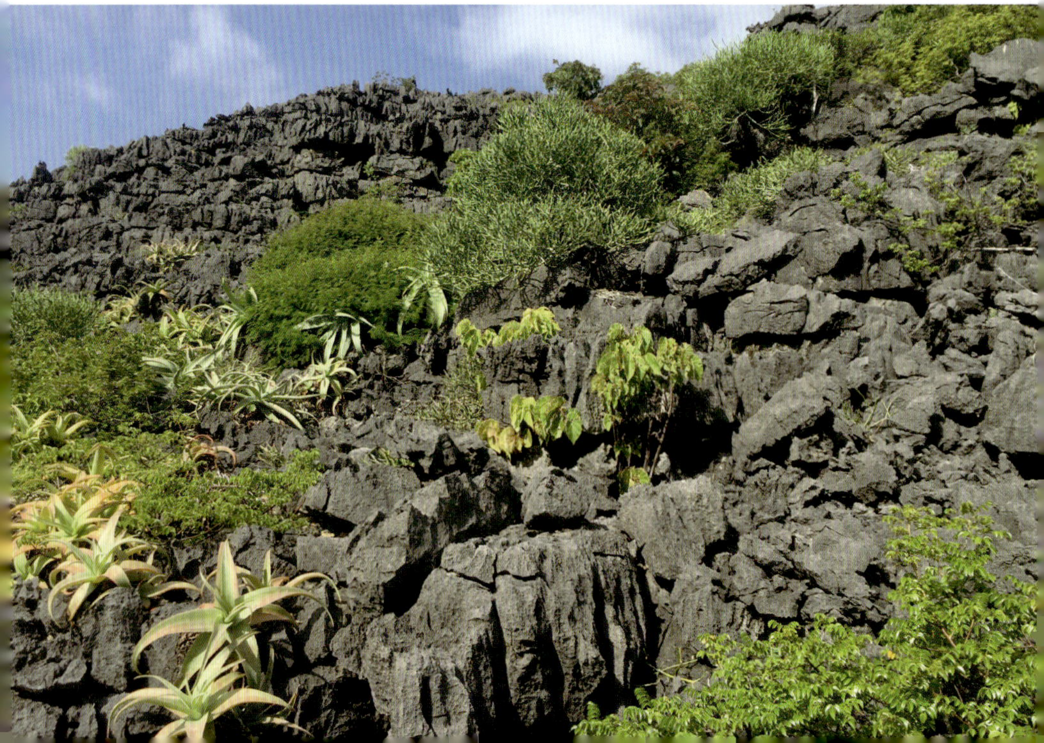

areas, the plant life has much in common with that growing on termite mounds—another instance of a very different soil type within a grass-dominated habitat. Vegetation on inselbergs, especially rock-splitting figs (*Ficus*), grows in cracks and crevices in the rocks, as well as on the shallow and rocky soils that form, though these sometimes slide off the rock faces into oblivion. Growing conditions are harsh, with blazing hot days exposed to the sun, often cold nights, and long periods with little water. As such, only a hardy subset of plants grows in this habitat, and some are specialized exclusively to grow in these harsh conditions. Some typical plants include blue ears (*Cyanotis*), monkeys tails (*Xerophyta*), aloes, a small selection of orchids, lantern flowers (*Ceropegia*), euphorbias, specialized groundsels (*Senecio*) and everlasting-flowers (*Helichrysum*), lip ferns (*Cheilanthes*), and Resurrection Plant (*Myrothamnus flabellifolius*). Trees tend to be a subset of what grows in surrounding habitats, though some species of tropical chestnuts (*Sterculia*) are partial to large rocks. Grasses are almost always a part of the vegetative mix; most are shared with surrounding habitats, especially in savanna, but some are specialized to inselbergs. Lichens and bryophytes are common and often blanket large portions of the exposed rock. The inselbergs of central and eastern Madagascar support a very low growth of plants, lacking the trees and larger shrubs typical in Africa. Classic plants of these Malagasy inselbergs include kalanchoe (*Kalanchoe*), pachypodium (*Pachypodium*), and *Myrothamnus moschatus*. The vegetation of western and northern Madagascar's limestone Tsingy is rich and distinctive and is essentially a limestone-adapted version of MALAGASY DECIDUOUS FOREST.

CONSERVATION: The rugged nature of this habitat tends to defend it from most human disturbance. Even in some heavily populated and human-modified areas, inselbergs remain as outposts of wild habitat.

WILDLIFE: Across the Afrotropics, inselbergs, koppies, and cliffs form the exclusive habitat for the cliff-jumping Klipspringer (IS) as well as several species of hyraxes (see sidebar 30, p. 272). In the Horn of Africa, rocky habitat is the home of Speke's Pectinator, an odd rodent, and Beira, an elegant small antelope. Along the Namib Escarpment, it holds the Black Mongoose and hyrax-like Dassie Rat. Leopards depend on cover, both to hunt and to avoid Lions and hyenas, and are very fond of rocky areas. They often prey on baboons, which range widely but also often find sanctuary in the same habitat. In Madagascar, the lush and inaccessible canyons within limestone Tsingy and

OPPOSITE PAGE:
Top: **Central Madagascar has some impressively vast inselbergs, including Tsaranoro Mountain.** © KEN BEHRENS, TROPICAL BIRDING TOURS

Bottom: **Eroded limestone habitats in Madagascar are rich in succulent plants. Nosy Hara National Park.** © KEN BEHRENS, TROPICAL BIRDING TOURS

THIS PAGE:
Klipspringer, the "Stone Jumper" is the quintessential inselberg mammal. Cape Fold Mountains, South Africa.
© KEN BEHRENS, TROPICAL BIRDING TOURS

SIDEBAR 30 HYRAX

Although elephants are often cited as the closest relatives of the hyraxes, the elephant-shrews, manatees, and Dugong are equal cousins that diverged some 60 MYA. This group, which originated in Africa, is the most primitive of all placental mammals (Eutheria) and is called the Afrotheria. Watching them scuttle around their rocky crevasses on an inselberg or koppie it is hard to believe that 30–25 MYA they were the most dominant herbivores in Africa. Then came the ungulates, which rose to claim that mantle, leaving the outcompeted hyraxes to retreat to marginal habitats. Rock Hyrax have their sweat glands on their feet; rather than a cooling mechanism, this allows the pudgy speedster increased grip on rocky surfaces.

Hyrax form 75% of Verreaux's Eagles' diet in most parts of their ranges.
© KEITH BARNES, TROPICAL BIRDING TOURS

Right: **Lushly vegetated inselberg and Plains Zebra in Kruger National Park, South Africa.** © KEN BEHRENS, TROPICAL BIRDING TOURS

eroded sandstone massifs are important habitats for wildlife. Crowned Lemurs can sometimes even be seen jumping cautiously across the jagged Tsingy.

Rocky habitat is the exclusive domain of several widespread birds, including Freckled Nightjar, Familiar Chat, White-winged and Mocking (IS) Cliff-Chats, Rock-loving Cisticola, rock-thrushes, and Cinnamon-breasted Bunting. Rocky mountains are patrolled by Verreaux's Eagles, which prey mainly

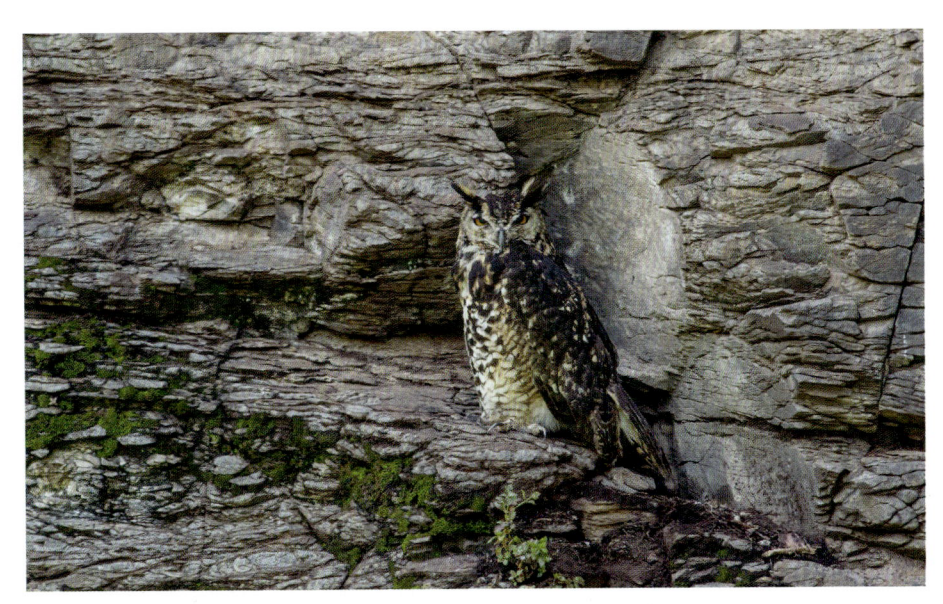

"MacKinder's" Cape Eagle-Owl nests and roosts on cliffs. © KEN BEHRENS, TROPICAL BIRDING TOURS

on hyraxes. In order to escape their nemesis predator, hyraxes post lookouts and have evolved "sunshade" eyebrows to help them look into the bright midday sun. Along the Namib-Angolan Escarpment, this habitat is the domain of two localized endemic birds: Hartlaub's Francolin and Rockrunner. In South Africa, it supports Ground Woodpecker, Buff-streaked Chat, and Kopje Warbler. In West Africa, there are three localized rock-loving species of firefinches. Cliffs and other rocky sites are crucial breeding habitat for many birds that feed primarily in other habitats. Classic examples are raptors, owls, swifts, ravens, and Rock Martin. Within **AFROTROPICAL LOWLAND RAINFOREST**, rocky caves and giant boulders are the breeding habitat of two of the continent's most bizarre birds, the White-necked and Gray-necked Picathartes (or Rockfowl), cave-dwelling, colonial-nesting passerines that look like no other birds and are notoriously difficult to find away from their nesting areas.

What wetlands are to birders, rocky habitats are to reptile enthusiasts: a habitat that covers a small portion of the landscape but is of disproportionate importance to the group of animals they seek. These habitats are crucial for many snakes and a high proportion of lizards. Examples include many skinks, girdled lizards (*Cordylus* and *Smaug*), flat lizards (*Platysaurus*), crag lizards (*Pseudocordylus*), agamas, flat geckos (*Afroedura*), leaf-toed geckos (*Hemidactylus*), dwarf geckos (*Lygodactylus*), thick-toed geckos (*Pachydactylus*), and Giant Plated Lizard (*Matobosaurus validus*). Many localized endemic reptiles in Madagascar are found exclusively on specific massifs or tracts of Tsingy. Some widespread Malagasy rock-loving reptiles include Dumeril's and Grandidier's Madagascar Swifts (*Oplurus quadrimaculatus* and *O. grandidieri*), Barbour's Day Gecko (*Phelsuma barbouri*), and some dwarf geckos.

DISTRIBUTION: Inselbergs, koppies, cliffs, and other rocky habitats are most frequently encountered in more arid zones, where rocks are less likely to be veiled by thick vegetation. They are especially common in the ancient arid corridor that runs all the way from the Horn of Africa to the Namib Desert and the Karoo Plateau in s. Africa. Occasional rocky hills and isolated boulders are found in central

SAVANNAS

Angola Cave-Chat is an Angolan near-endemic that is specialized to rocky habitats. © DUBI SHAPIRO

and west Africa, especially in the more arid Sahel zone but locally even in the rainforest. Some impressive granite inselbergs are found in Madagascar, especially on the deforested high plateau. There are a couple of vast sandstone massifs in sw. Madagascar and three major areas of Tsingy in the west and north.

WHERE TO SEE: Simba Koppies, Serengeti National Park, Tanzania; Spitzkoppe, Namibia; Tsingy de Bemaraha National Park, Madagascar.

Madagascar's "Tsingy", uplifted then eroded oceanic limestone, is one of the region's most striking and distinctive rocky habitats. It is home to the lovely Tsingy Plated Lizard (*Zonosaurus tsingy*) *(inset)*.
© KEN BEHRENS, TROPICAL BIRDING TOURS

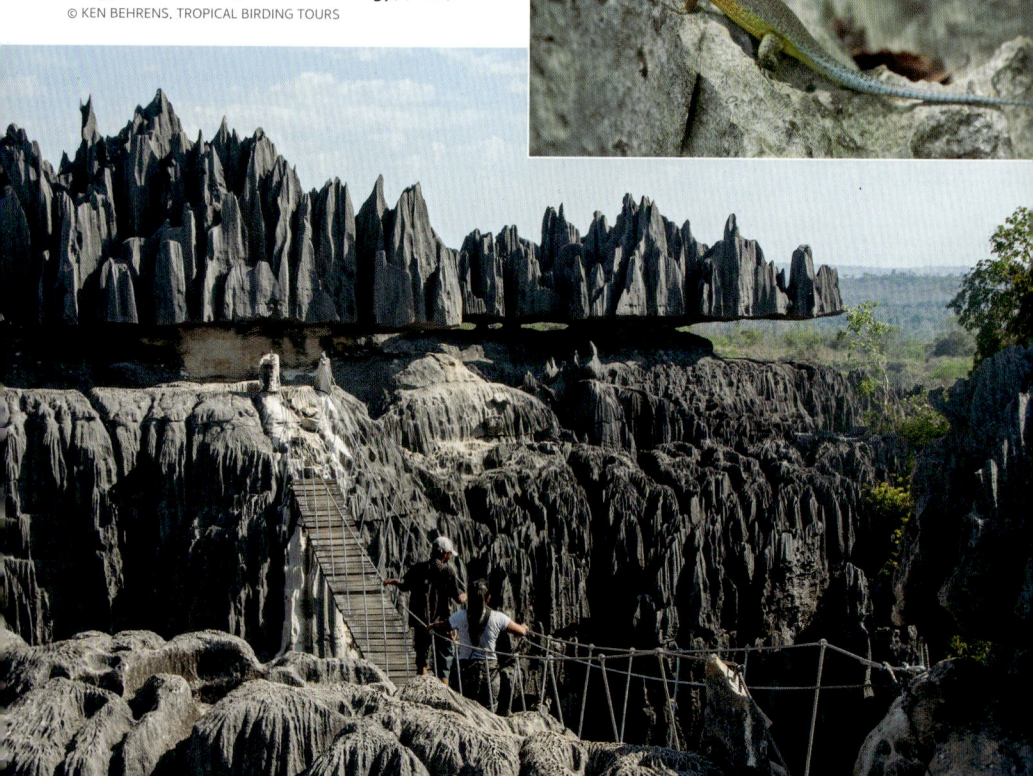

Af6H GUSU

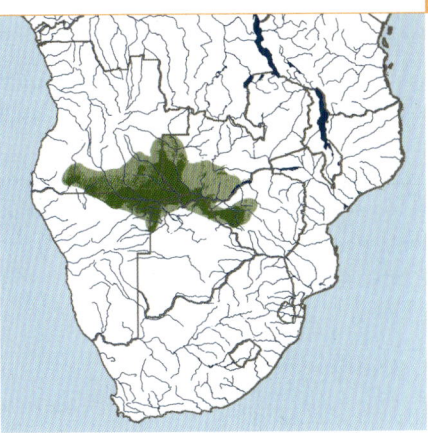

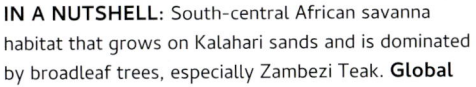

IN A NUTSHELL: South-central African savanna habitat that grows on Kalahari sands and is dominated by broadleaf trees, especially Zambezi Teak. **Global Habitat Affinities:** OPEN EUCALYPT SAVANNA; OAK DEHESA; TEAK FOREST. **Continental Habitat Affinities:** MIOMBO; MOPANE; MAVUNDA. **Species Overlap:** MIOMBO; MAVUNDA; MOPANE; MOIST MIXED SAVANNA; KALAHARI DRY THORN SAVANNA; GUINEA SAVANNA.

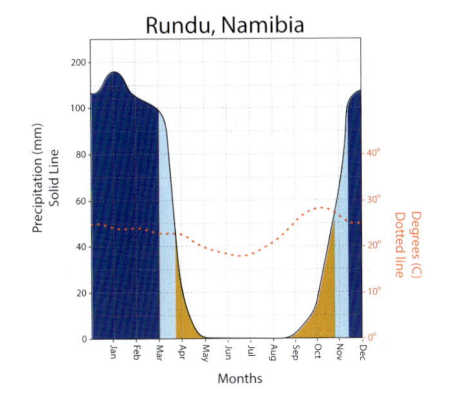

DESCRIPTION: Gusu is a fairly tall broadleaf woodland with abundant grass on the ground and distinct wet and dry seasons (Köppen **Awa**). It grows on deep Kalahari sands, with little clay or silt, that were primarily deposited by eolian processes (transported by wind). Trees that are able to penetrate the sand with their roots can access water that accumulates by lateral seepage. This "supplementary" water supply allows trees to grow as tall as 65 ft. (20 m), higher than the local level of rainfall would normally allow.

The dominant and characteristic tree is Zambezi Teak (*Baikaiea plurijuga*). This tree is similar in appearance to the dominant trees of MIOMBO, with a classic lollipop tree shape and drooping, pinnately compound leaves. The valuable, rich reddish-brown wood is often seen for sale along roadsides. Another important tree in Gusu is Angolan Teak (*Pterocarpus angolensis*), which can sometimes be the dominant tree, particularly in portions of sw. Zambia and Zimbabwe. Many other trees are common in Gusu, including Marula (*Sclerocarya birrea*), Large False-Mopane (*Guibourtia coleosperma*), monkey-oranges (*Strychnos*), Thorny Teak (*Pterocarpus lucens*), bushwillows (*Combretum*), Sand Camwood (*Baphia massaiensis*), terminalia (*Terminalia*), croton (*Croton*), donkeyberry (*Grewia*), shepherd's tree (*Boscia*), Mobola Plum (*Parinari curatellifolia*), Kalahari Podberry (*Dialium englerianum*), Peeling Plane (*Ochna pulchra*), Horn-pod Tree (*Diplorhynchus condylocarpon*), Wild Seringa (*Burkea africana*), and White Bauhinia (*Bauhinia petersiana*), which often form a midstory of smaller trees under the open canopy. *Brachystegia* (miombo) trees can be found locally, hinting at Gusu's affinity to MIOMBO. Trees other than teak tend to get the upper hand in heavily disturbed areas. Thick

Angolan Teak (*Pterocarpus angolensis*) is a classic Gusu tree, especially distinctive in the dry season when its winged seeds *(inset)* are conspicuous. Caprivi Strip, Namibia. © KEN BEHRENS, TROPICAL BIRDING TOURS

grass grows over most of the ground during the rainy season, and slowly dries up or is eaten by ungulates during the dry season.

Gusu has a similar feel and structure to Miombo, though it is generally more open. Miombo is more botanically diverse, with a mix of canopy trees, in contrast to the dominance of Zambezi Teak in typical Gusu. Half of the year, Gusu sees little rainfall, whereas heavy rains fall from November to March. Typical annual precipitation is 12–24 in. (300–600 mm). Gusu is generally easy to walk through, especially during the dry season, though in some areas the understory forms a dense thicket. Because of its open nature, there is moderate visibility at eye level. The canopy is usually discontinuous, typical of a savanna habitat.

As in Miombo, fire is important, and wildfires are common, especially at the onset of the rainy season. The smaller drainage lines support open grasslands, which are seasonally flooded. Larger drainage lines and rivers give rise to perennial wetlands, thickets, and MONSOON FOREST. These are most prominent and important along the Kavango River, Okavango Delta, and Zambezi River. A characteristic tree of the Monsoon Forest within Gusu is Manketti (*Schinziophyton rautanenii*).

CONSERVATION: Owing to its poor soil and lack of water, much of this habitat is sparsely populated by humans. Despite that, hunting is widespread. Fires are natural but have increased in frequency and severity because of human activity. Zambezi Teak is sensitive to fire and can be wiped out by frequent burning. Much of this woodland is also being overharvested for the valuable teakwood. Gusu is less resilient than other types of woodland; once land is cleared by a combination of cutting and fire, the hot sun destroys the organic matter in the soil, and the area tends to remain clear and never regenerate tall woodland.

WILDLIFE: Gusu supports a similar suite of species to MOPANE—essentially, a much-reduced selection of the wildlife of the more diverse adjacent MIOMBO and KALAHARI DRY THORN SAVANNA habitats. The dual influences of those habitats are evident in a list of some of the mammals of Gusu: Smith's Bush Squirrel, Rusty-spotted Genet, Roan and Sable Antelopes, Bushbuck, Greater Kudu, Steenbok, Puku, and African Buffalo. Primarily thanks to the existence of some well-protected areas, large predators persist, including Leopard, Lion, and African Wild Dog. This habitat is used by large numbers of African Savanna Elephants.

Although Gusu has much lower bird diversity than Miombo or Kalahari Dry Thorn Savanna, it does have abundant birdlife and usually feels more bird-rich than Miombo. Bradfield's Hornbill (IS) is the sole avian Gusu specialist, though it does also use adjacent habitats. Some of the most vocal birds, providing the background sound track, are Ring-necked Dove and Yellow-fronted Tinkerbird, an endlessly repeated "work harder" from the former and a high-pitched tooting from the latter. Gusu provides good refuge for large raptors, including Bateleur, Dark Chanting-Goshawk, Brown Snake-Eagle, and the rare Ovambo Sparrowhawk. Bird flocks are frequent, and common flock members are African Gray Hornbill, Chinspot Batis, African Paradise-Flycatcher, African Golden and

Zambezi Teak (*Baikaiea plurijuga*) is the most characteristic tree of Gusu. Caprivi Strip, Namibia. The distinctive leaves *(inset)* are usually easily recognized.
© KEN BEHRENS, TROPICAL BIRDING TOURS

African Black-headed Orioles, Rufous-bellied Tit, Southern Black-Tit, Greencap Eremomela, and Southern Yellow White-eye. Like many African savanna environments, there is a rich selection of bushshrikes, such as Brown-crowned Tchagra, Brubru, and Black-backed Puffback. The understory holds Fawn-colored Lark, Kurrichane Thrush, Red-backed Scrub-Robin, Piping Cisticola, and Tinkling Cisticola. Other typical birds of this habitat include Striped Kingfisher, Meyer's Parrot, Sharp-tailed and Violet-backed Starlings, Pale Flycatcher, White-breasted and Amethyst Sunbirds, Yellow-fronted Canary, Yellow-throated Bush Sparrow, and Golden-breasted Bunting. In ne. Namibia, this habitat holds Souza's Shrike, a major target for s. African birders, and one of the few places where this species is found outside Miombo.

Gusu supports many widespread snake and lizard species, including venomous snakes like Puff Adder (*Bitis arietans*), Black Mamba (*Dendroaspis polylepis*), Boomslang (*Dispholidus typus*), and Savanna Vine Snake (*Thelotornis capensis*). Other widespread reptiles that are often bumped into during the day are Mole Snake (*Pseudaspis cana*), Ground Agama (*Agama aculeata*), Wahlberg's Skink (*Trachylepis wahlbergii*), Bushveld Lizard (*Heliobolus lugubris*), Black-lined Plated Lizard (*Gerrhosaurus nigrolineatus*),

Top left: **Bradfield's Hornbill is a highly localized Gusu specialist bird.** © KEN BEHRENS, TROPICAL BIRDING TOURS

Left: **Souza's Shrike is mostly a Miombo bird, but it does also sneak into Gusu.** © KEITH BARNES, TROPICAL BIRDING TOURS

Right: **Common Ostrich creche attended by a male. Mahango Game Reserve, Namibia.** © KEN BEHRENS, TROPICAL BIRDING TOURS

Below: **South African Springhare is a nocturnal kangaroo-like rodent with a wide distribution that includes Gusu. The two springhares make up their own family Pedetidae, bizarrely related to the "flying squirrels," or anomalures, of the Central African rainforests.** © KEITH BARNES, TROPICAL BIRDING TOURS

and Rock Monitor (*Varanus albigularis*). Gusu comprises a major portion of the range of several much more localized reptiles, which may be encountered with luck. These include Anchieta's Cobra (*Naja anchietae*), Caprivi Rough-scaled Lizard (*Ichnotropis grandiceps*), Ceríaco's Tree Agama (*Acanthocercus ceriacoi*), and Chobe Dwarf Gecko (*Lygodactylus chobiensis*).

DISTRIBUTION: Gusu occurs in a fairly narrow zone from s. Angola to w. Zimbabwe and Zambia. Gusu and MOPANE together form a belt across s. Africa and provide a transition from the moister MIOMBO to their north and the drier and thornier savanna types to their south and west.

WHERE TO SEE: Kavango Region/Caprivi Strip, Namibia.

GRASSLANDS

Af7A AFROTROPICAL GRASSLAND

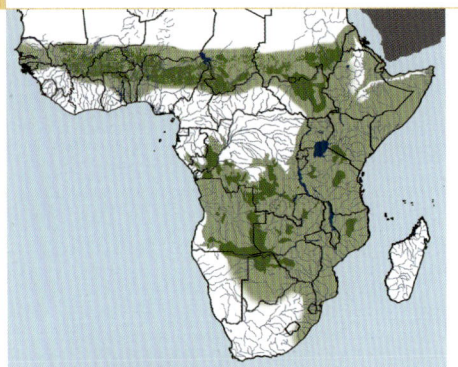

IN A NUTSHELL: Grass-dominated habitat with only occasional, scattered trees and bushes, found at low and middle elevations. **Global Habitat Affinities:** AUSTRALIAN TROPICAL TUSSOCK GRASSLAND; CAMPO; TERAI FLOODED GRASSLAND; MESOAMERICAN SAVANNA AND GRASSLAND. **Continental Habitat Affinities:** MONTANE GRASSLAND; Afrotropical savannas (seven different habitats). **Species Overlap:** Afrotropical savannas; MONTANE GRASSLAND.

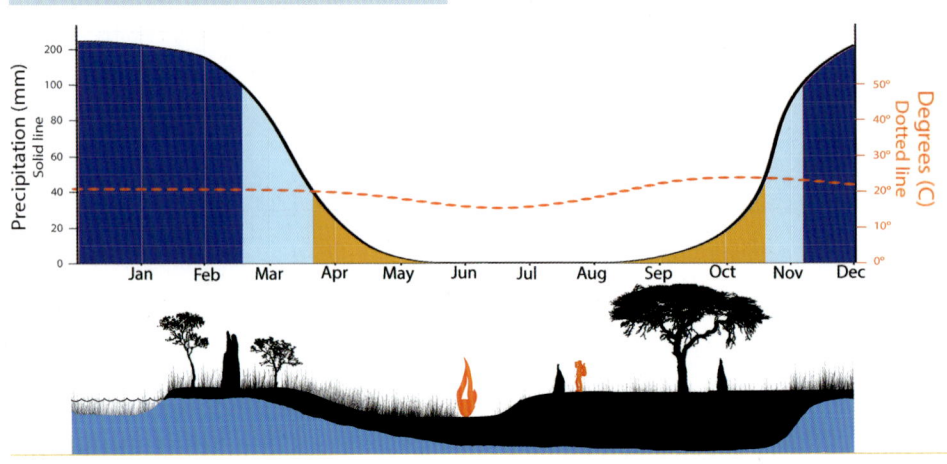

DESCRIPTION: Quite simply, grassland is habitat in which the dominant vegetation is grass and most of the wildlife is dependent on grass for its livelihood. Since savanna is by definition a grassland with some trees, there is obviously broad overlap between grassland and the various savanna habitats. Anyone who has visited a national park in eastern or southern Africa has seen that the interplay between tree-dominated and grass-dominated areas happens on a micro scale. One minute you are driving through a thicket-like area of acacia or Mopane, and suddenly you emerge onto a grassy and treeless plain. Part of what is complicated about grassland is that this habitat can be produced by different phenomena. One factor is a climate in which nearly all the precipitation occurs in half, or less than half, of the year. During the dry half, the habitat dries out and becomes prone to burning, especially due to lightning strikes at the beginning of the wet season. These fires wipe out tree seedlings and maintain the grassland. A related and often concurrent factor is the browsing of tree seedlings by herbivores, again preventing the growth of large trees. The other major influence that produces grassland is edaphic—related to soil types that don't allow the

Adok, South Sudan

Coffee Bay, South Africa

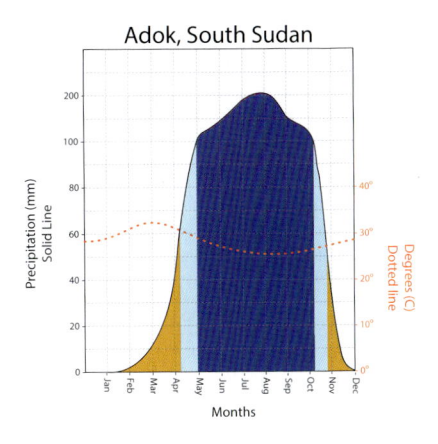

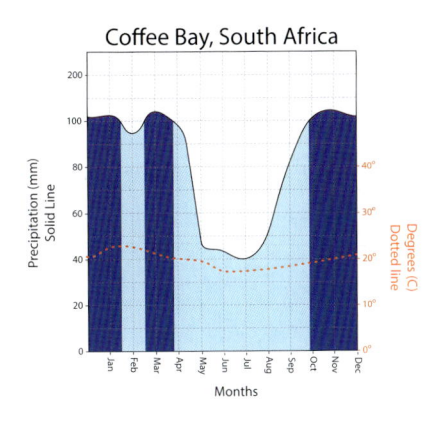

growth of trees. The most widespread example of such edaphic grassland is seasonally flooded areas (subhabitat Af7A-1) with heavy clay soils, such as the Dambos within the GUINEA SAVANNA and MIOMBO zones and the floodplains of the Okavango and Niger Deltas. Edaphic grasslands can also occur on hilltops or ridgelines with thin soils. Some such grasslands occur even deep in the rainforest zone of c. Africa. Because it can be produced by multiple different factors, this is not a habitat that is strictly tied to a specific climate or seasonality.

The precise composition of grassland varies considerably, depending on its type and location on the continent. Classic species of flooded grassland include Antelope Grass (*Echinochloa pyramidalis*), Wild Rice (*Oryza longistaminata*), Jaragua (*Hyparrhenia rufa*), and Common Russet Grass (*Loudetia simplex*). Just a few of the many grasses common in drier grassland (subhabitat

There are over 1 million migratory Blue Wildebeest in the Serengeti of Tanzania and Masai Mara of Kenya.
© KEN BEHRENS, TROPICAL BIRDING TOURS

Gemsbok or Southern Oryx in a pure grassland along the edge of the Etosha salt pan. Etosha National Park, Namibia. © KEN BEHRENS, TROPICAL BIRDING TOURS

Af7A-2) include Red Oat Grass (*Themeda triandra*), various *Andropogon*, Rhodes Grass (*Chloris gayana*), fountaingrasses (*Pennisetum*), and lovegrasses (*Eragrostis*).

Human disturbance also has a tendency to produce what are called "derived" grasslands in places that wouldn't typically have supported this habitat, such as the Western Cape of South Africa and the coast of East Africa. Cultivated fields of some crops have a grassland-like character and attract a few of the species of natural grassland. Remarkably, grasslands grow at both ends of the moisture gradient: into the Sahara and Namib Deserts after good rains and in seasonally flooded areas adjacent to permanent swamps in the wettest parts of the continent. Grassland is very much a boom-or-bust habitat. It grows with remarkable rapidity after good rains or recent flooding but can become stark and desertlike during the heart of the dry season. The burst of nutrients that can be produced by this habitat makes it very important for wildlife; it is at the heart of some of the greatest mammal concentrations and migrations left on Earth.

CONSERVATION: The frequent presence of water and good forage for domestic animals tends to attract humans to grasslands. Large wild mammals have been

Fire radically changes this landscape, but some wildlife, like this Crowned Lapwing, profit by eating roasted insects that are significantly easier to find. © KEITH BARNES, TROPICAL BIRDING TOURS

largely eliminated from the Niger Delta and Lake Chad floodplains by competition from domesticated animals and hunting. Massive water diversion schemes can have a drastic impact on floodplain environments. A dam on Zambia's Kafue River vastly reduced the seasonally flooded area and produced a 50% reduction in the local herd of Southern Lechwe, which formerly numbered over 100,000 individuals. Even worse results may follow from the still incomplete Jonglei Canal, which will divert water around the Sudd wetlands in South Sudan.

WILDLIFE: Grassland is of great importance to Africa's big mammals. Some species use this habitat throughout the year, while others migrate during the year between savanna and pure grassland habitats. Africa's grasslands support some of the greatest migratory mammal herds on Earth. The best-known of these is the Serengeti grassland of n. Tanzania, which hosts a circular annual migration of around 1.2 million Blue Wildebeests, 200,000 Plains Zebras, and 400,000 Thomson's Gazelles. The White Nile floodplains of the Sudd in South Sudan support around 1 million Kobs and 30,000 Nile Lechwes. The Liuwa Plain of w. Zambia supports 30,000 migratory Blue Wildebeests and 3000 Southern Lechwes.

For mammals, one major advantage of seasonally flooded grasslands is that they slowly expand and contract through the year, opening up new areas of ideal forage. This habitat is so attractive that several mammal species have specifically adapted to it, namely Bohor and Southern Reedbucks, Puku, Southern and Nile Lechwes, and Common Tsessebe. Many mammals found in savanna are fond of grassland. These include Cape Hare, Grévy's and Plains Zebras, White Rhinoceros, African Buffalo, several species of gazelles, Blue Wildebeest, Red Hartebeest, and Oribi. Grassland is an

Grassland that approaches steppe, with an abundant shrub component. Good habitat for Rosy-throated Longclaw *(inset)*. **Ndutu, Tanzania.** © KEN BEHRENS, TROPICAL BIRDING TOURS

White Rhino, a classic grassland grazing species, crops the vegetation into a fairway-esque aspect. © KEN BEHRENS, TROPICAL BIRDING TOURS

important grazing habitat for Hippopotamus when it emerges from its aquatic refuges to feed at night (see sidebar 40, p. 373). It is extensively used by predators such as Lion, Serval, African Wild Dog, and Cheetah. The soil types that tend to produce grassland also often attract termites, which in turn draw Aardvark, Aardwolf, and Ground Pangolin. Of course, grasslands are a rodent paradise and support a wide range of mice and rats.

Grasslands that occur in conjunction with trees, in a savanna habitat, attract a huge variety of birds, most of which are covered in the accounts of specific savanna types. This account will focus exclusively on grassland species. The vast majority of African birds are adapted to use trees to at least some extent, and only a limited, specialized subset can truly be considered grassland birds. These include some francolins, quail, buttonquail,

Many grasslands transform into wetlands when flooded. "Black" Lechwe and Red-billed Ducks in the Bangweulu Wetlands, Zambia.
© KEN BEHRENS, TROPICAL BIRDING TOURS

Af7A AFROTROPICAL GRASSLAND 285

Cheetah, often considered a classic grassland predator, also hunts in denser lightly wooded savanna vegetation. Masai Mara, Kenya. © HANH DUNG

This isolated grassland plain in the rain shadow of Mt. Meru in n. Tanzania comprises the entire range of "Beesley's" Spike-heeled Lark. The inset photo shows how drastically this habitat transforms in the dry season! MAIN PHOTO © KEITH BARNES; INSET © KEN BEHRENS, TROPICAL BIRDING TOURS

some bustards, several lapwings, some larks, most pipits and longclaws, several chats, and some widowbirds. Seasonally flooded grasslands attract a wide variety of wetland birds, both resident birds from adjacent permanent swamps and migratory ones that either travel from the Palearctic to spend the boreal winter in Africa or move around the continent of Africa in search of optimal habitat conditions (see sidebar 41, p. 378). A few species, such as Blue Quail, Black-rumped Buttonquail, and Streaky-breasted Flufftail, are highly specialized to seasonally flooded grassland. More general grassland species include Common Ostrich, Abdim's and White Storks, Black-chested Snake-Eagle, crowned-cranes, Temminck's Courser, Forbes's Plover, Yellow-throated Sandgrouse, Black Coucal, Flappet Lark (IS), Yellow-throated Longclaw (IS), African Pipit (IS), Banded Martin, Gray-rumped Swallow, Zitting and Desert Cisticolas, many migrant shrikes, wheatears, and queleas.

Pure grassland habitat is not particularly rich in reptiles. A few species, such as Grass Skink (*Trachylepis megalura*) and the grass lizards (*Chamaesaura*) are adapted to this habitat, and widespread species like rock pythons (*Python sebae* and *P. natalensis*) will use it.

DISTRIBUTION: Grassland is found throughout most of the Afrotropics, though is naturally lacking in the SUCCULENT KAROO and FYNBOS zones and lays dormant in the Namib Desert. However, even deep in the desert, in the occasional year with good rains, grassland can quickly and temporarily spring up, producing a quick flush of nutrients, including windblown seeds, which are of great importance to this resource-scarce ecosystem. Near the equator, tropical grasslands can be found from sea level up to around 6000 ft. (1800 m), where they give way to a mix of Afrotropical MONTANE GRASSLAND and montane forest. Away from the equator, the upper limit is lower, around 4000 ft. (1200 m).

The continent's largest stands of pure grassland occur on the Niger, Sudd, and Okavango floodplains, on the Serengeti, and in parts of w. Tanzania and Zambia. Elsewhere, such as throughout the MIOMBO and GUINEA SAVANNA zones and along the southeast coast, grassland is a common habitat but one that occurs in a fine matrix alongside more treed habitats. Grassland is rare in the heart of the West and Central African rainforest belt but becomes more common in the transition zones surrounding the Congo Basin and along the West African coast. At higher elevations, as in the mountains of Ethiopia and East Africa and the Highveld of South Africa, tropical grassland is replaced by Afrotropical Montane Grassland. The extent of grassland in Madagascar prior to the arrival of humans is disputed, but the general lack of grassland-adapted species seems to argue for this having been a rare natural habitat. Nonetheless, much of the island is now covered in derived grassland (see MALAGASY GRASSLAND AND SAVANNA).

WHERE TO SEE: Serengeti National Park, Tanzania; Masai Mara National Park, Kenya; Awash National Park, Ethiopia; Kafue National Park, Zambia.

SIDEBAR 31   A TALE OF TWO SEASONS

Lowland grassland may be the African habitat that shows the greatest variability across the year. After a heavy rain, grasses spring up at an incredible rate, the landscape is lush and green, and wetlands with shallow pools form. Waterbirds and large mammals flock to the area. But toward the end of the dry season, or during a period of drought, the same area resembles a desert. Not only is the grass dry and yellow, but much of it may simply disappear, consumed by herbivores, burned, or desiccated and blown away by the wind. In the most extreme cases, nothing but cracked and sun-baked soil remains, with a dormant seed bank awaiting the onset of the next rains. This radical transformation is one of the most dramatic and profound anywhere in the world.

Af7E MALAGASY GRASSLAND AND SAVANNA

IN A NUTSHELL: Grassland and savanna on Madagascar, originally native, but expanded and maintained by human-induced fires. **Global Habitat Affinities:** INDIAN TROPICAL GRASSLAND; SALT STEPPE; CAMPO SUJO. **Continental Habitat Affinities:** MOIST MIXED SAVANNA; AFROTROPICAL GRASSLAND (superficially similar but natural rather than anthropogenic). **Species Overlap:** SPINY FOREST; MALAGASY DECIDUOUS FOREST; INDIAN OCEAN LOWLAND RAINFOREST.

DESCRIPTION: Much of Madagascar was once forested. The dominant habitat in the east was rainforest, in the central and northern west was deciduous forest, and in the southwest SPINY FOREST. The pre-human nature of the High Plateau has been hotly debated and remains mysterious, but it likely included a mix of drier forest, woodland, and natural grassland habitats, which were maintained by natural fires and grazing from indigenous species. Across all these zones, forest has been greatly reduced, and grassland and savanna are now the most common habitats, covering at least two-thirds of Madagascar. The natural grasslands that did exist in Madagascar had endemic grass and ant species and have been greatly expanded by humans. While these habitats vary significantly across the island, which will be discussed below, they all share a biologically depauperate nature compared with Madagascar's forest environments. Anthropogenic savanna certainly occurs elsewhere in Africa but not as extensively as on Madagascar; and elsewhere it is far richer biologically, due to the abundant presence of natural savanna habitats and species on the continent.

The main human pressure that has turned forest and woodland into savanna and grassland over the last millennium, or perhaps longer, is frequent burning, mainly to create cattle pasture and secondarily to clear land for cultivation. This continues to be the case: virtually every corner of this habitat burns at least once every year. As such, most of the plants of Malagasy grassland are fire-resistant in various ways. Some grasslands are also maintained by grazing from zebu cattle, which replaced the natural grazers, such as extinct hippos, giant tortoises, and perhaps Elephant Bird.

The High Plateau is mostly covered in pure grassland, around 3 ft. (1 m) tall, vast stretches of which are dominated by a handful of species of exotic grass. Drainage lines and low wet areas often support dense growth of pandanus trees, and sometimes INDIAN OCEAN MONTANE RAINFOREST remnants. Smaller clefts are often choked with cosmos (*Cosmos*), an exotic aster from Mexico.

Right: **Pandanus swamp often persists along streams in fire-prone areas. Isalo, Madagascar.** © KEN BEHRENS, TROPICAL BIRDING TOURS

Below: **On Madagascar's fire-swept High Plateau, forest remains only in tiny patches along drainage lines and in other wet places. The dominant habitat is biologically impoverished anthropogenic grassland.** © KEN BEHRENS, TROPICAL BIRDING TOURS

In some areas, fire-resistant aloes persist within the grasslands. The indigenous Cerulean Flax-Lily (*Dianella ensifolia*) is also locally common. Pine and eucalyptus plantations have been planted widely, and these trees sometimes become naturalized.

Lower-elevation savanna covers most of western and northern Madagascar. Here the grassland is taller, up to 9 ft. (3 m), and generally dominated by grass species with broader, flatter leaves. Trees are much more common than on the High Plateau. These have maximum heights of 25–40 ft.

Baobabs, like these Grandidier's Baobabs (*Adansonia grandidieri*) at the famous "Allee des Baobabs" are often the sole remnants of formerly forested environs after other trees are cleared.
© KEN BEHRENS, TROPICAL BIRDING TOURS

(8–12 m), though most formations are shorter. One of the most common and striking types of savanna is that dominated by indigenous *Bismarckia nobilis* palms. A good example of this landscape is the iconic approach to Isalo National Park from the west. Baobabs (especially *Adansonia za* and *A. grandidieri*) are often left when other trees are cleared, making for an incredibly striking albeit tragic landscape in parts of the west. One well-known example is the "Alley of Baobabs" near Morondava. Other typical western savanna trees and bushes include Marula (*Sclerocarya birrea*), *Erythroxylum platyclados*, spikethorns (*Gymnosporia*), jujube (*Ziziphus*), Mango (*Mangifera indica*), *Dicoma incana*, physic nuts (*Jatropha*), *Acridocarpus excelsus*, Lala Palm (*Hyphaene coriacea*), *Stereospermum*, *Cloiselia oleifolia*, and hook-thorns (*Senegalia*). Some grassland occurs in the east, mostly in areas where persistent fire and woodcutting have destroyed secondary forest.

Some of the typical grasses of savanna and grassland include needlegrasses (*Aristida*), panicgrasses (*Panicum*), Cogongrass (*Imperata cylindrica*), thatching grasses (*Hyparrhenia*), *Loudetia*, Tanglehead (*Heteropogon contortus*), and Yellow Thatching Grass (*Hyperthelia dissoluta*). The plants of this habitat come from a variety of sources. Many are exotic, introduced from Africa or Eurasia. Some

A highland landscape of anthropogenic pine and eucalyptus savanna and cultivation.
© KEN BEHRENS, TROPICAL BIRDING TOURS

are indigenous. There are 18 species of indigenous trees from the western deciduous forest that also occur in savanna. A few species have also been contributed by the Spiny Forest. Around 40% of the grass species are also indigenous, a strong argument for the widespread presence of natural grasslands before the arrival of humans.

CONSERVATION: This biologically depauperate habitat is certainly not threatened and, in fact, is growing in extent every year because of ongoing deforestation.

WILDLIFE: While natural grassland habitats may once have held some of Madagascar's now-extinct megafauna, the much vaster grassland and savanna today are remarkably poor in wildlife. Despite this being the land of lemurs, most lemurs avoid this habitat. One exception is the widespread Gray Mouse Lemur, which can be found in western savanna habitats, sometimes adjacent to villages. The most common mammals are introduced rodents. Some bats occur, including fruit bats like Madagascar Fruit Bat and Madagascar Flying Fox, which range widely in search of fruit.

There are remarkably few grassland-adapted birds in Madagascar. The most common species in this habitat are Madagascar Lark, Madagascar Cisticola, and African Stonechat. Some raptors, including Black Kite, Madagascar Buzzard, Madagascar Kestrel, and the increasingly rare Madagascar Harrier, will hunt in grassland. There are some gamebirds: the endemic Madagascar Partridge and Madagascar Buttonquail and the widespread Harlequin Quail (lowlands) and Common Quail (highlands). Savannas are slightly richer in birds than pure grasslands and support species like Helmeted Guineafowl, Madagascar Hoopoe, Namaqua Dove, Madagascar Bee-eater, Madagascar Bulbul, Madagascar Magpie-Robin, Common Jery, and Chabert Vanga. The southwestern subspecies of Stripe-throated Jery readily uses more thickly treed savanna habitat. Malagasy Palm Swift is common in areas with palms.

Af7E MALAGASY GRASSLAND AND SAVANNA 291

Gray-headed Lovebird, a common species in w. Madagascar savannas.
© KEN BEHRENS, TROPICAL BIRDING TOURS

The endemic Madagascar Sandgrouse is increasingly rare but can be found locally in lowland western savanna and grassland environments. © KEN BEHRENS, TROPICAL BIRDING TOURS

Right: **Panther Chameleons (*Furcifer pardalis*)** are quite common in northern Malagasy anthropogenic savanna. © KEN BEHRENS, TROPICAL BIRDING TOURS

Below: **Madagascar Bullfrog (*Laliostoma labrosum*)** is one of a small set of amphibians that can survive in anthropogenic savanna. © KEN BEHRENS, TROPICAL BIRDING TOURS

This habitat supports a few reptiles. Some of the most common are Jeweled Chameleon (*Furcifer lateralis*), a couple species of skinks (*Trachylepis*), Bernier's Striped Snake (*Dromicodryas bernieri*), and Madagascan Whipsnake (*Thamnosophis lateralis*). Frogs do not thrive in this fire-swept habitat, but wet areas hold a few. By far the most common amphibian is the introduced Mascarene Ridged Frog (*Ptychadena mascareniensis*).

Endemism: There are around 20 plants endemic to grasslands on Madagascar.

DISTRIBUTION: Grassland and savanna dominate the High Plateau and most of the west. They are more locally distributed in the eastern lowlands, where secondary forest is the more common anthropogenic habitat. Above 6500 ft. (2000 m), there are montane grasslands that are botanically quite different from those below, and are more closely allied with Afrotropical **MONTANE GRASSLAND**.

WHERE TO SEE: RN7 highway, Madagascar.

The spectacular Malagasy endemic *Phymateus saxosus* grasshopper can be common in anthropogenic habitats. © KEN BEHRENS, TROPICAL BIRDING TOURS

Af7F MONTANE GRASSLAND

IN A NUTSHELL: Grass-dominated habitat found at middle and high elevations. **Global Habitat Affinities:** AUSTRALASIAN MONTANE GRASSLAND; SHOLA GRASSLAND; HAWAIIAN GRASSLANDS; HUMID PUNA. **Continental Habitat Affinities:** AFROPARAMO; MONTANE HEATH. **Species Overlap:** MONTANE HEATH; AFROPARAMO; AFROTROPICAL GRASSLAND; and Afrotropical savannas (seven different habitats).

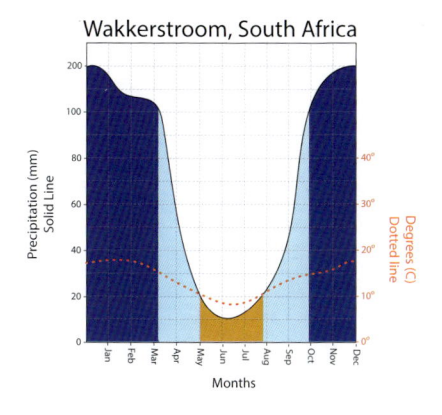

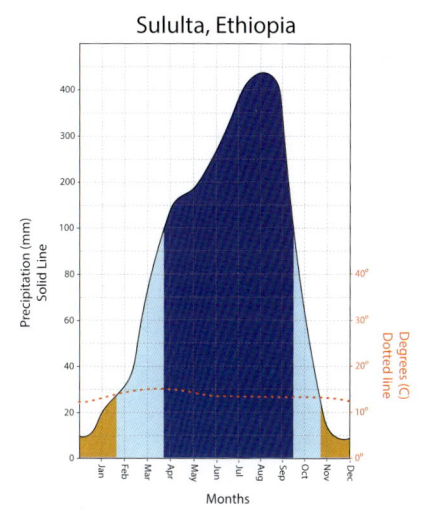

DESCRIPTION: While AFROTROPICAL GRASSLAND is widespread across the continent, Montane Grasslands are much more localized, confined to higher elevation and mountainous areas. Classic African Montane Grasslands are found in the same zone as montane forest. It is difficult to know the extent of these middle-elevation grasslands before the advent of human-caused fires, but it is certain that they have greatly expanded in modern times at the expense of MOIST MONTANE FOREST. This habitat can be distinguished from AFROPARAMO by its lower elevation, greater density and dominance of grass, different mix of grass species, and the central role of fire.

Wet-season grassland at Suikerbosrand Nature Reserve, South Africa *(left)* compared with dry-season grassland *(right)* at Ezemvelo Nature Reserve, South Africa. © KEN BEHRENS, TROPICAL BIRDING TOURS

The climate in this habitat is harsh, with warm temperatures during the day and cold, sometimes freezing, temperatures at night. Montane Grassland can develop across a broad range of rainfall levels, from the driest parts of the South African Highveld (Köppen **Cwb**), which receive only 16 in.

The Montane Grasslands of the high Simien Mountain of n. Ethiopia support the rare Walia Ibex, a biogeographic remnant of Palearctic influence. Several plants in the Ethiopian highlands also have close affinities to Europe. © MARIUS BURGER

White-tailed Gnu, or Black Wildebeest, was almost extinct in the wild, reduced to around 600 individuals; now numbering 18,000 its status has been dramatically improved. Mountain Zebra National Park, South Africa.
© KEITH BARNES, TROPICAL BIRDING TOURS

(400 mm) of annual rain, to the wettest parts of the Ethiopian Highlands (Köppen **Cwb**), which can receive 100 in. (2500 mm).

The precise composition of the grasses that dominate this habitat varies considerably depending on elevation and location on the continent. Although they look superficially similar, different areas have significantly different assemblages of plants and animals, forming distinctive subhabitats: South African Temperate "Highveld" (Af7F-1), Tropical "Nyasi" (Af7F-2), Abyssinian Grassland (Af7F-3), and in drainage areas and fire-protected gullies Montane Shrubland (Af7F-4). The most common grass in the Highveld of e. South Africa is Red Oat Grass (*Themeda triandra*). Other typical grasses include trident grasses (*Tristachya*), thatching grasses (*Hyparrhenia*), *Exotheca abyssinica*, fescues (*Festuca*), signalgrasses (*Brachiaria*), *Monocymbium ceresiiforme*, Common Russet Grass (*Loudetia simplex*), fountaingrasses (*Pennisetum*), foxtails (*Setaria*), and Bluestems (*Andropogon*). Wet areas often hold true sedges of the genus *Carex*. Herbaceous plants can be common, especially in moist and sheltered spots. A few examples are felworts (*Swertia*), mouse-ear chickweeds (*Cerastium*), lady's mantles (*Alchemilla*), *Helichrysum*, buttercups (*Ranunculus*), and Ouhout (*Leucosidea sericea*; see sidebar 32, p. 299).

CONSERVATION: Montane Grasslands are among the habitats that are most frequently converted to cultivation. This has happened to the majority of the grasslands of the Ethiopian Highlands and the South African Highveld. Even in places where natural grassland remains, this habitat is often overused for grazing.

WILDLIFE: Montane Grassland is an important habitat for many large mammals, while rodents are universally abundant. Typical larger mammals include baboons, Oribi, Mountain Reedbuck, and

THIS PAGE:
Left: **Southern Bald Ibis is a threatened endemic of Southern Africa's Highveld grasslands, eating insects and often frequenting recently burnt fields.** © KEN BEHRENS, TROPICAL BIRDING TOURS

Below left: **Veldfire at Dombeya, Johannesburg, South Africa.** © KEITH BARNES, TROPICAL BIRDING TOURS

OPPOSITE PAGE:
Top: **Montane Grassland is not as prone to flood as lower-elevation grassland, but marsh areas do develop along streams. Where streams form steep-sided gullies, key microhabitat for Blue Swallow is found.** MAIN PHOTO ANDRINGITRA NATIONAL PARK, MADAGASCAR, © KEN BEHRENS; INSET © KEITH BARNES, TROPICAL BIRDING TOURS

Bottom: **Yellow-breasted Pipit at Wakkerstroom, South Africa, where forbs are flowering in dense grassland. The appearance of the habitat and the plumage of the bird change drastically in the wet versus dry** *(inset)* **seasons!** © KEN BEHRENS, TROPICAL BIRDING TOURS

Common Eland. Lion, Leopard, and hyenas all sometimes use this habitat. Smaller predators include Black-backed Jackal and Serval. South Africa's Highveld once held a diverse set of big mammals, but most of them are now regionally extinct or confined to reserves. The Black Wildebeest was extinct in the wild but has now been reintroduced. Smaller mammals such as Cape Fox, Meerkat, and Gray Rhebok remain common. At the montane grassland-forest ecotone in the Albertine Rift mountains, "Mountain" Eastern Gorillas like eating plants from the arrowroot (Marantaceae) and ginger (Zingiberaceae) families.

Montane Grassland is a poor habitat for reptiles, though a few species have adapted to it. These include Ornate Sandveld Lizard (*Nucras ornata*), Zimbabwe Girdled Lizard (*Cordylus rhodesianus*), Bayon's Skink (*Trachylepis bayonii*), Grass Skink (*T. megalura*), and the grass lizards (*Chamaesaura*). The Highveld holds the remarkable Sungazer (*Smaug giganteus*), a big spiky lizard that is the largest member of the girdled lizard family. This is also the habitat for South Africa's recently discovered Cream-spotted Mountain Snake (*Montaspis gilvomaculata*), the only member of its genus.

The Montane Grasslands of West Africa and Angola are poor in birds, probably due to their isolation and perhaps to the scarcity of natural grasslands before frequent human-caused fires. The more contiguous archipelago of Montane Grassland from Eritrea to South Africa is somewhat richer in birds. Typical species include various harriers and other migrant raptors, Striped Flufftail, the

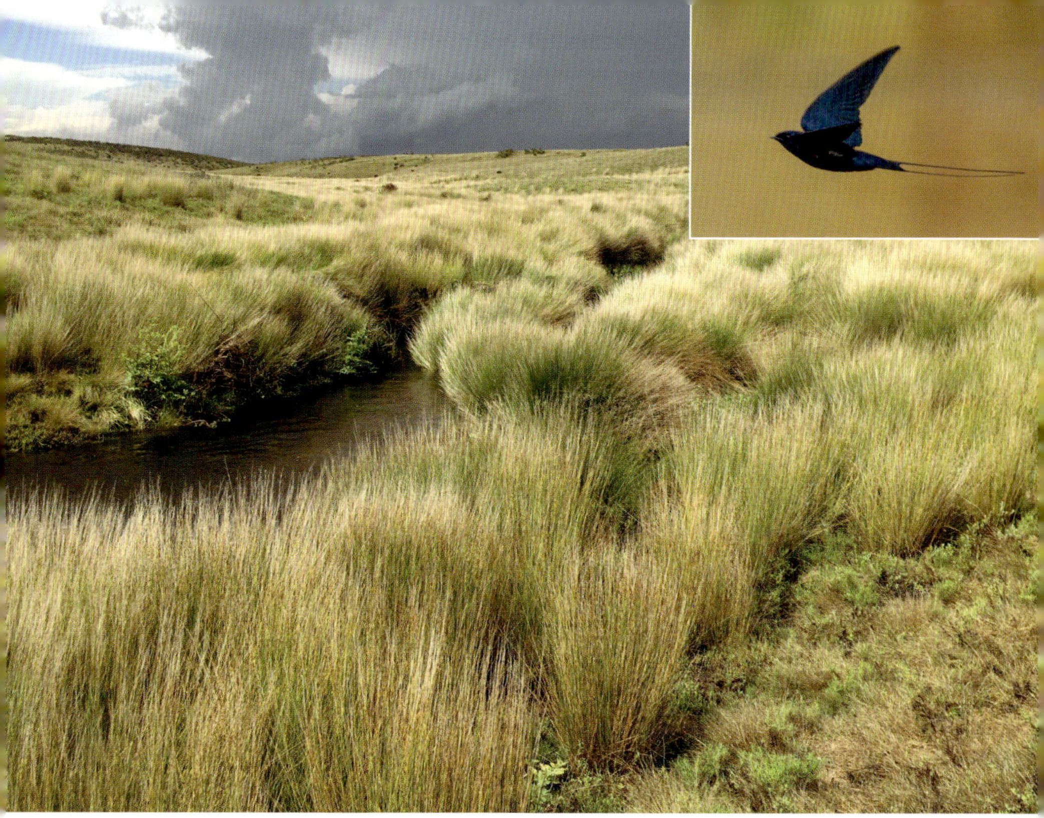
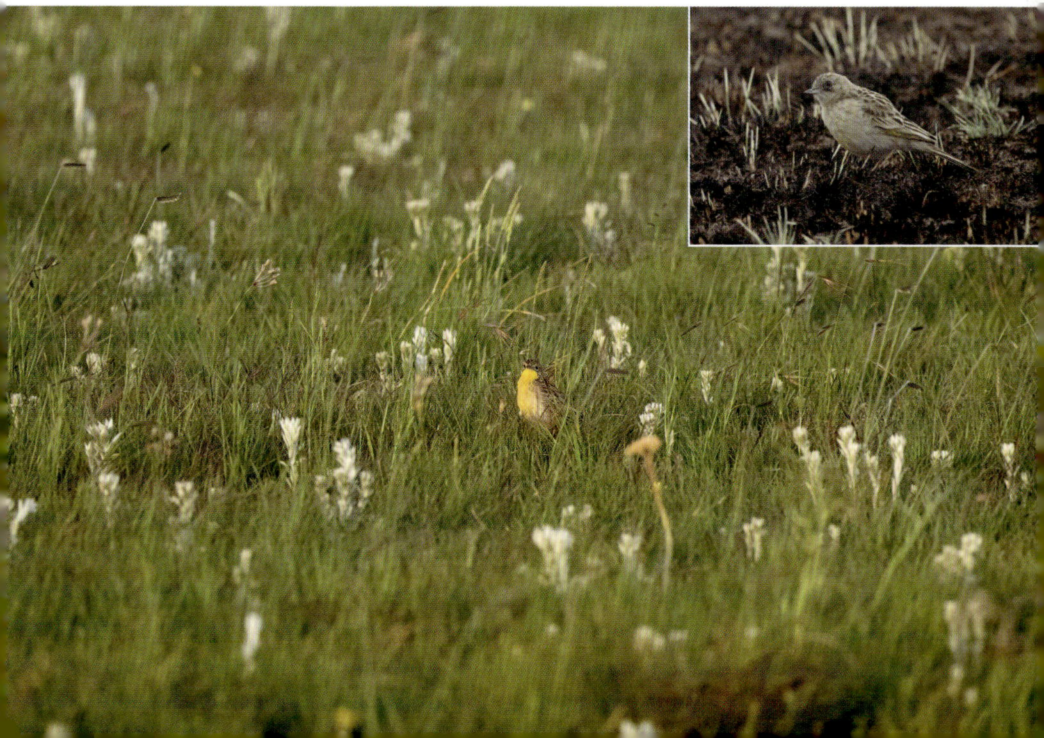

rare migratory Montane Blue Swallow, Malachite Sunbird, and several species of cisticolas. The most interesting birds of this habitat are the more localized endemics.

Endemism: The scattered, archipelago-like nature of this habitat in Africa has led to many fascinating local endemics. The two major centers of endemism are the Ethiopian Highlands and the Highveld of South Africa. Ethiopia is home to Gelada, a large primate that uses a complex vocabulary second only to that of *Homo sapiens*. It is also used by Ethiopian Wolf, the world's most endangered canid. The region's wonderful lineup of endemic birds includes Wattled Ibis, Spot-breasted Lapwing, and Abyssinian Longclaw (IS). South African Highveld is also rich in birds, including Blue Crane, Blue Bustard (IS), Buff-streaked Chat, Cape Grassbird, Botha's Lark, and

Left: **Rocky mountain grassland. Dedza, Malawi.** © KEN BEHRENS, TROPICAL BIRDING TOURS

Below: **Gelada (despite their appearance not closely related to baboons) on the Ankober Escarpment, Ethiopia. This is the world's only primate that is primarily a grazer, surviving mainly on grass blades, seeds, rhizomes, and roots. Insects are eaten only rarely.** MAIN PHOTO © KEN BEHRENS; INSET © KEITH BARNES, TROPICAL BIRDING TOURS

Yellow-breasted Pipit. The mountains of Kenya and Tanzania have a few localized grassland species, like Jackson's Francolin and Sharpe's Longclaw. The mountains between the Ethiopian Highlands and the South African Highveld have many localized endemic plants.

DISTRIBUTION: Montane Grasslands usually occur in a matrix alongside MONTANE HEATH, MOIST MONTANE FOREST, and AFROTROPICAL MONTANE DRY MIXED WOODLAND. Near the equator, Montane Grasslands can be found down to around 6000 ft. (1800 m). Away from the equator, Montane Grasslands are found at lower elevations; the Highveld grasslands of South Africa grow at 4000–7500 ft. (1200–2300 m). The largest areas of natural Montane Grassland were originally found in the Ethiopian Highlands and South African Highveld. These two major blocks were loosely connected by an archipelago of mountains. However, the grasslands on the mountains of Zimbabwe, Mozambique, and Malawi, in the Eastern Arc Mountains, and in far w. West Africa may have been very limited originally but have become much more widespread due to frequent human fires. The highest mountains in Madagascar support a grassland that is superficially similar to the MALAGASY GRASSLAND AND SAVANNA but has a different botanical composition and probably represents a natural rather than anthropogenic habitat.

WHERE TO SEE: Wakkerstroom, South Africa; Nyika Plateau, Malawi; Bale Mountains National Park, Ethiopia.

SIDEBAR 32   EDGE CASE STUDY: HIGHLAND OUHOUT SHRUBLAND

There is a distinctive type of shrubland growing locally in Southern Africa, primarily along the edges of Highveld Montane Grasslands. The most typical plant is Ouhout (*Leucosidea sericea*), a distinctive plant whose narrow leaves have serrated edges. Like Hagenia (*Hagenia abyssinica*), a classic plant of dry montane woodland, Ouhout belongs to the Rose family, and Ouhout is Hagenia's closest relative. This formation represents a transition from Montane Grassland to the pockets of MOIST MONTANE FOREST that lie below. In some areas it is restricted to gullies, while in others, Ouhout dominates large areas, sometimes choking out the understory grass. These large monotypic stands are likely results of overgrazing by domestic animals. The wildlife of this shrubland is essentially an extremely reduced subset of what is found in South African Moist Montane Forest, while the surrounding and interspersed grasslands hold typical Highveld species. These shrublands could be considered their own habitat, lumped into Moist Montane Forest, or, as in this book, considered as a subhabitat (Af7F-4) of Montane Grassland.

Ouhout shrubland, when dense enough, is a seasonal home to the Southern African endemic Bush Blackcap. Wakkerstroom, South Africa.
© KEN BEHRENS, TROPICAL BIRDING TOURS

MEDITERRANEAN FORESTS, WOODLANDS, AND SCRUBS

Af8A FYNBOS

IN A NUTSHELL: A botanically incredibly diverse heath habitat of s. South Africa, which is dominated by restios, ericas, proteas, and geophytes. **Global Habitat Affinities:** AUSBOS (AUSTRALIAN SOUTHERN LOWLAND HEATHLAND); NORTHWEST EUROPEAN COASTAL AND MONTANE HEATH; ASIAN TEMPERATE HEATHLAND; MAQUIS; PACIFIC CHAPARRAL. **Continental Habitat Affinities:** STRANDVELD; RENOSTERVELD; MONTANE HEATH; MAGHREB MAQUIS. **Species Overlap:** STRANDVELD; RENOSTERVELD; NAMA KAROO; SUCCULENT KAROO; MONTANE GRASSLAND.

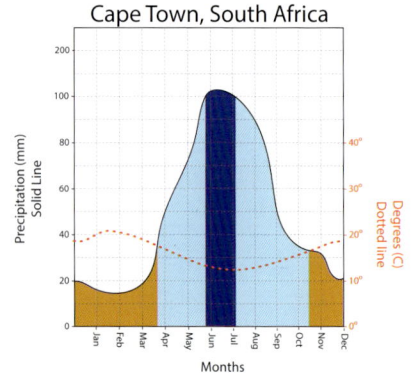

Fynbos burns regularly and then regenerates, from top to bottom: recent burn, 5 years post burn, 10 years post burn, 15-20 years post burn

DESCRIPTION: This hard-leaved (sclerophyllous) heath shrubland growing on nutrient-poor soil in a Mediterranean climate with wet winters and hot dry summers (Köppen **Csb**) is one of the Afrotropics' most extraordinary. Botanists have classified six floral kingdoms in the world, and Fynbos (along with RENOSTERVELD and "western" STRANDVELD) in this small corner of South Africa is one of them (The Greater Cape Floristic Region)— and by far the smallest. This treatment is merited by the monumental plant diversity and endemism, with 13 genera containing over 100 species! There are around 9000 plants in this floral kingdom, 69% of which are endemic. Compared with similar habitats growing in

The three most conspicuous groups of Fynbos plants: erica heath *(left)*, **sedge-like restio** *(middle)*, **and protea** *(right)*. © KEITH BARNES (LEFT AND RIGHT); © KEN BEHRENS (CENTER), TROPICAL BIRDING TOURS

Mediterranean climates, Fynbos is 1.7 times more diverse than sw. Australia, 2.2 times more diverse than the Mediterranean basin or California, and 3.0 times more diverse than Chile.

Four groups of plants dominate: restios, ericas, proteas, and geophytes. These plants all show adaptations that allow them to grow in an environment where fire and long dry periods are common. Restios, which make up their own family (Restionaceae), are evergreen, rush-like plants that may look like sedges or grasses to the untrained eye. However, these are the most characteristic plants, typically being present in almost all types of Fynbos. Members of the erica family, commonly known as heaths, are shrubs with fine, leathery, evergreen leaves. Within this family, the single genus *Erica* has over 600 representatives in the Fynbos! Proteas form a family of evergreen bushes that have cone-like flower heads that can be spectacular when blooming. The tallest shrubs are generally 3–10 ft. (1–3 m) tall. Geophytes are plants with underground storage organs, in this case usually bulbs.

Protea and restio-dominated Fynbos, 4–5 years after the most recent fire. Bredasdorp, South Africa. © KEN BEHRENS, TROPICAL BIRDING TOURS

Taller, bushier Fynbos 12–15 years old with dense protea and erica elements becoming moribund and in need of a burn. *Berzelia* (inset), a heath-like plant that belongs to the endemic Bruniaceae plant family, can be found in wetter Fynbos. Nature's Valley, South Africa. © KEITH BARNES, TROPICAL BIRDING TOURS

Moist seeps and river sources are often clogged with *Berzelia*, a genus of water-loving, flexible, fine-leaved shrubs with spherical bobbles of white flowers that is endemic to the Fynbos. Grasses and tall trees are both uncommon, though they occur locally, especially where there is more summer rainfall, east of Knysna. Although Fynbos appears inviting from a distance, the density of shrubs usually makes walking difficult, except in occasional open glades or postfire environments.

There are two extreme seasons in most Fynbos, a cooler, moist austral winter (May–Aug) and a hot dry austral summer (Nov–Mar). The climate is fairly mild, especially along the coast, though snowfall can occur on mountaintops in winter. Annual rainfall varies widely, from 10 to 80 in. (250–2000 mm), though a typical range is 12–30 in. (300–750 mm). Fynbos receives rain almost exclusively in winter in the west, but east of Knysna, rain is year-round or even more frequent in the summer. As suggested by its enormous plant diversity, the Fynbos is far from uniform. There is tremendous heterogeneity and complexity to the local plant communities, driven by the wide variety of rainfall regimes, elevations, soil types, and slope aspects. But the main driver of diversity and vegetation structure is fire (see sidebar 33, p. 303). Fynbos is fire-adapted, and it needs fire to thrive; without it, many plant species would go extinct and communities would become moribund, species-poor protea thicket. Fires, mostly during the hot, dry summer, which is paired with strong winds, burn hot and tend to denude huge swathes of vegetation. The result appears devastating, and can be for the human communities living near this habitat, but fire is essential, as the vegetation and wildlife have coevolved to need the renewal. In the immediate aftermath, the

| SIDEBAR 33 | FIRE: FRIEND OR FOE? |

Fire destroys. Fire renews. Fire creates. We often hear stories about disastrous blazes that wreak havoc on communities. The damage to property is real, and the human cost is huge. But most habitats in Africa need fire to varying degrees to keep their ecology in sync. Most savannas burn in the dry season. This kills young saplings (keeping the habitat more open) and also removes the moribund layer of grass that has accumulated, returning nutrients to the soil. Once the rains come, grasses sprout and provide new growth for grazers like zebras. Fire, in combination with mega- and mini-fauna (see sidebars 40, p. 373; 34, p. 306; and 17, p. 188), keeps many of those systems functioning. In other habitats like FYNBOS, fire is simply non-negotiable. It is everything. Without it, the heathlands get overgrown (moribund) and many of the understory components (where much of the diversity lies) die off by being smothered. Fynbos is so dependent on fire that its climax trees—proteas—often hold their seeds in fire resistant canopy capsules because dispersing them under normal conditions would see them decimated by predatory mice. When fire comes through, the capsule burns open and the seeds are dispersed and are able to germinate in a predator-free, low-competition environment. Many Fynbos plants are fire-adapted. But the Fire Lily (*Cyrtanthus ventricosus*) is perhaps the most extreme in a guild of pyrophytic geophytes (fire-adapted with an underground storage organ like a bulb). Within a couple of days of a devastating fire, the crimson flowers of this species erupt in a charred landscape, proving that there is life after death. Sunbirds and the Table Mountain Pride Butterfly (*Aeropetes tulbaghia*) pollinate this plant in what resembles a post-apocalyptic landscape. The flowers don't last long: the plant seeds, leafs out, then goes dormant, not flowering again until a fire repeats the process, a wait of perhaps 15–20 years!

Fynbos, burnt down to the sand. Fire Lily (*Cyrtanthus ventricosus*) is a bulb that flowers only after intense fire rids the environment of competition; it can then lie dormant for decades, waiting for the next "devastating" fire to bloom again.
© CALLAN COHEN, BIRDING AFRICA

seemingly barren, bleak, postfire environment will at first be inhabited by several highly specialized bulbs, which will flower briefly before lying dormant for another 15–20 years until the next fire event. After 2–3 years a low shrubland, 1.6–3 ft. (0.5–1 m) tall, develops and is easy to walk through; it is dominated by restios and ericas. After 5–15 years, protea stands are 3–5 ft. (1–1.5 m) tall, and large erica shrubs have grown up. The habitat becomes awkward to pass through, and plant species diversity peaks. After 15–20 years, protea stands become 7–10 ft. (2–3 m) tall and are often impenetrable thickets; species diversity slowly declines, awaiting the rejuvenating effects of the next fire.

CONSERVATION: Large areas of lowland Fynbos have been cleared for agriculture and development, notably around the sprawling city of Cape Town. The montane Fynbos has been less affected by humans, and many mountain ranges are formal conservation areas. However, all Fynbos is subject to a more insidious threat in the form of introduced invasive plants. Areas that have been converted to agriculture or treed suburbs and farmyard gardens can have a somewhat grassland or savanna-like character and attract species that historically wouldn't have been found in this zone.

WILDLIFE: Fynbos does not offer an abundance of easy nutrients to animals and therefore is a relatively low-density wildlife environment. Its mammal assemblage was further reduced by hunting after the arrival of Europeans. The impressive "Cape" Lion, potentially a distinct subspecies with a black mane and ears, larger than savanna lions, was extirpated in the late 1800s, along with African Savanna Elephant and African Buffalo; however, these animals probably mostly avoided using classic Fynbos, preferring the nutrient rich neighboring **RENOSTERVELD**. The local Bontebok and "Cape" Mountain Zebra have been reintroduced here in completely unsuitable habitats, where they are dependent on human-made grazing lawns! The Fynbos still holds an assemblage of smaller mammals, including an impressive lineup of diminutive antelopes. Cape Grysbok is near-endemic to this habitat and occurs alongside the more widespread Common Duiker, Steenbok, and Gray Rhebok. Smaller predators include the local Cape Genet, Cape Fox, and Cape Gray Mongoose. The "Cape" Leopard, smaller than its savanna cousins and with a much larger home-range, is much reduced in numbers but persists at low density. It prefers rugged mountain wilderness areas but is sometimes seen surprisingly close to humans in the Kogelberg Biosphere Reserve. One of the most important mammalian components, the rodents, are seldom seen or noticed but are responsible for a significant amount of pollination and seed dispersal of keystone plants like proteas. The rodent roster includes several local endemics like Verreaux's Mouse, Cape Spiny Mouse, and Grant's Rock Rat.

Although it is not Africa's most diverse birding habitat, the Fynbos is of great interest to birders because of its endemic and near-endemic birds, including two families that are near-endemic: rockjumpers and sugarbirds. Great-looking species of rocky outcrops (see **INSELBERGS AND KOPPIES**, p. 268) include Cape Rockjumper (IS), the flicker-like Ground Woodpecker, and Cape and Sentinel Rock-Thrushes. Thicker and wetter *Berzelia* seeps hold Cape Grassbird and Victorin's Warbler (IS) and the rarely seen Striped Flufftail. When ericas flower, pollinators can be abundant, including Orange-breasted (IS), Southern Double-collared, and Malachite Sunbirds. Flowering protea stands hold Cape Sugarbird (IS) and Protea Canary (IS). The Fynbos Buttonquail (IS) is perfectly adapted to the fire ecology of this habitat, specializing in restio-dominated postburn habitat 3–5 years old. As the habitat

Orange-breasted Sunbird is a Fynbos endemic, key pollinator of both ericas and proteas and is an indicator species for this habitat.
© KEN BEHRENS, TROPICAL BIRDING TOURS

Above: **Cape Sugarbird feeding on and pollinating a flowering pincushion plant (*Leucospermum* sp.—Proteaceae family), a classic Fynbos scene.** © KEITH BARNES, TROPICAL BIRDING TOURS

Below: **Cape Rockjumper in restio-dominated Fynbos. This species typically prefers the open nature of more recently burned environments. Rooiels, South Africa.** MAIN PHOTO © KEN BEHRENS; INSET OF MALE © KEITH BARNES, TROPICAL BIRDING TOURS

SIDEBAR 34 HABITAT HELPERS—POLLINATORS

Plants and wildlife exist in a complex dynamic whereby one is constantly affecting the other. This is clearly apparent in the mutualistic (equally beneficial) relationship between plants and animal pollinators. Sunbirds are the most famous of Africa's avian pollinators, and many flowers feed them delectable nectar in exchange for their help in transferring pollen. Many plants have developed long tubular flowers, and the sunbirds' bills have coevolved to be decurved to easily access the flowers. A feathery tongue means that nectar is quickly lapped up, and a high metabolism means that the birds keep feeding and transferring pollen. Because flowers flush and then vanish, nectarivorous birds like sunbirds are usually locally nomadic, appearing in large numbers when flowers are abundant and moving on as soon as the flowers wilt. But insects are arguably even more important pollinators, and they have evolved equally bizarre apparatus for the job. Not only that, but some plants like orchids have evolved to allow only a single species of insect pollinator, with complex apparatus restricting pollination to that species. At night, bats and moths take over and do an equally critical job for many species. Wildlife also acts as critical dispersers of fruit and seed. See sidebar 17, p. 188.

The stunning Golden-winged Sunbird enjoys *Leonotis* nectar on the Ngorongoro Crater rim. Insects like this bee fly (family Bombyliidae) *(inset)* with an incredibly long proboscis have evolved specialized equipment for the job of pollination. © KEITH BARNES, TROPICAL BIRDING TOURS

Above: **Several reptiles in the Fynbos have tiny ranges, some restricted to a single mountain massif. Hawequa Flat Gecko (*Afroedura hawequensis*) is a good example, found only in the Du Toitskloof and the Limietberg Mountains.** © KEIR AND ALOUISE LYNCH

Right: **Cape Mountain Rain Frog (*Breviceps montanus*), a perpetually "sulking" Fynbos endemic frog. Owing to their terrestrial nature, they live mostly underground far away from water; tailed froglets emerge from eggs, skipping the tadpole life cycle stage.** © KEIR AND ALOUISE LYNCH

thickens up, the nomadic buttonquail shifts to more suitable, recently burnt patches. Karoo Prinia, Neddicky, Cape Bunting, Cape Canary, and Cape Siskin are widespread. Lowland Fynbos on the Agulhas plain supports endemic specialties like Cape Clapper Lark and is perfect breeding habitat for low-density widespread species like Denham's Bustard. Mountain Fynbos provides refuge for Rock-Hyrax-hunting Verreaux's Eagle and Jackal Buzzard.

The Fynbos has a rich set of reptiles comprising more than 100 species. Specialties include Southern Rock Lizard (*Australolacerta australis*), Cape Mountain Lizard (*Tropidosaura gularis*), Black (*Cordylus niger*) and Oelofsen's (*C. oelofseni*) Girdled Lizards, and false girdled lizards (*Hemicordylus capensis* and *H. nebulosus*). Three species of charismatic chameleon occur, including some with miniscule ranges, such as Elandsberg Dwarf Chameleon (*Bradypodion taeniabronchum*). This region is also one of the most diverse in the world for tortoises, and Fynbos is home to such species as Parrot-beaked Padloper (*Homopus areolatus*) and Angulate Tortoise (*Chersina angulata*).

Endemism: The Fynbos encompasses countless areas of local interest for botanists. For example, 30 separate and discrete botanical complexes have been identified within mountain Fynbos alone. These tend also to be nodes for over 22 species of endemic frogs, with three endemic species of comical Jabba-the-Hutt-like *Breviceps* rain-frogs, seven endemic moss frogs (*Arthroleptella*) with miniscule ranges, and six endemic ghost-frogs (*Heleophryne*), the latter an ancient lineage found nowhere else on Earth. There are over a dozen endemic reptiles, and taxonomic changes suggest more, especially in complexes with cryptic species like *Afrogecko* and dward leaf-toed (*Goggia*) geckos. Eight bird species (discussed above) are endemic or near-endemic to Fynbos. The invertebrate fauna is extremely rich, with many endemics, including many ancient lineages, such as the stunning Cape Stag Beetles (*Colophon*). There is a slight difference between the wildlife of mountain and lowland Fynbos, with lowland Fynbos sharing more species with STRANDVELD and RENOSTERVELD.

DISTRIBUTION: Fynbos is found both along the coast and on the mountains of s. South Africa, reaching 7900 ft. (2400 m). There are some isolated montane Fynbos communities that are completely surrounded by KAROO habitat. Although there are similarities between Fynbos, RENOSTERVELD, and STRANDVELD, their structure and wildlife communities are different.

WHERE TO SEE: Cape of Good Hope Nature Reserve, South Africa; Kogelberg Biosphere Reserve, South Africa.

On the eastern side of False Bay the Kogelberg supports a full array of Fynbos-endemic birds and mammals. Some of these, like Victorin's Warbler and Protea Seedeater, are oddly absent on the Cape Peninsula, only 25 mi. (40 km) away, despite perfect habitat being found for them there. Rooiels, South Africa.
© KEITH BARNES, TROPICAL BIRDING TOURS

Af8B STRANDVELD

IN A NUTSHELL: Dense coastal scrub of evergreen hard-leaved (sclerophyllous) and fleshy drought-resistant shrubs and succulents with spring (Aug–Sep) displays of flowering herbaceous annuals. **Global Habitat Affinities:** WALLUM; NORTHWEST EUROPEAN COASTAL AND MONTANE HEATH; ASIAN TEMPERATE HEATHLAND; MAQUIS; PACIFIC CHAPARRAL. **Continental Habitat Affinities:** FYNBOS; SUCCULENT KAROO; RENOSTERVELD; MONTANE HEATH; MAGHREB MAQUIS. **Species Overlap:** FYNBOS; RENOSTERVELD; NAMA KAROO; SUCCULENT KAROO.

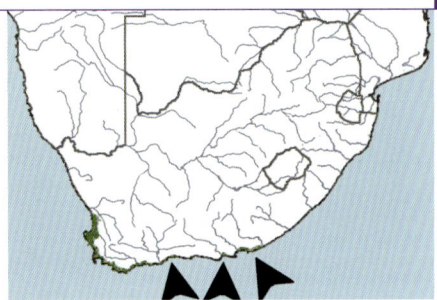

DESCRIPTION: Along with FYNBOS and RENOSTERVELD, botanists have incorporated "western" Strandveld into the Greater Cape Floristic Region, the smallest and most complex of Earth's six floral kingdoms. Meaning "beach vegetation'" in colloquial Afrikaans, this shrubland is virtually restricted to nutrient-poor sandy acidic soil or coastal dunes below 650 ft. (200 m). Although botanists consider much of it a type of Fynbos, it is much less diverse and structurally distinct and is virtually devoid of proteas and ericas. The daisy family (Asteraceae) forms a major component of all elements of Strandveld. This habitat is mostly a dense evergreen shrubland with a low canopy, 4–5 ft. (1.2–1.5 m) tall, populated by hard-leaved shrubs like Sea Guarrie (*Euclea racemosa*), Beach Salvia (*Salvia africana-lutea*), Tortoise Berry *(Nylandtia spinosa)*, Blue Kuni-rhus (*Searsia glauca*) and White Bristle-bush (*Metalasia muricata*) and shrublike daisies like West Coast Bitou (*Osteospermum moniliferum*) and salad thistle (*Didelta*). Occasionally isolated individual Milkwood (*Sideroxylon inerme*) and num-num trees (*Carissa*), which may have climbers, occupy wind-protected gullies. Stunted shrubby trees form thickets, especially in "eastern" Strandveld. Where these stands become dense enough, the habitat becomes SOUTH COAST FOREST MATRIX. A low understory comprises fat-leaved and thick-stemmed succulents mostly shorter than 19 in. (50 cm) and dominated by spurges, the distinctive stonecrop family (Crassulaceae), and aloes. There are some geophytes (bulbs), the most striking of which are bright pink amaryllis (Amaryllidaceae) flowers that paint areas cerise at times. The ground layer is covered with fat-leaved fig-marigolds (Aizoaceae), bean-capers (*Zygophyllum*), and a phenomenal diversity of perennials and annuals, many of which are daisies, including White Rain Daisy (*Dimorphotheca pluvialis*), Gousblom (*Gazania krebsiana*), Livingstone Daisy (*Cleretum bellidiforme*), and wild sorrels (*Oxalis*). There is a small grassy component, but it never dominates.

There are two extreme seasons in Strandveld (Köppen **Csb**): a cooler austral winter (May–Aug) and a hot dry austral summer (Nov–Mar). The climate is mostly mild, 44–86°F (7–30°C) with no snow, and

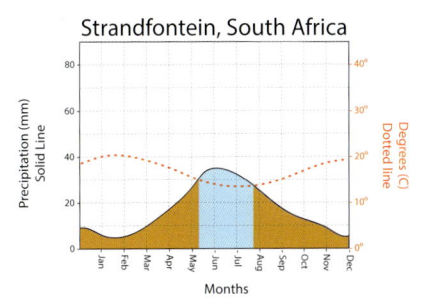

Profusion of annual and perennial spring wildflowers, mostly daisies (Asteraceae), carpet West Coast National Park, South Africa. © KEITH BARNES, TROPICAL BIRDING TOURS

Shrubby Aster-dominated Strandveld with restios (note the absence of ericas and proteas) in the foreground, with dune thicket behind. De Hoop Nature Reserve, South Africa. © KEN BEHRENS, TROPICAL BIRDING TOURS

Strandveld is a botanical wonderland! From the left: Black Flag (*Ferraria crispa*), Dainty Soldier-in-a-Box (*Albuca cooperi*), and two unidentified asters (Asteraceae). © KEN BEHRENS, TROPICAL BIRDING TOURS

frost and hail are rare. Annual rainfall is strongly seasonal in western Strandveld, falling mostly in the austral winter, but is more year-round farther east. Rainfall averages 5–30 in. (125–750 mm), though a typical range is 8–14 in. (200–350 mm). Mist and fog provide occasional precipitation year-round. Winds are strong throughout the year and stunt the growth of shrubs. In the western Strandveld, after late winter rains, the annuals flower, producing perhaps one of the most spectacular flowering displays (along with the SUCCULENT KAROO's Namaqualand) anywhere on Earth. This is somewhat unpredictable and may vary from year-to-year depending on timing and the total quantity of rain; some years are much better than others. Classic wisdom, however, suggests that late August to early September is your best bet at catching the finest flower displays.

CONSERVATION: The Strandveld is threatened, with massive areas having been cleared for agriculture and coastal development. In addition, alien vegetation, especially Australian Port Jackson Willow (*Acacia saligna*) and Coastal Wattle (*A. cyclops*), which were deliberately planted to stabilize the coastal dunes, have run wild and occupied massive sectors of Strandveld. Although most of the big mammals of this habitat were cleared out soon after European hunters arrived at the Cape, much of the remaining natural vegetation is now part of private "game farms" and conservation initiatives, with much wildlife having been restocked. Several very important reserves comprise primarily Strandveld, and important sectors are formally protected.

Strandveld can be locally dominated by grass. West Coast NP, South Africa. © KEN BEHRENS, TROPICAL BIRDING TOURS

SIDEBAR 35 — IS IT A MOLE OR IS IT A RAT?

Constituting their own family (Bathyergidae), the mole-rats are neither moles nor rats. In fact, they are more closely related to porcupines and Neotropical capybaras than either of the groups that give them their name! The "sand-puppy," or Naked Mole-rat, of the Horn of Africa is perhaps the most famous of these critters. With an appearance only a colonial mother (the Queen no less) could love, and along with the Damaraland Mole-rat, these are the only truly eusocial mammals, with social structure like bees and ants! The 22 species in five genera in this family are restricted to sub-Saharan Africa. They are a lineage of fossorial (digging) mammals that live underground, constructing complex burrow systems and foraging on bulbs, roots, and tubers. Long, strong claws and massive teeth are used for digging. Nostrils that constrict and a mouth that conveniently closes behind the teeth prevent dirt from entering. Tiny eyes and small ears also help with underground living. Although their mounds are abundant, seeing any fossorial creature is difficult, and perhaps your best bet to see one is to find a Cape Dune Mole-rat in the early spring when rains force them to disperse.

Cape Dune Mole-rat with powerful claws and teeth for digging and small ears and eyes to assist moving through sandy subterranean burrows. © KEITH BARNES, TROPICAL BIRDING TOURS

WILDLIFE: Strandveld, despite much of it superficially appearing to be a scrubby dry wasteland, supports a surprising array of wildlife. Although much was decimated after the arrival of European settlers, the establishment of formal conservation areas (especially West Coast National Park and De Hoop Nature Reserve), as well as private-lands conservation initiatives, have resulted in significant restocking of large ungulate populations, including: Springbok, Bontebok, Gemsbok, Eland, "Cape" Mountain Zebra, and Red Hartebeest. Small ungulates, which managed to escape the ravages of hunting, include the near-endemic Cape Grysbok and Gray Rhebok, which both thrive, alongside the widespread Steenbok and Common Duiker. Although "Cape" Lions were extirpated and "Cape" Leopards avoid the habitat, the Strandveld is particularly good for smaller cats, including Caracal (which enjoys the high density of gamebirds here) and African Wild Cat. Other small predators include Cape Genet, Yellow Mongoose, Cape Fox, Bat-eared Fox, and Honey Badger. The soft sands and dunes are suitable for a suite of rare and near-endemic burrowing (fossorial) mammals that mostly live underground, and Cape and Grant's Golden Moles and Cape Dune Mole-rat are all best found in Strandveld. During the winter rains these animals can sometimes be seen dispersing, scuttling across dunes or even lawns in coastal towns in broad daylight.

Birding Strandveld is productive, and communities are moderately rich, sharing more with Karoo habitats than FYNBOS. Pied Starling, Capped Wheatear, Karoo Scrub-Robin, and Bokmakerie bounce on the ground. Gray Tit, Cape Crombec, and Chestnut-vented Warbler glean small shrubs. Speckled and White-backed Mousebirds flutter between ridgeline dunes. The two most striking birds that make significant use of Strandveld are Black Harrier—which breeds in this habitat (and RENOSTERVELD) before dispersing elsewhere— and the strident Black Bustard. Gamebirds are abundant, with Cape Spurfowl and Helmeted Guineafowl omnipresent and Gray-winged Francolin more secretive but also common. Where Common Ostrich have been reintroduced, they breed well, and have become common. Southern Double-collared and Malachite Sunbirds are abundant, especially when aloes and crassulas are flowering. Seedeaters are also abundant with canaries featuring regularly: Yellow and White-throated in western Strandveld and Cape, Brimstone, and Streaky-headed in eastern Strandveld. The shrub thickets and Milkwood stands in eastern Strandveld support the local endemic

Top: **Caracal is perhaps more readily seen in Strandveld than in any other habitat.** © KEITH BARNES, TROPICAL BIRDING TOURS

Middle: **Black Bustard, locally known as Southern Black Korhaan. In spring, the loud calls of this bustard punctuate the Strandveld. West Coast National Park, South Africa.** © KEN BEHRENS, TROPICAL BIRDING TOURS

Bottom: **Strandveld is rich in tortoises, including the cute Parrot-beaked Padloper (*Homopus areolatus*).** © KEITH BARNES, TROPICAL BIRDING TOURS

Black Girdled Lizard (*Karusaurus niger*) is more typical of Fynbos, but there is also an isolated population in Strandveld near Langebaan, South Africa. © KEITH BARNES, TROPICAL BIRDING TOURS

Southern Tchagra and Knysna Woodpecker, along with a few species more typical of forests, such as Southern Boubou and Bar-throated Apalis. In summer, the intra-African migrant yellow-billed form of Black Kite and Common Buzzard are abundant predators.

Strandveld is rich in reptiles, especially snakes, supporting Puff Adder (*Bitis arietans*), Cape Cobra (*Naja nivea*), Boomslang (*Dispholidus typus*), and smaller endemic species, particularly Cape Sand Snake (*Psammophis leightoni*), Many-horned Adder (*Bitis cornuta*), and Southern Adder (*B. armata*)—first described only in 1997—which is globally vulnerable and virtually restricted to Strandveld. Some other charismatic reptiles include Cape (*Bradypodion pumilum*) and Namaqua (*B. occidentale*) Dwarf Chameleons and Large-scaled Girdled Lizard (*Cordylus macropholis*). This region is also one of the most diverse in the world for tortoises, and Strandveld is home to such species as Parrot-beaked (*Homopus areolatus*) and Speckled Padlopers (*Chersobius signatus*) and Angulate Tortoise (*Chersina angulata*). Some frogs also favor the soft-sand habitat, especially the odd-looking Sand Rain Frog (*Breviceps rosei*) and Cape Sand Toad (*Vandijkophrynus angusticeps*).

Endemism: There are a significant number of endemic plants in the Strandveld, but it is significantly less rich or diverse than the FYNBOS. Some of the vertebrates are endemic, but most are shared with the FYNBOS and SUCCULENT KAROO.

DISTRIBUTION: Restricted to coastal South Africa from around Eland's Bay to the Kei River, north of East London, reaching around 12 mi. (20 km) inland in places, especially along estuaries.

WHERE TO SEE: West Coast National Park and De Hoop Nature Reserve, South Africa.

Cape Longclaw in West Coast National Park, South Africa. Note the abundance of succulent ground cover plants. © KEN BEHRENS, TROPICAL BIRDING TOURS

Af8C RENOSTERVELD

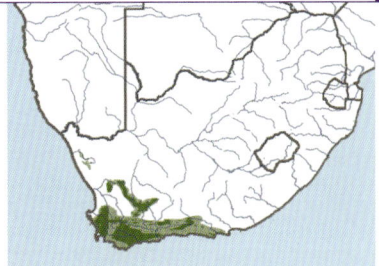

IN A NUTSHELL: A uniform low scrub dominated by Renosterbos shrubs and, in most places, significant grass cover, with a very high diversity of seasonally active geophytes, succulents, and legumes. **Global Habitat Affinities:** AUSTRALASIAN MONTANE HEATHLAND; NORTHWEST EUROPEAN COASTAL AND MONTANE HEATH; ASIAN TEMPERATE HEATHLAND; MAQUIS; PACIFIC CHAPARRAL. **Continental Habitat Affinities:** FYNBOS; STRANDVELD; SUCCULENT KAROO; MONTANE HEATH; MAGHREB MAQUIS. **Species Overlap:** FYNBOS; STRANDVELD; NAMA KAROO; SUCCULENT KAROO; MONTANE GRASSLAND.

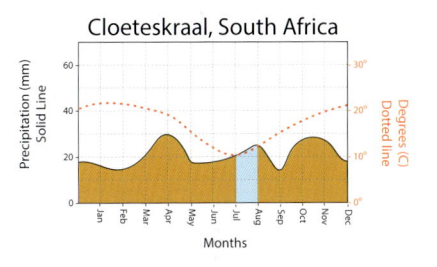

DESCRIPTION: Along with FYNBOS and western STRANDVELD, botanists have incorporated Renosterveld into the Greater Cape Floristic Region, the smallest and most complex of Earth's six floral kingdoms. This habitat is unfairly sometimes called Fynbos's ugly sister! Renosterveld means "rhinoceros vegetation" in colloquial Afrikaans, and this shrubland is strongly associated with nutrient-rich clay and silt soils derived from shales and granite. Although sometimes considered a type of Fynbos, it is structurally distinct, and Fynbos's most characteristic elements, restios, proteas and ericas, are absent or rare. Renosterveld is unusual in that it is often dominated by a single species. Renosterbos (*Dicerothamnus rhinocerotus*) is a smallish, round, rosemary-like shrub in the aster family that grows to be 3–6 ft. (1–2 m) tall and has small, triangular gray-green leaves growing directly off the stem of long, slender branches. Older plants are gnarled, with smooth, grayish bark. Not all Renosterveld is dominated by Renosterbos, as many versions of this habitat are primarily vegetated by grasses and other shrubs.

There are two seasons in Renosterveld (Köppen **Csb**): a cooler austral winter (May–Aug) and a hot austral summer (Nov–Mar). The climate is mostly mild 44–86°F (7–30°C) with little snow, frost, or hail. Annual rainfall is variable, sometimes year-round and sometimes seasonal, falling mostly in the austral winter (Jun–Aug) and averaging 12–24 in. (300–600 mm), although more summer rainfall occurs in the east and the Overberg. The transformation of Renosterveld from an "ugly sister" gray shrubland into a flower-festooned "Cinderella" when the bulbs, succulents, and legumes bloom in late winter and spring (Aug–Oct) is spectacular but ephemeral, only lasting 4–8 weeks. Renosterveld transitions fairly rapidly into Fynbos when soils become nutrient poor (e.g., sandstone) or NAMA KAROO when rainfall drops below 10 in. (250 mm).

Historically, this habitat was likely much grassier, with less Renosterbos and a greater diversity of shrubs, but the reduction in fire and replacement of wild ungulates with stock like sheep has led to massive overgrazing and likely transformed Renosterveld into a much shrubbier habitat. Other shrubs include Wild Rosemary (*Eriocephalus africanus*), Klein Perdekaroo (*Oedera genistifolia*), Jakkalsstert (*Anthospermum aethiopicum*), doll roses (*Hermannia*), currant-rhuses (*Searsia*), and members of the daisy and pea families. The vegetation seldom develops into thickets and is easy to pass through, except in valleys and on steep south slopes that haven't burned for a long time. The ground layer is fairly open and dominated by both C3 (temperate type) and C4 (tropical type) grasses, including Red Oat Grass (*Themeda triandra*) and genera like *Ehrharta*, *Pentaschistis*, and *Pentameris*. Other common understory components include daisies (Asteraceae) and a huge diversity of bulbs that flower in the late winter and spring (Aug–Oct), mostly iris (Iridaceae; e.g., *Gladiolus*, *Moraea*, *Babiana*), lilies (Liliaceae), *Ornithoglossum*, *Amaryllis*, and various terrestrial ground orchids. Renosterveld that burns in late summer or early autumn typically produces a proliferation of flowering bulbs and annuals the following spring.

CONSERVATION: Lowland Renosterveld is one of the most threatened and fragmented habitats on Earth, with over 90% of that in the Overberg converted to agriculture (see SOUTH AFRICAN TEMPERATE CULTIVATION, p. 418) and stock farming, and fewer than 50 parcels exceed 250 acres (100 ha) in size. Virtually none of the habitat is formally conserved by the state, and the vast majority of what remains is in private hands, complicating conservation issues. However,

Classic Renosterveld, dominated by Renosterbos (*Dicerothamnus rhinocerotus*; larger bushes in foreground), but note the significant component of grasses. Matjiesrivier, South Africa. © KEN BEHRENS, TROPICAL BIRDING TOURS

Quartz outcrops within Eastern Rûens Shale Renosterveld in the Overberg form microhabitats that support *Gibbaeum* succulents, or "Ostrich toes" *(inset)*, along with a suite of other highly localized endemic species. Haarwegskloof Renosterveld Reserve, South Africa. © KEN BEHRENS, TROPICAL BIRDING TOURS

conservation easements on private land coordinated and supported by organizations like the Overberg Renosterveld Conservation Trust attempt to keep what remains safe and well managed. Renosterveld is a fire-adapted habitat, and fire is a key management tool in this highly fragmented landscape. It also seems that there were originally more grasses in Renosterveld, and historical overgrazing by livestock may have resulted in it becoming shrubbier. Management practices could reverse this trend. Montane Renosterveld is in slightly better condition, though less diverse. Renosterveld in the Eastern and Northern Cape Provinces are less threatened as their value for agriculture is limited.

WILDLIFE: Owing to its high soil fertility, Renosterveld probably supported significant populations of African megafauna before they were extirpated by European hunters in the 1800s: Black Rhinoceros, African Buffalo, African Savanna Elephant, and myriad antelope and their predators (like African Wild Dog and Cheetah) were probably once common. This was the stronghold of Bluebuck, a stunning antelope closely related to Roan and Sable, and also occasional habitat for Quagga, both tragically hunted to global extinction. It was also the likely stronghold for Bontebok, which almost suffered the same fate, save for a merciful intervention (see sidebar 36, p. 318). Despite the fragmentation, smaller antelopes persist, including Common Duiker, Steenbok, Cape Grysbok, and Gray Rhebok. Cape Porcupines enjoy digging up and eating the profusion of bulbs. Small predators include Caracal, Cape Genet, Cape Gray, Egyptian, and Yellow Mongooses. Despite the loss of termitaria with land transformation, Aardwolf and Aardvark are still occasionally recorded in remnants.

| SIDEBAR 36 | WHERE DID THE SOUTHERN UNGULATES ROAM? |

Before their natural microhabitats could be properly documented, Bluebuck and Quagga were extinct, and Bontebok had dwindled to a few hundred animals. But they were likely most common in lush, grassy, nutrient-rich lowland RENOSTERVELD that has now been almost entirely and irrevocably converted to wheatfields. The few natural patches that do remain have been overgrazed by sheep and other domestic grazers. Bontebok were saved from extinction by the creation of Bontebok National Park near Swellendam, and once numbers increased, some were translocated to other state and private reserves. However, the FYNBOS habitat of most of these locations is nutrient-poor, and vegetation is difficult to eat and digest; in all likelihood, it is suboptimal habitat for these charismatic antelopes. Today they are supported by the creation of artificial grazing lawns, and they do not eat much of the natural Fynbos vegetation. So, although today we see Bontebok in Fynbos-dominated reserves, this was unlikely their preferred habitat.

Bontebok, which may have preferred lowland Renosterveld before its mass conversion to agriculture, came close to extinction before being reintroduced widely, though mostly in nonideal Fynbos habitat. Bontebok National Park, South Africa. © KEITH BARNES, TROPICAL BIRDING TOURS

For birds, the historical avifauna is unknown, but it may have been richer than in today's fragmented and modified Renosterveld. Larger patches support both breeding Black Harrier and Black Bustard; this is one of the strongholds of the Endangered harrier. Thicket shrubs, mostly growing in valleys, produce berries and attract frugivores like Cape Bulbul, Cape White-eye, and Red-winged and Pied Starlings. Cape Bunting, Cape Robin-Chat, Karoo Scrub-Robin, and Bokmakerie bounce across the ground, and generalists like Karoo Prinia and Grey-backed Cisticola occur throughout. Cape Spurfowl is common, and Gray-winged Francolin more secretive but also fairly common. Helmeted Guineafowl is abundant, though it prefers human-modified areas over natural

Bokmakierie is found throughout the Strandveld, Fynbos, and Renosterveld. It is the world's only largely terrestrial bushshrike. © KEN BEHRENS, TROPICAL BIRDING TOURS

Renosterveld. Seedeaters are also abundant with canaries featuring: Yellow, White-throated, Cape, Brimstone, and Streaky-headed. Along river courses the striking Cape Bishop occurs, and Pearl-breasted Swallows seasonally hawk. Larks are quite capable of foraging in the wheatbelt matrix that Renosterveld is embedded in, but in adjacent patches of more open Renosterveld, both Large-billed and the highly local and endemic Agulhas (Overberg) and Cape Clapper (Swartland) Larks breed. There are good numbers of reptiles and frogs that utilize Renosterveld.

Endemism: Renosterveld has a relatively low rate of endemic shrubs, with most species also occurring in adjacent FYNBOS, but there are hundreds of threatened endemic and near-endemic legume and geophyte (especially bulbs) species, and so, unfortunately, this habitat is ground zero for plant extinction on Earth. New species of plants are described from this habitat virtually every year. Renosterveld's endemic mammals have been hunted to extinction, while some of its endemic birds have fortunately adapted to the surrounding anthropogenic wheatbelt matrix (see SOUTH AFRICAN TEMPERATE CULTIVATION). It supports many species with ranges restricted to s. Africa and shared with Fynbos, STRANDVELD, SUCCULENT KAROO, and NAMA KAROO. The spectacular and Critically Endangered Geometric Tortoise (*Psammobates geometricus*) has a tiny global range and favors Renosterveld fragments, where it eats grass and geophytes. The bizarre Cape Caco (*Cacosternum capense*) frog is also endemic to the greater Renosterveld matrix but has adapted to using undulating pans in loamy or clay areas including in cultivated land.

DISTRIBUTION: Lowland Renosterveld is restricted primarily to two lowland plains areas close to Cape Town, sw. South Africa: (1) the northern Overberg Plain just south of the Langeberg mountains, extending from Caledon to Mossel Bay; and (2) the Swartland, west of the Cederberg and Koue Bokkeveld mountains, from Citrusdal south to Paarl. Although primarily found below 2800 ft.

Above: **Black Harrier over Renosterveld, with cultivated wheat behind. Renosterbos (*Dicerothamnus rhinocerotus*) detail** *(inset).* © KEITH BARNES, TROPICAL BIRDING TOURS

Right: **Renosterveld in the foreground (dominated by *Oedera squarrosa*), with the cultivated fields that have largely replaced it in the background. Private land not far from Haarwegskloof Reserve, South Africa.** © KEITH BARNES, TROPICAL BIRDING TOURS

(850 m), isolated patches occasionally occur in higher mountains up to 6200 ft. (1900 m). Montane Renosterveld occurs in highland areas of Namaqualand, especially the Kamiesberg highlands between Calvinia and Sutherland; on the fringes of the Little and Great Karoo basins; and across to the Eastern Cape (Baviaanskloof).

WHERE TO SEE: Haarwegskloof Renosterveld Reserve, South Africa.

Af8D MAGHREB MAQUIS

IN A NUTSHELL: Dry thicket heathland around the Mediterranean with trees that are probably not in their climax state. **Global Habitat Affinities:** AUSTRALIAN TEMPERATE HEATH THICKET; MAQUIS; MIDDLE EASTERN MAQUIS; PACIFIC CHAPARRAL. **Continental Habitat Affinities:** Has a superficial resemblance to tall FYNBOS but a very different ecology. **Species Overlap:** MAGHREB GARRIGUE; MAGHREB JUNIPER OPEN WOODLAND; MAGHREB BROADLEAF WOODLAND.

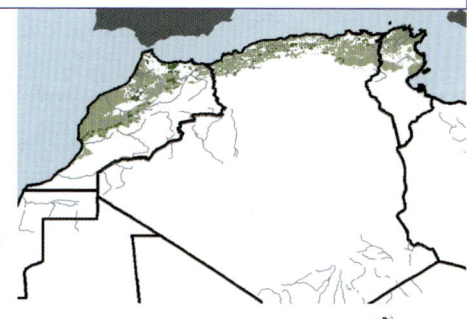

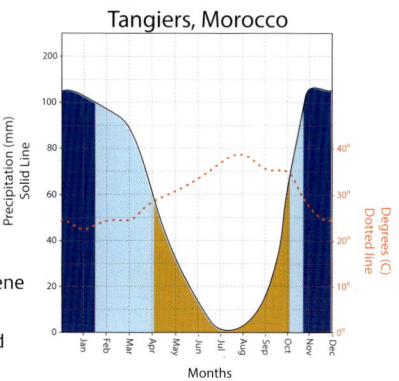

DESCRIPTION: The original early Holocene (12,000–4000 YBP) habitat of n. Africa is still open to dispute. Some suggest that it was predominantly a mixed evergreen woodland of cypress, juniper, and oak. Others think it was a fire-adapted wooded savanna, while others believe that the Maghreb Maquis that exists now is a natural response to the Mediterranean climate (Köppen **Csa**). Although Maghreb Maquis is often regarded as a human-induced, disturbed (plagioclimax) habitat, it is likely that in the early Holocene a Maghreb Maquis existed as an alternate stable state (climax as opposed to a human-induced climax, so called plagioclimax) in gaps within mixed woodland, oak woodland, conifer woodland, or a savanna-type habitat. An increase in fire frequency, overgrazing by domestic sheep and goats, clearing for agriculture, and the development of olive groves has all but eliminated the oaks and conifers. The plants native to the original (though very limited in extent), completely natural Maghreb Maquis have a massive advantage in this human-altered environment and in modern times have become the dominant vegetation around the Mediterranean basin and through the western parts of the Middle East.

Maghreb Maquis is a dense shrub or small tree community with a canopy between 3 and 20 ft. (1–6 m) high, most often around 10 ft. (3 m). It consists mostly of evergreen canopy plants like junipers, Aleppo Pine (*Pinus halepensis*), Carob (*Ceratonia siliqua*), Strawberry Tree (*Arbutus unedo*), and Wild Olive (*Olea oleaster*). Oleander (*Nerium oleander*) and Rock Rose (*Cistus albidus*) form a shrub layer, while in more open areas, many annuals grow until they are shaded out by sclerophyllous plants. Walking through Maghreb Maquis is not an easy task, as the canopy at eye level is nearly solid. The canopy is usually not tall enough to be able to walk under comfortably, and almost all canopy plants have shrubby growth forms, with small branches growing up the trunk and multiple trunks growing from one base. Because this environment gets most of its rain during the winter

Little Owl in Oued Massa Nature Reserve, Morocco. Flowers are abundant in springtime Maghreb Maquis, and the many sclerophyllous plants give off a wonderful, fragrant, herb garden-esque smell.
© KEN BEHRENS, TROPICAL BIRDING TOURS

months, and the summer drought months are brutal for plants, many plants are sclerophyllous, possessing leathery leaves with a waxy coating that helps limit transpiration. Plants also tend to have narrower leaves than those in broadleaf deciduous forest. Some non-sclerophyllous shrubs are deciduous, losing their leaves in summer to avoid desiccation.

A taller version of anthropogenic Maghreb Maquis occurs on steep slopes in the Atlas Mountains, where clearing of cedar and fir forests has allowed Aleppo Pines and Phoenician Juniper (*Juniperus phoenicea*) to become established as short canopy trees, with an understory of Mastic (*Pistacia lentiscus*), Holly Oak (*Quercus ilex*), and Wild Olive. Given time, this habitat in the Atlas Mountains would probably turn into MAGHREB JUNIPER OPEN WOODLAND rather than MAGHREB FIR AND CEDAR FOREST owing to a lack of moisture, but this potential woodland rarely passes the shrub regrowth stage because of grazing, lopping, and fire.

CONSERVATION: Because a lot of Maghreb Maquis is anthropogenic or fire-dependent, a product of the modification of MAGHREB PINE FOREST and MAGHREB BROADLEAF WOODLAND, it can revert to these forests in reserves where fires are prevented or grazing reduced. Outside these reserves, change to MAGHREB GARRIGUE habitat is likely through continued fires and overgrazing. Much of the coastal Maghreb Maquis has been turned over to cultivation, and further pressure is added by urban development and tourism infrastructure. In the more arid areas, Maghreb Maquis is being degraded into MAGHREB HOT SHRUB DESERT and even SAHARAN REG DESERT. Higher temperatures are likely with climate change, but the real unknown is the possible change of precipitation cycles. These overgrown heathlands are often on poorly developed soils with little capacity to hold moisture, so a decrease in rainfall over the already very dry summers will result in some shrubs not being able to survive. Given that this habitat is vitally important for migrating bird species, after crossing the Sahara

Maghreb Maquis is a crucial part of the habitat mix used by the Endangered Northern Bald Ibis. © KEN BEHRENS, TROPICAL BIRDING TOURS

on their northerly journey or as often the last real vegetation they encounter on the southern journey, protection of Maghreb Maquis is vital for the survival of many European migrant birds.

WILDLIFE: Animal assemblages in the Maghreb Maquis have been drastically affected by 4000 years of intense human pressure, resulting in the extinction of many mammals (see the **MAGHREB FIR AND CEDAR FOREST**, p. 54). Barbary Macaque is now very rare in this habitat. Both Algerian Hedgehog and Long-eared Hedgehog occur alongside Crested Porcupine, Wild Boar, and African Wildcat.

Maghreb Maquis holds many birds and is key for the small Palearctic *Curruca* warblers that reach their greatest diversity around the Mediterranean. This habitat supports several species, including Dartford, Western Subalpine, Spectacled, Tristram's, and Sardinian Warblers. Melodious and Western Olivaceous Warblers

Tristram's Warbler, a specialist bird of montane Maghreb Maquis.
© DANIELE ARDIZZONE, FLUYENDO PHOTOGRAPHY

also breed in Maghreb Maquis. In winter it supports Eurasian Blackcap, Little Owl, Common Nightingale, and Song Thrush. In spring and summer, Maghreb Maquis hosts breeding migrants such as Red-necked Nightjar, Rufous-tailed Scrub-Robin, Woodchat Shrike, Western Orphean Warbler, and Marmora's Warbler. This habitat also becomes vitally important as the first feeding place for spring migrant passerines that have crossed the Sahara on their way to the mixed forests of northern Europe. Following these birds are numerous accipiters and falcons, along with other raptors such as Short-toed, Booted, and Bonelli's Eagles.

Maghreb Maquis has many more reptiles than other habitats of the Palearctic, with chameleons as well as other lizards and snakes. Streams and seeps provide habitat for Berber Toad (*Sclerophrys mauritanica*) and Stripeless Tree Frog (*Hyla meridionalis*).

Rufous-tailed Scrub-Robin migrates north from sub-Saharan Africa and breeds in Maghreb Maquis during the boreal summer. © DANIELE ARDIZZONE, FLUYENDO PHOTOGRAPHY

DISTRIBUTION: Maghreb Maquis is widespread and common below 3500 ft. (1100 m) around the s. Mediterranean coast from Morocco to Tunisia, with small, isolated patches across to the Red Sea that are almost unmappable at a regional scale. Maghreb Maquis–type habitat is found in Macaronesia and on the Azores, Madeira, and the Canaries.

WHERE TO SEE: Sierra Bullones, Morrocco; Casablanca, Morocco.

This patch of Maghreb Maquis near Nador, Morocco, lacks larger bushes and trees and is best regarded as an ecotone between Maghreb Maquis and Maghreb Garrigue. © CARLOS N. G. BOCOS

Af8D MAGHREB MAQUIS

A rugged swath of Maghreb Maquis near Alhucemas, Morocco. Note the typical mix of small trees, shrubs, and rocky open ground. © CARLOS N. G. BOCOS

Af8E MAGHREB GARRIGUE

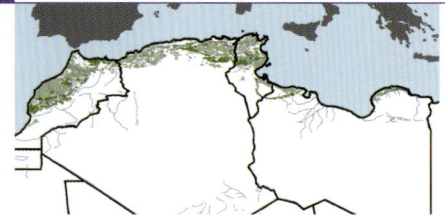

IN A NUTSHELL: An open scrub habitat found around much of the Mediterranean Sea; akin to a depauperate MAGHREB MAQUIS. **Global Habitat Affinities:** GARRIGUE; ASIAN GARRIGUE. **Continental Habitat Affinities:** Has a superficial resemblance to FYNBOS but a very different ecology. **Species Overlap:** MAGHREB MAQUIS.

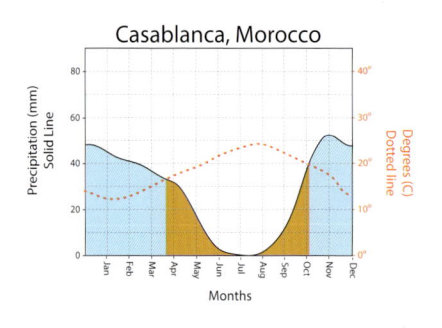

DESCRIPTION: Maghreb Garrigue is regarded as both a natural habitat and more commonly an anthropogenic habitat because if protected from fire and overgrazing, it would likely return to MAGHREB MAQUIS or even to dry conifer or oak forests. This same habitat (or something extremely similar) also occurs in Europe, where it is also referred to as Garrigue. It is a very depauperate, short, and scrubby counterpart of Maquis, and it is usually found in parts of this region that are hotter and have a drier microclimate than Maghreb Maquis (though still Köppen **Csa**) or on nutrient-deficient soils developed over limestone, marble, quartzite, or granite. Where soils are deeper and more nutrient rich from bedrocks such as sedimentary mudstone, igneous basalt, and more rarely gabbros, Maghreb Maquis is more resistant to degradation, but these richer soils are usually turned over to the cultivation of olive groves.

This shrubland is less than 6 ft. (2 m) high, with a mix of trees that are also characteristic of Maghreb Maquis, though these are usually overgrazed and much shorter and shrubbier than in Maghreb Maquis. Common species include Wild Olive (*Olea europea*), Strawberry Tree (*Arbutus unedo*), Libyan Strawberry Tree (*A. pavarii*), Lentisk (*Pistacia lentiscus*), and Spiny Burnet (*Poterium spinosum*). In the eastern part of North Africa, where Spiny Burnet becomes more dominant, the Maghreb Garrigue can start to resemble a very short thornscrub and is called

Maghreb Garrigue is similar to Maghreb Maquis but even more scrubby and open, with few or no tall trees. Tamari National Park, Morocco. © NICK ATHANAS, TROPICAL BIRDING TOURS

Af8E MAGHREB GARRIGUE

Together, Maghreb Maquis and Maghreb Garrigue are paradise for *Curruca* warblers, such as Western Subalpine Warbler. © KEITH BARNES, TROPICAL BIRDING TOURS

"Phrygana" and "Batha." Many of the common household herbs originated from this environment in Europe, and many have very closely related species in Africa, such as Dandelion-leaved Sage (*Salvia taraxacifolia*), Wild Thyme (*Thymus serpyllum*), and Egyptian Lavender (*Lavandula multifida*). Given that this habitat is the end result of human overuse, it is no wonder that it is found in disturbed areas all around the Mediterranean.

CONSERVATION: As explained above, Maghreb Garigue is normally a result of MAGHREB MAQUIS degradation so has similar problems of fires and overgrazing. These problems combined with higher temperatures through climate change, and the possible change of precipitation cycles, will place serious pressure on many of the shrub species, almost certainly leading to local extinctions.

WILDLIFE: For a discussion of the now-extinct megafauna of North Africa see the MAGHREB FIR AND CEDAR FOREST (p. 54). Because this is primarily an anthropogenic habitat, mammals are limited, though the Western Gerbil inhabits small patches of natural Maghreb Garrigue when scattered Argan trees (*Argania spinosa*) are present. Both Hoogstraal's Gerbil and Occidental Gerbil, along with Barbary Striped Grass Mouse, occur in dunal Maghreb Garrigue of the Sabka Plains.

Birdlife in Maghreb Garrigue is similar to that of the MAGHREB MAQUIS from which it derives but is much reduced, both in species and numbers. The more open nature of Maghreb Garrigue has allowed species from drier and sparser habitats to become more widespread. Birds that take advantage of Maghreb Garrigue's open nature include Small Buttonquail, Dupont's Lark, a variety of other larks and pipits, Black-eared and Black Wheatears, Spectacled Warbler, and Masked Shrike. Short-toed, Booted and Bonelli's Eagles and Black Kite are all regularly sighted.

DISTRIBUTION: This habitat is very widespread on Mediterranean islands and all around the Mediterranean coast. It extends east through much of Turkey into Syria, where it merges into West Asian Semi-desert Thornscrub.

WHERE TO SEE: Cap Ghir, Morocco.

Maghreb Garrigue's open and often rocky nature makes it attractive to open-country birds like Black Wheatear. © DANIELE ARDIZZONE, FLUYENDO PHOTOGRAPHY

Af8F MAGHREB BROADLEAF WOODLAND

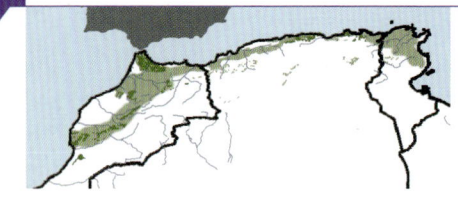

IN A NUTSHELL: Mixed deciduous and evergreen broadleaf forests of oaks and wild olive found in North Africa. **Global Habitat Affinities:** MEDITERRANEAN OAK FOREST. **Continental Habitat Affinities:** TAPIA. **Species Overlap:** MAGHREB GARRIGUE; MAGHREB MAQUIS; MAGHREB FIR AND CEDAR FOREST.

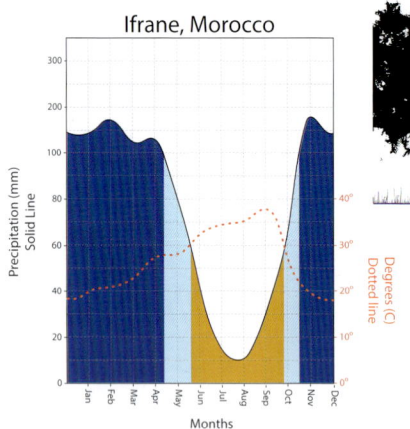

DESCRIPTION: The Maghreb region has evergreen oak forest and woodland that are dominated by three sclerophyllous (thick, waxy-leaved) species: Holly, Cork, and Kermes Oaks (*Quercus ilex*, *Q. suber*, and *Q. coccifera*; subhabitat Af8F-1). Though all mix with other species such as Wild Olive (*Olea oleaster*), they tend not to grow together. The remnants of

Higher-elevation oak woodland in the Anti-Atlas Mountains, Morocco. The structure of this habitat is oddly reminiscent of the Pinyon-Juniper Woodland of North America. © CARLOS N. G. BOCOS

once-widespread Argan (*Argania spinosa*) woodlands (subhabitat Af8F-2) are now virtually extinct and only exist as tree plantations. Holly Oak and Kermes Oaks grow in areas between 1300 and 7900 ft. (400–2400 m) with a wide range of rainfall and are common trees in n. Africa. The formerly massive Cork Oak forests of the Maghreb, which were the largest in the world until 1900, have been decimated, and now only small fragments exist.

In the early Holocene, oak and conifer forests would have extended down to the Mediterranean coast in areas that are now MAGHREB GARRIGUE. The upper-elevation Holly Oak forests blend in with the MAGHREB FIR AND CEDAR FOREST, and there can be extensive ecotones between these habitats. Even when the canopies of the two forest types are distinct, the understories change at a different rate. Holly Oak forests can be seen as an ecotone between drier Cork Oak forests and moister fir and cedar forests, in a similar way that the Cork Oak forests at lower elevations are an ecotone between drier Aleppo Pine (*Pinus halepensis*) woodlands and moister Holly Oak forests.

Holly Oak forests can be very tall, with a canopy of up to 60 ft. (20 m). In contrast, Cork Oak forests are usually uneven, and trees have crooked branches starting at 10 ft. (3 m) and rounded crowns between 20 and 30 ft. (6–9 m). Occasionally, Cork Oak forest canopies even out and become fairly uniform, with trees up to 50 ft. (15 m). In the vast majority of Cork Oak forests, the bark is stripped for commercial cork production, and it is rare to find a mature Cork Oak without scaring, making it easy to identify this tree species. The very thick, vertically cracked bark can be 2 in. (5 cm) thick and is removed from much of the trunk and lower branches of the trees. The exposed trunk quickly turns from red to black. The harvested (and therefore protected) trees can be hundreds of years old.

Because Cork Oak forests have a more open canopy than Holly Oak and Kermes Oak forests, they tend to have a more pronounced shrub layer when not heavily grazed; although this shrub layer is shorter than that of the Holly Oak forests, it tends to be thicker with less deciduous species and therefore more difficult to walk through. With that said, the shrub layer in both types of forest tends

Argan woodland in the Sous Region, Morocco. Though natural Argan has been virtually eliminated, Argan plantations remain, and are still rich in birdlife. © CARLOS N. G. BOCOS

to have the same species, and in the wetter winter months the understories can look identical. Things change in the hotter summer months, since the more open Cork Oak forests dry out more, putting more water stress on the shrub layer. Where large oak forests still exist with a closed canopy, the subcanopy is open, and there is a significant low shrub layer. Walking through these forests is not difficult. Common understory species of Cork Oak forest include European Fan Palm (*Chamaerops humilis*), Irish Strawberry Tree (*Arbutus unedo*), Mediterranean Broom (*Genista linifolia*), Wooly Rock-Rose (*Halimium lasianthum*), Spanish Lavender (*Lavandula stoechas*), and Moroccan Chamomile (*Ormensis multicaulis*). Holly Oak forests also contain Scorpion Vetch (*Coronilla valentina*), Spurge-laurel (*Daphne laureola*), which is neither a spurge nor a laurel, Pineapple Broom (*Cytisus battandieri*), and more widespread European plants such as Common Hawthorn (*Crataegus monogyna*).

The soils of this habitat remain moist and do not have the intense red staining from iron oxyhydroxides of the soils of the surrounding MAGHREB GARRIGUE. This suggests that the original forests maintained a more humid microclimate in the subcanopy and undergrowth than surrounding habitats despite intense summer droughts.

The climate of this type of broadleaf forest is not surprisingly called "Mediterranean" (Köppen **Csa**) and is similar to s. Europe and the coast of s. California. Winters are mild (rarely getting frost) and fairly humid, but the summers are brutal, with scorching temperatures and no rain. These are tough growing conditions, and for that reason most of the plants are sclerophyllous to handle the drought conditions for half of the year.

CONSERVATION: Throughout North Africa, most oak forests have been converted to cultivation because they tend to grow on better soils than do pine or juniper woodlands and are ideal for olive production. Of the once-massive oak forests only a few extensive forests remain in protected areas. Smaller patches that are crucial to migrating bird species are widespread, and Cork Oak plantations are great bird habitat. As long as olive and other crop farming practices remain artisanal, with small plots owned by local farmers, the birds are fairly secure. However, if industrial monoculture takes hold as it has in Europe, these remaining forests will be decimated as will the migrating bird populations that use them.

WILDLIFE: For a discussion of the now-extinct megafauna of North Africa see the MAGHREB FIR AND CEDAR FOREST (p. 54). Cuvier's Gazelle—endemic to the Maghreb—persists in these open forests, especially in the Tunisian Atlas, although it will use shrubby edges of the deserts too. This species underwent a dramatic decline and was thought extinct until small numbers were rediscovered. Barbary Macaques persist in this woodland, alongside African Wolf and several species of gerbil and other small mammals.

There are several resident birds that are common throughout broadleaf forest and woodland, and most of these are shared with Europe. Examples include Great Spotted Woodpecker, Short-toed Treecreeper, Eurasian Jay, European Robin, Mistle Thrush, Common

Maghreb Broadleaf Woodland supports a suite of widespread Palearctic species such as Eurasian Nuthatch. © KEITH BARNES, TROPICAL BIRDING TOURS

Great Spotted *(left)* **and Lesser Spotted** *(right)* **Woodpeckers are two of the Palearctic's most widespread woodpeckers, using a broad range of wooded and forested habitats.** © KEITH BARNES, TROPICAL BIRDING TOURS

Nightingale, Black Redstart, Coal and Great Tits, and Eurasian Nuthatch. Afrotropical flavor is added by the widespread Common Bulbul. Many species are summer-breeding migrants, such as Booted Eagle, Common Nightingale, and Spotted Flycatcher. Rocky areas hold breeding European Stonechat and Blue and Rufous-tailed Rock-Thrushes. Some localized specialties that use these woodlands include Maghreb Owl, Red-necked Nightjar, Atlas Flycatcher, and Maghreb Magpie. In drier and more open areas, there is Levaillant's Woodpecker, and in scrubby Holm Oak with scattered trees the local Moussier's Redstart and Sardinian Warbler occur. In scrub along rivers there is Rufous-tailed Scrub-Robin and Black-eared Wheatear. The rare endemic Algerian Nuthatch prefers forest dominated by Atlas Oak (*Quercus tlemcenensis*) but also moves into cedar and fir elements. The endemic *buvryi* subspecies of Hawfinch uses Cork Oak and alders at moderate elevations. Warblers do well in broadleaf woodland, with species including Melodious, Western Olivaceous, and Western Bonelli's Warblers, alongside Common Firecrest and Woodchat Shrike. In North Africa, Dark Chanting-Goshawk, Western Orphean Warbler, and Black-crowned Tchagra are especially fond of Argan plantations, where they are found alongside Eurasian Thick-knee, Red-rumped Wheatear, and a variety of larks. In winter Argan supports migrant Song Thrush and Redwing that visit from their European breeding grounds.

North African Fire Salamander (*Salamandra algira*) and Sharp-ribbed Newt (*Pleurodeles waltl*) occur under logs and in moist microclimates alongside Coastal Skink (*Chalcides mionecton*). Streams and seeps provide habitat for Berber Toad (*Sclerophrys mauritanica*) and Stripeless Tree Frog (*Hyla meridionalis*).

DISTRIBUTION: Oak forests remain in limited patches in the Atlas Mountains of Morocco and the mountains of Algeria. Formerly widespread coastal and lower-montane forests from sw. Morocco to Libya have largely been decimated. Argan forests are now mainly anthropogenic but occur locally in Morocco.

WHERE TO SEE: Tazekka National Park, Morocco; Talassemtane National Park, Morocco; Arganeraie Biosphere Reserve, Morocco; Chréa National Park, Algeria.

Af8G ALBANY THICKET

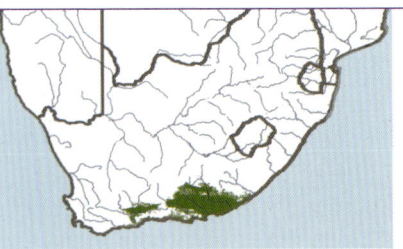

IN A NUTSHELL: Dense semi-succulent thorny scrub of subtropical se. Africa. **Global Habitat Affinities:** CARRASCO; SERTAO CAATINGA; DRY CHACO. **Continental Habitat Affinities:** SPINY FOREST. **Species Overlap:** MOIST MIXED SAVANNA; SOUTH COAST FOREST MATRIX; NAMA KAROO; KALAHARI DRY THORN SAVANNA; FYNBOS; MONTANE GRASSLAND.

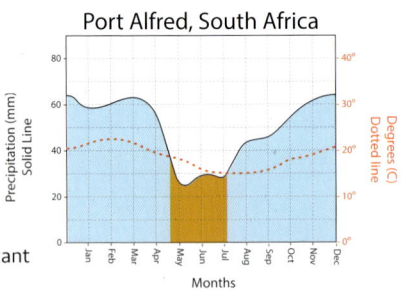

DESCRIPTION: Albany Thicket is a dense, woody, impenetrable, semi-succulent spiny habitat of south-central South Africa that is 6–13 ft. (2–4 m) tall with some emergents up to 23 ft. (7 m). It occurs patchily, often adjacent various other habitats. Overall, it is something of a transitional habitat between NAMA KAROO and SOUTH COAST FOREST MATRIX. Although sometimes considered a Mediterranean habitat, it most resembles dry deciduous woodlands that have a significant succulent component, like the South American Sertao Caatinga and Dry Chaco. With the skyline dominated by fat-leaved Porkbush (*Portulacaria afra*) and emergent candelabra-like tree euphorbias, much of the Albany Thicket is visually very distinctive. It occurs almost exclusively on shallow to deep loamy clay soils derived from siltstone and shales but also locally on sandstone and limestone. This is a dry and harsh environment (Köppen **Cfb**), with huge fluctuations in temperature, especially in the arid interior: summer maxima can exceed 104°F (40°C), while nights below freezing are common in the winter. The temperature averages 60–77°F (15–25°C). Although it is considered to be semi-xeric, rainfall is moderate throughout, ranging from 4 to 35 in. (105–900 mm) but generally 16–24 in. (400–600 mm) annually. Rain occurs year-round, with a slight increase in summer (Nov–Mar) in the northeast and winter (Jun–Aug) in the southwest, but the quantity of rainfall is highly variable, with some years being much drier than others. Unpredictable seasonal droughts are regular. The coast is climatically more moderate than the interior, and frost is common in winter at higher elevations. The predominance of stem and leaf succulents, lacking dry woody kindling, means fire is rare in Albany Thicket.

Albany Thicket has around 1600 plant species. The dominant plants are mostly succulent or sclerophyllous (possessing small, hard leaves), giving them considerable drought resistance, and the seasonal appearance of this habitat is remarkably uniform irrespective of seasonal drought stresses. This habitat may represent a relict form of vegetation that was more widespread in Africa during the Eocene (56–34 MYA). Larger trees and shrubs are deep rooted or have water storage organs. Some succulents are able to use unique biochemical pathways to enhance their drought resistance (see sidebar 10, p. 111). One of the most characteristic plants is Porkbush, which has

Hoodia in the foreground, aloes in the background, and Porkbush (*Portulacaria afra*) in the inset. Typical succulent components of Albany Thicket. © KEITH BARNES, TROPICAL BIRDING TOURS

dense small succulent leaves and soft fatty stems. It grows as a small shrub or tree, usually 8–14 ft. (2.5–4.5 m) tall. Porkbush is a member of the Didiereaceae and a relative of the famous octopus trees of Madagascar's SPINY FOREST. Another striking element of much Albany Thicket is tree-form spurges (e.g., *Euphorbia triangularis* and *E. grandidens*) that grow up to 23 ft. (7 m) in height and which have cactus-like form, with gnarled grey bark and fleshy trunk-like arms reaching for the sky. These are filled with a creamy latex comprising alkaloids and terpenoid compounds that is distasteful and toxic, and although spurges may get damaged by large herbivores pushing them around, they are free from herbivore attention as food plants. On deeper soils and in more protected gullies, trees grow, such as Jacket Plum (*Pappea capensis*) and Karoo Boer-bean (*Schotia afra*), which has bright red flowers that attract sunbirds and butterflies.

The midstory has a wide diversity of easily recognized aloes (Asphodelaceae), with thick, succulent spiky leaves arranged in a whorl. Aloes often grow in clumps, and seasonally, long spikes of spectacular flowers transform hillsides into yellow or orange wonderlands. Aloe leaves are filled with a colorless gel, and one aloe species native to Arabia (*Aloe vera*) is now cultivated as a major commercial medicinal product. Shrubs, including several spiny ones, grow in the midstory. Examples include num-nums (*Carissa*), Kooboo Berry (*Mystroxylon aethiopicum*), Common Gwarrie (*Euclea undulata*), fragrant Camphor Bush (*Tarchonanthus camphoratus*), donkeyberry (*Grewia*), a variety of rhus (*Searsia*) bushes, and members of the daisy family. Shrubs tend to grow into one another leaving little to no intervening space. The spectacular bird-of-paradise flowers in the

Left: **Note the characteristic emergent *Euphorbia* "candelabra trees" in the dense Albany Thicket, as well as the sharp ecotone to grassland patches where most of the ungulates graze. Kariega Valley, Eastern Cape, South Africa.** © KEITH BARNES, TROPICAL BIRDING TOURS

Below: **Albany Thicket near Grahamstown, South Africa. The emergent tree euphorbias are visually striking.** © KEITH BARNES, TROPICAL BIRDING TOURS

genus *Strelitzia*, with giant broad strap-like leaves betraying their affinity to the banana family, are also characteristic elements. Albany Thicket supports four of the world's five members of the bird-of-paradise family (Strelitziaceae). This is a Gondwanaland relict family whose final member is the amazing Traveler's Palm (*Ravenala madagascariensis*), another biogeographic connection between Albany Thicket and Madagascar. The shift between the upper canopy and lower shrub layer is sometimes obscured by climbers and lianas, which include succulent, herbaceous, and woody climbers, many in the genus *Asparagus*.

The ground cover is also thick, often with a variety of stonecrops, especially pig's-ears in the genera *Cotyledon* and *Crassula*. Other ground covers include fleshy zebra cactus (*Haworthia*) and sour-fig (*Carpobrotus*). Although not common, an interesting group in Albany Thicket is carrion flowers (*Stapelia*), a genus of low-growing spineless stem succulents. Their weird red-and-flesh-colored hairy flowers smell like rotten meat, attracting pollinators like blowflies. The grass component of Albany Thicket is poorly developed, but some bush clumps transition to AFROTROPICAL GRASSLAND via a very sharp boundary.

Albany Thicket is one of the preferred habitats of the Southern Tchagra, endemic to South Africa and Eswatini. The num-nums (*Carissa*) *(inset)* are distinctive dense shrubs that are a key element of the midstory of this habitat. © KEITH BARNES, TROPICAL BIRDING TOURS

SIDEBAR 37 — WHAT'S IN A THICKET?

Webster's dictionary defines thicket as "a dense growth of shrubbery or small trees." Essentially, the term refers to a structural quality of dense vegetation that makes it almost impossible to move through. Many different and unrelated types of vegetation may form "thicket-type" physiognomy, with examples including vine tangles in lowland East Africa, montane Afrotropical scrub, or palm thickets in lowland rainforest. But sometimes, thickets also represent a cohesive and distinctive habitat with a unique floral composition and biogeography, such as the ALBANY THICKET in se. South Africa. Another interesting and similar system is the localized Itigi Thicket (sidebar 13, p. 153) of ne. Zambia and c. Tanzania, that is included within Monsoon Forest (p. 151). This thick, semiarid shrubland is 10–15 ft. (3–5 m) tall and is home to several endemic plants and reptiles, even though the bird and mammal assemblages are not particularly distinctive.

Many areas of the East Coast Forest Matrix have a thicket-like character. Save, Mozambique.
© KEN BEHRENS, TROPICAL BIRDING TOURS

CONSERVATION: The removal of larger habitat-modifying herbivores in the 1800s may have resulted in significant bush encroachment in Albany Thicket. This thickening trend was countered by habitat destruction and degradation by humans, as much of the area was converted to agriculture and pastoralism, creating more open grassland areas. Natural herbivores were largely replaced by domestic ones, like goats and sheep, which browse differently, in a more damaging fashion, and do not disperse seeds as effectively. Despite Albany Thicket having significant biomass, it is not very productive and recovers slowly after grazing. There is evidence that grazing by domestic stock has led to ecosystem-level degradation of over 92% of the habitat. Urbanization (especially in the Nelson Mandela Bay Municipality) and alien plants are two additional threats. Although it remains poorly protected, several conservation areas hold important elements of this habitat, most notably Mountain Zebra and Addo Elephant National Parks. The latter has been expanded to be the third largest conservation area in South Africa.

WILDLIFE: Although botanically Albany Thicket has more in common with NAMA KAROO and SUCCULENT KAROO, the wildlife is really a combination of MOIST MIXED SAVANNA and SOUTH COAST FOREST MATRIX habitats. The botanical uniqueness of Albany Thicket would suggest a higher concentration of endemic vertebrates, but it is hypothesized that significant contractions during Pleistocene glacial cycles may have eliminated many of them. High nutrient value and proximity to permanent water suggest that historically Albany Thicket supported a high density and diversity of herbivores including abundant African Savanna Elephants, African Buffalo, and rhinos, which may have been key habitat modifiers. Although these were largely removed by hunting in the 1800s, they are being returned to significant portions of Albany Thicket that exist within protected areas. Lions, Black and White Rhinoceroses, and Spotted Hyena were reintroduced to Addo Elephant National Park. Many plants are endowed with thorns, spines, and toxic latex, suggesting that these defenses evolved because of considerable natural browsing pressure by large herbivores. Larger antelope include Red Hartebeest and Eland. Mountainous areas have Smith's Red Rock Rabbit, "Cape" Mountain Zebra, and Mountain Reedbuck. Antelope such as Common Duiker and Greater Kudu have persisted, as have Warthog, although they are much reduced. More open, forest-like areas support Bushpig and Bushbuck. Vervet Monkeys and Chacma Baboon are abundant.

The birdlife is an unusual mix of open savanna, dense woodland, and even some primarily forest-dwelling birds. Typical savanna species include Yellow-breasted Apalis, Tawny-flanked Prinia, White-browed Scrub-Robin, Emerald-spotted Wood-Dove, Black-collared Barbet, Gray-headed Bushshrike, Black-backed Puffback, Dark-capped Bulbul, Chinspot Batis, Southern Black-Tit, Black Cuckooshrike, Black-crowned Tchagra, and Cape Starling. But there are also typical species of denser vegetation, some of which are shared with thicker areas of STRANDVELD. These include

Red-fronted Tinkerbird is one of many species that are equally at home in Albany Thicket and adjacent habitats. © KEITH BARNES, TROPICAL BIRDING TOURS

Angulate Tortoise (*Chersina angulata*) uses Albany Thicket, Fynbos, and other shrub-dominated habitats in the region. The male has an extended bowsprit under the chin, which they use to fight other males for territory. © KEITH BARNES, TROPICAL BIRDING TOURS

South African endemics such as Knysna Woodpecker and Southern Tchagra but also the more widespread Green-backed Camaroptera, Red-fronted Tinkerbird, Bar-throated Apalis, Southern Boubou, and Sombre Greenbul. Perhaps the most interesting bird is the "Cape" Barred Owlet, a taxon hundreds of kilometers away from the next population and, pending an ongoing taxonomic investigation, perhaps the only bird endemic to the Albany Thicket! Along rivers and in protected gullies, typical dense forest species can be found, including Crowned Eagle, Knysna Turaco, Scaly-throated Honeyguide, Tambourine Dove, Forest Weaver, Olive Bushshrike, African Crested Flycatcher, Terrestrial Brownbul, Olive Woodpecker, and Crowned Hornbill. At transitional grassy edges there are local endemics like Swee Waxbill and Forest Buzzard, as well as Red-necked Spurfowl. And whenever there are flowers, especially aloes, these are attended by nectarivores like double-collared, Amethyst, and Malachite sunbirds and Black-headed Oriole. Interestingly, bird abundance and diversity are higher in Albany Thicket that has had African Savanna Elephants present for longer periods; however, this seems to be the case only where elephant densities are low and cover of Albany Thicket is high. Elephants reduce the canopy and tend to create more open pathways through this otherwise impenetrable habitat. Unsurprisingly, frugivores and bark-gleaners are more common in dense thickets, and insectivores, nectarivores, and hawking species more common in more open thickets. The nature of Albany Thicket means that a greater proportion of birds are small bodied, able to negotiate dense tangles, but also, a high proportion of species are frugivores, as many Albany Thicket plants produce berries. In adjacent grassy patches, open country specialists like Red-collared and Fan-tailed Widowbirds and Pied Starling can be seen.

Among reptiles, there are many species that are regionally endemic that are shared with surrounding habitats, such as Parrot-beaked Padloper (*Homopus areolatus*). There are only a few endemic reptiles, but some are quite spectacular, like Southern Dwarf Chameleon (*Bradypodion ventrale*) and Albany Adder (*Bitis albanica*), which was only discovered in 1997.

Endemism: Some 200 plant species are endemic and another 165 near-endemic, including a high percentage in the following families: milkweeds (Asclepiadaceae), stonecrops (Crassulaceae), spurges (Euphorbiaceae), and several families of Asparagales. Flagship species include some rather spectacular endemic cycads, including Karoo and Eastern Cape Blue Cycads (*Encephalartos lehmannii* and *E. horridus*). Faunal endemism is low.

DISTRIBUTION: Mostly distributed in South Africa's Eastern Cape Province, it also extends into the eastern reaches of the Western Cape Province. Although some Albany Thicket is coastal, much of it occurs in the interior away from the coast, in harsher inland terrain. It is adjacent to, and marginally intergrades with, almost every other habitat type in South Africa in a variety of transitions.

WHERE TO SEE: Addo Elephant National Park, South Africa; Kariega Game Reserve, South Africa.

Af8H TAPIA

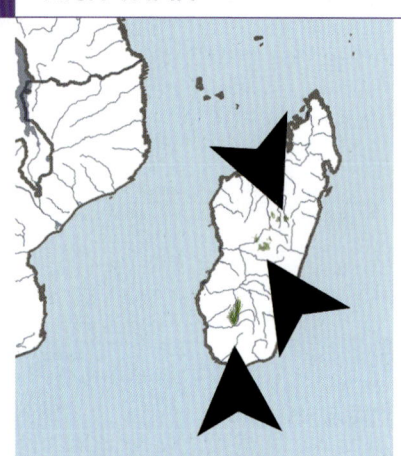

IN A NUTSHELL: Park-like woodland of sclerophyllous Tapia trees, growing on the High Plateau of Madagascar. **Global Habitat Affinities:** OAK DEHESA; CALIFORNIA OAK SAVANNA; GRASSY MULGA; AUSTRALIAN TEMPERATE HEATH THICKET. **Continental Habitat Affinities:** MAGHREB BROADLEAF WOODLAND. **Species Overlap:** MALAGASY GRASSLAND AND SAVANNA; MALAGASY DECIDUOUS FOREST; SPINY FOREST.

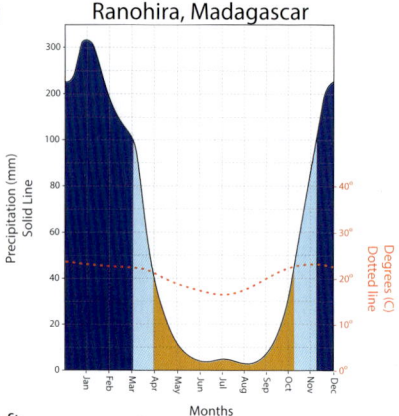

DESCRIPTION: Central Madagascar holds a fascinating and unique habitat called "Tapia," the Malagasy name for the *Uapaca bojeri* tree. This habitat defies easy classification alongside the rest of the continent's habitats (see sidebar 38, p. 341), and its degree of anthropogenicity is also controversial. Not surprisingly, the overwhelmingly dominant tree in this habitat (95% or more of the trees) is the Tapia, which has a rounded canopy supported by trunks that are sometimes straight but are more typically twisted and angular. The thick and corky bark, which can be an inch (2.5 cm) thick, gives Tapia a fire resistance that has proven key to its persistence on the frequently fire-swept High Plateau. The oval-shaped leaves are thick and leathery (sclerophyllous). The canopy height varies from 13 to 40 ft. (4–12 m), with the lower end of this range being most common. Whatever its height, the canopy tends to be uniform, with few emergents. Trees are well spaced, giving Tapia a park-like feeling.

Alongside the eponymous Tapia, some of the most common trees and larger shrubs are members of the Madagascar-endemic Asteropeia and Sarcolaenaceae families: *Xerochlamys itremoensis*, *Sarcolaena oblongifolia*, *Leptolaena pauciflora*, and *Asteropeia densiflora*. Other common woody species include *Aphloia theiformis*, *Coptosperma*, *Syzigium*, and *Weinmannia lucens*. The understory usually includes heath (Ericaceae) species, which like Tapia are resistant to fire and dehydration. During the wet season, a lush ground cover of herbaceous plants grows, mostly grasses and asters (Asteraceae). Other understory plants include aloes, *Pachypodium*, *Xerophyta*, euphorbias, Flax Lily (*Dianella ensifolia*), and *Kalanchoe* (in the Stonecrops)—plants typical of Africa's arid zone and hinting at an ancient link. Tree ferns and epiphytes, common features of the rainforests farther east, are lacking and scarce, respectively. Small lianas do occur.

Like most of Madagascar, Tapia experiences two seasons: a dry cold winter and a hot wet summer (Köppen **Cwa**). Winter temperatures can dip to around freezing, and those in summer can

Pleasingly park-like Tapia woodland. Col des Tapias, Madagascar. © KEN BEHRENS, TROPICAL BIRDING TOURS

reach 95°F (33°C), but average temperatures are 62–72°F (17–22°C). The rainy season runs from October to April, with the majority of rainfall, averaging 40 in. (1000 mm) annually, occurring from December to February. The vast majority of the High Plateau, including the Tapia, burns during the dry season. So this habitat cycles annually between having a lush, grassy understory during the rainy season to being quite stark and open after the winter burns.

Tapia is found from 2600 to 5250 ft. (800–1600 m). It grows exclusively on rocky substrates, which can be gneissic, granitic, or quartzitic. The rugged and rocky highland landscapes where Tapia grows, in combination with this pleasingly park-like woodland, make for some beautiful bucolic landscapes.

There is major controversy about the character of the Malagasy High Plateau before the arrival of humans (which itself is a hotly debated issue!). Some experts have asserted

Tapia tree (*Uapaca bojeri*) has thick, leathery leaves and thick, fire-resistant bark. Isalo, Madagascar.
© KEN BEHRENS, TROPICAL BIRDING TOURS

that all Madagascar was forested, whereas more recent work suggests that the High Plateau was a complicated mosaic that included natural savanna and woodland. It was often believed that Tapia formerly covered vast parts of Madagascar, but more recent studies suggest that its extent may not have changed significantly. Another controversy is the degree to which humans have shaped this habitat. Undoubtedly humans have hugely increased the frequency with which Tapia is burned and preferentially cut other trees growing within Tapia woodlands. Both behaviors tend to weed out other large woody plants and produce monotypic stands of Tapia. This woodland has been valued by people for a very long time, and they have likely shaped it to their needs and desires, much as humans shaped the Mediterranean Oak Forest of Europe. So while it is certain that Tapia woodland existed long before humans arrived in Madagascar, it is likely that Tapia today is different from what it would have been thousands of years ago—less diverse and perhaps more open. Another question is whether Tapia is a stable habitat in the absence of human disturbance or whether it represents a regressive stage in the degradation and aridification of an area that formerly would have supported rainforest.

CONSERVATION: On an island with many conservation woes, Tapia stands out as an example of people living in harmony with nature. Tapia woodland has long been valued as a source of firewood, fruit, and mushrooms and, perhaps most importantly, as the main source of the "lanibe" silk cocoons (see Wildlife section). As such, it has received protection and sustainable management over hundreds of years. Nonetheless, there are conservation concerns, including overharvesting for wood, clearing for cultivation, and perhaps most seriously, encroachment by exotic pine and eucalyptus trees.

WILDLIFE: On an island of incredible biodiversity and endemism, Tapia is mostly notable for its lack of wildlife! Tapia and adjacent rocky areas are used occasionally by the "face of Madagascar's biodiversity," the Ring-tailed Lemur. Other lemurs do not seem to use this habitat. The birds are very similar to those of the woodier versions of **MALAGASY GRASSLAND AND SAVANNA**. Typical species include Helmeted Guineafowl, Malagasy Turtle-Dove, Madagascar Hoopoe, Common Jery, Malagasy Bulbul, Malagasy White-eye, Souimanga and Malagasy Sunbirds, and Red Fody. One slightly scarcer species that uses Tapia is Forest Rock-Thrush, especially the *bensoni* subspecies in the Isalo region. Grassy glades and more open surrounding areas hold widespread open-country species like Madagascar Lark and Madagascar Cisticola.

The star animal of Tapia is actually an insect, a Malagasy endemic moth called "Landibe" (*Borocera cajani*). This is a silk moth whose

Ring-tailed Lemur is widespread in sw. Madagascar, and can sometimes be found in Tapia. © KEN BEHRENS, TROPICAL BIRDING TOURS

The birds of Tapia are widespread habitat generalists like the Madagascar Sunbird.
© KEN BEHRENS, TROPICAL BIRDING TOURS

Tapia woodland in Isalo National Park, Madagascar. The canopy is entirely made up of Tapia trees, but there is a wider diversity of woody shrubs in the midstory and understory.
© KEN BEHRENS, TROPICAL BIRDING TOURS

preferred food plant is the Tapia tree. Since ancient times, highlanders have been collecting the cocoons of this moth, which they use to produce silk. The silk fibers are dyed with various plant-based pigments, charcoal, and mud, then woven into beautiful "lambas." These lengths of cloth have long been central to many aspects of Malagasy culture, used to mark group identity, as marriage gifts, and most importantly, to shroud the dead.

Endemism: No species are endemic to this habitat, but most of its wildlife is endemic to Madagascar.

DISTRIBUTION: Found in scattered patches on the central and southern portions of the Malagasy High Plateau, from west of Antananarivo, south and west to the Isalo region. The total extent of Tapia woodland is estimated at around 300,000 acres (133,000 ha).

WHERE TO SEE: Isalo National Park, Madagascar; Col des Tapia, between Ambositra and Antsirabe along the RN7 highway.

SIDEBAR 38 — MALAGASY TAPIA, A GLOBAL CLASSIFICATION ENIGMA

Part of what makes Tapia woodland interesting is that it is a classification challenge. Fundamentally, as a woodland dominated by a leathery-leaved tree, it is unlike anything else in Africa. Structurally, it is very similar to oak woodlands from North America and Europe, but unlike those woodlands, its rainfall occurs almost entirely during the hot summer months. It is probably best considered a sclerophyll woodland, alongside several other habitats in this category from Australia, which are dominated by leathery-leaved trees and receive summer rainfall. Finding a habitat like this, belonging to an entirely different "family" of habitats, hidden away on the Central Plateau of Madagascar, is a wonderful anomaly of the sort that makes the "Eighth Continent" such a biogeographic wonderland. Further spice is provided by the question of whether Tapia is a fundamentally natural formation or whether it is profoundly anthropogenic, like the Mediterranean Oak Forest, a habitat formed by and for human use.

African Heathlands Dendrogram

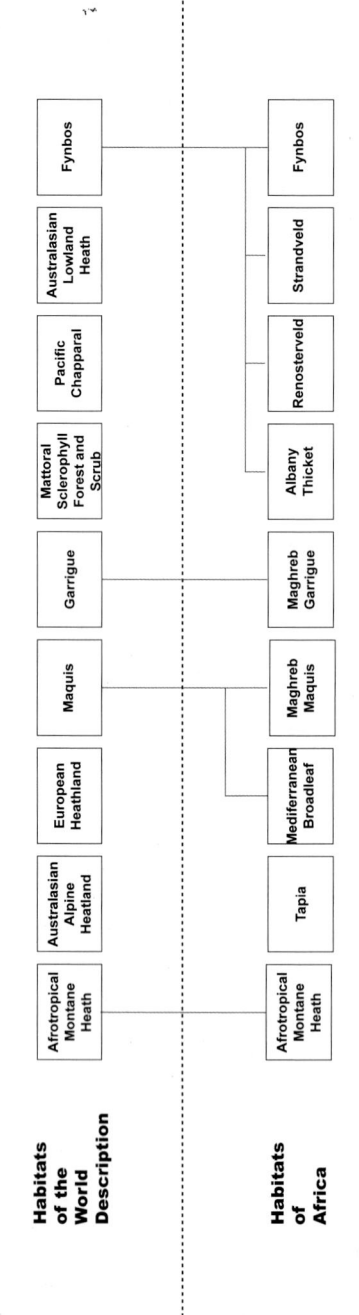

Habitats of the World Description	Afrotropical Montane Heath	Australasian Alpine Heatland	European Heathland	Maquis	Garrigue	Mattoral Sclerophyll Forest and Scrub	Pacific Chapparal	Australasian Lowland Heath	Fynbos
Habitats of Africa	Afrotropical Montane Heath	Tapia	Mediferranean Broadleaf	Maghreb Maquis	Maghreb Garrigue	Albany Thicket	Renosterveld	Strandveld	Fynbos

ALPINE TUNDRAS AND MONTANE HEATHS

Af10A AFROPARAMO

IN A NUTSHELL: Habitat found at the very highest elevations in Africa, sparsely vegetated with grass, herbs, low heath, and/or succulents. **Global Habitat Affinities:** ANDEAN CUSHION PARAMO; ANDEAN GRASSY PARAMO; AUSTRALASIAN ALPINE TUNDRA; HUMID PUNA. **Continental Habitat Affinities:** HIGH ATLAS ALPINE MEADOW. **Species Overlap:** MONTANE HEATH; MONTANE GRASSLAND.

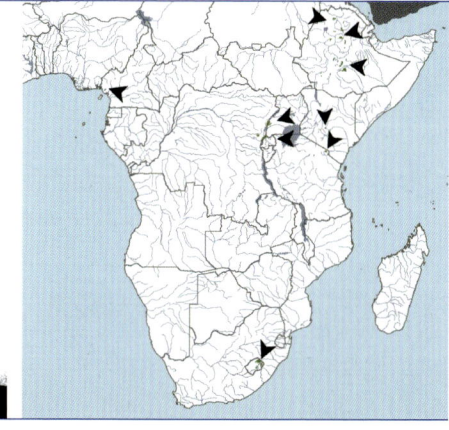

DESCRIPTION: This is the habitat on the top of Africa's highest mountains. Afroparamo is found above 12,800 ft. (3900 m) on tropical massifs, such as Kilimanjaro and Ethiopia's Bale Mountains. In temperate Lesotho's Drakensberg mountains, it starts much lower, at 8900 ft. (2700 m). This is an extreme and relatively impoverished habitat, due to both its inherent harshness and its archipelago-like formation, in which islands of Afroparamo habitat are separated by vast intervening areas. Although appearing similar, the species composition and wildlife are distinct in four different geographic areas, which are deserving of subhabitat status: Abyssinia (Af10A-1), East Africa (Af10A-2), Lesotho (Af10A-3), and Mt. Cameroon (Af10A-4).

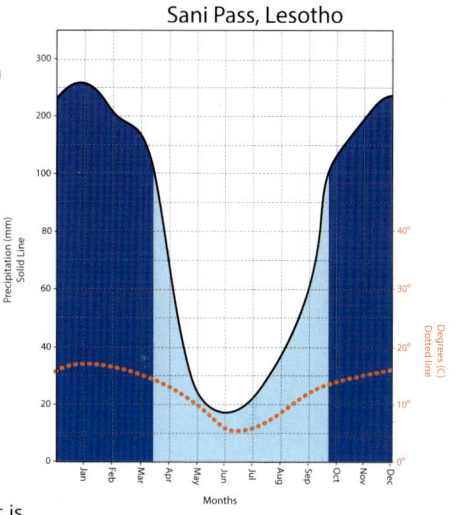

Afroparamo habitat varies a lot, both on a continental scale and more locally. What defines it is that it grows at high elevations, where only low vegetation is able to survive. It is closely allied with both MONTANE HEATH and Afrotropical MONTANE GRASSLAND, shares many plants with both, and is sometimes lumped into one of these habitat categories. In this book, we distinguish Afroparamo habitat from heath because heath vegetation, while present in Afroparamo, is much less dominant and/or takes on a stunted growth form. Afroparamo habitat is distinguished from grassland by the more scattered nature of the grass present and its different mix of grass species.

While Montane Grassland is maintained and often created by fire, most of the Afroparamo grass species are intolerant of fire. The transition between Afroparamo and heath or grassland habitats can be nebulous and difficult to pinpoint. The same is true of the transition from high-elevation NAMA KAROO to Afroparamo that occurs in Lesotho.

The substrate on most mountains is rocky, with shallow soil that, in part, is formed by freezing and thawing action. The vegetation can take many forms. In some areas it is dominated by grasses, in some by miniature heath bushes, in some there are meadows of herbaceous plants, and often, low succulent shrubs are the most common vegetation. Sometimes these succulents grow as pincushions or low mats, making some versions of Afroparamo habitat strongly resemble the Paramo of South America. The tops of the highest mountains, such as Kilimanjaro, are rocky Alpine deserts that are virtually unvegetated save for some hardy lichens and mosses.

The climate in this habitat is extremely harsh (Köppen **Dfc**)—some mountains have recorded a temperature fluctuation of 72°F (40°C) in a single day. Warm temperatures often occur during the day, while cold, often freezing, temperatures are typical at night. One of the characteristics of Afroparamo is that freezes can occur anytime of year. Temperatures as low as −5°F (−20°C) have been recorded in Lesotho. Afroparamo is dependent on elevation rather than rainfall, so naturally is found across a broad range of rainfall levels, from the desert-like high Kilimanjaro to the wettest

Red-hot poker (*Kniphofia*) is often present in Afroparamo. Simien Mountains, Ethiopia.
© KEITH BARNES, TROPICAL BIRDING TOURS

Af10A AFROPARAMO

Above: **Sani Pass, Lesotho, in winter. This far south, Afroparamo includes a Nama Karoo-linked element of its shrub growth, and birds like Sickle-winged Chat, Gray Tit, and Layard's Warbler can be found here at their eastern extent.** © KEITH BARNES, TROPICAL BIRDING TOURS

Below: **Ethiopian Wolf, a rodent-hunting canid of the Ethiopian highlands. Sanetti Plateau, Bale Mountains National Park.** © KEN BEHRENS, TROPICAL BIRDING TOURS

Afroparamo moonscape, with giant lobelias on the Sanetti Plateau, Bale Mountains National Park, Ethiopia. Habitat for Ethiopian Highland Hare *(inset).* MAIN PHOTO © IAIN CAMPBELL; INSET © KEN BEHRENS, TROPICAL BIRDING TOURS

parts of the Ethiopian Highlands, which can receive 100 in. (2500 mm) of precipitation. Snowfall is frequent, and snow blankets large portions of the Drakensberg during the austral winter (Jun–Aug).

It is hard to generalize about the plants of this habitat, as its botanical composition varies greatly; only around 20 vascular plants are found both in Lesotho and the mountains of East Africa. But some typical plants include lady's mantles (*Alchemilla*); heath (*Erica*); succulent *Helichrysum* species; mouse-ear chickweeds (*Cerastium*); *Carex* sedges; and *Agrostis, Deschampsia, Festuca, Pentaschistis,* and *Poa* grasses.

The sclerophyllous shrub *Chrysocoma ciliata* is common in Lesotho, especially in heavily grazed areas. In Lesotho, the most common grass is Mountain Wire Grass (*Tenaxia disticha*). Other frequent grasses there include *Merxmuellera drakensbergensis*, Goat-beard Grass (*Festuca caprina*), and *Pentaschistis*. Ethiopia has *Koeleria* and *Aira* grasses and a few other plants that are shared with Europe, evidence of the continents' links during the ice ages.

Afroparamo habitat in Ethiopia and East Africa boasts some of the continent's most striking plants. There are several species of giant lobelias, which grow massive flowers up to 16 ft. (5 m) tall. Equally strange are the giant groundsels (*Dendrosenecio*), which can have a thick, corky trunk and sometimes punctuate vast stretches of landscape.

CONSERVATION: This habitat has largely been protected by its inhospitality. Overgrazing, which can drastically modify the plant composition of Afroparamo areas, is a growing problem in Lesotho and Ethiopia.

WILDLIFE: This habitat generally lacks big mammals, but it is rich in rodents. In the Bale Mountains of Ethiopia, there is a remarkable density of Big-headed African Mole-rats. These huge rodents spend most of their time in extensive burrow systems but cautiously poke their big tawny heads out of the ground during the day. The Bale Mountains also have several species of grass rats and the endemic Ethiopian Highland Hare. This lineup of smaller mammals has proven a sufficient food source for an endemic species of canine, the Ethiopian Wolf. This beautiful animal looks like a huge long-snouted and long-legged red fox. It spends most of its time hunting rodents in Afroparamo habitat, though it also ventures lower, into AFROTROPICAL MONTANE DRY MIXED WOODLAND, MONTANE HEATH, and MONTANE GRASSLAND. Ethiopia's northern Afroparamo zone, notably in the Simien Mountains, supports a gorgeous endemic wild goat, the Walia Ibex.

The Afroparamo habitat of West Africa and East Africa is poor in birds, probably due to the mountains' isolation. The only habitat specialist in East Africa is Scarlet-tufted Sunbird, an iridescent emerald-green sunbird with extended central tail feathers. Ethiopia and Lesotho each support slightly higher bird diversity, mostly of endemic species.

Endemism: Ethiopia's Afroparamo habitat supports several endemics, including

Afroparamo is rich in rodents like Sloggett's Ice-Rat, an endemic of Lesotho and adjacent South Africa. © KEN BEHRENS, TROPICAL BIRDING TOURS

Orange-breasted Rockjumper is a striking species of high elevations and one of only two members of a Southern African endemic family.
© KEITH BARNES, TROPICAL BIRDING TOURS

Moorland Francolin, Wattled Ibis, Spot-breasted Lapwing, and Moorland Chat, which is shared with Kenya and n. Tanzania. Ethiopia has isolated resident populations of several more typically Eurasian species, namely Ruddy Shelduck, Golden Eagle, and Red-billed Chough, and also supports a small population of Wattled Cranes. The high mountains of Ethiopia are used as wintering habitat by many Palearctic-breeding birds between September and March, including various ducks, shorebirds, harriers, and eagles. Afroparamo habitat in Lesotho is the breeding area for Mountain Pipit and the year-round home of Drakensberg Rockjumper and Drakensberg Siskin.

There are a several specialist reptiles in Lesotho: Common Crag Lizard (*Pseudocordylus melanotus*), Lang's Crag Lizard (*P. langi*), Essex's Mountain Lizard (*Tropidosaura essexi*), Cottrell's Mountain Lizard (*Tropidosaura cottrelli*), and Drakensberg Rock Gecko (*Afroedura nivaria*). Though amphibians are rare in such cold places, the Drakensberg has an endemic frog with a great name: Ice Frog (*Amietia vertebralis*). Drakensberg Frog (*Amietia hymenopus*) can also be found in streams through Afroparamo habitat. This habitat comprises many areas of low plant diversity but high endemism. The Lesotho highlands especially stand out as a hotspot for localized plants.

DISTRIBUTION: Afroparamo habitat is found on the highest mountains from Ethiopia to Lesotho and e. South Africa. These include the highest mountains of Ethiopia (the tallest of which is Ras Dejen in the Simien Mountains, at 14,928 ft./4550 m); Mt. Kenya (17,057 ft./5199 m) and the highest portions of the Aberdares (13,127 ft./4001 m) in Kenya; Mt. Kilimanjaro (19,341 ft./5895 m) and Mt. Meru (14,968 ft./4562 m) in Tanzania; the highest parts of the Ruwenzori Range, on the border between Uganda and DRC (16,762 ft./5109 m); the highest parts of the Virunga Range, on the Rwanda-DRC-Uganda border (14,787 ft./4507 m); Mt. Elgon on the Uganda-Kenya border (14,177 ft./4321 m); and the highest parts of eastern Lesotho (reaching a maximum height of 11,424 ft./3482 m). The summits of some of the lower mountains in East African, Malawi, Zimbabwe, and Mozambique also have pockets of habitat in particularly cold and exposed enclaves that resemble Afroparamo. Elsewhere in Africa, the only true Afroparamo habitat is found on the top of Mt. Cameroon (13,250 ft./4040 m). This habitat generally occurs immediately above MONTANE HEATH. In some places, it directly abuts MONTANE GRASSLAND and/or the stunted and heath-rich upper reaches of MOIST MONTANE FOREST.

WHERE TO SEE: Above Sani Pass, Lesotho; Bale Mountains National Park, Ethiopia.

Despite its harshness, Afroparamo holds a few cool reptiles, like Common Crag Lizard (*Pseudocordylus melanotus*).
© KEN BEHRENS, TROPICAL BIRDING TOURS

| SIDEBAR 39 | AFRICA'S GREAT RIFT VALLEY |

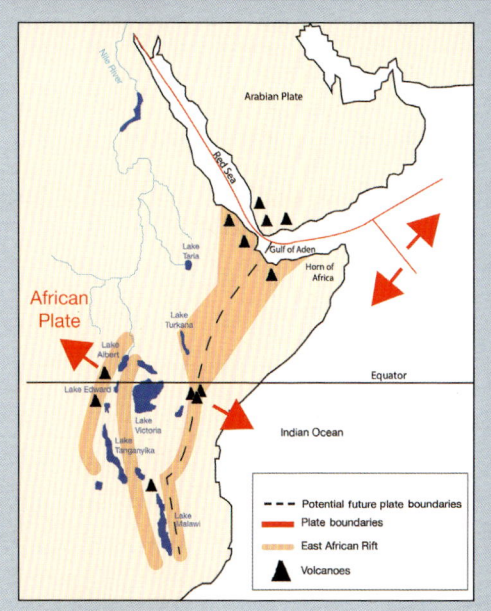

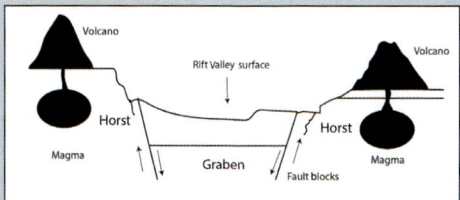

The Great Rift Valley is a remarkable geological feature, where you can stand in what may eventually become a new sea between a big chunk of East Africa and the rest of the continent. It is actually part of a three-branched system, with sister rifts under the Red Sea and the Gulf of Aden, yet unlike the other two arms of the system, it did not fill with ocean water. As the Nubian (the main African) Plate and the new Somalian (East African) Plate drift apart, the land surface ruptures along fault lines, and the area between these faults drops down. The area that drops down is called a graben, and the areas that are lifted up are called horsts. To make the whole scenario even more dramatic, areas of thinner crust on either side of the rift valley are prone to volcanism, giving us Mt. Kilimanjaro and many other spectacular peaks. These mountains, in turn, receive a lot more rain than the surrounding flatlands. So this rifting is the reason you can be in a montane forest on Tanzania's Mt. Meru while looking down at semi-desert thornscrub and alkaline lakes in the bottom of the Great Rift Valley.

Side of the Great Rift Valley, covered in a dense form of Moist Mixed Savanna. © KEN BEHRENS, TROPICAL BIRDING TOURS

Af10B HIGH ATLAS ALPINE MEADOW

IN A NUTSHELL: The areas of bare ground and tundra at very high elevations in the mountains of North Africa. **Global Habitat Affinities:** EUROPEAN ALPINE TUNDRA; NEARCTIC ALPINE TUNDRA; ASIAN ALPINE TUNDRA. **Continental Habitat Affinities:** AFROPARAMO. **Species Overlap:** MAGHREB JUNIPER OPEN WOODLAND.

DESCRIPTION: Above tree line in the Atlas Mountains of the Maghreb there is alpine tundra or meadow, where despite growing in a Mediterranean climate of dry summers and wet winters, the temperatures are so low that the area can be snow covered from November to April. As a result, these habitats have much more in common with the alpine tundras of Europe (Köppen **Dfc**) than the mountaintops of the rest of Africa. This habitat does differ from European alpine tundra in that the soils are poorly developed lithosols and aridisols originally developed under drier, harsher conditions than exist today and even at present have very dry conditions for much of the year. Life here is tough.

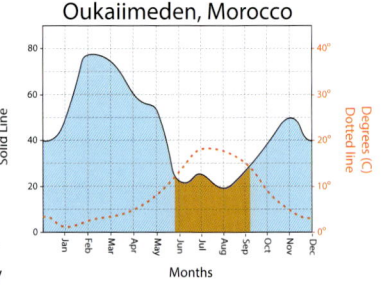

At lower elevations, from 8200 to 9840 ft. (2500–3000 m), High Atlas Alpine Meadow is typically a low shrubland, with dense shrubs about 2–3 ft. (0.6–1 m) tall. It seems to be migrating lower because of the clearing of MAGHREB FIR AND CEDAR FOREST and MAGHREB JUNIPER OPEN WOODLAND. This is interesting because in its purest and most diverse form, alpine meadow is expected to move to higher elevations because of climate change. Fortunately, there are still some high peaks, including Jbel Toubkal at 13,670 ft. (4167 m), offering high-elevation refuge for this habitat. Above 9800 ft. (3000 m) this habitat gradually changes to feature cushion plants alongside dwarf shrubs, such as Snowy Mespilus (*Amelanchier ovalis*), Mountain Currant (*Ribes alpinum*), and even Common Juniper (*Juniperus communis*) in a dwarf form of a species found throughout colder parts of the Northern Hemisphere. Spiny Madwort (*Hormathophylla spinosa*) is common in shrubbier tundra. Where the tundra is rocky, with many areas of bare ground, it is dominated by cushion plants such as Drinn Sandwort (*Arenaria pungens*), interspersed with sedges such as the widespread Hair-like Sedge (*Carex capillaris*) and grasses such as Atlas Fescue (*Festuca mairei*) and Alpine Fescue (*F. brachyphylla*).

CONSERVATION: Although, like most habitats in North Africa, protection is limited and human pressures are great, the tundra regions are known for their beauty, and many are therefore protected in national parks such as Toubkal National Park in Morocco, which contains the country's highest peak. Some other areas have ski resorts, especially near Marrakech, but although they seem to place stress on the accessible environments (and therefore the ones that most tourists see), they have limited impact on a larger scale. The single greatest threat to these ecosystems is

Af10B HIGH ATLAS ALPINE MEADOW

A lush alpine meadow in its green summertime aspect with flowering forbs. Oukaïmeden, Atlas Mountains, Morocco. © KEITH BARNES, TROPICAL BIRDING TOURS

climate change, as these habitats are limited to the highest slopes. Interestingly though, some species move downslope with climate change in other parts of the world. This habitat currently persists at only the very highest elevations of the Middle Atlas and other lower ranges. We should expect to lose this habitat from many peaks.

WILDLIFE: Little persists in the way of large mammals, although in the past these peaks probably supported Barbary Sheep (Aoudad) and the occasional "Barbary" Leopard. These beasts are all but gone now. Small mammals include mice and gerbils, which provide food for a small population of Red Fox, but overall this habitat is remarkably mammal free. There are not many birds up here either, but those that do occur are very special. Perhaps the most characteristic bird is the Alpine Accentor. From the Hedge Sparrow family (Prunellidae), this is a Palearctic species that bridged the gap to Africa. The Crimson-winged Finch is a lovely suffusion of buffs and pinks, and this high-elevation specialist persists year-round, though it descends in winter when snow covers the ground. The African subspecies (*alienus*) is near-endemic to this habitat. The Horned Lark is one of the most variable species on the planet, with 47 races, but the subspecies endemic to the Atlas Mountains is sometimes considered a separate species and is most at home in the alpine tundra. The Atlas Wheatear is an endemic summer breeder restricted to higher mountains and favors tundra meadows, although in winter it departs to the Sahel. Two specialist high-elevation Palearctic corvids also

352 ALPINE TUNDRAS AND MONTANE HEATHS

High Atlas Alpine Meadow in spring retains a wintry aspect and can be harsh and uninviting. Oukaïmeden, Morocco. © NICK ATHANAS, TROPICAL BIRDING TOURS

"Atlas" taxon of the widespread Horned Lark. © DANIELE ARDIZZONE, FLUYENDO PHOTOGRAPHY

Af10B HIGH ATLAS ALPINE MEADOW

Above: **Atlas Wheatear breeds in High Atlas Alpine Meadows, then migrates to winter in the lowlands of West Africa.** © DANIELE ARDIZZONE, FLUYENDO PHOTOGRAPHY

Right: **Crimson-winged Finch, a specialist of the highest portions of the Atlas Mountains.** © CARLOS N. G. BOCOS

exist here, Alpine and Red-billed Choughs, probing the grassy meadows with their long, slender bills and giving their piercing calls. The high peaks are patrolled by that most characteristic avian mountain hunter, Golden Eagle, and the occasional Lammergeier, though the latter is becoming increasingly rare.

In the High Atlas there is a good variety of herps, including some localized ones like Atlas Mountain Skink (*Chalcides montanus*), Andalusian Wall Lizard (*Podarcis vaucheri*), Atlas Day Gecko (*Quedenfeldtia trachyblepharus*), and Atlas Dwarf Lizard (*Atlantolacerta andreanskyi*). Along small creeks one can find Common Toad (*Bufo bufo*) and Spiny Toad (*B. spinosus*), as well as the more local Mediterranean Tree Frog (*Hyla meridionalis*). Rocky alpine meadows attract Moroccan Painted Frog (*Discoglossus scovazzi*) and rockier areas with scrub support endemic Mountain Skink (*Chalcides montanus*) and the beautiful Mountain Viper (*Vipera monticola*).

DISTRIBUTION: Generally limited to mountains above 9200 ft. (2800 m), though a less diverse form is moving lower with the destruction of the forested zone. This high-elevation limitation means that this habitat is most prevalent in the High Atlas and elsewhere is localized to the summits of the Middle Atlas and Aurès.

WHERE TO SEE: Oukaïmeden, Morocco.

Af10C MONTANE HEATH

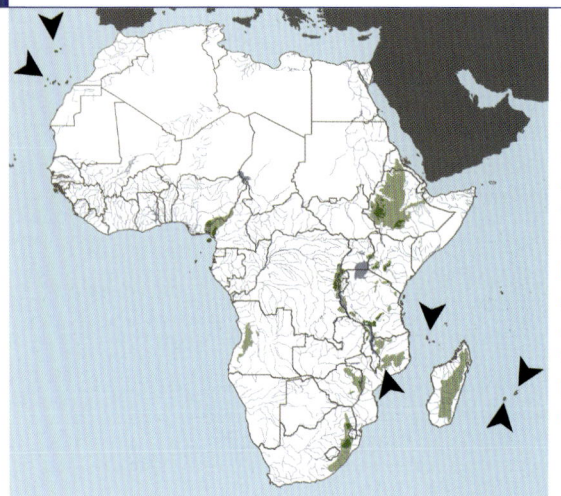

IN A NUTSHELL: A montane heath-dominated shrub habitat that is fire-prone and grows on poor soils, mainly at high elevations but locally down to moderate elevations. **Global Habitat Affinities:** AUSTRALASIAN MONTANE HEATHLAND; WEST EUROPEAN COASTAL HEATH. **Continental Habitat Affinities:** FYNBOS; RENOSTERVELD; STRANDVELD; AFROPARAMO. **Species Overlap:** AFROTROPICAL MONTANE DRY MIXED WOODLAND; AFROPARAMO; NAMA KAROO; MONTANE GRASSLAND; MOIST MONTANE FOREST.

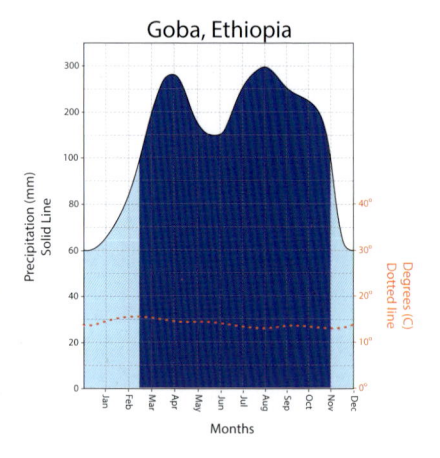

DESCRIPTION: Heath is a shrubland dominated by members of the heath family (Ericaceae), which includes heathers. These plants tend to have fine, leathery leaves and a brownish or off-green coloration, giving the habitat a distinctive character even at a distance. Heath usually grows on poor, acidic, well-drained soil and is highly prone to fire. In the Afrotropics, the most diverse heath habitat is the FYNBOS and related habitats of sw. South Africa. The continent's only other heath, the Montane Heath covered here, seems to act as a buffer between the frequently burned MONTANE GRASSLAND on mountaintops and the lower-lying and generally unburned MOIST MONTANE FOREST. At the transition from montane forest to heath, there can be large ericas, up to 24 ft. (7 m) tall, heavily swathed in mosses and other epiphytes. Hagenia trees and bushes (*Hagenia abyssinica*) are also often part of this mix. Higher up, the moorland (a wetter kind of heath) gradually becomes shorter, down to 3–7 ft. (1–2 m), then gives way to Afroparamo, though even Afroparamo includes some dwarf heaths and has been regarded as an impoverished version of Montane Heath. Between these transitions and extremes, a typical Montane Heath has a height of 6–13 ft. (2–4 m). Heath is a habitat of highly variable temperatures, with cold nights and warm days (Köppen **Cfb**). Temperatures can be well below freezing, especially during winter in s. Africa, while daily temperature swings of 50°F

Montane heathland and grassland. Andringitra National Park, Madagascar. © KEN BEHRENS, TROPICAL BIRDING TOURS

(30°C) have been recorded. Rainfall is highly variable—scant in some places but as high as 100 in. (2500 mm) annually in the wettest montane heathlands.

As suggested by the name, Montane Heath is indeed dominated and characterized by members of the heath family. The mix of species varies regionally. On Mt. Karthala, on Grande Comore, the common heath is the Comorian endemic *Erica comorensis*. Madagascar has a rich assemblage of heath species, including many endemics. Typical species of African mainland Montane Heath include *Erica*

Classic high mountain heathland at 6995 ft. (2132 m) on top of Marojejy, Madagascar.
© KEN BEHRENS, TROPICAL BIRDING TOURS

trimera and *E. silvatica*. Another typical heath, the Tree Heath (*E. arborea*) is found both in the mountains of sub-Saharan Africa and in MAGHREB MAQUIS habitats surrounding the Mediterranean, as well as in the Macaronesian islands. Non-heath plants often found in Montane Heath habitat include hardleaves (*Phylica*) and a variety of asters (Asteraceae).

WILDLIFE: Montane Heath is a poor habitat for wildlife. In general, it has a small subset of the species of adjacent MOIST MONTANE FOREST, AFROTROPICAL MONTANE DRY MIXED WOODLAND, and MONTANE GRASSLAND habitats. Black Rhinoceros is one mammal that readily uses heath, though it has been eliminated from most of its former range. A few rodents are restricted to high mountain habitats, including Montane Heath. There are two such species in Malagasy heath.

Typical birds include Abyssinian and Rwenzori Nightjars, Cape Eagle-Owl, African Stonechat, Moorland Chat, Cinnamon Bracken-Warbler, and a variety of sunbirds, including Tacazze and Malachite Sunbirds and several double-collared sunbirds. Ethiopia has perhaps the richest Montane Heath in the Afrotropics, which corresponds to the richness of the adjacent Afrotropical Montane Dry Mixed Woodland. The Indian Ocean island of Grande Comore has the distinction of supporting the only Afrotropical bird species that is entirely restricted to heath habitat: Comoro White-eye. This habitat is also used by the island endemic Grande Comoro Brush-Warbler. On Réunion, Montane Heath is inhabited by the endemic Réunion Stonechat, while in Madagascar, heath species include Madagascar Flufftail, Madagascar White-eye, and Madagascar Sunbird.

Above: **Gelada troop in Montane Heath. This extraordinary primate has the second-largest vocabulary after humans. Ankober Escarpment, Ethiopia.** © KEN BEHRENS, TROPICAL BIRDING TOURS

Left: **Jackson's Francolin is a Montane Heath specialist, but it will enter juniper, yellowwood, and Hagenia forests and woodlands as well. Aberdare National Park, Kenya.** © KEN BEHRENS, TROPICAL BIRDING TOURS

Left: **Rouget's Rail** is an extroverted rail that readily uses upland habitats, including heath (*Erica*). *Inset:* detail of a heath plant from Réunion. © KEN BEHRENS, TROPICAL BIRDING TOURS

Below: **Mountain Nyala** is an Abyssinian endemic mammal that uses a mix of montane habitats, which includes heath. Bale Mountains National Park, Ethiopia. © KEN BEHRENS, TROPICAL BIRDING TOURS

Endemism: In Montane Heath habitats on the Indian Ocean islands, which have very different wildlife from those on mainland Africa, endemism is driven more by the high endemism of adjacent forest habitats than by the heath itself. On the continent, this habitat is readily used by several of the endemic mammals and birds of the Ethiopian Highlands: Gelada, Mountain Nyala, Ethiopian Wolf, and Moorland and Chestnut-naped Francolins. Elsewhere, it is an important habitat for the localized Albertine Rift endemic Rwenzori Red Duiker and for the Kenyan specialties Jackson's Francolin and Aberdare Cisticola.

DISTRIBUTION: Found exclusively in the higher mountains of the Afrotropics, this habitat typically occurs in a narrow elevational range, above MOIST MONTANE FOREST but below AFROPARAMO. Near the equator, this zone is between 10,000 and 11,500 ft. (3000–3500 m). Away from the equator, as in central and south Madagascar, heath can naturally occur down to 5900 ft. (1800 m). This habitat can also be produced at lower elevations by frequent, human-caused fires, which transform montane forest into heath, which itself is often a transitory step on the way to conversion into anthropogenic montane grassland. Such artificial heaths are common in Madagascar, down to at least 3300 ft. (1000 m). Heath habitat is also found on some of the islands that surround Africa and Madagascar. It is present in Macaronesia, on the Azores and Madeira, and in the Indian Ocean islands on Grande Comore and Réunion.

WHERE TO SEE: Bale Mountains National Park, Ethiopia; Aberdare National Park, Kenya.

FRESHWATER HABITATS

Af11A AFROTROPICAL DEEP FRESHWATER MARSH

IN A NUTSHELL: Habitat permanently flooded with fresh water, vegetated by hydrophilic reeds, shrubs, and floating plants. **Global Habitat Affinities:** AUSTRALIAN TROPICAL WETLAND; NEOTROPICAL TROPICAL WETLAND; ASIAN TROPICAL WETLAND; NEARCTIC REEDBED MARSHES; EUROPEAN REEDBEDS. **Continental Habitat Affinities:** SOUTH AFRICAN TEMPERATE WETLAND. **Species Overlap:** AFROTROPICAL SHALLOW FRESHWATER MARSH; SOUTH AFRICAN TEMPERATE WETLAND; RIVERS; FRESHWATER LAKES AND PONDS; TIDAL MUDFLATS AND ESTUARIES.

DESCRIPTION: Freshwater wetlands are among the most productive and diverse habitats in Africa and are critical for a vast array of wildlife. They are also crucial to humans, watering most of the food they consume, such as rice, which is the staple food for half the Earth's human population. Various types of wetlands, some of which are covered separately in this book, tend to occur in close association with each other. These include RIVERS, FRESHWATER LAKES AND PONDS, seasonally flooded AFROTROPICAL GRASSLAND, AFROTROPICAL SHALLOW FRESHWATER MARSH, and this habitat. Much of what is described in this account applies more broadly to freshwater wetlands, which are given their most lengthy coverage here. Subsequent accounts will cover other freshwater habitats more briefly, mostly focusing on their distinguishing characteristics and specialist wildlife.

Deep freshwater wetland can take many forms. The most common is tall reedbed marshes, which often include or are dominated by Papyrus (subhabitat Af11A-2). But this habitat is complex in its own right and can include fine-leaved grasses, sedges, floating vegetation, shrubs, and small trees. Freshwater SWAMP FOREST is covered separately, as its primary function is that of a thickly forested environment, though it often occurs in close conjunction with small patches of marsh.

Deep marshes support plants that can grow in the

A classic deep freshwater wetland at Mabamba Swamp, Uganda. The deepest areas have floating vegetation used by jacanas, pygmy-geese, and lapwings. © KEN BEHRENS, TROPICAL BIRDING TOURS

Vast marshes in Botswana. Papyrus *(inset)* lines the river channels, while shallower marsh is found at the fringes, where the marsh transitions to grassland. MAIN PHOTO © GRANT ATKINS, ATKINSON PHOTOGRAPHY AND SAFARIS; INSET © KEN BEHRENS, TROPICAL BIRDING TOURS

anaerobic conditions produced by perennial flooding. These plants have a diverse set of forms, making this an endlessly fascinating habitat to explore. Grasses are some of the most important and prominent wetland plants. Widespread and common species include the common reeds (*Phragmites*), cattails (*Typha*), Wild Rice (*Oryza longistaminata*), and Hippo Grass (*Vossia cuspidata*). Africa's most famous wetland grass is Papyrus (*Cyperus papyrus*), which can form vast monotypic stands, especially between Botswana and Uganda. Elsewhere it is often found in a mélange alongside other marsh plants. Papyrus towers remarkably high for a marsh plant, with a mature height of 16 ft. (5 m). It is also incredibly fast growing and can reach this height in only 10 weeks! Other plants that often grow alongside Papyrus include the fern *Cyclosorus striatus*, *Melastomastrum*, morning glory (*Ipomoea*), *Limnophyton*, Swamp Hibiscus (*Hibiscus diversifolius*), *Impatiens irvingii*, primrose-willow (*Ludwigia*), and knotweeds (*Polygonum*).

Floating plants tend to grow in slow-flowing areas with deeper water and sometimes blanket huge areas. Common species include water lilies (Nymphaeaceae), mosquito ferns (*Azolla*), pondweeds (*Potamogeton*), Water Snowflake (*Nymphoides indica*), duckweeds (*Lemna*), and Water Cabbage (*Pistia stratiotes*). Although Water Cabbage is native to the continent, it has spread far beyond its original range and is a serious invasive species. Common Water Hyacinth (*Pontederia crassipes*) and Giant Salvinia (*Salvinia molesta*) are invasive introduced species that are now common. Though they are usually anchored to the soil below, Papyrus and Hippo Grass will also grow in floating mats, extending out into deeper water. When portions of these mats become dislodged in river systems, they form mini islands that float downstream. In the Congo Basin, most marshes are dominated by Hippo Grass. Openings within SWAMP FOREST have thick growth of *Lasimorpha senegalensis*, a member of the arum family that has arrow-shaped leaves and a flower reminiscent of a pitcher plant, which locally seems to fill the niche of Papyrus. Although Madagascar has Papyrus, it is less common than in Africa, and many deep marshes are dominated by another arum, the Water Banana (*Typhonodorum lindleyanum*). Deep marshes are sometimes fringed by swamp-adapted shrubs and small trees, especially when occurring in forest zones. More frequently, deep marshes are fringed by or occur in a mélange alongside shallower marsh of shorter stature, comprising fine-leaved grass and/or sedge (see AFROTROPICAL SHALLOW FRESHWATER MARSH). These shallower areas may dry out during the dry season. Although the water levels in deep marshes do fluctuate significantly, especially in the monsoonal climate of the savanna zone, this habitat grows exclusively in areas that are permanently flooded and never dry out completely. The vegetation is typically dominated by tall reeds and Papyrus, as opposed to shorter and finer-leaved grasses and sedges.

CONSERVATION: Although Africa still contains far more pristine marshes than most of the other continents, this habitat has nonetheless been heavily affected by humans. Marshes are often destroyed to make way for agriculture or human settlement. The marshes of w. Madagascar support several endemic species that are endangered due to habitat destruction and human disturbance. Wetlands easily collect pollutants, as human contaminants naturally flow downhill. Invasive plants such as Common Water Hyacinth, an Amazonian species, now clog huge areas that were formerly far more diverse. Conserving marshes and other wetlands has to be considered one of the continent's greatest conservation challenges, and this uniquely vulnerable habitat may serve as a sort of litmus test for the wider state of conservation.

WILDLIFE: This is one of the most exciting habitats for wildlife watching, and any safari will spend much time in and around wetland habitats. There are some species that specialize in more shallow sedge and fine-grass marsh, and those will be discussed in that habitat account (see p. 356). But most wildlife will readily use both deep and shallow marshes, so these two are broadly addressed together below.

Water draws wildlife like a magnet, and there is no surer way to liven up a dull midday in the savanna than to head for a river or marsh. While virtually all larger mammal species visit wetlands in order to drink, only a small set of species use this as their primary habitat. Hippopotamuses spend their days in wetlands, then range widely through surrounding habitats at night in search of fodder. The Sitatunga is a remarkable marsh-dwelling antelope that has long, narrow, widely splayed hooves that allow it to traverse boggy ground. Several other

Grosbeak Weaver weaves a gorgeous nest. It is just one of a large array of weavers that prefer to nest in deep freshwater wetlands. © KEN BEHRENS, TROPICAL BIRDING TOURS

antelopes are found mainly at the edges of wetlands. These include reedbucks, Puku, lechwes, and Waterbuck. Other mammals often found in and around wetlands are Serval, Side-striped Jackal, Marsh Mongoose, and African Buffalo. In e. Madagascar, one remarkable species of lemur has adapted to life in reedbeds: the Critically Endangered Lac Alaotra Bamboo Lemur.

While wetlands are relatively poor in reptiles, they do host the continent's biggest and most fearsome species, Nile Crocodile. They are also prime habitat for Nile Monitor (*Varanus niloticus*) and several water-adapted snakes. Wetlands are the primary habitat of many of continental Africa's 800-some species of amphibians and some of Madagascar's hundreds of endemic frogs. Africa is home to around 20% of the world's approximately 15,000 freshwater fish species; the Congo River alone has more than 700 species.

Goliath Heron versus African Fish-Eagle. Epic scenes like this sometimes play out at wetlands, where many different species come into contact.
© KEITH BARNES, TROPICAL BIRDING TOURS

Wetlands are paradise for birds and birders alike. Anyone with ambitions of checking off a long bird list must spend lots of time in this habitat. Many entire groups of birds are found almost exclusively in wetlands. These include herons, egrets, ibises, geese, shorebirds, terns, and gulls. Many members of these groups will use both freshwater and saltwater wetlands. This is the habitat of some of the continent's most superb birds, such as Goliath Heron, African Darter, Yellow-billed and Saddle-billed Storks, Hamerkop, Wattled Crane, Gray Crowned-Crane, and Giant Kingfisher. Shoebill, which may be Africa's most sought-after bird, makes its home in Papyrus swamps, along with a range of other

Saddle-billed Stork. © KEN BEHRENS, TROPICAL BIRDING TOURS

Shoebill is one of Africa's most sought-after birds. It is found in very low densities in deep marshes, especially Papyrus. © KEN BEHRENS, TROPICAL BIRDING TOURS

Papyrus Gonolek is a skulking though vocal bushshrike that is entirely confined to the interior of Papyrus marshes. © KEN BEHRENS, TROPICAL BIRDING TOURS

specialists, like Papyrus Gonolek, Papyrus Yellow-Warbler, and White-winged Swamp Warbler. Wetlands are hunted by African Fish-Eagle, harriers, and African Grass-Owl.

Although many of this habitat's birds are big and flashy, it also supports smaller birds, including kingfishers, swallows, wagtails, stonechats, swamp warblers, reed warblers like the widespread African Reed Warbler (IS), cisticolas, prinias, weavers, bishops, widowbirds, and waxbills. Deep marshes provide excellent cover for skulking birds like White-backed Night-Heron, Little Bittern, Rufous-bellied Heron, Greater Painted-Snipe, flufftails, African Swamphen, and a variety of rails. Areas of floating vegetation are the preferred habitat of African Jacana, White-backed Duck, Blue-billed Teal, and African Pygmy-Goose.

Some species readily use both deep marshes and adjacent savanna and forest habitats. Good examples are White-browed and Coppery-tailed Coucals, Hartlaub's Babbler, Gray-capped Warbler, and Red-chested Sunbird. Reedbeds are the breeding habitat for several species of weavers and bishops, such as Cape, Southern Masked-, Southern Brown-throated, Northern Brown-throated, and Thick-billed Weavers and Southern and Northern Red Bishops (see sidebar 14, p. 163).

African Jacana, trotting across a floating deep marsh of water lily pads. © KEN BEHRENS, TROPICAL BIRDING TOURS

Deep reed marshes are found locally throughout the Indian Ocean islands (subhabitat Af11A-1) but most commonly in w. Madagascar, where they are associated with broad, slow-flowing river systems. These Malagasy wetlands strongly resemble those of the African mainland. They are the

Long-toed Lapwing deftly trots across floating marsh vegetation.
© KEITH BARNES, TROPICAL BIRDING TOURS

exclusive habitat of two endemic Malagasy birds, namely Madagascar Jacana and Sakalava Rail. They are also an important part of the habitat mix of the Critically Endangered Madagascar Fish-Eagle, Endangered Malagasy Pond-Heron, and the much more common Madagascar Swamp-Warbler.

Most wetland species are widespread, probably due to the far-flung nature of this habitat, as only species that are able to find new habitat when local conditions change will survive. Indeed, many wetland birds are migratory or nomadic and move around to take advantage of changing water conditions across the continent (see sidebar 41, p. 378). Many Palearctic-breeding wetland birds migrate to Africa between September and April. Examples are ducks, harriers, shorebirds, various warblers, and wagtails.

Madagascar Swamp Warbler is a classic reed marsh species, endemic to Madagascar.
© KEN BEHRENS, TROPICAL BIRDING TOURS

Endemism: Most wetland species are widespread. But within the Afrotropics' wetland zone, there are minor centers of bird endemism in Botswana and Zambia and around Lake Victoria. Three bird species, Kilombero Cisticola, White-tailed Cisticola, and Kilombero Weaver, are endemic to a wetland area in central Tanzania. Many of the localized endemic birds are species virtually confined to pure stands of Papyrus, such as Papyrus Gonolek, Papyrus Yellow-Warbler, and Papyrus Canary. The endemic Cape Verde Swamp Warbler has adapted to a variety of anthropogenic habitats but was probably originally confined to deep marshes of Giant Reed (*Arundo donax*).

DISTRIBUTION: Deep marshes are found locally across the Afrotropics, usually nested within other habitats, ranging from desert to AFROTROPICAL LOWLAND RAINFOREST. They are most common in areas of intermediate rainfall, typical of the savannas of much of East Africa and Southern Africa. In the forest zone of the Congo Basin and West Africa, there is abundant water, but flooded areas tend to grow SWAMP FOREST rather than marsh. In the drier parts of the continent, there is rarely sufficient water to support deep marshes. Papyrus is found up to 7500 ft. (2300 m) elevation, while other types of deep marsh can be found slightly higher.

The wetlands in the southern parts of South Africa, extending onto the higher elevations of e. South Africa, are temperate rather than Afrotropical wetland. Although they look superficially similar, they lack the floating vegetation common in Afrotropical deep marshes and also lack many widespread Afrotropical species, so they are covered in a separate habitat account (see SOUTH AFRICAN TEMPERATE WETLAND). Strings of rich lakes and associated marshes run down the two arms of the Great Rift Valley, from Ethiopia to Tanzania and from Uganda to Malawi. Western Tanzania and Zambia hold some important wetlands, including the Bangweulu Wetlands and the Kafue Flats. The continent's most famous wetland is the Okavango Delta, where a broad river flows out of the mountains of Angola, forms an expansive delta, and then seeps into the sands of the dry Kalahari in n. Botswana. One of the largest wetlands on Earth is the Sudd of South Sudan, which covers some 22,000 mi.^{2} (57,000 km^2). Madagascar has vast reedbed marshes at Lac Kinkony and Lac Alaotra.

WHERE TO SEE: St. Lucia, South Africa; Okavango Delta, Botswana; Lake Awassa, Ethiopia; Queen Elizabeth National Park, Uganda; Bangweulu Wetlands, Zambia.

Af11B AFROTROPICAL SHALLOW FRESHWATER MARSH

IN A NUTSHELL: Habitat that is permanently or seasonally flooded with shallow fresh water, mostly vegetated by hydrophilic sedges and fine grasses. **Global Habitat Affinities:** AUSTRALIAN TROPICAL WETLAND; NEOTROPICAL TROPICAL WETLAND; ASIAN TROPICAL WETLAND; NEARCTIC SEDGE AND GRASSLAND MARSHES. **Continental Habitat Affinities:** SOUTH AFRICAN TEMPERATE WETLAND; AFROTROPICAL GRASSLAND. **Species Overlap:** SOUTH AFRICAN TEMPERATE WETLAND; AFROTROPICAL GRASSLAND; AFROTROPICAL DEEP FRESHWATER MARSH; FRESHWATER LAKES AND PONDS; RIVERS; TIDAL MUDFLATS AND ESTUARIES; ROCKY SHORELINE.

DESCRIPTION: There is so much internal complexity within wetland habitats that it is tricky to start parsing out separate habitats. But there is a striking and consistent difference between marshes growing in areas that are permanently and deeply flooded and those that are ephemeral, growing in areas of shallower water, which often partially dries out during the dry season. Shallow marshes are much shorter and more uniform in stature and are dominated by fine-leaved sedges and grasses. They feel almost like a grassland and quite unlike the diverse and obviously waterlogged deep marshes, with their tall reeds, floating vegetation, and open pools. The height of a shallow marsh is typically 1.5–3 ft. (0.5–1 m).

Highland shallow reed and fern marsh in Kamiranzovu Swamp, Nyungwe National Park, Rwanda. This is the key habitat for the Albertine Rift endemic Grauer's Swamp Warbler *(inset)*. © KEN BEHRENS, TROPICAL BIRDING TOURS

A classic seasonally flooded grassland from the Miombo zone, with Miombo woodland and inselberg in the distance and a core of permanently moist Dambo wetland along drainage lines. This whole foreground area might flood after a heavy rain. These are Roan Antelope. Mutinondo, Zambia. © FRANK WILLEMS

In some places, deep and shallow marsh grow side by side. A typical example is the progression from an open slow-flowing river to a tall reed and/or papyrus marsh, to a shorter, shallow sedge and grass marsh, and then to a dry grassland and/or savanna habitat. Shallow marsh very often borders deep marsh, providing the transition to a grassland habitat. In many cases, shallow marsh blends very gradually into seasonally flooded grassland. After heavy rains, large portions of grassland can temporarily function as shallow marshes. The subtle distinction between a grassland that seasonally floods and a shallow marsh is that a grassland dries out completely during the dry season, whereas portions of a shallow marsh dry out but wet drainage lines, often with slow-flowing water, remain even in the heart of the dry season. The wetter areas tend to be dominated by sedges, while the drier fringes comprise fine-leaved grasses. Even at the height of the rainy season floods, the water rarely covers the tallest plants. During the dry season, shallow wetlands often burn, especially at their fringes, as fires sweep into them from adjacent savanna and grassland habitats.

While shallow marsh occurs locally throughout the continent, it is more common at middle and high elevations. There are large tracts of shallow marsh in the MIOMBO zone, which are often referred to as Dambos (subhabitat Af11B-2). These marshes tend to form in broad shallow basins, nested within Miombo woodland. While their individual scale is small, they cumulatively cover a surprisingly large portion of the Miombo zone: 13% of Zambia is covered in Dambo. There are also large tracts of shallow marsh associated with some of Africa's great floodplains, including the Okavango, Lake Chad, and the White Nile's Sudd wetlands.

Above: **Shallow marsh within montane rainforest in e. Madagascar. Marshes like this often include abundant ferns. This is the habitat of some Malagasy endemic birds, including Madagascar Rail** *(inset).*
© KEN BEHRENS, TROPICAL BIRDING TOURS

Below: **This photo shows how shallow and deep marsh interlace. Most of this is shallow marsh, but in the top left corner is a deeper area that supports Papyrus. Moist Montane Forest abuts the wetlands. Arusha National Park, Tanzania.** © KEN BEHRENS, TROPICAL BIRDING TOURS

Sedges and grasses are the most typical plants of this habitat and make up the majority of the vegetation. Common sedges include Swamp Sawgrass (*Cladium mariscus*), true sedges (*Carex*), flatsedges (*Cyperus*), hairsedges (*Bulbostylis*), fringe rushes (*Fimbristylis*), *Scirpus*, umbrella-sedges (*Fuirena*), spikerushes (*Eleocharis*), and bulrushes (*Typha*). Common grasses are *Loudetia phragmitoides*, silvergrasses (*Miscanthus*), Antelope Grass (*Echinochloa pyramidalis*), Wild Rice (*Oryza longistaminata*), Thatching Grass (*Hyparrhenia rufa*), and Ditch Millet (*Paspalum scrobiculatum*). Ferns are common shallow marsh components in some areas, especially at higher elevations in Africa and in e. Madagascar. Other plants typical of this habitat include yellow-eyed grasses (Xyridaceae), St. John's worts (*Hypericum*), knotweeds (*Polygonum*), and Humped Bladderwort (*Utricularia gibba*). Sphagnum moss is found in some marshes, especially at higher elevations. Flowering herbs are common.

CONSERVATION: The endemic-rich sedge marshes of e. Madagascar (subhabitat Af11B-1) have been so widely destroyed that they probably deserve to be considered the island's most-threatened habitat, despite the better-known plight of the island's forests. Shallow marshes on the continent are subject to similar pressures as deep marshes.

WILDLIFE: The majority of wetland wildlife remains the same across various types of marshes and is addressed under AFROTROPICAL DEEP FRESHWATER MARSH. This section will focus on the wildlife that is more characteristic of shallow marshes.

The Dambos in the MIOMBO zone are very important to local mammal populations as sources of both food and water. Many species range freely between Miombo woodland and Dambo wetland. Hartebeest are especially fond of the edges of shallow marshes. These marsh edges also provide excellent open hunting areas for predators such as Lions and Cheetahs.

Shallow wetlands are of great importance for many birds and provide ideal foraging conditions. Many species nest in the thicker cover provided by deep wetlands, but prefer to forage in shallow wetland. Several secretive species of rails are virtually confined to this habitat. These include Chestnut-headed and Streaky-breasted Flufftails and Striped Crake. Shallow wetland is also the

Above: **Dwarf Bittern is a good example of an intra-African migrant that opportunistically takes advantage of local flooding conditions, specializing in Afrotropical Shallow Freshwater Marsh.** © KEITH BARNES, TROPICAL BIRDING TOURS

Left: **Serval, an elegant predator that often hunts rank grassland or adjacent shallow wetlands.** © HANH DUNG

Yellow-billed Stork uses a wide variety of wetland habitats and is found across most of sub-Saharan Africa, even including Madagascar. © KEITH BARNES, TROPICAL BIRDING TOURS

preferred habitat of the dainty Lesser Jacana and gorgeous Blue Quail. It is readily used by many other skulking aquatic birds, including Dwarf Bittern, Red-chested Flufftail, African Crake, Lesser Moorhen, African Snipe, and Black Coucal. Although it is not as rich in weaver species as deep marshes, it is favored by a variety of weavers, including Yellow-crowned Bishop, Yellow Bishop, and Yellow-mantled Widowbird. During the breeding season, the skies above many shallow wetlands are full of the songs of cisticolas. Several of these small, cryptic birds breed almost exclusively in shallow wetland and give a variety of high-pitched metallic songs while on the wing over their marshy territories. Marsh Owl hunts this habitat and can often be seen perching at dusk on fence posts or other convenient perches. During the rainy season, shallow marshes produce a bounty of food that attracts migrating birds, both Palearctic and intra-African, such as White and Abdim's Storks, respectively.

Shallow wetlands within Madagascar's eastern rainforest zone are the exclusive habitat of Gray Emutail and Slender-billed Flufftail. They are also used frequently by Madagascar Rail, Madagascar Flufftail, Meller's Duck, and Madagascar Snipe.

Endemism: The Dambos of the MIOMBO zone form a minor center of endemism. There is one species (Grauer's Rush-Warbler) endemic to shallow highland marshes along the Albertine Rift.

DISTRIBUTION: Shallow marshes are found locally across the Afrotropics, usually nested within other habitats. They are one of the most common habitats in the MIOMBO zone. They are common within the grass-dominated habitats that cover much of East and Southern Africa and are much more scarce in the forest zone of Central and West Africa. The Highveld of e. South Africa contains an abundance of shallow marsh; these are covered under SOUTH AFRICAN TEMPERATE WETLAND, though they occur in the area of transition from temperate to tropical marsh. The montane forest zone along the eastern escarpment in Madagascar supports a distinctive type of shallow marsh (subhabitat Af11B-1).

WHERE TO SEE: Kasanka National Park, Zambia; Ranomafana National Park, Madagascar.

Af11C NORTH AFRICAN TEMPERATE WETLAND

IN A NUTSHELL: Permanent and ephemeral wetlands similar to those found throughout the colder parts of the world. **Global Habitat Affinities:** AUSTRALASIAN TEMPERATE FRESHWATER WETLAND; NEOTROPICAL TEMPERATE WETLAND; ASIAN TEMPERATE WETLAND. **Continental Habitat Affinities:** SOUTH AFRICAN TEMPERATE WETLAND. **Species Overlap:** FRESHWATER LAKES AND PONDS; RIVERS; TIDAL MUDFLATS AND ESTUARIES; ROCKY SHORELINE.

DESCRIPTION: The broad term "wetland" encompasses a vast and complicated set of microhabitats in which water is the key determinant of the environment. This habitat's vast array of microhabitats have a cohesion and a similar set of wildlife that leads us to treat them as one broader habitat. Some of the microhabitats considered under wetlands include bogs, reed islands and reedbeds, flooded grassland, and marshes. In contrast to the wetlands in tropical or southern temperate Africa, the climate surrounding these wetlands is either a dry Mediterranean (Köppen **Csa**) or desert climate (Köppen **Bsa**). Although this is predominantly a freshwater system, almost all the existing wetlands in North Africa are near the coast and contain some brackish water. Most freshwater wetlands are ephemeral, filling with winter rains and early spring runoff from the Atlas

A wetland teeming with terrapins, Northern Shovelers, Marbled Ducks, a Black-winged Stilt, and a Green Sandpiper. Oued Massa, Morocco. © KEN BEHRENS, TROPICAL BIRDING TOURS

A wetland on Gran Canaria, Spain. © VINCENT LEGRAND

Mountains, then drying out through the summer months. Historically there were massive annual floods down the Nile, so extensive reed and Papyrus wetlands lined the floodplains. In contrast to the Maghreb, the Egyptian wetlands did not only rely on winter rainfall to fill the marshes but also received water at other seasons from other rainy periods in Ethiopia, South Sudan, and Uganda, so many of the wetlands of the Nile Delta retained water throughout the year.

Dayet Aoua Lake is one of the few wetlands within the Atlas Mountains. Because the surrounding forests received orographic rainfall, this wetland is fed by local runoff and snow melt. It is highly altered with almost no surrounding marsh vegetation and also highly prone to drastic changes in water level. From an ecological perspective, it acts more as an artificial dam than a natural lake.

CONSERVATION: Wetlands in this region have been under intense pressure for centuries, from activities such as overexploitation of the water through large-scale irrigation projects, which affects water tables; drainage for other land use; abandonment of grazing, which leads wetlands to become overgrown and choked with other vegetation; and pollution from industrial and agricultural practices. The Maghreb has had dramatic declines in wetlands over the latter half of the 20th century, with almost no extensive wetland protected north of the Sahara, but this pales in comparison to the changes along the Nile, where practically all wetlands of any significance have been turned over to agriculture and the Aswan Dam has stopped the annual flooding of the Nile. The Sahara used to have large marshes associated with oases (pp. 83–86), but these have all been modified by human use and no longer function as wetland environments.

WILDLIFE: Some of the largest concentrations and variety of birdlife in North Africa can be seen in wetland areas, which provide breeding sites for some species, wintering sites for others, and critical stopover sites for other species. In the deeper open waters, diving ducks, pelicans, loons, and grebes can be found. Shallow waters at the edges support dabbling ducks, swans, and geese. Muddy edges may be inhabited by spoonbills, shorebirds, bitterns, crakes, and rails.

The ephemeral nature of the Maghreb wetlands suits the timing of wintering migrants well. Wetland areas attract large numbers of wintering birds, such as ducks, shorebirds, and geese, which winter here but breed in Northern Europe and Western Asia. But they also hold large

FRESHWATER HABITATS

Right: **Zitting Cisticola**, the "default" cisticola on a continent where this family reaches incredible diversity— there are 128 African members— making this the continent's most diverse bird family.
© KEN BEHRENS, TROPICAL BIRDING TOURS

North African wetlands support the farthest westerly population of the widespread **Ruddy Shelduck**. © KEN BEHRENS, TROPICAL BIRDING TOURS

colonies of breeding waterbirds. This bird migration means that the composition of species and the numbers of individuals are often in flux within these dynamic microhabitats. In general, the wetland birds of any area do not really relate to the surrounding vegetation types, so the same overall assemblage of bird species can be found in different North African Temperate Wetlands, regardless of whether they are surrounded by desert, steppe, or deciduous forest.

Many of the wetlands in the region are on important flyways, such as the East Atlantic Flyway, which links migratory birds from nw. Europe with the west coast of Africa, in particular Mauritania. Eastern Africa has an equivalent, and even today, millions of birds migrate along the remnant river edge marsh vegetation of the Nile on their migration between Africa and Eurasia.

Spectacular aggregations often occur when birds gather to prepare to migrate or at large-scale winter feeding areas. Such concentrations of birds can include shorebirds (stilts, snipes, sandpipers, avocets, plovers), pelicans, ducks, geese, cranes, rails, gulls, terns, swans, grebes, storks, ibises, herons, egrets, and cormorants.

DISTRIBUTION: These wetlands used to dot the North African coast and, as recently as 5700 YBP (see sidebar 7, p. 85), the Sahara held the largest lakes in the world. Wetlands are now mainly found along the Nile and in a few coastal reserves, as this habitat has been converted to managed lakes and dams. Most of the oasis marshes have been drained, while a few mountain wetlands remain in the Atlas and other highlands. The Macaronesian Islands hold some wetlands.

WHERE TO SEE: Along the edge of the Nile River, Egypt; Al Qattamiya Wetland, Egypt; Dayet Aoua Lake, Morocco; Sidi Moussa-Oualidia Wetland, Morocco.

Eurasian Wigeon migrates from n. Eurasia to spend the winter in North African Temperate Wetlands, among other places. © KEITH BARNES, TROPICAL BIRDING TOURS

SIDEBAR 40 — HABITAT ARCHITECTS

There are three African animals that are capable of completely transforming habitats. The first is obvious—after all it is the largest terrestrial mammal on the planet. Elephants are big, and they throw their weight around, literally. They are capable of felling large trees with a pair of tusks, some bulk, and a little belligerence. Herds of elephants can turn dense woodland into open savanna surprisingly quickly. Savannas have evolved in sync with elephants and often "need" their bolshiness to maintain maximum species diversity. Elephants open up savannas, which keeps fires more regular and grass and herbs a key component. This maintenance also allows in mammalian grazers and keeps the whole system ticking over.

The other two architect animals are less obvious, the first because it does its job where it cannot easily be seen. By day Hippopotamuses build water channels throughout their aquatic habitat, creating corridors for other aquatic animals to move. Their feces alter the chemical composition of the water, enriching it and maintaining high fish diversity. Owing to their sensitive skin, which is easily sunburnt, hippos emerge from the water only after dark. They range widely, swinging their heads back and forth close to the ground, cropping vegetation closely with their lips. They transform thick riverine grassy fringes into cropped grazing lawns, leaving woody vegetation nearby untouched. Whales and dolphins are their closest living relatives, but there were several other hippo species that went extinct as recently as the last ice age. When they come into contact with humans, hippos are not well liked, as they are the most dangerous mammal in Africa. But their vital role in keeping riverine fisheries active means that people would be well served to work a compromise to keep them around.

The final habitat architect seems to be too small to matter, the humble termite (*Odontotermes*). There is hardly any soil in Africa that has not been shaped by termites, especially in the savannas, and some mounds are centuries old. Termites aerate the soil, increasing water penetration. Soil aeration, together with food and droppings accumulation, creates patches of elevated fertility, rich in nitrogen and phosphorus, encouraging growth of grasses, shrubs, and trees. Trees grow better and bear more fruit near mounds, attracting more insects, which bring more lizards. At the same time, improved grass growth is a magnet for grazers such as zebra and African Buffalo, which fertilize the area further with dung! It is an endless positive feedback cycle. Termites also space their mounds evenly, 65–400 ft. (20–120 m) apart in a "polka dot" pattern. Computer models suggest that this pattern perfectly utilizes the savanna's resources, attracting more plants, insects, and reptile communities than randomly spaced mounds. After rains, winged termite alates can swarm spectacularly, providing a glut of food for everything from impalas to eagles and people.

Hippos are perfectly adapted to water. Only ears, eyes, and nostrils break the water's surface while 99% of the animal remains submerged. © KEITH BARNES, TROPICAL BIRDING TOURS

Af11D SOUTH AFRICAN TEMPERATE WETLAND

IN A NUTSHELL: Habitat that is permanently waterlogged or temporarily flooded with freshwater. Vegetated by temperate hydrophilic reeds, rushes, and sedges, often with peat formations, including high-elevation mires. **Global Habitat Affinities:** EUROPEAN REEDBEDS; AUSTRALASIAN TEMPERATE FRESHWATER WETLAND; NEOTROPICAL TEMPERATE WETLAND; ASIAN TEMPERATE WETLAND. **Continental Habitat Affinities:** NORTH AFRICAN TEMPERATE WETLAND; AFROTROPICAL DEEP FRESHWATER MARSH; AFROTROPICAL SHALLOW FRESHWATER MARSH. **Species Overlap:** AFROTROPICAL DEEP FRESHWATER MARSH; AFROTROPICAL SHALLOW FRESHWATER MARSH; FRESHWATER LAKES AND PONDS; RIVERS; TIDAL MUDFLATS AND ESTUARIES.

DESCRIPTION: Various types of wetlands, which are covered separately in this book, occur in close association. These include RIVERS, FRESHWATER LAKES AND PONDS, AFROTROPICAL DEEP FRESHWATER MARSH; AFROTROPICAL SHALLOW FRESHWATER MARSH; seasonally flooded AFROTROPICAL GRASSLAND, and this habitat. Vleis (subhabitat Af11D-1) are catchment areas, fringes or floodplains of streams and lakes where the soil remains moist without being eroded by flowing

South African Shelduck in a classic South African Temperate Wetland, fringed by *Bulbinella*.
© KEITH BARNES, TROPICAL BIRDING TOURS

water. In areas where some seasonal flooding occurs, austral winter (May–Aug) in the Cape and austral summer (Nov–Mar) elsewhere, wetlands may have temporarily or permanently flooded restio and sedge marshes. Where water is more permanent, rushbeds and reedbeds include Common Reed (*Phragmites australis*), Clustered Flat-sedge (*Cyperus congestus*), and Sea Clubrush (*Bolboschoenus maritimus*)—a widespread temperate rush that is largely absent from tropical Africa but reappears in Europe—and Bulrush (*Typha capensis*). Small sedges include genera *Carex* and *Cyperus*, and there are myriad graminoids (plants with a grass-like morphology); on the fringes of the moist soil, a handful of geophytic bulbs and herbs, including orchids, geraniums, and the spectacular red-hot pokers (*Kniphofia*) occur. Owing to salt leaching or proximity to the coast, some wetlands may have a brackish character.

Lesotho has two types of high-elevation mires (subhabitat Af11D-2) above 9000 ft. (2750 m): bogs on cooler, steeper, and moister slopes with highly organic soils; and larger fens on warmer, gentler, drier slopes with mineral soils. Both are dominated by short sedges and grasses and the occasional moss—which is not as prevalent as in the Northern Hemisphere. The main sedge is *Isolepis angelica*; rushes include *Juncus inflexus* and *J. oxycarpus*; other plants include rare endemic daisies like *Cotula paludosa* and grasses like *Merxmuellera disticha*. Bogs are permanently wet and fens mostly have standing water only in the austral summer. In both, small hummocks and tarns (pools) occur; neither have an obvious

Reedbeds in a typical South African Temperate Wetland in Knysna, South Africa. © KEITH BARNES, TROPICAL BIRDING TOURS

Exotic Australian wattles (*Acacia*) choke wetlands and soak up precious freshwater, a major conservation challenge in the Cape Provinces of South Africa. © KEITH BARNES, TROPICAL BIRDING TOURS

stream channel, and they both form peat (partially decayed organic matter)—a feature of precipitation-rich cold climates.

South African Temperate Wetlands differ most markedly from those farther north by what they lack. There is very little Papyrus and a vastly less diverse community of wetland plants, and although there are aquatic herbs, the large floating mats of water lilies (*Nymphaea*) so typical of tropical systems are absent. These wetlands occur where mean temperatures span 48–64°F (9–18°C), but in winter the temperature often drops below freezing and frost is frequent at higher elevations. Rainfall ranges around 20–47 in. (500–1200 mm).

CONSERVATION: There are some significant protected areas of this habitat, including Agulhas, Wilderness, uKhahlamba (South Africa), and Sehlabathebe (Lesotho) National Parks. In addition, there are numerous bird sanctuaries and local reserves and several Ramsar Convention sites. Introduced Australian wattles (*Acacia*) threatens these wetlands, occupying habitat and sucking up perennial water. Lesotho mires are heavily affected by stock trampling and overgrazing, which degrade wetlands and result in erosion.

WILDLIFE: Most of the big mammals that would have used these wetlands were driven locally extinct after European hunters decimated them, but some have been locally reintroduced, like Hippopotamus. Among the smaller mammals Cape Clawless and Spotted-necked Otters are still common, if rarely seen, and there are many small mammals, like vlei rats (*Otomus*). Wetland birds are mobile and cosmopolitan, and those using this habitat are mostly a subset of Afrotropical wetland birds save for some important groups that are rare or absent: birds associated with floating mats of vegetation, such as jacanas and pygmy-goose, and wetland nomads, such as Dwarf Bittern, Rufous-bellied Heron, Allen's Gallinule, Striped Crake, and Streaky-breasted Flufftail.

Africa's frightfully tricky flufftails, which look like crakes but are more closely related to finfoots, like using this habitat. This is prime habitat of one of Africa's most enigmatic birds, the Critically Endangered White-winged Flufftail. Long thought to be an intra-African migrant to Ethiopia, it is now known to breed in South Africa, where it is restricted to a handful of sites and is hardly ever seen. Red-chested and Striped Flufftails will also use this habitat, although they are also widespread outside it. These wetlands are also good for some rare breeding species, including the *capensis* subspecies of Great Bittern in taller reedbeds and Wattled Crane in more open marshland. Black Storks fly from nearby cliffs to forage in these wetlands. Although ducks are mobile across the continent, the mix in this habitat tends to be unique, with Cape Shoveler, Maccoa Duck, and South African Shelduck often present. African Rail and Black and Baillon's Crake patrol the marsh edges along with other common Rallidae. Among the shorebirds, common residents include African Snipe, Black-winged Stilt, and Pied Avocet in shallow water; Three-banded Plover and Blacksmith and African

Otters like Cape Clawless Otter remain common, though shy and elusive, in South African Temperate Wetlands. © ANGELA KEY

Af11D SOUTH AFRICAN TEMPERATE WETLAND

African Quailfinch visits wetland verges to drink, which is when you have the best chances to see this shy and cryptic member of the waxbill family. © KEN BEHRENS, TROPICAL BIRDING TOURS

Wattled Lapwings on the shores; and Palearctic breeding shorebirds that abound in the austral summer (Sep-Mar), including Ruff, Little Stint, Common Greenshank, and Marsh and Wood Sandpipers. Overhead White-winged and Whiskered terns—the so-called marsh terns—hunt. Pied and Malachite Kingfishers are frequent hunters of small fish. Flooded fringes often have Cape Wagtail and the bright Orange-throated Longclaw patrolling and may attract drinking African Quailfinch. Reedbeds are haunted by the dour Little Rush Warbler, Lesser Swamp Warbler, and migrant Marsh Warbler. In the summer the reedbeds are transformed by scarlet, golden, and velvety-black breeding plumage of Yellow, Yellow-crowned, and Southern Red Bishops and, in the east, Fan-tailed and Red-collared Widowbirds along with nesting Cape Weaver and Southern Masked-Weaver.

Many frogs are found in these wetlands, including a few spectacular endemics such as Long-toed Tree Frog (*Leptopelis xenodactylus*), Western Leopard Toad (*Sclerophryspantherina*), Cape Platanna

Above: **Pied Kingfisher is one of the world's most widespread kingfishers, and readily uses both tropical and temperate wetlands in sub-Saharan Africa.** © KEN BEHRENS, TROPICAL BIRDING TOURS

Right: **Male bishops like this Southern Red Bishop radically transform in the breeding season, from brown streaky "LBJs" into velvet-and-scarlet beauties.** © KEN BEHRENS, TROPICAL BIRDING TOURS

(*Xenopus gilli*), Montane Marsh Frog (*Poyntonia paludicola*), and the stunning but diminutive 0.6 in. (2 cm) long and Critically Endangered Micro Frog (*Microbatrachella capensis*).

Endemism: Most wetland birds are widespread. A variety of amphibians are endemic to these wetlands as are several plants.

DISTRIBUTION: From sea level around Cape Town to Port Elizabeth and then gradually higher in the mountains of eastern South Africa and Lesotho, reaching 8200 ft. (2500 m) in the Highveld as the latitudes become subtropical. In the north their limit stretches as far as the upper plateau of the Leolo Mountains south of Polokwane. This habitat grades into AFROTROPICAL DEEP FRESHWATER MARSH (see p. 358) and AFROTROPICAL SHALLOW FRESHWATER MARSH in a mélange along the coastal plain of eastern South Africa north of Port Elizabeth and in northern Gauteng, where the South African escarpment starts to descend into the subtropics north of Tshwane (Pretoria). South African Temperate Wetlands usually occur as a mosaic with other habitats, ranging from FYNBOS and RENOSTERVELD to MONTANE GRASSLAND.

WHERE TO SEE: Wilderness National Park, South Africa; Sehlabathebe National Park, Lesotho.

SIDEBAR 41 — INTRA-AFRICAN MIGRANTS

Long-distance Palearctic migrants are the classic and well-known migrants in Africa (see sidebar 6, p. 76). But Africa also has a significant but poorly studied amount of both nomadism and internal bird migration. The intra-African migrants include some herons, storks, raptors, quail, flufftails, crakes and rails, bustards, pratincoles, plovers, cuckoos, nightjars, bee-eaters, swallows, pipits, and weavers, plus African Golden-Oriole and African Paradise-Flycatcher; many dozens of species in total. Most of these birds fall into one of two categories. The first is waterbirds that are seeking optimal water levels and feeding conditions and move around the continent taking advantage of the regional timing of rainy seasons or of unusually wet conditions that happen unpredictably. Oftentimes, within days of an unusually heavy rain, migrant waterbirds begin appearing, begging the still largely unanswered question of how they knew to migrate! The second category of intra-African migrants are mostly species that breed during summer in Southern Africa, then migrate to more equatorial parts of the continent during the frigid austral winter. Many species in this category are resident in equatorial regions but become migrants in the south of their range.

Brown-chested Lapwing has an odd east–west migration pattern across the continent, unusual for an intra-African migrant. © KEITH BARNES, TROPICAL BIRDING TOURS

Af11E FRESHWATER LAKES AND PONDS

IN A NUTSHELL: A freshwater habitat with permanent still water that is too deep for the growth of emergent vegetation. **Global Habitat Affinities:** FRESHWATER LAKES, DAMS, AND PONDS (European); NEARCTIC OPEN WATER. **Species Overlap:** RIVERS; AFROTROPICAL DEEP FRESHWATER MARSH; AFROTROPICAL SHALLOW FRESHWATER MARSH; SOUTH AFRICAN TEMPERATE WETLAND; NORTH AFRICAN TEMPERATE WETLAND; SALT PANS AND LAKES; TIDAL MUDFLATS AND ESTUARIES.

DESCRIPTION: Lakes and ponds are important components of wetland systems. They store large amounts of precious fresh water and provide habitat for some specialized terrestrial wildlife plus a rich set of aquatic species. Africa's Great Lakes hold 25% of the world's unfrozen surface fresh water; a remarkable wealth of the element central to life.

Natural lakes vary a great deal, from shallow, murky, and nutrient rich to deep, clear, and nutrient poor. Ponds are far more common now than they would have been naturally because of the human construction of dams, especially to hold water in arid areas, such as the Karoo, Horn of Africa, and Sahel. Some lakes, such as the deep Rift Valley lakes, change little throughout the year, whereas shallow lakes in more arid country expand and contract dramatically when they receive rainy season inflow. Vast areas can be transformed from grassland or savanna into functionally being parts of a huge lake, until the waters recede again. Saline lakes look superficially similar but function very differently and are covered under SALT PANS AND LAKES.

Most lakes and ponds are associated with rivers and fringed by wetlands. The typical pattern is that the shoreline is lined with deep reedbed marsh, comprising Papyrus, *Phragmites*, or both. Shallower lakes may be bordered by shorter sedge and grass marsh. Ponds are often vegetated with the same floating vegetation described under AFROTROPICAL DEEP FRESHWATER MARSH. Some benthic (growing on the lakebed), nonemergent, aquatic vegetation does grow in areas too deep for emergent vegetation but receiving enough light to allow plant growth. Depending on

Shorebirds, ducks, flamingoes, herons, and terns teeming at a lake in the Great Rift Valley of Ethiopia.
© KEN BEHRENS, TROPICAL BIRDING TOURS

Small ponds in savanna environments act as magnets for mammals, like these Plains Zebra in Tarangire National Park, Tanzania. The surrounding habitat is Moist Mixed Savanna. © KEN BEHRENS, TROPICAL BIRDING TOURS

water clarity this zone is generally 3–12 ft. (1–4 m) deep. Many types of algae are also present in open water. In the forest zone of Central and West Africa, lakes may be bordered directly by **SWAMP FOREST** without an intervening layer of marsh. Those various types of marginal wetlands are addressed elsewhere; this account primarily covers the surface of open water. Some lakes also have muddy shorelines, forming the inland equivalent of the bird-rich coastal **TIDAL MUDFLATS AND ESTUARIES**.

CONSERVATION: Throughout the Afrotropics, the water in lakes can be overutilized, leading them to slowly shrink. The opposite sort of problem can also occur, where owing to human causes, silt accumulates in the bottom of lakes, pushing their water levels up and flooding the previous shoreline and adjacent habitats. This can be seen in some of Kenya's Rift Valley lakes, most famously Lake Nakuru, which was abandoned for several years by its famous flamingo population owing to the lack of appropriate open shoreline habitat. One of the continent's greatest conservation tragedies occurred in the 1950s when the Nile Perch was introduced into Lake Victoria, resulting in the extinction or near extinction of hundreds of endemic cichlids. Although it is a growing concern, Africa hasn't experienced as many problems with fertilizer and sewage runoff, and subsequent algal blooms, as many other parts of the world.

WILDLIFE: Wetlands as a whole are incredibly important wildlife habitat. Their general role as such is discussed in **AFROTROPICAL DEEP FRESHWATER MARSH**. This section will focus on wildlife that uses areas of open water and mudflats, as opposed to adjacent marshes and swamps. In terms of megafauna, lakes and ponds are favored by Hippopotamus, African Savanna Elephant (a capable swimmer!), and Nile Crocodile. They are also ideal habitat for several species of otters, Nile Monitor (*Varanus niloticus*), terrapins, and many species of snakes and frogs. Quite a few birds will feed in areas of open water. Some species float on the open water and either dabble or dive into the water

Af11E FRESHWATER LAKES AND PONDS

to feed. Classic examples include Little and Great Crested Grebes, Great White and Pink-backed Pelicans, Great and Long-tailed Cormorants, a variety of ducks, Spur-winged Goose, and Red-knobbed Coot. Other species tend to hover or swoop over the water in order to feed. Good examples are Gray-headed Gull and White-winged, Whiskered, and Gull-billed Terns. Pied Kingfishers hover high, then plunge into the water to pluck fish. Muddy shorelines are especially bird rich. They are generally less productive than coastal mudflats but can still have very high densities of birds, including a variety of shorebirds, storks, herons, and egrets. African Quailfinch is an odd and beautiful little short-tailed member of the waxbill family that often leaves drier habitats to drink at the muddy edges of lakes and ponds. Some long-legged birds will readily hunt along the edges of lakes and ponds, even in areas where they are not marshy or muddy, with Gray Heron a classic example.

African Fish-Eagle, one of the continent's most iconic and best-known bird species. © KEN BEHRENS, TROPICAL BIRDING TOURS

Although this book does not focus much on aquatic wildlife, this is where the real richness of lakes and ponds is to be found. There is an extraordinary assemblage of fishes and mollusks in Africa's Great Lakes. Lake Tanganyika alone has around 250 species of cichlids and 100 snails and bivalves, nearly all of which are endemic. Overall, the Great Lakes boast around 10% of the world's fish species. Its richest group is a massive radiation of cichlids, totaling around 1500 species.

DISTRIBUTION: Africa is gifted with a remarkable set of lakes. The grandest are all part of the African Great Lakes region, between northern Kenya and Malawi. Most of the Great Lakes lie in the Great Rift Valley, which in East Africa splits into the w. Albertine Rift and the Eastern Rift. The largest lake of all, however, is situated between these two rifts. The huge but shallow Lake Victoria is the world's second-largest lake by surface area, only bested by North America's Lake Superior. Lake Tanganyika is the world's longest freshwater lake and is also notable as the second-deepest, second-oldest, and second-most voluminous (after Russia's Lake Baikal). Africa's other largest lakes, in order of decreasing size, are Nyassa, Turkana, Albert, Mweru, Kivu, and Edward, all part of the Great Lakes system. West Africa's Lake Chad was once enormous, likely among the world's largest lakes, but has shrunk to a much smaller size due to the drying of the local climate in combination with human pressures. Smaller lakes and ponds are found throughout Africa. Like other wetlands, they can act as attractive oases for both humans and wildlife in dry areas.

WHERE TO SEE: Lake Victoria, Uganda, Kenya, and Tanzania; Lake Naivasha, Kenya; Lake Awassa, Ethiopia.

Terrapins like Serrated Hinged Terrapin (Pelusios sinuatus) are at home in African lakes and rivers, where they can often be seen basking on the banks, or sometimes even on the back of a Hippo! © MARIUS BURGER

Af11F RIVERS

IN A NUTSHELL: A freshwater habitat with running water that is mostly unvegetated. **Global Habitat Affinities:** AUSTRALASIAN SANDY RIVERBEDS; NEOTROPICAL FLOWING RIVERS; FLOWING RIVERS; LOWLAND RIVERS; UPLAND RIVERS. **Continental Habitat Affinities:** FRESHWATER LAKES AND PONDS. **Species Overlap:** FRESHWATER LAKES AND PONDS; AFROTROPICAL DEEP FRESHWATER MARSH; AFROTROPICAL SHALLOW FRESHWATER MARSH; SOUTH AFRICAN TEMPERATE WETLAND; NORTH AFRICAN TEMPERATE WETLAND; TIDAL MUDFLATS AND ESTUARIES; SALT MARSH.

DESCRIPTION: Rivers are intimately connected with Africa's other wetlands. In particular, they occur in close association with FRESHWATER LAKES AND PONDS and AFROTROPICAL DEEP FRESHWATER MARSH. Slow-moving rivers often have Afrotropical Deep Freshwater Marsh along their banks, while in the rainforest zone of Central and West Africa, rivers are typically fringed by SWAMP FOREST. In much of the MIOMBO zone, and along the Indian Ocean coast, rivers are fringed with various semi-evergreen forests (MONSOON FOREST, SOUTH COAST FOREST MATRIX, or EAST COAST FOREST MATRIX). In general, the presence of rivers allows the linear penetration of drier habitats by moister habitat (see Fig. 4, p. 13). As examples, a river running through w. Somalia, through a semi-desert thornscrub environment, will be lined with MOIST MIXED SAVANNA, and a river running through the GUINEA SAVANNA of West Africa will be lined with a lush forest resembling AFROTROPICAL LOWLAND RAINFOREST. So rivers serve to add a great deal of diversity and complexity to the landscape. That aside, this account mainly treats the open water portion of rivers rather than the

Hamerkop is an African and Malagasy near-endemic monotypic family that barely sneaks into the s. Arabian Peninsula. It is an amphibian-eating specialist that uses a variety of wetlands, especially rivers and streams. Its nest *(inset)* is extraordinarily huge, proportionately one of the largest among birds! MAIN PHOTO © KEITH BARNES; INSET © KEN BEHRENS, TROPICAL BIRDING TOURS

SIDEBAR 42 — IN DENIAL: AN AFRICAN CORRIDOR TO THE MED

The Nile is one of the longest rivers in the world, at 4130 mi. (6500 km). It starts at Lake Victoria in the sweltering Afrotropics, surrounded by lowland rainforest in Jinja, Uganda, and spills into the Mediterranean Sea in Egypt, having crossed one of the driest deserts on Earth. The Nile offers a pathway for African species to penetrate and extend from the tropics to a Mediterranean climate and spill over into the Levant and Arabia. Several mammal species atypical of the surrounding deserts can thus be found along the Nile and its alluvial plains, especially in the southern stretches of the Sahara, including Savanna Path Shrew, Lesser Leaf-nosed Bat, and Honey Badger. There are only two main north–south corridors through the Sahara—the nw. African coast and the Nile Valley—and as these become drier with climate change, Mediterranean North Africa and sub-Saharan Africa will become increasingly isolated.

---

associated riparian habitats. It is also restricted to flowing water significant enough to characterize as a river; for the purposes of this book, streams and smaller areas of flowing water are viewed as microhabitats of the habitat through which they flow.

Africa is mostly characterized by slow-flowing rivers. It lacks huge mountain ranges like the Himalayas and Andes and their associated fast-flowing rocky torrents and cascades. Africa's rivers tend to have a winding course, changing slowly over time, periodically leaving oxbow lakes along their former course. The Nile River of the northeast was historically regarded as the world's longest river, but that title may ultimately go to the Amazon. Trivia aside, it is an

Above: **The Ogooue River carves its way through Lope National Park in Gabon. Rock Pratincole and other riverine specialists like the rocky outcrops. Note the lowland forest-grassland mosaic on the nearby hills.** © CHRISTIAN BOIX, AFRICA GEOGRAPHIC

Right: **Africa is strangely lacking in the waterfall-associated bird species found on other continents. Calandula Falls, Angola.** © KEN BEHRENS, PROMISED LAND VENTURES

impressive 4130 mi. (6650 km) long! The Congo River is somewhat shorter, at 2922 mi. (4700 km), but is the third largest river in terms of discharge volume, after the Amazon and Ganges, and the second largest in terms of drainage area. A little-known fact is that the Congo is the world's deepest known river, with a maximum depth of 720 ft. (219.5 m). Other major rivers, in order of decreasing length, are the Niger, Zambezi, Orange, Limpopo, Senegal, Kavango, and Volta. The world-renowned Victoria Falls, on the Zambezi River, is one of the largest falls on Earth, over a mile (1.6 km) wide. Muddy banks, sandbanks, and emergent rocks are all important riverine microhabitats. Unlike other continents, Africa lacks species closely associated with waterfalls.

CONSERVATION: Rivers easily collect pollutants, as human contaminants naturally flow downhill, and water is a perfect conductor. Invasive plants such as Common Water Hyacinth (*Pontederia crassipes*), an Amazonian species, can clog slow-moving rivers and spread easily.

WILDLIFE: For an overview of wetland wildlife see **AFROTROPICAL DEEP FRESHWATER MARSH**. Rivers are a major part of the landscape in most of Africa's premier mammal-watching locations, such as the Serengeti, Masai Mara, and Kruger National Park. They are magnets for big mammals, which visit for both food and water. One of the world's great wildlife spectacles occurs on the Mara River of Kenya and Tanzania, when huge herds of Blue Wildebeest and Plains Zebra cross the river during their annual migration. Most swim across and survive, but many drown in panicked stampedes, and others fall prey to the massive Nile Crocodiles that lurk at crossing points. In general, rivers are great places to stop and scan for both mammals and birds. They not only attract wildlife but also allow good visibility, especially in more closed habitats.

Pod of Hippopotamuses in Serengeti National Park, Tanzania. Rivers provide daytime refuge for easily sunburned Hippos, which leave the river and range widely to feed at night.
© KEN BEHRENS, TROPICAL BIRDING TOURS

Egyptian Plover, a monotypic endemic African family that prefers sandy riverbanks and islands. © KEN BEHRENS, TROPICAL BIRDING TOURS

Southern Carmine Bee-eaters nest colonially in riverbanks, one of the most incredible birding spectacles on the continent. Kavango River, Namibia. © KEN BEHRENS, TROPICAL BIRDING TOURS

Open rivers are not nearly as rich in birds as the wetlands associated with them, but they do provide habitat for quite a few species, including some riverine specialists. Rivers through forested environments are the preferred habitat of the extremely elusive African Finfoot, as well as African Black Duck, Half-collared Kingfisher, and Cassin's Flycatcher. More open rivers hold Hamerkop, Saddle-billed Stork, thick-knees, a variety of boldly-colored and aggressive lapwings, Giant and Pied Kingfishers, and Madagascar and African Pied Wagtails. Exposed river rocks are the breeding microhabitat of Rock Pratincole in Africa and Madagascar Pratincole in Madagascar. Sandy river islands are the domain of one of Africa's most sought-after birds, the Egyptian Plover. This is a strikingly handsome shorebird with a black collar and saddle, blue-gray back, white eyebrow, and peachy throat and belly. Part of its appeal for birders is that it makes up its own monotypic (single species) bird family. Gray Pratincole and African Skimmer also nest on sandy river islands in loose colonies. African River Martin nests in sandy riverbanks along the Congo and Ubangi Rivers, as well as along the adjacent coast. Bee-eaters also key into vertical riverbank habitat, which is perfect for their nests: a very long entrance tunnel and a nesting chamber deep inside the soil. Bee-eaters that nest along rivers include Rosy, Northern Carmine, Southern Carmine, and White-fronted. Though Africa almost completely lacks the large suite of river island specialists found in the Amazon basin, a handful of species, such as Gosling's Apalis, prefer riverside scrub and forest.

DISTRIBUTION: Rivers, both large and small, are found throughout Africa and Madagascar, except for the true deserts: the Sahara and the Namib. However, even in the arid Namib, the ephemeral Kuiseb River, which only flows once in a blue moon, plays a key role in keeping its two main habitats NAMIB ROCK DESERT and NAMIB SAND DESERT distinct (see pp. 77–93). Although the waters seldom flow, the Western Riparian Woodlands (subhabitat Af 2–1) along the river margins form a key subhabitat in the desert.

WHERE TO SEE: Murchison Falls National Park, Uganda; Zambezi River at Victoria Falls, Zambia / Zimbabwe.

SALINE HABITATS

Af12A AFROTROPICAL MANGROVE

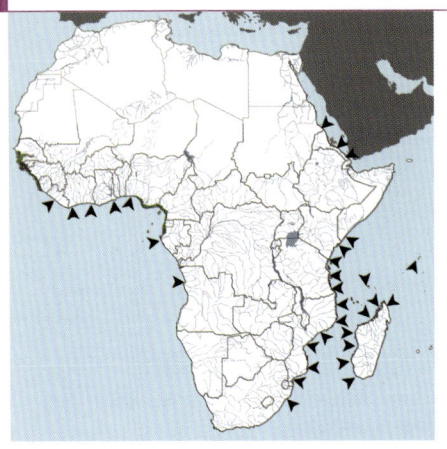

IN A NUTSHELL: Forest of salt-tolerant trees that grows coastally in the intertidal zone. **Global Habitat Affinities:** AUSTRALASIAN MANGROVE FOREST; NEOTROPICAL MANGROVES; INDO-MALAYAN MANGROVE FOREST. **Continental Habitat Affinities:** SWAMP FOREST. **Species Overlap:** SWAMP FOREST; AFROTROPICAL LOWLAND RAINFOREST; EAST COAST FOREST MATRIX; SOUTH COAST FOREST MATRIX; TIDAL MUDFLATS AND ESTUARIES; SALT MARSH; SANDY BEACH AND DUNES.

DESCRIPTION: Mangroves are trees that have a high tolerance for salt, allowing them to grow in the intertidal zone along the seashore and on brackish estuaries. Heavy surf prevents the growth of mangrove seedlings, so this habitat is confined to more sheltered areas, such as estuaries, lee shores of islands and sandbars, and the interior of barrier reefs. Although the species composition of mangrove forest varies throughout the world, its structure and function remain much the same. Among the biogeographic realms, the Afrotropics holds the third-most extensive mangrove habitat, after Australasia and Indo-Malaysia. Australian and Indo-Malaysian mangroves hold a diverse and specialized set of wildlife, whereas Afrotropical Mangrove does not, for reasons that are somewhat mysterious.

Trees that qualify in the broad sense as mangroves are found in many different families. The pressures of living in an exacting environment have forced these unrelated trees to evolve a set of similar adaptations, an example of convergent evolution. One is the ability to survive in low-oxygen soil. Some species have stilt roots to lift them out of the water, and others push "breathing roots" (pneumatophores) out of the mud to give them access to oxygen. Another mangrove adaptation is thick and leathery leaves to limit water loss.

Botanically, the East African, Arabian Peninsula, and Indian Ocean Island mangroves are similar to those elsewhere around the Indian Ocean (subhabitat Af12A-1) and are completely different from the mangroves of West Africa (subhabitat Af12A-2), which are closely allied to those along the Atlantic coast of the Americas. The most common and widespread species in the east are Red Mangrove (*Rhizophora mucronata*), Black Mangrove (*Bruguiera gymnorrhiza*), Indian Mangrove (*Ceriops tagal*), Cannonball Mangrove (*Xylocarpus granatum*), Gray Mangrove (*Avicennia marina*), Apple Mangrove (*Sonneratia alba*), and White-flowered Black Mangrove (*Lumnitzera racemosa*). Most of these species are found all the way to the western Pacific. In the poorer west, there are only six major

Malachite Kingfisher in mangroves. Many mangrove species have stilt roots. Cuanza River mouth, Angola.
© KEN BEHRENS, PROMISED LAND VENTURES

species: three known as Red Mangrove (*Rhizophora racemosa*, *R. harrisonii*, and *R. mangle*), Black Mangrove (*Avicennia germinans*), White Mangrove (*Laguncularia racemosa*), and the introduced but quickly spreading Nipa Palm (*Nypa fruticans*). Note that the same common name is used for multiple species of mangroves, some from both the Indian and Atlantic, meaning that common names are often not helpful when discussing mangrove trees! Three factors determine the exact mix of mangrove

Below: **Mangroves in Madagascar sometimes support the endemic White-throated Rail. Cap d'Ambre, Madagascar.**
© KEN BEHRENS, TROPICAL BIRDING TOURS

species in any single location: (1) soil, whether sand or clay; (2) duration and frequency of flooding by sea water; and (3) salinity of the water, whether brackish or pure seawater.

Typical mangrove forest is around 30 ft. (10 m) tall, though well-developed stands can reach 150 ft. (45 m). The trees are normally stunted along the shoreline but become taller inland, where they give way to SWAMP FOREST, AFROTROPICAL LOWLAND RAINFOREST, EAST COAST FOREST MATRIX, SOUTH COAST FOREST MATRIX, or savanna habitat. This azonal habitat is not directly dependent on rain for its sustenance, so mangroves grow in areas with a wide range of precipitation. However, areas without either sufficient rainfall or fresh groundwater, such as the coast of the Namib Desert, lack mangroves.

West African mangroves can grow quite tall and often transition into lowland rainforest on their inland side. Amansuri Conservation Area, Ghana. © KEITH BARNES, TROPICAL BIRDING TOURS

CONSERVATION: Human populations tend to be dense along the coast, heavily affecting adjacent mangroves. Although this habitat may seem inaccessible at first glance, mangroves are cut for firewood, charcoal, and building material. Areas of mangroves are often converted into shrimp farms, saltworks, or rice paddies. Their coastal location also makes them susceptible to oil spills and river pollution. Some mangroves can store ten times more carbon than terrestrial forests, making them a vital conservation priority.

WILDLIFE: Considering how long mangroves have been present in the Afrotropics, this is a remarkably poor habitat for terrestrial birds and mammals compared with the mangroves of the Indo-Malayan or Australasian regions. For the most part, mangroves host a much-reduced subset of the species of adjacent habitats. Of course, mangroves are crucial to marine environments, harboring a huge and diverse array of species, and are especially crucial as a nursery for juvenile fish. This habitat is used by charismatic aquatic beasts like West African Manatee, Dugong, Nile Crocodile, and marine turtles.

Although virtually no species are restricted to this habitat, the assemblages of wildlife in West Africa, East Africa, the Malagasy region, and around the Red and Arabian Seas are all quite different.

West African mangrove is botanically poorer but richer in wildlife, perhaps because it occurs alongside freshwater SWAMP FOREST and AFROTROPICAL LOWLAND RAINFOREST. The muddy substrate covered in dense roots seems to preclude the presence of most ground-dwelling mammals, but the canopy holds primates including Western Red Colobus, Angolan and Gabon Talapoins, Sooty Mangabey, and Mona and Campbell's Monkeys. A fair number of West African rainforest birds will use mangroves, and this is the primary habitat for Mouse-brown (IS) and Carmelite Sunbirds.

The mangroves along the Indian Ocean coast are much poorer in terms of terrestrial wildlife. The one bird strongly associated with this habitat is Mangrove Kingfisher. Around the Red and

Mangroves are almost always associated with tidal mudflats. Belalands Flats, Madagascar. © KEN BEHRENS, TROPICAL BIRDING TOURS

Arabian Seas, Collared Kingfisher and Clamorous Reed Warbler (IS) are found almost exclusively in mangroves. In Madagascar, several endemic species such as White-throated Rail, Madagascar Swamp Warbler, and Malagasy Kingfisher use this habitat. It is also important for the Critically Endangered Madagascar Fish-Eagle and provides daytime refuge for Madagascar Flying Fox.

DISTRIBUTION: Mangrove forest is found along coasts in warm tropical and subtropical regions. It is almost universally found alongside TIDAL MUDFLATS AND ESTUARIES, and in many places, especially the drier parts of the mangrove zone, there is also SALT PAN on the inland side of the mangroves—an area that floods then evaporates only occasionally, giving the soil a very high saline content. In the Afrotropics, the largest tracts are at river mouths, including the Zambezi, Rufiji, Niger, and Gambia Rivers, where the presence of brackish water allows mangroves to penetrate up to 120 mi. (190 km) inland. In West and Central Africa, mangroves are found from Mauritania south to Angola. Along Africa's Indian Ocean coast, they occur locally from Somalia to South Africa. In the Indian Ocean, mangroves are common along the west coast of Madagascar and very local on the east coast and on other islands. Mangroves can also be found locally around the Red and Arabian Seas.

WHERE TO SEE: Saloum National Park, Senegal; Mida Creek, Kenya; between Tulear (Toliara) and Ifaty, Madagascar.

Af12B SALT PANS AND LAKES

IN A NUTSHELL: Sparsely vegetated or unvegetated habitat on highly alkaline soils, usually evaporation pans. **Global Habitat Affinities:** AUSTRALIAN SALT PAN; SALINE ANDEAN LAKES; SODA PANS AND INLAND SALT MARSHES; ASIAN SALT PAN; PLAYAS. **Continental Habitat Affinities:** SALT MARSH; TIDAL MUDFLATS AND ESTUARIES. **Species Overlap:** TIDAL MUDFLATS AND ESTUARIES; SALT MARSH; ROCKY COASTLINE; SANDY BEACH AND DUNES; AFROTROPICAL DEEP FRESHWATER MARSH; AFROTROPICAL SHALLOW FRESHWATER MARSH; SOUTH AFRICAN TEMPERATE WETLAND; NORTH AFRICAN TEMPERATE WETLAND.

DESCRIPTION: An extremely hostile environment in which very few plants will grow, this habitat is created and defined by alkaline soil that is incredibly rich in sodium carbonate. Pans form when large lakes evaporate, when seawater is trapped in a coastal lagoon and then evaporates, or when salt-rich deposits of volcanic material accumulate in lake basins and along river courses. The two main subhabitats are coastal (Af12B-1) and inland (Af12B-2). This habitat might be considered little more than a microhabitat within arid environments if not for its huge extent in the Etosha Pan in Namibia, Makgadikgadi Pans in Botswana, and a scattering of large soda lakes in the Great Rift Valley, including Tanzania's Lakes Natron and Manyara, Kenya's Lake Nakuru, and Ethiopia's Lake Abiata. Despite its frequent aridity, the life of this habitat is water, which can be direct rainfall or inflow from rivers and streams. Although dry pans are virtually barren, blue-green algae will grow and small shrimp quickly hatch after a pan is flooded, providing a source of food in the ecosystem. Despite their apparent inhospitable nature, soda lakes are highly productive, with high rates of photosynthesis enabled by the huge amount of dissolved carbon dioxide. Flooded salt pans rapidly become wetlands bustling with feeding and nesting birds, most famously flamingos. There is a small set of halophytic plants that will grow on salt pans, usually along the edges, forming grassland, shrubland, or marsh. These plants include Salt Grass (*Sporobolus spicatus*) and other *Sporobolus*, Smooth Flatsedge (*Cyperus laevigatus*), *Odyssea paucinervis*, chenopods (Chenopodioideae), and seepweeds (*Suaeda*). Away from the pans, the habitat tends to transition via a sharp boundary to NORTHERN or KALAHARI DRY THORN SAVANNA or MOPANE. Artificial commercial salt pans can replicate the conditions of natural ones and be highly attractive to waterbirds.

CONSERVATION: This inhospitable habitat is generally not directly affected by human activity, though it is sensitive to changes in the larger landscape. Kenya's Lake Nakuru experienced excessive freshwater runoff from

Lake Tsimanampetsotsa, Madagascar, a classic saline lake, and the Malagasy stronghold of flamingoes. © KEN BEHRENS, TROPICAL BIRDING TOURS

adjacent agricultural areas during the early 2010s. The excessive water, nutrients, siltation, and eutrophication caused flamingos to temporarily abandon the area, though flamingos returned in 2020.

WILDLIFE: Most mammals rarely frequent this habitat, except when forced to cross it on the way to somewhere else. Scavengers like Black-backed Jackals occasionally raid waterbird colonies in the pans.

Birds are by far the most abundant and conspicuous form of wildlife. Most famous among these are flamingos, which breed exclusively in this habitat. While Greater Flamingo breeds more widely, the only historically consistent breeding site for Lesser Flamingo (IS) on the whole continent has been Tanzania's Lake Natron. When not breeding, flamingos move all around the continent. The creation and enhancement of Kamfers Dam in South Africa has boosted the declining Lesser Flamingo population, providing another dependable breeding site. When Namibia's Etosha Pan floods, every 7–10 years, Greater and Lesser Flamingos flock there to breed. But oftentimes, the pan dries too quickly to allow the young birds to be ready to migrate, stranding huge numbers of them, which either die or require a massive rescue operation. Another bird that nests colonially on flooded salt pans is Great White Pelican. The sight of many thousands of flamingos and pelicans standing virtually shoulder to shoulder on a salt pan is one of Africa's great wildlife spectacles.

Right: **The edges of Namibia's Etosha Pan support the endemic Etosha Agama (***Agama etoshae***), which is perfectly colored to blend in with its pale, salty environment.** © KEN BEHRENS, TROPICAL BIRDING TOURS

Below: **Greater and Lesser Flamingos often feature on African safaris and briefly pull attention away from big mammals. Lake Abiata, Ethiopia.** © KEN BEHRENS, TROPICAL BIRDING TOURS

Above: **Chestnut-banded Plover is a salt pan specialist.** © KEN BEHRENS, TROPICAL BIRDING TOURS

Right: **Unlike all other dabbling ducks, Cape Teal frequently use salt pan habitat.** © KEITH BARNES, TROPICAL BIRDING TOURS

Salt pans are frequented by a range of shorebirds, both residents, such as Pied Avocet and Kittlitz's Plover, and migrant species, like Ruff and Curlew Sandpiper. The dapper Chestnut-banded Plover (IS) is virtually restricted to this habitat. Other birds partial to salt pans include terns and gulls, Eared Grebe, and Cape Teal. One odd sight that may greet a visitor to a dry salt pan is Common Ostrich in the distance, distorted by the heat haze. Amazingly, these hardy birds sometimes choose to nest in this habitat because of the absence of predators.

There are a few reptiles that have adapted to this harsh habitat; Etosha and Makgadikgadi Pans each have an endemic agama.

DISTRIBUTION: Salt pans are found in arid environments where evaporation generally outpaces precipitation, so they are confined to southwestern, northeastern, and northern Africa. The two largest pans are in the Kalahari of the southwest, while the Great Rift Valley has a string of soda lakes, the largest of which is Lake Natron. A scattering of smaller salt pans, including artificial ones, may be found elsewhere, both inland and along the coast. Arid sw. Madagascar has one major saline lake, Tsimanampetsotsa. There are some vegetative communities growing on highly saline soils that we do not include in this habitat because of their lack of visual diagnosability or distinctive set of wildlife. Examples include portions of the Karoo, some seasonally flooded grasslands in East Africa, and some desert shrublands in n. Africa.

WHERE TO SEE: Etosha National Park, Namibia; Lake Manyara, Tanzania; Lake Abiata, Ethiopia.

Af12C TIDAL MUDFLATS AND ESTUARIES

IN A NUTSHELL: Intertidal coastal areas with rich soil that is largely unvegetated. **Global Habitat Affinities:** AUSTRALASIAN MUDFLAT; NEOTROPICAL TIDAL MUDFLAT; MEDITERRANEAN TO CASPIAN TIDAL FLATS; ASIAN TEMPERATE TIDAL MUDFLAT; NEARCTIC TIDAL MUDFLAT. **Continental Habitat Affinities:** SALT MARSH; SALT PANS AND LAKES. **Species Overlap:** SALT MARSH; ROCKY COASTLINE; SANDY BEACH AND DUNES; SALT PANS AND LAKES; AFROTROPICAL DEEP FRESHWATER MARSH; AFROTROPICAL SHALLOW FRESHWATER MARSH; SOUTH AFRICAN TEMPERATE WETLAND; NORTH AFRICAN TEMPERATE WETLAND; OFFSHORE ISLANDS; PELAGIC WATERS.

A classic estuary, at the mouth of the Bushman's River, Eastern Cape, South Africa.
© KEITH BARNES, TROPICAL BIRDING TOURS

DESCRIPTION: Coastal mudflats mainly occur where rivers flow into the ocean, carrying sand, silt, and clay, which are deposited along the estuary. These soils are much richer than the nutrient-poor substrates of sandy and rocky coastlines found away from estuaries and support a bounty of invertebrates, which in turn support other wildlife, mainly migratory birds. Some mudflats are tidal, exposed at low tide. Others are exposed only when the winds correctly align. This is a harsh environment, due to wave action, constantly changing water levels, and the high salt content of ocean water.

SALINE HABITATS

Constant wave and wind action and high salt content make this a tough environment in which little or no vegetation grows. Knysna Estuary, South Africa. © KEITH BARNES, TROPICAL BIRDING TOURS

WILDLIFE: This habitat is very poor in mammals, although at high tide it is an important component of the habitat of Dugong and West African Manatee. Mudflats are a birder's paradise due to the range of shorebirds they attract. Many widespread species, such as Gray Heron, Little Egret, Whimbrel, Bar-tailed Godwit (IS), and Black-winged Stilt are found along coasts throughout the Afrotropical region. There are some 20 species of Palearctic-breeding shorebirds that migrate to Africa's highly productive mudflats during the non-breeding season (Sep–Mar), including some 2 million individual birds at Banc d'Arguin in Mauritania, one of the highest shorebird densities anywhere on Earth! Some shorebirds are more localized to the Indian or Atlantic Ocean side of the continent: Indian Ocean shorebirds include Crab-Plover, sand-plovers, and Terek Sandpiper, while Red Knot is largely restricted to Atlantic coast.

DISTRIBUTION: Found locally all around the continent, including on the Indian Ocean islands. In slightly more sheltered areas, this basically unvegetated habitat is replaced by SALT MARSH, a low habitat of halophytic (salt-tolerant) plants.

WHERE TO SEE: West Coast National Park, South Africa; Walvis Bay, Namibia; Banc d'Arguin National Park, Mauritania.

Crab-Plover makes up its own family and is found along the Indian Ocean coastline.
© KEN BEHRENS, TROPICAL BIRDING TOURS

Af12D SALT MARSH

IN A NUTSHELL: Coastal areas that support low, salt-tolerant marsh vegetation, mainly grasses and/or succulents. **Global Habitat Affinities:** AUSTRALIAN COASTAL SALT MARSH; NEOTROPICAL COASTAL LAGOONS; EUROPEAN COASTAL SALT MARSH; NEARCTIC SALT MARSH. **Continental Habitat Affinities:** SALT PANS AND LAKES; TIDAL MUDFLATS AND ESTUARIES. **Species Overlap:** TIDAL MUDFLATS AND ESTUARIES; ROCKY SHORELINE; SANDY BEACH AND DUNES; SALT PANS AND LAKES; AFROTROPICAL DEEP FRESHWATER MARSH; AFROTROPICAL SHALLOW FRESHWATER MARSH; SOUTH AFRICAN TEMPERATE WETLAND; NORTH AFRICAN TEMPERATE WETLAND; AFROTROPICAL MANGROVE; OFFSHORE ISLANDS; PELAGIC WATERS.

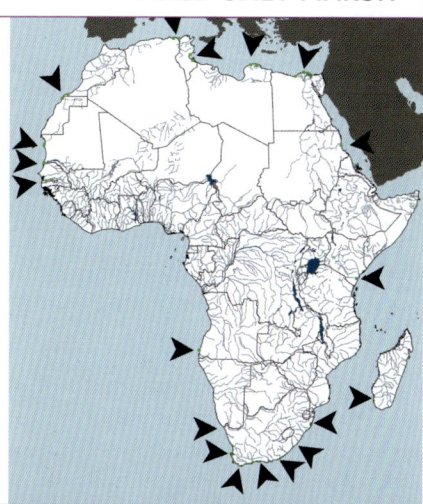

Salt Marshes often have a meadow-like aspect. Knysna Estuary, South Africa.
© KEITH BARNES, TROPICAL BIRDING TOURS

DESCRIPTION: Salt Marshes grow mainly in sheltered coastal spots such as bays and estuaries, where wave action is lessened. They thrive in the intertidal zone but sometimes grow above the normal high tide line, in areas up to 8 ft. (2.5 m) above mean sea level. The height of the vegetation gradually increases: shortest at the water's edge and tallest in the supratidal zone (above the normal high tide line), where it transitions to other habitats.

In the temperate and subtropical zone of South Africa, Salt Marshes are often grass dominated, resembling those in North America. Cordgrass (*Spartina maritima*) is often found in large monotypic stands. Succulents such as Marsh Samphire (*Salicornia meyeriana*), Perennial Glasswort (*Salicornia perennis*), and *Sarcocornia tegetaria*, *S. decumbens*, and *S. capensis* are also common and sometimes dominate large tracts of marsh. Other widespread species include Salt Marsh Rush (*Juncus kraussii*),

SALINE HABITATS

A close-up look at Salicornia, a typical component of an African Salt Marsh. © KEITH BARNES, TROPICAL BIRDING TOURS

arrowgrasses (*Triglochin*), introduced St. Augustine Grass (*Stenotaphrum secundatum*), Sea Lavender (*Limonium scabrum*), Dwarf Eelgrass (*Zostera capensis*), and Seashore Dropseed (*Sporobolus virginicus*). Shrubs like *Chenolea diffusa* sometimes grow in the higher reaches. Overall, the halophytic plants of this environment are similar to those of SALT PANS AND LAKES. In the tropics, Salt Marshes are much less studied and less common. These occasional tropical Salt Marshes usually comprise succulent species rather than grasses. As would be expected, Salt Marshes in the Red Sea have a very different botanical composition from those of Southern Africa, holding species such as *Arthrocaulon macrostachyum*, Nitre Bush (*Nitraria retusa*), and *Tetraena alba*. Some species, like *Dactyloctenium geminatum* and *Suaeda monoica*, are found all along the Indian Ocean from South Africa up to the Red Sea. Salt Marshes along the s. Mediterranean have a mix of plant species similar to those from s. Europe.

WILDLIFE: Afrotropical Salt Marshes are not particularly rich in larger wildlife but do support a few widespread bird species, such as African Pied Wagtail. The richest marshes in South Africa hold a slightly larger variety of species, including Cape Teal, Levaillant's Cisticola, and Orange-throated Longclaw. Salt Marshes provide a refuge for shorebirds and wading birds at high tide when mudflats are submerged. Shorebirds like Kittlitz's Plover, that prefer drier areas, find permanent refuge in Salt Marsh. Longer-legged species such as Black-winged Stilt and Common Greenshank feed in deeper marsh pools. The remarkable mini-kangaroo-like Four-toed Jerboa is a nocturnal Salt Marsh specialist from North Africa about which precious little is known; its conservation status is Data Deficient.

Four-striped Grass Mouse is one of a handful of rodents that will feed in Salt Marshes. West Coast National Park, South Africa. © KEN BEHRENS, TROPICAL BIRDING TOURS

DISTRIBUTION: Primarily a temperate and subtropical habitat that is largely, but not completely, replaced in similar ecological conditions by AFROTROPICAL MANGROVE in the tropics. As such, Salt Marshes are common along the coast of South Africa, with only a smattering elsewhere. Strangely, although Salt Marsh is common along the northern coast of the Mediterranean, it remains rare along the southern coast. Usually it occurs just inland from TIDAL MUDFLATS AND ESTUARIES, in slightly more sheltered but still highly saline areas. In the subtropics, it often occurs alongside AFROTROPICAL MANGROVE.

WHERE TO SEE: West Coast National Park, South Africa.

Af12E SANDY BEACH AND DUNES

IN A NUTSHELL: Coastal areas with poor sandy soil, either along a wave-swept beach or slightly inland in a sand dune system. **Global Habitat Affinities:** AUSTRALIAN SANDY BEACH; EUROPEAN DUNES AND SANDY SHORES; NEARCTIC SAND BEACH AND DUNES. **Continental Habitat Affinities:** NAMIB SAND DESERT; SAHARAN ERG DESERT. **Species Overlap:** TIDAL MUDFLATS AND ESTUARIES; SALT MARSH; ROCKY SHORELINE; SALT PANS AND LAKES; OFFSHORE ISLANDS; PELAGIC WATERS.

DESCRIPTION: Most of Africa's coastline consists of sandy beaches and rocky areas that range from flat rock to abrupt cliffs (see ROCKY SHORELINE). Sand dunes accumulate just inland of the intertidal zone in some areas, creating a localized desertlike environment. The areas between lines of sand dunes are often wet, holding marshes, most typically AFROTROPICAL SHALLOW FRESHWATER MARSH but sometimes also AFROTROPICAL DEEP FRESHWATER MARSH. Rare natural saline lagoons, along with more common human-created ones, can also be found along sandy stretches of shoreline (see SALT PANS AND LAKES).

A few hardy plants grow above the high tide line. These mostly consist of widespread grasses, creeping vines including Beach Morning Glory (*Ipomoea pes-caprae* and *I. imperati*), and succulents like Sea Purslane (*Sesuvium portulacastrum*) and Beach Soutbos (*Chenolea diffusa*). Small woody shrubs are able to take hold in some areas. Trees are generally lacking, though the superficially pine-like Australian *Casuarina* has been widely introduced and thrives along beaches. Along the South African coast, there are some members of the aster and ice plant families that have adapted to sandy beaches, such as Cape Beach Daisy (*Arctotheca populifolia*), Blue-and-white Daisybush (*Dimorphotheca ecklonis*), and Purple Dewplant (*Disphyma crassifolium*).

Classic sandy beach and dune habitat at Dunes de Dovela, Mozambique. © KEN BEHRENS, TROPICAL BIRDING TOURS

398 SALINE HABITATS

A small sandy beach on an island covered with Indian Ocean Lowland Rainforest. Nosy Mangabe, Madagascar.
© KEN BEHRENS, TROPICAL BIRDING TOURS

WILDLIFE: Although beach is the favorite habitat of sun-seeking vacationers around the world, it is generally a poor habitat for wildlife. In sub-Saharan Africa, the richest shoreline habitats are those of w. South Africa and Namibia, where the cold waters of the Benguela Current bring abundant nutrients close to the shore, which is a mix of sandy and rocky areas. This area has some massive Cape Fur Seal colonies, some on the mainland and some on offshore islands. These colonies attract scavengers like Brown Hyena and Black-backed Jackal.

Birds of the cold coastline of sw. Africa include African Oystercatcher and Bank, Cape, and Crowned Cormorants. More widespread birds found along much or all of the continent's sub-Saharan shorelines are Western Reef-Heron, White-fronted Plover, Great Crested Tern, and Sandwich Tern. Some species, like Lesser Crested Tern, are restricted to the Indian Ocean coast,

African Penguin is a somewhat shocking beach denizen of South Africa's Western Cape Province. These birds have started nesting at famous mainland sites like Boulder's Bay and Stony Point since human-populated areas have created predator-free safe zones. © KEN BEHRENS, TROPICAL BIRDING TOURS

Left: **White-fronted Plover** is the quintessential sandy beach bird throughout the Afrotropics, though it will also use larger rivers and can be found on inland sandbanks in places like Kruger National Park.
© KEN BEHRENS, TROPICAL BIRDING TOURS

Below left: **Beach crab in Gabon.** © CHRISTIAN BOIX, AFRICA GEOGRAPHIC

Below: **Huge coastal dunes at Lac Anony, Madagascar.**
© KEN BEHRENS, TROPICAL BIRDING TOURS

while others, like West African Crested Tern and Damara Tern, are found mainly along the Atlantic. There are many generalist coastal birds that use both sandy and rocky shorelines and TIDAL MUDFLATS AND ESTUARIES. But even these species tend to prefer mudflats because of the greater abundance of food. Although sandy beaches are far poorer than mudflats for shorebirds, this habitat does attract some migrant species, like Sanderling (IS), and is the preferred habitat of the resident White-fronted Plover. The sandy shores of North Africa are closely aligned with those of Europe and host a variety of migrating plovers and sandpipers. North Africa does have much higher gull diversity than sub-Saharan Africa, with common species including Mediterranean, Yellow-legged, and Lesser Black-backed Gulls.

Sandy beaches provide breeding habitat for Green (*Chelonia mydas*), Loggerhead (*Caretta caretta*), Olive Ridley (*Lepidochelys olivacea*), Hawksbill (*Eretmochelys imbricata*), and Leatherback (*Dermochelys coriacea*) Sea Turtles, which are found throughout most of the region but are classified as Vulnerable, Endangered, or Critically Endangered. Their decline is a result of human activities, including beach development, hunting, and dumping trash into the ocean.

DISTRIBUTION: Sandy beach is Africa's most common coastal habitat, though stretches of rocky shoreline and coastal cliffs are also frequent. These two common habitats are occasionally broken by TIDAL MUDFLATS AND ESTUARIES, AFROTROPICAL MANGROVE, and sometimes SALT MARSH in estuarine areas. Tall stabilized dunes along the s. Indian Ocean coast support a distinctive type of forest (see SOUTH COAST FOREST MATRIX, p. 164).

WHERE TO SEE: Pemba, Tanzania; Nosy Be, Madagascar.

Af12F ROCKY SHORELINE

IN A NUTSHELL: Rocky coastal areas, lacking vegetation or sparsely vegetated, varying from low rocky beach to towering cliffs. **Global Habitat Affinities:** AUSTRALIAN ROCKY HEADLAND; NEOTROPICAL ROCKY COASTLINE; ASIAN TEMPERATE ROCKY COASTLINE AND SANDY BEACH; NEARCTIC ROCKY COASTLINE; EUROPEAN ARCTIC AND TEMPERATE ROCKY HEADLAND. **Species Overlap:** SANDY BEACH AND DUNES; TIDAL MUDFLATS AND ESTUARIES; SALT MARSH; SALT PANS AND LAKES; AFROTROPICAL DEEP FRESHWATER MARSH; AFROTROPICAL SHALLOW FRESHWATER MARSH; OFFSHORE ISLANDS; PELAGIC WATERS.

DESCRIPTION: Rocky formations are Africa's second-most common type of coastline, after sandy beach. These formations vary a great deal, from beaches of jumbled boulders; to beaches of fine, rounded rocks; to low, fixed, wave-eroded rocks; to towering cliffs (subhabitat Af12F-1). The geological origins of these various Rocky Shorelines around the continent are diverse.

There are not many plants that can survive in this harsh environment with its salt spray and rocky soil. Most of the plants are widespread beach-adapted species, though there are a few plants specialized to cliff niches. Succulent shrubs sometimes form a low, dense thicket on coastal headlands. These come from genera including *Euphorbia*, *Tetragonia*, and ice plants (*Carpobrotus*). Rock pools hold various marine algae, anemones, and small fish.

WILDLIFE: Rocky coastal habitats are generally poor in terrestrial wildlife, though they do provide critical breeding and resting grounds for some coastal and marine species, especially in the cold waters of sw. Africa. There are massive Cape Fur Seal colonies in that area, which attract mammalian scavengers like Brown Hyena and Black-backed Jackal. African Penguin, Cape Gannet, and Bank, Cape, and Crowned Cormorants are all endemic to the Benguela Current zone, and all nest on rocky substrates. Small, rocky islands are the favored roosting sites of gulls and terns, such as Great Crested, Lesser Crested, Roseate, Common, and Antarctic Terns and Kelp Gull. Rocky shorelines are very poor in shorebirds but are the favored habitat of the long-distance migrant Ruddy Turnstone and are key for the endemic African Oystercatcher, which feeds on bivalves like mussels along the marine fringes. Coastal cliffs provide breeding habitat for White-tailed Tropicbirds in the Gulf of Guinea and on the Indian Ocean islands. Along the North African

Right: Cape Cormorant at Kommetjie, Western Cape, South Africa. Note the vast kelp beds in the fringing ocean, a key microhabitat requirement for the endemic Bank Cormorant that also roosts on the same rocks.
© KEN BEHRENS, TROPICAL BIRDING TOURS

Below: Rocky tidepools support mussels and anemones *(insets)*, urchins, and a bounty of other sea life. Knysna Heads, South Africa.
© KEITH BARNES, TROPICAL BIRDING TOURS

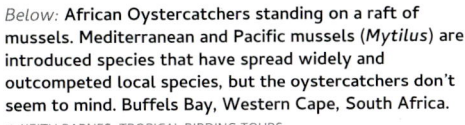

Below: African Oystercatchers standing on a raft of mussels. Mediterranean and Pacific mussels (*Mytilus*) are introduced species that have spread widely and outcompeted local species, but the oystercatchers don't seem to mind. Buffels Bay, Western Cape, South Africa.
© KEITH BARNES, TROPICAL BIRDING TOURS

402   SALINE HABITATS

Right: **Cape Gannets against the rugged cliffs of the Cape Peninsula. Western Cape, South Africa.** © KEN BEHRENS, TROPICAL BIRDING TOURS

Below: **Sooty Gull, a species typical of both rocky and sandy shoreline along the Indian Ocean and Red Sea coasts.** © KEN BEHRENS, TROPICAL BIRDING TOURS

Bottom right: **Rocky Shoreline at Kommetjie, Western Cape, South Africa.** © KEITH BARNES, TROPICAL BIRDING TOURS

coast, rocky areas are the preferred habitat for Great Cormorant and European Shag. They provide breeding areas for Eleonora's Falcon, which specializes in hunting the migratory birds that seasonally flood the region.

DISTRIBUTION: Rocky areas and seaside cliffs are found locally all along the African coast and around the Indian Ocean islands. They are especially common in w. South Africa and on the Comoros, Seychelles, and Mascarene Islands.

WHERE TO SEE: Cape Peninsula, South Africa.

Af12G OFFSHORE ISLANDS

IN A NUTSHELL: Offshore Islands, either sandy or rocky, usually covered in grass or low scrub, that provide refuge for oceanic wildlife and a small selection of other species. **Global Habitat Affinities:** INDO-MALAYAN OFFSHORE ISLANDS. **Species Overlap:** PELAGIC WATERS; ROCKY COASTLINE; SANDY BEACH AND DUNES; TIDAL MUDFLATS AND ESTUARIES; SALT MARSH; AFROTROPICAL MANGROVE.

DESCRIPTION: While Offshore Islands look insignificant on a map, they provide crucial habitat for breeding seabirds, sea turtles, and seals because they are generally free of the predators of the mainland and are also less likely to have been pillaged by humans. They also give refuge to migrating birds, such as shorebirds, gulls, and terns, which also use beaches and mudflats along the mainland.

Most small Offshore Islands are sandy, formed by the breakdown of coral reefs, though some are sedimentary, produced by the outflow of sediment from large rivers; some are volcanic; and others are rocky. An example from the last category is the Inner Seychelles, which are granitic fragments of Gondwana.

Aldabra lies in the w. Indian Ocean, fairly close to n. Madagascar, though it is politically part of the Outer Seychelles. This island is the world's second-largest coral atoll and one of its best preserved. Aldabra is covered in a diverse set of microhabitats, including open scrub, dense thickets, and a low mix of herbs and grasses that is sometimes referred to as "tortoise turf"

Nosy Ve, Madagascar, a typical small offshore tropical island, with a mix of sandy and rocky shoreline. Islands like Europa in the Mozambique channel are used for nesting by Red-footed Booby *(inset)*.
MAIN PHOTO © KEN BEHRENS; INSET © KEITH BARNES, TROPICAL BIRDING TOURS

since it is heavily grazed by the island's huge population of around 100,000 Aldabra Giant Tortoises (*Aldabrachelys gigantea*). Aldabra supports almost 300 species of plants, of which dozens are endemic, either to Aldabra itself or to the Outer Islands.

CONSERVATION: As a crucial refuge to nesting seabirds and breeding marine mammals, these islands are of outsized conservation importance. Despite their remoteness, most islands have still been heavily affected by hunting and egg collection, but the most drastic impact has been the introduction of mammals such as rats, goats, and rabbits. The rats prey on seabird eggs, especially those of burrow-nesting shearwaters and petrels. Goats and rabbits overgraze the vegetation and drastically alter the natural landscape. In some places, such as Round Island, off Mauritius, there has been a concerted effort to rid islands of exotic animals and restore them to their natural state.

WILDLIFE: Although there are only a few rocky islands off the coast of w. South Africa and Namibia, they are heavily used by wildlife as they provide secure breeding sites in the nutrient-rich, cold waters of the Benguela Current. Sites such as South Africa's Dassen and Dyer Islands and Namibia's Penguin Islands provide breeding habitat for most of the world's Cape Fur Seals, African Penguins, Cape Gannets, Kelp and Hartlaub's Gulls, and Cape, Crowned, and Bank Cormorants.

The remote Subantarctic Islands of the African sector of the Southern Ocean include Tristan and Gough (British Overseas Territories); Bouvetoya (Norway); Prince Edward Islands (South Africa); and the Crozets, the Kerguelens, Amsterdam, and the St. Paul islands (all French Southern Territories). Between them they support tens of millions of seabirds and a handful of endemics and are a major hotbed for seabird diversity. They provide refuge for Southern Elephant Seal, fur seals, 7 species of penguin, 10 species of albatross including rare, breeding-endemic species like Tristan and Amsterdam Albatross, and 30 species of shearwater, prion, petrel and storm-petrel. Many of these birds feed in the pelagic oceans around Southern Africa in the austral winter (Apr–Sep), when most of them are not breeding. The Subantarctic Islands' rocky coastlines have localized species like Imperial and Kerguelen Shag. Brown Skua and Subantarctic and Kerguelen Terns patrol the nearshore waters and the bizarre Black-faced Sheathbill forages for scraps in penguin colonies. Other island endemics range from the Eaton's Pintail to Tristan Thrush and Inaccessible Island Rail. But perhaps the coolest endemics are the collection of *Neospiza* finches on Tristan. Their lineage arrived from South America, and they have since undergone an evolutionary radiation similar to the Galápagos's Darwin's finches. Inaccessible Island Finch is the weirdest species; three morphs (genetically almost

Rocky Offshore Islands in the cold Benguela Current zone provide breeding sites for Cape Gannet. © KEITH BARNES, TROPICAL BIRDING TOURS

identical) occupy the same island where they behave as different species ecologically, with two small-billed forms and one 30% larger that has a large bill and forages almost exclusively on higher-elevation *Phylica* bushes. Two other islands lie farther north, in the tropical South Atlantic: Ascension and St. Helena, both British territories. These islands host nesting seabirds, including the endemic Ascension Frigatebird, and St. Helena has the endemic Saint Helena Plover.

Islands in warm tropical waters of both the Atlantic and Indian Oceans support breeding colonies of tropicbirds, frigatebirds, Brown Booby, Brown Noddy, and several terns, including Sooty Tern (IS). Aldabra, in the Seychelles, has the world's second-largest frigatebird colony, hosting both Great and Lesser Frigatebirds. Aldabra also has two endemic landbirds (Aldabra Drongo and Fody) and a flightless subspecies of White-throated Rail— the last flightless bird left in the w. Indian Ocean. Bird Island, also in the Seychelles, hosts more than a million breeding Sooty Terns. Europa Island, in the Mozambique Channel between Madagascar and Africa, is the largest breeding site for Green Sea Turtle (*Chelonia mydas*) in the Indian Ocean and also hosts three-quarters of a million pairs of breeding seabirds. Black Noddy breeds on islands only in the Atlantic portion of the Afrotropical region. Red-tailed Tropicbirds breed on some Indian Ocean islands, including Nosy Ve off Madagascar. Breeding colonies of Wedge-tailed and Tropical Shearwaters, Red-tailed Tropicbird, Masked and

Red-footed Boobies, Lesser Noddy, and Roseate Tern are found only on islands on the Indian Ocean side of the continent. Islands of the n. Indian Ocean and Red Sea are home to White-eyed Gull and Lesser Crested, Sandwich, and White-cheeked Terns. Round Island, off Mauritius, is one of the few relatively high-elevation (rather than low and sandy) tropical islands that lacks rats and as such is a critical breeding refuge for several seabirds, such as Trindade, Kermadec, and Herald Petrels (these three are mixed up in a complicated hybrid swarm), and also holds Bulwer's and Black-winged Petrels and Wedge-tailed Shearwater. Interior Réunion is the only breeding location for Barau's and Mascarene Black Petrels.

Macaronesia (Cape Verde, Azores, Madeira, and Canary archipelagos) in the nw. Atlantic is a crucial area for breeding seabirds. Most nest on small, rugged, uninhabited rocky islets, some volcanic, that are often devoid of vegetation or have scattered grass and shrubs. Seven species are breeding endemics to one or more of the archipelagos: Cape Verde, Boyd's, and Barolo Shearwaters; Zino's and Fea's Petrels; and Cape Verde and Monteiro's Storm-Petrels. The latter two are cryptic species that were recently discovered in colonies of almost identical-looking Band-rumped Storm-Petrel. Other breeding seabirds in Macaronesia include European and White-faced Storm-Petrels, Manx and Cory's Shearwaters, Bulwer's Petrel, Red-billed Tropicbird, Brown Booby, and Sooty and Roseate Terns.

Sandy Offshore Islands provide crucial breeding habitat for Green, Loggerhead (*Caretta caretta*), Olive Ridley (*Lepidochelys olivacea*), Hawksbill (*Eretmochelys imbricata*), and Leatherback (*Dermochelys coriacea*) Sea Turtles, all of which are threatened or endangered.

Endemism: Many island groups have a selection of breeding endemic birds, and there are major differences between the species breeding in the nw. Atlantic, tropical Atlantic, southwest side of Africa, the Subantarctic Islands, and the tropical Indian Ocean. A few species and subspecies of lizards are endemic to small Offshore Islands. Round Island has an exceptional assemblage of endemic reptiles: Round Island Keel-scaled Boa (*Casarea dussumieri*), Round Island Day Gecko (*Phelsuma guentheri*), and Round Island Ground Skink (*Leiolopisma telfairii*).

Rocky islands are haul-out sites for Cape Fur Seals. Table Bay, Western Cape, South Africa.
© KEN BEHRENS, TROPICAL BIRDING TOURS

DISTRIBUTION: The subantarctic zone to the south of the African continent has many islands, which vary from large to small (subhabitat Af12G-1). The largest concentrations of small tropical islands are found in the Red Sea, in the Seychelles, and off Madagascar, Mauritius, and Rodrigues (Tropical Indian Ocean, subhabitat Af12G-3). There are a few Offshore Islands in the cold water of the Benguela Current off sw. Africa (subhabitat Af12G-2). The coast of West Africa has relatively few islands, though there are some in the Gulf of Guinea (subhabitat Af12G-4). Macaronesia has a large number of islets (subhabitat Af12G-5).

WHERE TO SEE: Ile aux Cocos, Rodriguez, Mauritius; Nosy Ve, Madagascar; Bird Island, Lambert's Bay, South Africa.

Af12H PELAGIC WATERS

IN A NUTSHELL: The deep waters lying off the continental shelf surrounding Africa and the Indian Ocean islands. **Global Habitat Affinities:** AUSTRALIAN TEMPERATE PELAGIC WATERS; SOUTH AMERICAN TEMPERATE PELAGIC WATERS; MEDITERRANEAN TO CASPIAN PELAGIC WATERS; ASIAN TEMPERATE PELAGIC; NEARCTIC PELAGIC WATERS. **Species Overlap:** OFFSHORE ISLANDS; ROCKY COASTLINE; SANDY BEACH AND DUNES; TIDAL MUDFLATS AND ESTUARIES; SALT MARSH.

DESCRIPTION: The Atlantic Ocean lies west of Africa, and the Indian Ocean lies to its east. Cape Agulhas, South Africa, the continent's southernmost point, is somewhat arbitrarily considered the border between the two. The Mediterranean borders Africa to the north and is connected to the Atlantic, while the Red Sea and Gulf of Aden surround ne. Africa and are connected to the Indian. Although the ocean is contiguous and appears more-or-less uniform to the eye, much is happening in the water columns that differentiates stretches of ocean that are easily defined as subhabitats: Southern Ocean (Af12H-1); Benguela Temperate Current (Af12H-2); Tropical Indian Ocean (Af12H-3); Tropical Atlantic Ocean (Af12H-4); and Northern Temperate Waters (Af12H-5).

Although the waters surrounding Africa, save for those of the Mediterranean, are generally poorly known, both oceans are locally rich in marine mammals and seabirds. The edges of the continental shelf, where there is a drop-off between a relatively shallow seabed and much deeper waters, tend to have nutrient-rich upwellings and thus tend to be the best areas for seeing oceanic wildlife. In some places, such as off the island of Mauritius and off Dakar, Senegal, this drop-off is very close to shore. In general, the n. Indian Ocean is fairly poor in oceanic wildlife, and the Southern and Atlantic oceans are a bit richer. The richest portion of the Atlantic is off sw. South Africa and Namibia, where the cold Benguela Current brings abundant nutrients from the Southern Ocean. Cape Town, South Africa, is the one place in the continent where boat trips regularly go offshore in search of pelagic birds.

WILDLIFE: Some widespread marine mammals such as Humpback Whale, Blue Whale, Orca, and Common Bottlenose Dolphin are found throughout the oceans of the region. Other species are divided between the warmer, more tropical waters (Short-finned Pilot Whale, Pygmy Killer Whale, Melon-headed Whale, Spinner Dolphin, and Pantropical Spotted Dolphin) and the cooler waters around the southwest side of Africa (Southern Right Whale and Heaviside's Dolphin).

For birds, the major divide is between the Atlantic and Indian Oceans, though a few common birds, like Wilson's Storm-Petrel, are found in both. Widespread Atlantic species include Cory's and Sooty Shearwaters and European and Leach's Storm-Petrels. The Mediterranean generally has a reduced subset of Atlantic seabirds but is the heart of the range of Yelkouan and Balearic Shearwaters.

Above: **Giant Petrels battling for food in Southern Ocean waters.** © KEITH BARNES, TROPICAL BIRDING TOURS

Below: **Great White Shark is found in all the oceanic waters surrounding Africa, ranging from warm and tropical to cool and temperate.** © KEN BEHRENS, TROPICAL BIRDING TOURS

Black-browed Albatross is a large seabird species of the cool water of the southern oceans. It is regularly seen off sw. Africa and occasionally is even spotted from shore around Cape Town, South Africa. © KEN BEHRENS, TROPICAL BIRDING TOURS

Typical Indian Ocean pelagic species are Wedge-tailed, Tropical, and Flesh-footed Shearwaters and Great and Lesser Frigatebirds. The richest and best-known African waters for seabirds are those off Cape Town, South Africa. Here the rich, cold waters of the Benguela Current attract a great diversity of pelagic birds. Different seasons bring strikingly different birds. Many species of the far Southern Oceanic Islands move north during the austral winter, roughly between May and October; these include Black-browed and Wandering Albatrosses, giant-petrels, and Cape Petrel. During the austral summer, different birds visit from the north during their nonbreeding season; these include European Storm-Petrel, Sabine's Gull, and Parasitic, Pomarine, and Long-tailed Jaegers.

WHERE TO SEE: Pelagic birding boat trip out of Cape Town, South Africa.

Orcas are the apex predator throughout the entire realm of Africa's offshore waters. © KEITH BARNES, TROPICAL BIRDING TOURS

ANTHROPOGENIC

Af13A HUMID LOWLAND CULTIVATION

IN A NUTSHELL: Humid lowland areas dominated by human-planted tree, root, and tuber crops. **Global Habitat Affinities:** PADDY FIELDS; NEOTROPICAL OIL PALM. **Continental Habitat Affinities:** SAVANNA CULTIVATION; TROPICAL MONTANE CULTIVATION. **Species Overlap:** AFROTROPICAL LOWLAND RAINFOREST; TROPICAL MONTANE CULTIVATION; GUINEA SAVANNA; MOIST MIXED SAVANNA; MALAGASY GRASSLAND AND SAVANNA; INDIAN OCEAN LOWLAND RAINFOREST; INDIAN OCEAN MONTANE RAINFOREST.

Cultivated "farmbush" area at the edge of remnant rainforest. Kumbira, Angola.
© KEITH BARNES, TROPICAL BIRDING TOURS

DESCRIPTION: This human-created habitat occurs within Africa (subhabitat Af13A-1) and the Indian Ocean islands' lowland rainforest zone (subhabitat Af13A-2). In the Upper Guinea region, from Guinea to Cameroon, this habitat has largely replaced rainforest. Cultivated areas occur alongside villages and cities, with adjacent still-forested areas usually degraded into "farmbush." Intact rainforest is restricted to more remote or nationally protected areas. Far more forest remains in the Congo Basin region, though wherever people are found, they use an agricultural system similar to that of Upper Guinea. Humid Lowland Cultivation is also the dominant habitat in

Humid cultivation, mostly bananas, with a forest fragment behind. Southwest Ghana. © GRAY TAPPAN

east and north Madagascar and on the other Indian Ocean islands. The climate parallels that of lowland rainforest: hot and humid with frequent year-round rainfall. The growing season is 9–12 months long.

Two main types of plants are cultivated in this habitat: (1) tree crops primarily grown for cash and connected with international markets; and (2) roots and other food crops grown mainly for subsistence or sold locally. The most widespread and important tree crops are oil palm and cacao. Rubber, cloves, citrus, and robusta coffee are of secondary importance. The main source of palm oil, the African Oil Palm (*Elaeis guineensis*), is indigenous to West Africa and as such its plantations are far richer in wildlife than oil palm plantations elsewhere in the world. Oil palm is usually grown at a small scale, part of a mixed agricultural system, but there are larger industrial plantations in some places. Rubber is also sometimes cultivated in industrial plantations. Cacao is a crucial cash crop in most of West Africa, Cameroon, and nw. Madagascar. While some cacao is grown in open plantations, it is more typically grown in shaded plantations with a tree canopy, which can be quite rich in birds and, in Madagascar, in lemurs. Cloves and vanilla are grown frequently in e. Madagascar and the Comoros.

The majority of crops in the second category, food crops, are roots and tubers, the most important of which are cassava, yam, cocoyam, and sweet potato. Other important food crops, which are often part of mixed plantations, include plantains, maize, rice, and various fruit trees such as mango, African Breadfruit (*Treculia africana*), papaya, and avocado. Litchi and jackfruit are common on the Indian Ocean islands. Maize is a common crop, and small amounts of other cereals are grown, though they are more characteristic of SAVANNA CULTIVATION. A variety of domestic animals are kept as part of small-scale mixed farms. Mauritius is an outlier as a humid cultivated area where sugarcane is the dominant crop. In e. Madagascar, rice is grown wherever possible, since it is the preferred local food.

WILDLIFE: In Africa, humid cultivated areas tend to have a subset of the birds of AFROTROPICAL LOWLAND RAINFOREST, similar to the mix of species found in Anthropogenic Rainthicket (see p. 146) but still further reduced. In addition, they are used by some birds of natural savanna habitats, such as GUINEA SAVANNA and MOIST MIXED SAVANNA. While some mammals persist in the low-intensity agricultural areas of the Congo Basin, mammals have been virtually wiped out in the Upper Guinea region. Lemurs are largely absent from humid cultivated areas in Madagascar, save for the northwestern cacao plantations, which have proven attractive to several species, including the localized Pariente's Fork-marked Lemur and Giant Mouse Lemur.

Naturally this habitat holds some birds that occur in a broad range of habitats, species such as Black Kite, Red-chested Cuckoo, Speckled Mousebird, African Pied Wagtail, Common Fiscal, Green-backed Camaroptera, Collared Sunbird, and Bronze Mannikin. Forest and forest edge species that sometimes use cultivated areas include Western Gray Plantain-eater, Simple Greenbul, Chattering Cisticola, Dusky-blue Flycatcher, Splendid Starling, Splendid and Olive-bellied Sunbirds, and Northern Yellow White-eye. Savanna birds that have taken advantage of human disturbance to enter the traditional forest realm include African Thrush, Northern Puffback, Violet-backed Starling, and Copper Sunbird. Oil palm plantations attract frugivores like hornbills and Gray Parrot plus Swamp Palm Bulbul. Moist oil palm plantations can even hold the secretive White-spotted Flufftail.

Humid cultivated areas on the Indian Ocean islands tend to be much poorer in wildlife than those on the continent, as these originally heavily forested islands lack a reservoir of indigenous species that are adapted to open habitats. The starkest example of this is the island of Mauritius, the former refuge of some fabulous forest birds, the most famous of which was the Dodo. Today Mauritius is nearly entirely deforested, the Dodo and many other forest species are extinct, and about one-third of the island is covered in vast sugarcane monocultures. Cultivated areas in eastern Madagascar support widespread Malagasy birds such as Madagascar Coucal, Madagascar Bulbul, Madagascar Magpie-Robin, Madagascar White-eye, Red Fody, and the ubiquitous introduced Common Myna.

WHERE TO SEE: Kakum National Park, Ghana (cultivated areas in the periphery); Ambanja, Madagascar (shade-grown cacao).

Northern Giant Mouse Lemur is a restricted-range species of n. Madagascar that locally thrives in shade-grown cacao plantations, in the northwest. © KEN BEHRENS, TROPICAL BIRDING TOURS

Af13B SAVANNA CULTIVATION

IN A NUTSHELL: Lowland and middle-elevation areas of sub-Saharan Africa, with a moderately moist, highly seasonal climate, dominated by human-planted cereal, root, and tuber crops. **Global Habitat Affinities:** None. **Continental Habitat Affinities:** HUMID LOWLAND CULTIVATION. **Species Overlap:** AFROTROPICAL GRASSLAND; MONTANE GRASSLAND; Afrotropical savannas (seven different habitats); MALAGASY GRASSLAND AND SAVANNA.

DESCRIPTION: Huge portions of Africa's vast savanna zones are used by humans primarily as grazing grounds for domestic animals. However, this chapter is focused on the portions of the savanna zone that are cultivated, in general the moister areas. The natural habitats in this cultivated zone are AFROTROPICAL GRASSLAND, MOIST MIXED SAVANNA, GUINEA SAVANNA, MIOMBO, MOPANE, and GUSU. Some portions of dry thorn savanna areas are cultivated, taking advantage of a brief abundance of moisture during the short dry season. The growing season in Savanna Cultivation habitats ranges from 2.5 to 6 months, and they receive a similar amount of rainfall to adjacent natural savanna habitats.

What characterizes this habitat is the preeminent importance of cereal crops. In n. sub-Saharan Africa, the dominant crops are millet and sorghum, with maize and groundnuts also crucial (subhabitat Af13B-1). From s. South Sudan and s. Ethiopia southwest to Angola and south to e. South Africa, maize (corn) is the staple crop (subhabitat Af13B-2). Other common savanna cultivation crops include beans,

Cultivated savanna areas in Ethiopia seem to be a preferred habitat of a couple scarce endemics, such as the Stresemann's Bush-Crow.
© KEN BEHRENS, TROPICAL BIRDING TOURS

tobacco, vegetables, sunflower, sesame, fonio (a small indigenous African millet), and watermelon. Rice is the preeminent crop in Madagascar and is found locally in Africa. Cassava and other root crops are common, especially in the area of transition to the humid cultivation zone. Cotton is grown in some drier savanna environments. Domestic animals, especially cattle, goats, and chickens, are common

Dry versus wet season comparisons of cultivated savanna areas. Senegal. © GRAY TAPPAN

throughout and come to be the dominant form of livelihood in the drier parts of the savanna zone, where agriculture is only supplemental.

Population density, and associated density of cultivated areas, varies greatly. The northern savanna has long been heavily populated and historically acted as a sort of highway across the whole continent, south of the parched Sahara but north of the forbidding rainforest. Human populations are much lower in most of East and Southern Africa, though there are pockets of dense settlement and cultivation, such as much of Malawi and parts of Zimbabwe and Tanzania. The most densely populated portions of the Horn of Africa and East Africa are supported by TROPICAL MONTANE CULTIVATION. Through most of Africa, savanna cultivation is undertaken at small scale: family farms that grow a mix of crops and also hold livestock and forage for wild foods. The exception to this is e. South Africa, and some other parts of Southern Africa, where farms are larger and more industrialized. There are huge sugarcane plantations in e. South Africa, often in conjunction with large pine, eucalyptus, and wattle tree plantations (see TREE PLANTATIONS).

WILDLIFE: Larger mammals have been virtually eliminated from cultivated areas in North Africa. Elsewhere in the continent, some mammals of savanna habitats will venture into adjacent cultivated areas. In some areas, such as parts of Kenya and Tanzania, this actually becomes more common as humans have densely colonized parts of the migratory ranges of big mammals such as Plains Zebra and Blue Wildebeest.

The dual facts that most farming is mixed and small scale, and that a large proportion of African birds are adapted to moderately open savanna-type habitats (see sidebar 4, p. 62) mean that this is an excellent habitat for birds. In some cases, lightly farmed areas seem to have more birds than adjacent natural savannas. And in general, there are few savanna birds that aren't sometimes found in cultivated areas. Much of n. Sub-Saharan Africa is so densely populated that most birding takes place in this habitat, rather than in unmodified savanna habitat, and despite that the birding is excellent. Human-populated areas offer rich pickings for scavengers like Marabou Stork, Black Kite, and Hooded Vulture. Open fields are excellent hunting grounds for a wide variety of raptors, including Red-necked and Grasshopper Buzzards in the north and kestrels throughout. Baobabs are often spared, even when other trees are cut, leaving nesting habitat for Mottled Spinetail. Grainfields are used by Common Quail and some larks and pipits. Fallow fields are favored by Forbes's Plover, a variety of lapwings, some sandgrouse, and migrant wagtails and wheatears. Nocturnal birds of this habitat include Fiery-necked and Plain Nightjars and Barn Owl. Common birds in areas with at least some remaining indigenous trees include Ring-necked Dove, mousebirds, Eurasian Hoopoe, Little and White-throated Bee-eaters, some hornbills, Fork-tailed Drongo, Gray-backed Fiscal, several species of starlings and sunbirds, Northern and Southern

Manioc cultivation at the edge of gallery Spiny Forest in sw. Madagascar. Beza Mahafaly Reserve. © KEN BEHRENS, TROPICAL BIRDING TOURS

Af13C TROPICAL MONTANE CULTIVATION   415

Secretarybird is an Endangered bird that is the only member of an African endemic family. In some areas, where it isn't persecuted, it will forage in cultivated areas. © KEN BEHRENS, TROPICAL BIRDING TOURS

Gray-headed Sparrows, a bounty of weavers, Red-cheeked Cordonbleu, and paradise-whydahs. Northern Savanna Cultivation has Rose-ringed Parakeet, Vinaceous Dove, Little Green Bee-eater, and Piapiac. East Africa has Fischer's Lovebird and Superb Starling. The birds of cultivated areas in west and central Madagascar are the same as those in MALAGASY GRASSLAND AND SAVANNA and are covered in that account (see p. 287).

DISTRIBUTION: This habitat is common in a vast swath across n. sub-Saharan Africa, much of East and Central Africa, and in the northern and eastern parts of Southern Africa. It is also found in west and central Madagascar, alongside larger tracts of MALAGASY GRASSLAND AND SAVANNA. The transition to TROPICAL MONTANE CULTIVATION occurs at approximately 3300 ft. (1000 m).

Af13C TROPICAL MONTANE CULTIVATION

IN A NUTSHELL: Montane areas that are dominated by a wide variety of human-planted crops. **Global Habitat Affinities:** TEA PLANTATION; COFFEE AND CARDAMOM. **Continental Habitat Affinities:** HUMID LOWLAND CULTIVATION; SOUTH AFRICAN TEMPERATE CULTIVATION; SAVANNA CULTIVATION. **Species Overlap:** AFROTROPICAL MONTANE DRY MIXED WOODLAND; MOIST MIXED SAVANNA; MOIST MONTANE FOREST; MONTANE GRASSLAND; AFROTROPICAL GRASSLAND; MALAGASY GRASSLAND AND SAVANNA.

DESCRIPTION: Sub-Saharan Africa's mountains hold some of the continent's most fertile and densely populated areas. Humans have largely converted AFROTROPICAL MONTANE DRY MIXED WOODLAND, MOIST MONTANE FOREST, and MONTANE GRASSLAND into varied cultivation. Growing seasons are long in most mountains, and even the drier Abyssinian mountains have two rainy seasons a year. There are three main divisions within montane cultivation: cereal-dominated areas, complex mixed shambas, and rice cultivation.

Cereal crops (subhabitat Af13C-2) tend to predominate in higher, cooler, and drier areas, such as the higher parts of the Abyssinian highlands, Lesotho and adjacent parts of South Africa, and parts of the Angolan highlands. The mix of cereals changes locally. In large parts of Ethiopia, tef (a tiny indigenous grain), barley, and wheat predominate. Elsewhere, the preferred montane cereal is maize. Sorghum is common in Lesotho. Beans and groundnuts are grown alongside maize in West Africa and Angola. Fruits such as apples and peaches, which are typically associated with temperate areas, are grown locally. Nyger seed, legumes, oats, flax, and potatoes are important crops in parts of Ethiopia. Livestock are widely held in cereal-growing areas and are used both for labor and for meat. As a broad generalization, cereal-growing areas have replaced natural grasslands, and they have also served as an intrusion of a grassland ecological-equivalent habitat into areas that were formerly well wooded. Highland cereal-growing areas are allied with the wheatlands of the SOUTH AFRICAN

ANTHROPOGENIC

Despite their harshness and ruggedness, most of the Abyssinian highlands are cultivated with grains like wheat, barley, and tef. Simien Mountains, Ethiopia.
© KEITH BARNES, TROPICAL BIRDING TOURS

TEMPERATE CULTIVATION

zone but do not have that habitat's Mediterranean climate.

"Shamba" is a Swahili word that is used generally for any farm but which tends to be applied to the complex, well-watered farms that are common surrounding Lake Victoria (subhabitat Af13C-1). Much of this area has rich volcanic soil. Shamba-style farms are also found in the Eastern Arc Mountains, the mountains of s. Kenya and ne. Tanzania, s. Ethiopia, and in the Cameroon-Nigeria highlands. Part of what characterizes shambas is their complexity of elevation, aspect, rainfall regimen, soil, and associated crops. Farmers here have long used this rich tapestry of niches to grow a diverse assemblage of crops. Shambas generally replace and approximately replicate the structure of montane forests. The typical farms found on the slopes of Mt. Kilimanjaro are a perfect example of the complexity of shamba cultivation. They have canopy trees that are often indigenous species, bananas grown in the shade of the canopy trees, coffee grown in the shade of the bananas,

Rice is the preferred crop in the highlands of Madagascar, grown both in paddies (shown here) and as an upland crop. Soarano Valley, Madagascar. © KEN BEHRENS, TROPICAL BIRDING TOURS

legumes below the bananas, and root crops and vegetables at the bottom. All are fertilized by manure produced by domestic animals. This multistory arrangement approximates the complexity of a humid forest. In most shamba areas, bananas are the staple food. The common variety of bananas is *Musa acuminata*, but the indigenous Enset (*Ensete ventricosum*) predominates in Ethiopia. Coffee and tea are key cash crops. Other common plants include maize, beans, vegetables, rice, potatoes, and other root crops.

The highlands of Madagascar are mainly covered in MALAGASY GRASSLAND AND SAVANNA, but rice cultivation occurs wherever there is sufficient water. Rice is grown both in paddies and in upland areas. Other crops on the High Plateau include apples, peaches, grapes, and a wide variety of vegetables.

WILDLIFE: In general, montane cultivated areas are used only fleetingly by larger mammals that venture out of adjacent natural habitats. Human conflict with wildlife is becoming increasingly common in places where cleared and cultivated land lies adjacent to natural habitats. Examples are c. Kenya, where African Savanna Elephants venture out of Mt. Kenya and Aberdare National Parks, and the mountains of Rwanda and Uganda, where "Mountain" Eastern Gorillas sometimes raid farms.

Tropical Montane Cultivation is still rich in birdlife. Areas with a patchwork of natural or only slightly modified habitat, alongside human cultivation, can be incredibly rich in birds. The birds of this habitat largely fall out as preferring cereal-growing areas or complex shambas, with the former approximating natural grassland and the latter forest.

Widespread species of montane cereal cultivation include Black-shouldered Kite, Cape Crow, Rufous-naped and Red-capped Larks, African and Plain-backed Pipits, Capped Wheatear, Common Fiscal, Zitting Cisticola, Black-winged Bishop, and Red-collared Widowbird. The granaries of the Ethiopian highlands are especially rich in grassland birds. These include local endemics like Wattled Ibis, Swainson's Sparrow, and Ethiopian Siskin as well as widespread migratory species like White Stork, Common Quail, Common Crane, and a variety of harriers, kestrels, and wheatears.

Shambas can be wonderfully rich in birds, and spending time birding these farms can provide an excellent complement to birding the interior of forest. Typical widespread species are Red-eyed Dove, Dideric and Klaas's Cuckoos, Speckled Mousebird, Blue-breasted and Cinnamon-chested Bee-eaters, Rufous-necked Wryneck, African Thrush, White-browed Robin-Chat, African Paradise-Flycatcher, African Blue Flycatcher, Tropical Boubou, several species of white-eyes, Baglafecht Weaver, Bronze Mannikin, and Collared, Variable, and Scarlet-chested Sunbirds. East African shambas can be further enlivened by the likes of Gray Parrot, Ross's Turaco, Chubb's Cisticola, and Gray-capped Warbler. Shade-grown coffee plantations along the Angolan escarpment hold some of that country's endemic birds, including Red-crested Turaco and Gabela Akalat. Southern Ethiopian shambas have Bruce's Green-Pigeon, Ethiopian Bee-eater, and Ethiopian Boubou. Southern African montane farms can hold Cape Canary and Southern Masked-Weaver.

Malagasy rice paddies support a few birds such as Great, Little, and Black Egrets; Hamerkop; Madagascar Snipe; Madagascar Kingfisher; Madagascar Cisticola; and African Stonechat.

DISTRIBUTION: The cutoff between lowland and montane habitats is rarely clear, and cultivated habitats are no exception. The approximate watershed between lowland cultivation types and montane cultivation occurs around 3300 ft. (1000 m) in the tropics and lower in Southern Africa. The area around Lake Victoria is closely aligned with the HUMID LOWLAND CULTIVATION found just to its west but has a higher diversity of crops and a cooler climate and generally lacks the tree crops of the lowlands. Shamba-**type** cultivation generally shifts to cereal cultivation around 5600 ft. (1700 m). There are small areas of shamba-type cultivation on some of the islands surrounding Africa, namely São Tomé off Central Africa, on the slopes of Mt. Karthala on Grande Comore, and on the French territory of Réunion.

Af13D SOUTH AFRICAN TEMPERATE CULTIVATION

IN A NUTSHELL: Temperate areas of South Africa dominated by cereals and perennial fruits. **Global Habitat Affinities:** AUSTRALIAN TEMPERATE CROPLAND; NEARCTIC CROPLAND. **Continental Habitat Affinities:** NORTH AFRICAN TEMPERATE CULTIVATION; TROPICAL MONTANE CULTIVATION. **Species Overlap:** RENOSTERVELD; STRANDVELD; FYNBOS; MONTANE GRASSLAND; AFROTROPICAL GRASSLAND.

DESCRIPTION: At the far southern end of Africa is an area with a temperate, Mediterranean-type climate, with cold, wet winters and warm, dry summers. This is the Cape Floral Kingdom, where the primary natural habitats are FYNBOS, RENOSTERVELD, and STRANDVELD. This area has had a large human population since European colonization began in the 17th century. Most of the land capable of being cultivated was converted into farms long ago. This arable land was mostly found in Renosterveld, which grows on nutrient-rich clay and silt soils derived from shales and granite. This natural habitat has been far more affected by humans than any other habitat on the continent; only a tiny percentage of natural Renosterveld remains.

The primary crops growing in this temperate climate are perennial fruits (subhabitat Af13D-1): pome fruits (apples and pears), stone fruits (including peaches, plums, prunes, nectarines, and apricots), and grapes, some of which are turned into world-class wines. The Mediterranean climate is also good for citrus, olives, and a variety of vegetables. Finally, this area holds some important pockets of industrialized wheat production (subhabitat Af13D-2). Most wheat farms also grow canola and alfalfa and browse sheep.

WILDLIFE: This habitat's perennial plantations hold very little wildlife, though a few birds such as Cape Bulbul and Red-winged Starling can be found in fruit plantations. Wheat fields are the most interesting subhabitat for wildlife. Some mammals of adjacent natural habitats will forage in wheat

The wheatfields of South Africa's Overberg are a paradise for Blue Crane, South Africa's national bird. This species moved here en mass from its traditional range and habitat in the eastern grasslands.
© KEN BEHRENS, TROPICAL BIRDING TOURS

A classic scene from the Western Cape of South Africa: a Jackal Buzzard, cultivated fields, and sheep, with the Cape Fold Mountains in the background. © KEN BEHRENS, TROPICAL BIRDING TOURS

fields. These include Common Duiker and Steenbok. Small predators like Cape Gray and Yellow Mongooses will often hunt in this habitat. Grainfields replicate grassland and attract a subset of the birds of natural AFROTROPICAL GRASSLAND and MONTANE GRASSLAND. Examples include Common Quail, Helmeted Guineafowl, Denham's Bustard, Red-capped Lark, Cape Crow, Yellow and Southern Red Bishops, and African Pipit. Cultivated fields are one of the favorite habitats of South Africa's gorgeous national bird, the Blue Crane, which essentially shifted its global distribution into this habitat from the natural grasslands farther east. Natural RENOSTERVELD probably included a large grassy component, and perhaps for this reason cereal fields have proven a good habitat for the regional endemic Large-billed and "Agulhas" Cape Larks. Surprisingly, they are also used by a species that elsewhere is restricted to Karoo habitats, the Karoo Bustard, which like the Blue Crane has "colonized" this habitat.

Af13E NORTH AFRICAN TEMPERATE CULTIVATION

IN A NUTSHELL: Temperate areas of North Africa dominated by human-planted crops. **Global Habitat Affinities:** ASIAN TEMPERATE CROPLAND. **Continental Habitat Affinities:** SOUTH AFRICAN TEMPERATE CULTIVATION; TROPICAL MONTANE CULTIVATION. **Species Overlap:** MAGHREB MAQUIS; MAGHREB GARRIGUE; MAGHREB JUNIPER OPEN WOODLAND; SAHARAN REG DESERT; ROCKY HAMADA AND MASSIF.

DESCRIPTION: Along the far northern edge of Africa, north of the vast Sahara Desert, lies a strip of land that is closely aligned with the Mediterranean and quite distinct from Africa south of the Sahara. While the level of rainfall varies greatly throughout this region, the pattern is Mediterranean throughout: a cool winter when most rain occurs and a hot, dry summer. Mediterranean North Africa has been heavily populated for a long time, and most arable land has been cultivated for thousands of years. Grains, mainly wheat, barley, and rice, are the staple crops. As expected in the Mediterranean, olives, figs, and grapes are also important. Other common crops include maize, sugar beet, sugarcane, almonds, walnuts, potatoes, tomatoes, eggplant, peppers, carrots, citrus, onion, melons, apples, and peaches. This area benefits economically from being close to the rich markets of Europe, unlike the rest

Cultivated area on the fringes of the Sahara. Date palms are a crucial crop here. Todra Gorge, Morocco. © KEN BEHRENS, TROPICAL BIRDING TOURS

Intricate and incredibly labor-intensive terrace systems in the Atlas Mountains of Morocco have been inhabited and cultivated since antiquity. © KEN BEHRENS, TROPICAL BIRDING TOURS

Northern Lapwings are partial to agricultural fields, especially recently plowed ones. © KEITH BARNES, TROPICAL BIRDING TOURS

of the African continent. Agriculture in this region varies from small-scale and mixed to large-scale and industrial. Egypt's Nile Valley has a fascinating, ancient irrigation system that provides water to the farms of millions of small landholders. Livestock is important throughout and takes precedence in more arid areas, such as Morocco south of the Atlas Mountains and Egypt away from the Nile floodplain. In the western portions of this habitat, there is a mix of winter rainfall-watered farming and mostly small-scale irrigated farming.

WILDLIFE: Mammals have fared poorly in North Africa, and larger mammals have been virtually eliminated from cultivated areas. Birds have survived in better numbers. North African cultivated areas support a fairly diverse set of open-country and woodland species. Most of these are also widespread in Eurasia. Examples include Common Crane, Northern Lapwing, White Stork, Barn and Little Owls, Eurasian Hoopoe, Eurasian Bee-eater, Woodchat Shrike, Eurasian Greenfinch, and European Serin. Cultivated fields are excellent for the region's rich assemblage of larks, including Eurasian Skylark and Crested, Thekla, and Calandra Larks. Cultivated areas also hold some species with more limited s. European and n. African distributions, such as Cattle Egret, Egyptian Vulture, Spotless Starling, and Spanish Sparrow. Fields along the Moroccan coast are used by the Endangered Northern Bald Ibis. North Africa, including its cultivated areas, provides a crucial stopover point for many species that migrate from the Palearctic to spend the winter in sub-Saharan Africa. Just a few of many such species include Lesser Kestrel, Common Quail, Common Cuckoo, Eurasian Wryneck, Red-throated Pipit, Yellow Wagtail, Northern Wheatear, Greater Whitethroat, and Eurasian Golden Oriole.

DISTRIBUTION: This anthropogenic habitat is found from Morocco east to Egypt. It is broadest in the Atlas Mountains and narrows along the arid coast of n. Libya. In Egypt, south into Sudan, the Nile River allows a band of cultivation to pass all the way through the e. Sahara Desert. Oases (see sidebar 9, p. 100) also provide localized sources of water that have permitted cultivation deep into the Sahara. Cultivated areas are found high into the Atlas Mountains, up to 7000 ft. (2100 m) or higher.

Maghreb Magpie is a nw. African endemic, which is often found in cultivated areas. © ZSOMBOR KÁROLYI

Af13F TREE PLANTATIONS

IN A NUTSHELL: Monoculture tree plantations, mostly of various pines and eucalyptus but also other species. **Global Habitat Affinities:** AUSTRALIAN TREE PLANTATIONS; NEARCTIC TREE PLANTATIONS. **Species Overlap:** AFROTROPICAL MONTANE DRY MIXED WOODLAND; MOIST MONTANE FOREST; MAGHREB PINE FOREST.

DESCRIPTION: Trees are planted commercially and managed across the continent. These are typically same-age monocultures that can blanket huge stretches, though many subsistence and small-scale farmers also maintain small copses of tree crops. Although the vast majority of plantations comprise various species of pine (subhabitat Af13F-1) and eucalyptus (subhabitat Af13F-2), other plantation species include Black Wattle (*Acacia mearnsii*); Mexican Cypress (*Cupressus lusitanica*) in Libya and Ethiopia; Teak (*Tectona grandis*); Kashmir Tree (*Gmelina arborea*); Rubber Tree (*Hevea brasiliensis*); Terminalia in Nigeria, Ghana, and Ivory Coast; Cashew (*Anacardium occidentale*) mainly in East Africa, Mozambique, and Nigeria; Cork Oak (*Quercus suber*) in Algeria (subhabitat Af13F-3); Argan (*Sideroxylon spinosum*) in Morocco (subhabitat Af13F-4); Coconut Palm (*Cocos nucifera*) along the Indian Ocean coast, especially in Mozambique and e. Madagascar; Date Palm (*Phoenix dactylifera*) and Olive (*Olea europaea*) in North Africa; and Coastal She-oak (*Casuarina equisitifiolia*) and Gum Acacia (*Senegalia senegal*) in Senegal. Plantations are frequent at all elevations and in a variety of climates across the continent, although there is a tendency to grow them at moderately high elevations above 2000 ft. (670 m) as here the soils, climate, and rainfall are most suitable for rapid growth.

Trees selected for timber, like pines and eucalypts, are planted as evenly spaced small saplings, and as pioneer trees they tend to grow tall, straight, and quickly. Although these trees can reach a staggering 260 ft. (80 m), they are mostly harvested when they are 50–150 ft. (15–45 m) tall. Plantations are designed so that the trees' canopies barely interlock, to avoid damage during

Pine plantation in the Outeniqua Mountains of South Africa. This habitat is largely devoid of wildlife, but does provide good breeding structure for many species of raptors. © KEITH BARNES, TROPICAL BIRDING TOURS

harvesting. Once older than 5–10 years, most timber plantations have little to no understory and are easy to walk through, with a broad grid of open space beneath the towering canopy, but it is gloomy with the canopy preventing much light from reaching the plantation floor. Plantations have an unsurprisingly unnatural feel. Although from a distance, they appear superficially similar to Eurasian pine forests or Australian dry sclerophyll, on closer inspection the paucity of understory herbs, shrubs, or epiphytes lends a sterile quality. The main native plant that is able to survive on the fringes of these gloomy tree farms is bracken (*Pteridium*), sometimes forming dense clumps. Some plantations are grown for nontimber products: Gum Acacia produces gum used as a food additive, adhesive, and pharmaceutical; Rubber Trees are harvested for rubber and Cork Oak for cork. These plantations tend to be shorter and less gloomy but remain uniform and sterile. Plantations tend to transition to adjacent natural habitats with a sharp boundary.

CONSERVATION: The creation of tree plantations has caused myriad conservation problems across Africa, including habitat loss, mostly of grassland and montane forest, resulting in local extinctions, such as the elimination of Blue Swallow from most of its South African range. Replacing native plants with alien ones also results in changes to hydrology and soil properties, including soil structure and pH, which affect ecosystem functioning and nutrient recycling and can cause the local extinction of vital service providers like earthworms. However, plantations also supply crucial wood products for the human population, aiding economic development. A growing population and economy are likely to increase demand further and result in the expansion of this anthropogenic habitat. The availability of timber products from plantations sometimes means that subsistence harvesters are less likely to chop down native forest trees to meet their need for wood. This is a complex dynamic, but it is predicted that if plantations are not expanded, more indigenous forests are likely to be lost! The tree farm industry also contributes to local employment (e.g., 720,000 people in South Africa alone), providing income in more rural communities, which reduces their reliance on natural resources.

WILDLIFE: Tree plantations are almost devoid of wildlife, and calling them ecological deserts is doing deserts a severe injustice. Some small mammals are recorded using young plantations, but these vanish once the trees approach maturity. Walking through the average plantation, the silence is deafening save for the s. Cape Peninsula, where a small population of Common Chaffinch, introduced by none other than influential colonialist Cecil John Rhodes, give their musical rattle— a case of an introduced bird in introduced trees! Virtually no sub-Saharan African birds are adapted to use most of these plants for food, although sunbirds and white-eyes will occasionally feed on eucalyptus blooms. However, it is not food but rather the towering trees that make ideal nest sites that draw in birds of prey. Across the continent, many species of sparrowhawks, eagles, and even rarities like Bat Hawk now have their most-famous stakeouts near nests in plantations. Copses of trees in semiarid regions attract raptors formerly thought to be forest specialists, such as Red-breasted Sparrowhawk, which seem quite happy to hunt over open rangelands provided they have trees nearby for nesting. North of the Sahara, where there is natural MAGHREB PINE FOREST (see p. 67), pine plantations attract a very depauperate version of that wildlife community.

DISTRIBUTION: Some 38 million acres (15.4 million hectares) of plantation were spread across Africa in 2010, of which around 9.8 million acres (4 million hectares) is commercial, much of that in South Africa but also in Eswatini, Zimbabwe, Tanzania, Uganda, Ethiopia, and Mozambique. Most nations on the continent and Indian Ocean islands have some plantations, whether commercial or for subsistence purposes in local communities. Eucalyptus and pine plantations are now by far the most common type of "forest" on the High Plateau of Madagascar.

WHERE TO SEE: Sabie, South Africa.

Af13G CITIES AND VILLAGES

IN A NUTSHELL: Areas of concentrated human habitation. **Global Habitat Affinities:** AUSTRALIAN URBAN; ASIAN CITIES; NEARCTIC URBAN. **Species Overlap:** Smaller villages retain some similarity to surrounding natural habitats, but cities have very little in common with any natural areas.

DESCRIPTION: Humans have heavily altered the natural landscape of Africa in countless ways but nowhere more profoundly than in and immediately surrounding their settlements, from sprawling cities (subhabitat Af13G-1) to tiny and remote villages (subhabitat Af13G-2). These areas contain a bounty of different microhabitats that provide different niches for wildlife, including gardens of exotic flowers, stands of trees planted for shade and shelter, granaries, trash dumps, ornamental ponds, urban parks, manicured lawns, and concrete canyons between skyscrapers. Indeed, humans have created such a diverse and complex set of wildlife habitats that a whole book could easily be devoted just to them. Grazing lands and cultivated areas are covered in other sections; this section is devoted to the areas immediately associated with human habitation.

In forests, human settlements tend to be cleared to some extent and thus more open than the surrounding natural habitat. The opposite is true in more open environments, which characterize the majority of the Afrotropics; there, areas of human habitation tend to be lusher than their adjacent natural environs as people access water in order to grow ornamental plants and soften their surroundings, creating islands of lusher habitat. In areas of dry thorn savanna, towns tend to take on the character of MOIST MIXED SAVANNA, while in an area of Moist Mixed Savanna, a well-established town can have a character similar to MONSOON FOREST. Habitat islands like this can be very attractive to birds migrating through dry country. In some cases, human-modified habitats have allowed large range extensions of certain wildlife species. The most striking example of this is in the Cape area of South Africa, where many bird species have extended their ranges hundreds of miles to the west by using the new habitats created by humans. Large cities tend to be vegetated by ornamental trees and weeds, both of which are nonindigenous. Smaller villages may have some indigenous vegetation alongside garden plants.

Morarano, a typical small village in the Betsileo country on Madagascar's High Plateau. © KEN BEHRENS, TROPICAL BIRDING TOURS

The small African village of Kanjonde, Angola. Found in an area where the natural habitats are Miombo and Montane Grassland. © KEN BEHRENS, PROMISED LAND VENTURES

WILDLIFE: An amazing variety of larger mammals can survive alongside humans in areas where their presence is tolerated, as occurs in most of Ethiopia. Lush gardens can hold Gureza Colobus and other monkeys, squirrels, and Bushbuck. Treed urban areas frequently provide daytime refuge for Straw-colored Fruit Bat and other large bats. Predators like jackals, African Wildcat, Common Genet, and even Leopard can survive alongside large human populations. Spotted Hyenas are even known to live in the sewers of Addis Ababa, Ethiopia! In parts of South Africa, Hippopotamuses live in ponds and wetlands adjacent to towns and emerge to "mow" the lawns at night. In parts of southern and eastern Africa, herbivores like Impala, Thomson's Gazelle, Blesbok, and Plains Zebra have been introduced to the grounds of gated estates, where they munch the lush lawns. In contrast, the towns and cities of heavily hunted West Africa are virtually devoid of larger mammals.

Urban areas can provide excellent habitat for birds and are the main habitat of several introduced species, such as feral Rock Pigeon, Rose-ringed Parakeet (away from n. Africa), House Crow, European Starling, Common Myna, and House Sparrow (IS). They are favored in Ethiopia by White-collared Pigeon and throughout the continent by several species of indigenous sparrows. Some of these species are now so well adapted to human-created habitats that it is difficult to imagine where they lived before the ascendancy of human beings. In many parts of Africa, scavenging Black Kites and Marabou Storks are trash collectors and can be found in large numbers around dumps or perched ghoulishly on street lights. Hooded Vulture was formerly common in and around urban areas but has largely disappeared, part of the tragic Eastern Hemisphere crash in vulture populations. Lawns and artificial wetlands are favored by Hadada Ibis and Egyptian Goose. Mowed grass is attractive to Helmeted Guineafowl and several species of francolin and thrush. Tall buildings seem to function much as cliffs for some birds and even provide nesting habitat for Eurasian and Rock Kestrels and Red-winged Starling. Highway underpasses and culverts are colonized by Little and

Above: **Honey Badgers and many other animals can cozy up to humans when easy pickings are to be had.**
© KEN BEHRENS, TROPICAL BIRDING TOURS

Right: **The town of Wakkerstroom, in the Highveld of South Africa, originally was grassland, but the human-created habitat with abundant trees now attracts a new mix of savanna and forest species.**
© KEN BEHRENS, TROPICAL BIRDING TOURS

White-rumped Swifts and many species of swallows. As suggested by their names, Village Indigobirds and Village Weavers are often seen in towns. Some species that favor the MOIST MIXED SAVANNA–analogous habitat of lush gardens include Red-eyed and Laughing Doves, Dideric and Klaas's Cuckoos, bulbuls, and robin-chats. These habitats also attract mousebirds, white-eyes, and many species of sunbirds, which are especially partial to the flowers of introduced Australian bottlebrushes (*Callistemon*).

Urban habitats support many reptiles, including smaller snakes, skinks, and geckos. The most common gecko is the Tropical House Gecko (*Hemidactylus mabouia*; IS), which is easily found at night on walls and around lights across most of the Afrotropics. In reptile-rich Madagascar, human structures are often inhabited by Common House Gecko (*Hemidactylus frenatus*), several beautiful day geckos (*Phelsuma*), fish-scale geckos (*Geckolepis*), and sometimes even huge Madagascar velvet geckos (*Blaesodactylus*).

Top right: **Small village in a rice cultivation highland landscape. Ambositra, Madagascar.** © KEN BEHRENS, TROPICAL BIRDING TOURS

Above right: **"Coke's" Hartebeest with the Nairobi skyline behind. Nairobi National Park, Kenya.** © KEN BEHRENS, TROPICAL BIRDING TOURS

Right: **Humans and huge birds like these Marabou Storks coexist peacefully in towns where they are not persecuted, like on the shores of the Rift Valley lakes of Ethiopia.** © KEN BEHRENS, TROPICAL BIRDING TOURS

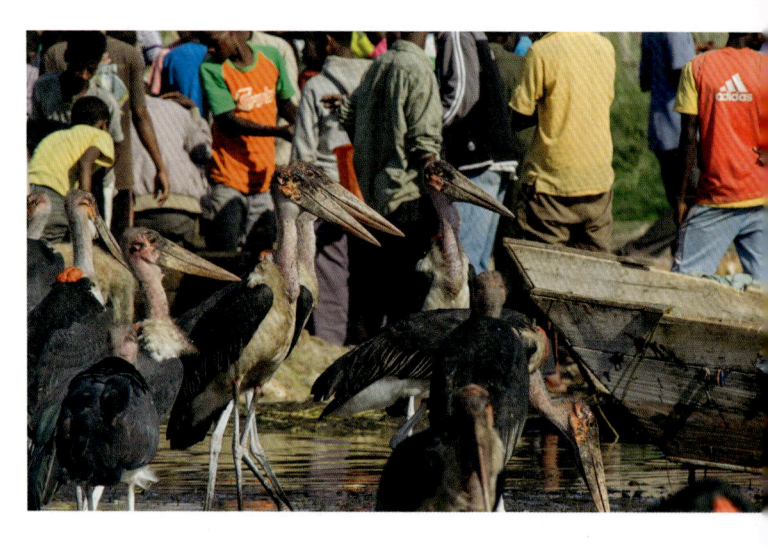

Af13H GRAZING LAND

IN A NUTSHELL: Areas grazed by domestic animals. **Global Habitat Affinities:** AUSTRALIAN OPEN GRAZING LAND; NEOTROPICAL COW PASTURE; ASIAN GRAZING LAND; NEARCTIC PASTURELAND AND RANGELAND. **Species Overlap:** Afrotropical savannas (seven different habitats); AFROTROPICAL GRASSLAND; MONTANE GRASSLAND.

DESCRIPTION: Grazing lands with minimal impact on the original habitats include those of the NORTHERN DRY THORN SAVANNA of the Horn of Africa, virtually all of which is grazed, and the Karoo shrublands and Highveld MONTANE GRASSLAND of Southern Africa. A more human-modified but still largely natural area is the vast swath of the Sahel, where the original habitat is Northern Dry Thorn Savanna and AFROTROPICAL HOT SHRUB DESERT. The story has been quite different on the Indian Ocean islands, especially Madagascar. There, vast areas of forest were burned to create cattle pasture, which is shockingly devoid of indigenous plants or wildlife. The major reason for the poverty of these habitats seems to be the lack of a reservoir of Malagasy savanna- or grassland-adapted species that can use them. On the African continent, forested areas that are destroyed to create pasture or for cultivated crops are quickly colonized by the species of adjacent grassland and savanna habitats.

WILDLIFE: Grazing lands generally have far fewer big wild mammals than do natural habitats because of the competition for food resources with domestic animals. When you see a herd of 100 goats somewhere in a dry thorn savanna, you can assume that they are essentially replacing the natural herbivores of that habitat. While herbivorous big mammals usually remain in reduced numbers, except in areas with lots of hunting, larger predators are rarely tolerated. Lion, Cheetah, Leopard, and hyenas have been wiped out across much of Africa in order to protect domestic animals. Even the smaller predators like Caracal, foxes, and jackals are often killed.

The situation is much better with birds. Grazing lands generally support a nearly complete set of the birds of the habitat in which they lie. A few species are even closely associated with domestic animals, often in lieu of the presence of wild mammals. Cattle Egret, Fork-tailed Drongo, Piapiac, and Wattled Starling all follow domestic animals, sometimes even perching on them and catching

This photo shows how grazing domestic animals shape even a fairly natural landscape. Spiny trees and bushes are cut to create a "kraal" enclosure, which in turn provides a new sort of habitat for some wildlife. Caraculo, Angola. © ARNON DATTNER, ECOPLANET FILMS

Oxpeckers (Yellow-billed on a donkey; Red-billed on a cow) will readily switch between wild and domestic animals. © KEN BEHRENS, TROPICAL BIRDING TOURS

the insects they disturb. Hornbills eat insects such as dung beetles that they extract from the dung of both wild and domestic animals. Oxpeckers eat ticks and other parasites off their host animals, drink their blood, and even use their hair to line their tree cavity nests. Certain groups of birds seem to thrive in grazed areas, occurring perhaps at even higher densities than in natural areas. These include some species of starlings, waxbills, and weavers. Two unusual birds, Stresemann's Bush-Crow and White-tailed Swallow, are restricted to a tiny and heavily grazed area of s. Ethiopia, where they remain in good numbers.

DISTRIBUTION: Vast areas of the Afrotropics are grazed at least occasionally by domestic animals: goats, sheep, cattle, donkeys, and camels. Thankfully, most of these grazing areas retain much of their natural character and wildlife and still function as savanna or grassland habitats.

Until the cows come home. Liben Plain, Ethiopia. © KEN BEHRENS, TROPICAL BIRDING TOURS

INDEX

Page numbers in **bold** indicate photographs.

Species mentioned as part of a 'list' are not indexed unless they are Indicator Species (IS), or there is significant information. The Introduction, and geographical features/countries have not been indexed, unless discussed in a Sidebar.

A

Aardvark 240, **240**, 256, 284, 317
Aardwolf 103, **123**, 240, 284, 317
Abura tree 170
Acacia 64
    Flat-topped 247, **247**
    Gum 422, 423
    Lahai 247
    Umbrella Thorn 83, 100, 122, 228, 237
    Whistling Thorn 228, **230**, 246
acacias 64, 122, 228, 237, **238**, 246
Accentor, Alpine 351
Addax **72**, 73, 122
Adder
    Albany 337
    Desert Mountain 81
    Horned **106**, 241
    Many-horned 314
    Peringuey's 91, **92**
    Puff 241, 278, 314
    Red 105
    Southern 314
African Dry Deciduous Forests 208, 209
African Heathlands, dendrogram 342
African Humid Forests, dendrogram 142
Afrogecko 308
Afroparamo 34, 35, 40, 46, 53, 293, 343–348, **344**, **345**, **346**, **347**, **348**, 350, 354, 357
Afrotropical Deep Freshwater Marsh 214, 244, **358**, 358–364, **360**, **361**, **362**, **363**, **364**, 365, 368, 374, 378, 379, 380, 382, 384, 390, 393, 395, 397, 400
Afrotropical Grassland 45, 121, 145, 150, 164, 167, 203, 227, 235, 256, 261, 280–286, **281**, **282**, **283**, **284**, **285**, 287, 293, 334, 358, 365, 374, 412, 415, 418, 419, 427
Afrotropical Hot Shrub Desert 77, 101, 107, 115, 118, 119, 121–125, **122**, **123**, **124**, **125**, 133, 136, 227–228, 233, 427
Afrotropical Lowland Rainforest 62, 142, 143–150, **144**, **145**, **146**, **147**, **148**, **149**, **150**, 151, 154, 156, 157, 160, 169, 170, 173, 174, 177, 180, 181, 197, 203, 205, 215, 244, 251, 258, 260, 261, 266, 267, 273, 364, 382, 386, 388, 409, 411
Afrotropical Mangrove 39, 143, 161, 167, 169, 386–389, **387**, **388**, **389**, 395, 396, 399, 403
Afrotropical Monsoon Forest *see* Monsoon Forest

Afrotropical Montane Dry Mixed Woodland 57, 61–66, **63**, **64**, **65**, **66**, 174, 176, 179, 247, 251, 299, 347, 354, 356, 415, 422
Afrotropical Montane Grassland *see* Montane Grassland
Afrotropical Shallow Freshwater Marsh 244, 255, 358, 360, **365**, 365–369, **366**, **367**, **368**, **369**, 374, 378, 379, 382, 390, 393, 395, 397, 400
Afrotropical Swamp Forest 39, 142, 143, 147, 152, 159, 169–173, **170**, **171**, **172**, **173**, 180, 358, 360, 364, 380, 382, 386, 388
Afrotropics, description 53
Agama
    Anchieta's 105
    Ceríaco's Tree 279
    Common 266
    Desert 86
    Etosha **391**
    Ground 278
    Mali 99
    Mucoso 219
    Mwanza Flat-headed **250**
    North African Rock 75, 120
    Tropical Spiny 226
agamas 231
Akalat
    East Coast 161
    Gabela 417
Albany Thicket 45, 106, 153, 164, 167, 243, 249, 332–337, **333**, **334**, **335**, **336**, **337**
Albatross
    Amsterdam 404
    Black-browed **408**
    Tristan 404
    Wandering 408
Albertine Rift montane forests 41
Albizia 210
    Moluccan 199, **200**
Aloe, Socotrine 135
aloes 333, 337
Alpine Tundras and Montane Heaths (Africa) 343–357
Afroparamo 34, 35, 40, 46, 53, 293, 343–348, 343–348, **344**, **345**, **346**, **347**, **348**, 350, 354, 357
High Atlas Alpine Meadow 55, 350–353, **351**, **352**, **353**
Montane Heath 38, 40, 46, 61, 62, 140, 141, 176, 196, 293, 299, 300, 309, 315, 343, 347, 348, 354–357, **355**, **356**, 357
Alstonia congensis 170
Amaryllis 309, 316
Amazon Terrafirma 143, 180
Ana Tree 90
Andean Cloudforest 174, 189
Andean Cushion Paramo 343
Andean Grassy Paramo 343
Angolan Deciduous Forest 151, 203, 209, 215–219, **216**, **217**, **218**, 219

Ant, Saharan Silver 86
Antelope
    Dwarf 148
    Roan 264, **264**, 277
    Royal 148
    Sable 256, **256**, 277
Anthropogenic Habitats (Africa) 290, 409–428
    Cities and Villages 424, 424–426, **425**, **426**
    Grazing Land **427**, 427–428, **428**
    Humid Lowland Cultivation 409, 409–411, **410**, **411**, 412, 415, 417
    North African Temperate Cultivation 418, 419–421, **420**, **421**
    Savanna Cultivation 290, 409, 410, **412**, 412–415, **413**, **414**, 415, **415**
    South African Temperate Cultivation 316, 319, 415–416, **418**, 418–419, **419**
    Tree Plantations 414, **422**, 422–423
    Tropical Montane Cultivation 409, 414, 415–417, **416**, 418, 419
Anthropogenic Rainthicket 146, 149
Apalis
    Bar-throated 314, 337
    Gosling's **385**
    Rudd's 161, 169
    Yellow-breasted 336
    Yellow-throated 65
Apple, Green 166
Arartree 67, 68, 117
Argan woodlands 329, **329**
Arid Tussock Acacia Shrubland 121
Ascarinopsis coursii 191
Asian Alpine Tundra 350
Asian Cities 424
Asian Cold Erg Desert 88
Asian Garrigue 326
Asian Grazing Land 427
Asian Peat Swamp Forest 169
Asian Salt Pan 390
Asian Temperate Cropland 419
Asian Temperate Heathland 300, 309, 315
Asian Temperate Pelagic 406
Asian Temperate Rocky Coastline and Sandy Beach 400
Asian Temperate Tidal Mudflat 393
Asian Temperate Wetland 370, 374
Asian Tropical Wetland 358, 365
asities 32
Asity
    Common Sunbird- **32**
    Schlegel's 213, **214**
    Yellow-bellied Sunbird- 192
Asphodelaceae **109**
Ass
    African Wild 97, 124
    Atlas Wild 56
Assegai Tree 177
asters 102, **310**, 311
Atacama Desolate Desert 82, 88

Auroch 118
Ausbos (Australian Southern Lowland Heathland) 300
Australasian Alpine Tundra 343
Australasian Littoral Rainforest 157, 164
Australasian Mangrove Forest 386, 388
Australasian Monsoon Vineforest 151, 157, 164
Australasian Montane Grasslands 293
Australasian Montane Heathland 315
Australasian Mudflat 393
Australasian Sandy Riverbeds 382
Australasian Swamp Forest 169
Australasian Temperate Freshwater Wetland 370, 374
Australasian Tropical Lowland Rainforest 143, 180
Australasian Tropical Montane Rainforest 174, 189
Australian Bottlebrush 426
Australian Coastal Salt Marsh 395
Australian Dry Vineforest 203
Australian Open Grazing Land 427
Australian Rocky Headland 400
Australian Salt Pan 390
Australian Sandy Beach 397
Australian Southern Lowland Heathland 300, 309
Australian Temperate Cropland 418
Australian Temperate Heath Thicket 321, 338
Australian Temperate Pelagic Waters 406
Australian Tree Plantations 422
Australian Tropical Tussock Grassland 280
Australian Tropical Wetland 358, 365
Australian Urban habitat 424
Avocet, Pied 376, 392
Aye-aye 183, **184**

B

Babbler
  Bare-cheeked 225
  Dapple-throat 178
  Gray-chested 178
  Hartlaub's 362
  Red-collared Mountain- 178
  Southern Pied- 241
  Spot-throat 178
  White-throated Mountain- 179
Baboon (primate)
  Chacma 104, 160, 217, 336
  Olive 217, 264
  Yellow 160, 256
Baboon, Socotra Island Blue (tarantula) 136
Badger, Honey 75, 231, 240, 312, **425**
Bald Ibis
  Northern **323**, 421
  Southern **296**
Bamboo 191
  African Alpine 175
Bamboo Lemur
  Gray **183**

Lac Alaotra 361
Northern **183**
Western Lesser 211
Banana, Water 360
bananas **410**, 416, 417
Baobab 247
  African 217
  Fony **130**
  Grandidier's 210, 289, **289**
baobabs 127, 128, 210, **211**, 289, 414
Barbet
  African **30**
  Anchieta's 257
  Banded 65, **65**, 250
  Black-collared 336
  "Brown-faced" Black-backed **152**
  Crested 250
  Miombo 257
  Pied **30**, 105
  Vieillot's 266
  White-eared 167
  Whyte's 257
  Yellow-breasted 233
Bat
  Madagascar Fruit 290
  Seychelles Flying Fox 200
  Straw-colored Fruit 188, 425
  Yellow-winged **266**
Bateleur 249, 277
"Batha" 327
Batis
  Chinspot 277, 336
  Leisler's 140
  Margaret's **205**, 206
  Pririt 105, 241
  Tenerife Long-eared 140
  Woodward's 161, 168
batises **30**
Bay-Owl, African 178
beach(es)
  Asian Temperate Rocky Coastline and Sandy 400
  Australian Sandy 397
  Nearctic Sand Beach and Dunes 397
  Sandy Beach and Dunes 386, 390, 393, **397**, 397–399, **398**, **399**, 400, 403, 406
Bear, Atlas Brown 56
Bee-eater
  Blue-breasted 417
  Cinnamon-chested 417
  Ethiopian 417
  Eurasian 421
  Little 414
  Little Green 415
  Madagascar 290
  Northern Carmine **232**, 385
  Red-throated 266
  Rosy 385
  Southern Carmine 385, **385**
  White-fronted 385
  White-throated 414
bee-eaters 385
bee fly **306**

Beech Forest 138
Beetle, Cape Stag 308
beetles, darkling 92–93
Beira 271
Bernieria, Long-billed 213
Berzelia 302, **302**, 304
biogeography 234
bird migration *see* migration, bird
Bishop
  Black-winged 417
  Cape 319
  Northern Red 362
  Southern Red 362, 377, **377**, 419
  Yellow 369, 377, 419
  Yellow-crowned 369, 377
Bismarckia nobilis 289
Bittern
  Dwarf **368**, 369, 376
  Great 376
  Little 362
  White-crested 171
Blackbird, Eurasian 141
Blackcap
  Bush 299
  Eurasian 65, 141, 324
Blackthorn 122
Black-Tit, Southern 278, 336
Blesbok 425
Blue Baboon, Socotra Island (tarantula) 136
Bluebuck 317, 318
Bluebush and Saltbush 107, 115, 121
Blue-eared Starling
  Greater 225
  Lesser 257, 266
Blue-Pigeon
  Madagascar **181**, 192
  Seychelles 201, **202**
Boa
  Dumeril's 132
  Round Island Keel-scaled 405
Boar, Wild 323
Boer-bean, Karoo 333
Bokmakerie 105, 312, 318, **319**
Bongo 147
Bonobo 147, **147**
Bontebok 304, 312, 317, 318, **318**
Booby
  Brown 404, 405
  Masked 404
  Red-footed **403**, 405
Boomslang 278, 314
Bou Amama 94–95
Boubou
  Ethiopian 417
  Southern 314, 337
  Tropical 155, 206, 417
  Zanzibar 161
Brachiaria **104**
Brachystegia (Miombo) 159, 204, 252, 253, **255**, **258**, 261, 275
bracken 423
Bracken-Warbler, Cinnamon 356

INDEX

Brigalow 252, 260
Bristlebill, Red-tailed 207
Broadbill
  African 155, 206
  Grauer's **32**, 178
Brownbul, Terrestrial 337
Brubru 241, 278
Brush-Warbler
  Grande Comoro 356
  Subdesert 129
Buffalo
  African 65, 167, 177, 206, 217, 240, 249, 264, 277, 283, 304, 317, 336, 361
  "Forest" African 147, 170
Bulbinella **374**
Bulbul
  Cape 318, 418
  Common 331
  Dark-capped 336
  Madagascar 290, 411
  Malagasy 340
  Red-eyed 91
  Seychelles 201
  Swamp Palm 171, 411
Bullfinch, Azores 141
Bullfrog
  Giant 106
  Madagascar **292**
Bunting
  Cape 307, 318
  Cinnamon-breasted 272
  Golden-breasted 241
  House 99
  Lark-like 81, 104, 105
  Socotra 136
  Somali 233
Burnet, Spiny 326
Bushbaby, Mohol 173
Bushbuck 65, 167, 177, 188, 250, 264, 277, 336, 425
Bush-Crow, Stresemann's **412**, 428
Bushpig 167, 206, 336
Bushshrike **30**
  Black-fronted 155
  Four-colored 155, 168, 206
  Gray-headed **30**, 336
  Green-breasted 179
  Many-colored 206
  Mount Kupe 179
  Olive 337
  Perrin's 206
  Rosy-patched 233
  Sulphur-breasted 155
  Uluguru 179
Bush Squirrels
  Ochre 160
  Red 167
  Smith's 160, 167, 277
bushwillows 229, 239, **245**, 261
Bustard
  Arabian **124**
  Black 313, **313**, 318
  Blue 298

Buff-crested 233
  Denham's 307, 419
  Heuglin's 125
  Karoo **103**, 104, 419
  Kori 231, **232**
  Little 60
  Ludwig's 104
  Nubian 124
  Rüppell's **78**, 79
Butterspoon Tree 177
Buttonquail 307
  Black-rumped 286
  Fynbos 304, 307
  Madagascar 290
  Small 327
Buzzard
  Common 314
  Forest 337
  Grasshopper 414
  Jackal 105, 307, **419**
  Long-legged 60, 69
  Madagascar 290
  Red-necked 414
  Socotra 136

C

Cabbage, Water 359
cacao 410, 411
Caco
  Cape 319
  Karoo 106, 114
  Namaqua 114
Cactus, Mistletoe 190
calcrete 126
California Oak Savanna 338
Camaroptera, Green-backed 337, 411
camels 85
Campo 280
Campo Cerrado 221
Campo Sujo 227, 235, 287
Canary
  Black-faced 155
  Black-headed 81, 104
  Black-throated 241
  Brimstone 313, 319
  Cape 307, 319, 417
  "Damara" Black-headed 114
  Island 141
  Lemon-breasted 168
  Papyrus 364
  Protea 304
  Streaky-headed 313, 319
  White-throated 313, 319
  Yellow 313, 319
Canaryberry, Sand 166
Canary Island Pine 68
"candelabra trees" 217, **218**, 334
Canegrass Dunes 82, 88
Cane Rat, Greater 146
Caracal 103, 118, 231, 240, 312, **313**, 317
Carapa 169
Carex 295
carrion flowers 334

Cashew 422
Caspian Wormwood Desert 101, 107, 115, 121
Casuarina 397
Cat
  African Golden 147
  African Wildcat 97, 312, 323, 425
  Black-footed 104, 240
  Sand 75, **75**, 97
Catbird, Abyssinian 65
"cathedral Mopane" 221, **223**
Caucasian Shrub Desert 71, 77, 82, 94, 101, 107, 115, 118, 121, 124
Cave-Chat, Angola **274**
Cedar
  Atlas 54, 55, 56, 59
  Clanwilliam 63
  Lebanese 54
  Mulanje 64
Central Africa, habitats 39–41
cereal crops 410, 412, 415, **416**, 418, 419
Cerradão 203
Cerrado Sensu Stricto 243
Cerrasco 332
Chaffinch
  Common 141, 423
  Gran Canaria Blue 141
  Tenerife 141
Chameleon
  Cape Dwarf 314
  Elandsberg Dwarf 307
  Elongate Leaf **195**
  Jeweled 292
  Karoo Dwarf 105
  Labord's 213
  Mediterranean 69
  Namaqua 92, **92**, 105
  Namaqua Dwarf 314
  Oustalet's 186
  Panther **292**
  Parson's 186
  Seychelles Tiger 202
  Socotra 136
  Southern Dwarf 337
  Uluguru Pygmy 179
  Usambara Three-horned **175**
  Usambara Two-horned 179
chameleons, dwarf 186–187
Chanting-Goshawk
  Dark 277, 331
  Eastern 234
  Pale 234, 241
Chaparral, Pacific 300, 309, 315, 321
Chat
  Angola Cave- **274**
  Arnold's 225
  Buff-streaked 273, 298
  Cape Robin- 318
  Familiar 81, 272
  Herero 269
  Karoo 104
  Mocking Cliff- 272
  Moorland 348, 356

Chat *continued*
    Mountain 81
    Red-capped Robin- 206, **207**
    Sickle-winged 104, **345**
    Tractrac 79, 91
    White-browed Robin- 417
    White-throated Robin- **166**, 250
    White-winged Cliff- 272
Chatterer, Fulvous **86**
Cheetah 79, 240, 264, **285**, 317, 427
    "Saharan" 97
Chenolea diffusa 396
Cherry, Mountain 59
chestnuts, tropical 271
Chevrotain, Water 148
Chihuahuan Desert 115
Chimpanzee 147
"Chipya" 153, 203, 204, **206**
Chough
    Alpine 353
    Red-billed 348, 353
Chrysocoma ciliata 346
cichlids 380, 381
Cistanche phelypaea **87**
Cisticola
    Aberdare 357
    Chatterine 411
    Chubb's 417
    Desert 286
    Grey-backed 318
    Kilombero 364
    Levaillant's 396
    Madagascar 290, 340, 417
    Piping 278
    Rock-loving 272
    Socotra 136
    Tinkling 278
    White-tailed 364
    Zitting **372**, 417
Cities and Villages **424**, 424–426, **425**, 346
Civet, African 147
Cliff-Chat
    Mocking 272
    White-winged 272
cliffs *see* Inselbergs and Koppies
Cloudforest 174, **190**
    Andean 174, 189
    Central American 174, 189
Clubrush, Sea 375
Coastal Heath, West European 354
coastline
    Asian Temperate Rocky, and Sandy Beach 400
    Nearctic Rocky 400
    Neotropical Rocky 400
    Rocky 390, 393, 403, 406
Cobra
    Anchieta's 279
    Cape 314
    False 120
    Forest 162, 168
    Mozambique Spitting **161**
    Shield 241

Snouted 241
Coffee and Cardamon 415
coffee plantations 416, 417
Cola, Hairy 166
Collared-Dove, African 231
Colobus
    Angola 64, 160, 177
    Guereza 64, **66**, 177, 425
    Niger Delta Red 170–171
    Tana River Red 160
    Uganda Red **147**
    Western Red 388
    Zanzibar Red 161
commiphoras **229**
Conifers (habitats) 54–69
    Afrotropical Montane Dry Mixed Woodland 57, 61, **61**, 61–66, **63**, **64**, **65**, **66**, 174, 176, 179, 247, 251, 299, 347, 354, 356, 415, 422
    Maghreb Fir and Cedar Forest **54**, 54–57, 59, 60, 69, 322, 323, 327, 328, 329, 330, 350
    Maghreb Juniper Open Woodland 54, 55, **57**, 57–60, **58**, **60**, 61, 67, 321, 322, 350, 419
    Maghreb Pine Forest 54, 57, **67**, 67–69, **68**, **69**, 138, 322, 422, 423
conifers, montane *vs* lowland forest 176
Coot, Red-knobbed 381
Cordgrass 395
Cordonbleu
    Red-cheeked 415
    Southern 241
Cormorant
    Bank 398, 400, **401**, 404
    Cape 398, 400, **401**, 404
    Crowned 398, 400, 404
    Great 381, 403
    Long-tailed 381
cosmos 287
Coua
    Blue 192
    Coquerel's 213
    Giant **212**, 213
    "Green-capped" Red-capped 131
    Red-breasted 185
    Red-fronted 192
    Running 129
    Verreaux's 131
Coucal
    Black 286, 369
    Coppery-tailed 362
    Madagascar 411
    White-browed 362
Courser
    Burchell's 79, 105
    Cream-colored **73**, 124
    Double-banded 105, 123
    Somali 124
    Temminck's 286
crab, beach 399
Crab, Coconut 202
Crab-Plover 394, **394**

Crag Lizard
    Common 348
    Drakensberg **348**
    Lang's 348
Crake
    African 369
    Baillon's 376
    Black 376
    Striped 368, 376
Crane
    Blue 298, **418**, 419
    Common 417, 421
    Gray Crowned- 361
    Wattled 348, 361, 376
Crassulacean acid metabolism (CAM) 111
Creeper, African Spotted **257**, 266
Crested Guineafowl
    Southern **167**, 168
    Western 155, 206
Crested Tern
    Great 398, 401
    Lesser 398, 401, 405
Crocodile
    Central African Slender-snouted **171**
    Nile 99, 361, 380, 384, 388
Crombec
    Cape **53**, 105, 241, 312
    Green 219
    Northern 231
    Red-capped **207**, 257
Cropland
    Asian Temperate 419
    Australian Temperate 418
    Nearctic 418
    see also Cultivation
Crossbill, Red 69
Crow
    Cape 417, 419
    House 425
    Stresemann's Bush- **412**, 428
Crowned-Crane, Gray 361
Cryptosepalum exfoliatum 203
Cuckoo
    Common 421
    Dideric 417, 426
    Klaas's 417, 426
    Levaillant's **66**
    Olive Long-tailed 207
    Red-chested 411
Cuckoo-Roller **32**, 44, 192
Cuckooshrike
    Black 336
    Gray 178
    Petit's 219
    Purple-throated 206
    Réunion 192
    White-breasted 225, 266
Cucumber, African Horned 239
Cucumber Tree 135
Cultivation
    Humid Lowland **409**, 409–411, **410**, **411**, 412, 415, 417

INDEX

North African Temperate 418, 419–421, 420, 421
Savanna 290, 409, 410, 412, 412–415, 413, 414, 415, 415
South African Temperate 316, 319, 415–416, 418, 418–419, 419
Tropical Montane 409, 414, 415–417, 416, 418, 419
cultivation, "shamba"/shamba 416, 417
Curruca warblers 323, 327
Cyanometra webberi 159
Cycad
  Eastern Cape Blue 337
  Karoo 337
Cyperus conglomeratus 122
Cypress
  Mexican 422
  Saharan 96
cypresses 55

D

daisies (Asteraceae) 109, 110, 309, 310, 316
Dambo 366, 366, 368, 369
Darter, African 361
Deccan Thornscrub 227
deciduous forest
  African Dry 208, 209
  Angolan 151, 203, 209, 215–219, 216, 217, 218, 219
  Asian Hyrcanian Temperate 138
  dendrogram 208
  European 138
  Indian Dry 209, 252
  Malagasy *see* Malagasy Deciduous Forest
  Pacific Dry 209, 215
  Southeast Asian Dry 209, 215, 260
  tropical *see* Tropical Deciduous Forests
Dehesa, Oak 275, 338
dendrograms
  African Deserts 70
  African Dry Deciduous Forests 208
  African Heathlands 342
  African Humid Forests 142
  African Savannas 220
"derived" grasslands 282
desert(s) 33, 53, 74, 77
  Asian Cold Erg 88
  Atacama Desolate 82, 88
  Caspian Wormwood 115, 121
  Chihuahuan 115
  cold 70
  dendrogram 70
  Dragon's Blood Tree Semi- 34, 133–137, 134, 135, 136, 137
  Galápagos Lowland 133
  hot 70
  hyperarid 77, 88
  Nearctic Desolate 82, 88
  Palearctic Shrub 71, 77, 82, 94, 101, 107, 115, 121, 124
  Sonoran 133
  Turanian Cold Erg 82, 88

see also Deserts and Arid Lands (Africa)
Desert Cauliflower 94–95
desertification 126, 230
Deserts and Arid Lands (Africa) 70, 71–137
  Afrotropical Hot Shrub Desert 77, 101, 107, 115, 118, 119, 121–125, 122, 123, 124, 125, 133, 136, 227–228, 233, 427
  Dragon's Blood Tree Semi-Desert 34, 133–137, 134, 135, 136, 137
  Maghreb Hot Shrub Desert 71, 73, 77, 82, 83, 94, 100, 101, 107, 115, 115–120, 116, 117, 118, 119, 119, 120, 121, 123, 124, 322
  Nama Karoo 72, 77, 79, 88, 95, 101, 101–106, 102, 103, 104, 105, 106, 107, 112, 114, 115, 121, 235, 300, 309, 315, 319, 332, 336, 344, 354
  Namib Rock Desert 71, 77–81, 78, 79, 80, 81, 88, 90, 94, 101, 106, 107, 114, 123, 385
  Namib Sand Desert 77, 81, 82, 88–93, 89, 91, 92, 114, 123, 385, 397
  Rocky Hamada and Massif 62, 71, 73, 77, 82, 83, 94–99, 95, 96, 97, 98, 99, 100, 115, 419
  Saharan Erg Desert 71, 73, 82–87, 83, 85, 86, 88, 94, 100, 115, 121, 397
  Saharan Reg Desert 71, 71–76, 72, 73, 74, 75, 76, 77, 82, 83, 84, 85, 94, 97, 100, 115, 118, 121, 322, 419
  Spiny Forest 44, 127–132, 128, 129, 130, 131, 132, 133, 180, 189, 209, 211, 213, 214, 287, 332, 333, 338
  Succulent Karoo 77, 88, 101, 102, 106, 107–114, 108, 109, 110, 111, 112, 113, 114, 121, 122, 133, 286, 300, 309, 311, 314, 315, 319, 336
Desert Steppe
  Temperate 101, 107
  Western 101, 107
Dibatag 233
Dik-Dik, "Damara" Kirk's 240
dispersers, seed 188
Dodo 182, 186, 411
Dog, African Wild 74, 225, 242, 249, 256, 264, 265, 277, 317
Doka 253, 261, 266
Dolphin, Common Bottlenose 406
Dormouse, Black-tailed Garden 56
Dove
  African Collared- 231
  Cloven-feathered 181
  Emerald-spotted Wood- 336
  European Turtle- 60
  Laughing 426
  Lemon 207
  Malagasy Turtle- 201, 340
  Namaqua 231, 290
  Red-eyed 417, 426
  Ring-necked 277, 414
  Rock 60
  Stock 60

Tambourine 337
Vinaceous 415
Dragon's Blood Tree 134, 135, 137
Dragon's Blood Tree Semi-Desert 34, 133–137, 134, 135, 136, 137
Dragon Tree 162
Drill (monkey) 147
Drongo
  Aldabra 404
  Fork-tailed 251, 414, 427
  Glossy-backed 427
  Square-tailed 168, 206
Dry Chaco 221, 243, 332
Dry Deciduous Yungas 209, 215
dry montane woodland 61–62, 63, 64, 65, 66
Duck
  African Black 385
  Hartlaub's 171
  Maccoa 376
  Marbled 370
  Meller's 369
  Red-billed 284
  White-backed 362
Dugong 388, 394
Duiker
  Blue 65, 167, 177, 205
  Common 104, 217, 250, 264, 304, 312, 317, 336, 419
  Harvey's 177
  Natal Red 167
  Rwenzori Red 357
  Weyns's 177
  Yellow-backed 205
dune(s)
  Canegrass Dunes 82, 88
  European Dunes and Sandy Shores 397
  Nearctic Sand Beach and Dunes 397
  Sandy Beach and Dunes 386, 390, 393, 397, 397–399, 398, 399, 400, 403, 406
Dune Forest 164, 165, 168
Dune Sharks (Grant's Golden Mole) 90
dung beetles 226, 226
duricrusts 126
Dwarf Chameleon
  Cape 314
  Elandsberg 307
  Karoo 105
  Namaqua 314
  Southern 337
dwarf chameleons 186–187
Dwarf Gecko
  Angola 257
  Chobe 279
dwarf geckos 179, 273

E

Eagle
  African Fish- 361, 362, 381
  Black-chested Snake- 251, 286
  Bonelli's 60, 69, 324, 327
  Booted 60, 69, 324, 327, 331

Eagle *continued*
  Brown Snake- 277
  Crowned 167, 178, **178**, 337
  Fasciated Snake- 167
  Golden 60, 69, 348, 353
  Madagascar Fish- 214, 364, 389, **389**
  Madagascar Serpent- 185, 192
  Martial 105, **251**
  Short-toed 60, 69, 324, 327
  Tawny 105, 231, 241
  Verreaux's 272–273, 307
Eagle-Owl
  Cape 65, 356
  "MacKinder's" Cape 273
  Pharaoh 99
  Shelley's 148, 149
  Usambara 179
East Africa, habitats 35–38
East Coast Forest Matrix 61, 62, 147, 151, 156, 157–163, **158**, **159**, **160**, **161**, **162**, **163**, 164, 168, 173, 174, 203, 219, 243, 244, 248, 252, 258, **335**, 382, 386, 388
edaphic grassland 280–281
Eerica comorensis 355
Egret
  Black 417
  Cattle 421, 427
  Great 417
  Little 394, 417
Eland 312, 336
  Common 240, 295–296
  Giant 264
Elephant
  African Forest 147, 170, 177
  African Savanna 64, 154, 167, 177, 206, 217, **222**, 225, 240, **248**, 249, 256, 264, 277, 304, 317, 336, 337, 380, 417
  "Desert" African Savanna 79, **80**
  "North African" Savanna 56
elephants, as habitat architects 337, 373
Elephant Shrew
  Golden-rumped **160**
  North African 118, 119
elephant shrews 160, **160**
emperor moth, caterpillar of 226
Emutail
  Brown 192
  Gray 369
Enset 417
Ephedra tilhoana 95
Eremomela
  Black-necked 257
  Greencap 278
  Yellow-bellied 241
  Yellow-rumped 104
Erg 49, 53, 71, 82
Erg Desert
  Asian Cold 88
  Saharan 71, 73, 82–87, **83**, **85**, **86**, 88, 94, 100, 115, 121, 397
  Turanian Cold 82, 88
Erg seas 82
ericas 301, **301**, 303, 354

Erica spp. 140, 301, 355–356
eucalyptus 288, **290**, 422
Euphorbia (euphorbias) **109**, 117, 128, **134**, **246**, 333, **334**
  King Jubia 117
Euphorbia candelabrum 217, **218**, **334**
Euphorbia stenoclada 128, **129**
European Alpine Tundra 350
European Arctic and Temperate Rocky Headland 400
European Coastal Salt Marsh 395
European Deciduous Rainforest 138
European Dunes and Sandy Shores 397
European Mixed Broadleaf Conifer Forest 76
European Reedbeds 358, 374

F

Falanouc, Eastern 185
Falcon
  African Pygmy 234
  Eleonora's 403
  Pygmy 241
Fanaloka 185
"farmbush", cultivated 409, **409**
ferricrete 126
Fever Tree **246**, 247
fig marigolds (Aizoaceae) 107, **109**
figs, rock-splitting 271
Finch
  African Crimson-winged 351
  Crimson-winged 351, **353**
  Inaccessible Island 404
  Trumpeter **75**
finches, *Neospiza* 404
Finfoot, African 171, **172**, 385
Fir
  Algerian 55
  Moroccan 55
fire 217, 237, 276, 287, 302, 303, 317, 366
  resistance 59, 62, 223, 338
Firecrest 56
  Common 331
  Madeira **139**, 141
firefinches 273
Fire Lily 303, **303**
Fiscal
  Common 411, 417
  Gray-backed 414
  Newton's 148
Fish-Eagle
  African **361**, 362, **381**
  Madagascar 214, 364, 389, **389**
Fishing-Owl
  Pel's 155, **155**
  Rufous 149, 171
  Vermiculated 171, **171**
Fish Poison Tree **198**
Flag, Black **311**
Flamingo
  Greater 391, **391**
  Lesser 391, **391**
flamingos 390, 391

Flax-Lily, Cerulean 288
Flowing Rivers 382
Flufftail **29**
  Buff-spotted 168
  Chestnut-headed 368
  Madagascar **29**, 192, 356, 369
  Red-chested 369, 376
  Slender-billed 369
  Streaky-breasted 286, 368, 376
  Striped 296, 304, 376
  White-spotted 171, 411
  White-winged 376
Flycatcher
  African Blue 417
  African Crested 168, 206, 337
  African Paradise- 277, 417
  Atlas 56, 69, 331
  Böhm's 257
  Cassin's 171, 385
  Dusky-blue 411
  Grand Comoro 186, 192
  Livingstone's 161
  Mariqua 225
  Seychelles Paradise- 201, **202**
  Spotted 65, 331
  Yellow 161
Flying Fox (bat)
  Madagascar 290, 389
  Seychelles **200**
flying foxes (bats) 161, 185
Fody 404
  Forest 192
  Red 200–201, 340, 411
  Seychelles 201
food crops 410, 412–413, 415, 416–417, 419
Forest
  African Humid Forests, dendrogram 142
  Afrotropical Montane *see* Moist Montane Forest
  Afrotropical Swamp 39, 142, 143, 147, 152, 159, 169–173, **170**, **171**, **172**, **173**, 180, 358, 360, 364, 380, 382, 386, 388
  Australasian Mangrove 386, 388
  Central American Semi-evergreen 151
  deciduous *see* deciduous forest
  Dune 164, **165**, 168
  European Mixed Broadleaf Conifer 76
  granite 197, 199, **199**
    see also Seychelles Granite Forest
  Groundwater 153, **156**
  Himalayan Pine 67
  igapó flooded 169
  Indo-Malayan Limestone 197, 268
  Indo-Malayan Mangrove 386, 388
  Jack Pine 67
  Laurel 50, 138–141, **139**, **140**, **141**
  Lodgepole Pine 67
  Maghreb Fir and Cedar **54**, 54–57, 59, 60, 69, 322, 323, 327, 328, 329, 330, 350
  Maghreb Pine 54, 57, **67**, 67–69, **68**, **69**, 138, 322, 422, 423
  Malabar Semi-evergreen 157, 164, 203

INDEX

Mediterranean *see* Mediterranean Forests, Woodlands and Scrubs
Mediterranean Dry Pine 67
Mediterranean Fir and Cedar 67
Mediterranean Juniper and Cypress 54
Mediterranean Oak 57, 328
Mediterranean Pine 54
Mesoamerican Dry Deciduous 209, 215
Middle Eastern Juniper 57
Moist Montane *see* Moist Montane Forest
Monsoon *see* Monsoon Forest
Nearctic Montane Mixed-Conifer 54, 67
Neotropical Semi-evergreen 151, 157, 164
rainforest *see* rainforest
Sand 164, 165, 168
Seychelles Granite 143, 180, 189, 197–202, **198**, **199**, **200**, **201**, **202**
Southeast Asian Semi-evergreen 151
Spiny *see* Spiny Forest
Swamp *see* Swamp Forest
Teak 275
Tropical Deciduous *see* Tropical Deciduous Forests
Várzea flooded 169
Warm Humid Broadleaf *see* Warm Humid Broadleaf Forests
see also Woodland
Forest Matrix
East Coast *see* East Coast Forest Matrix
South Coast *see* South Coast Forest Matrix
Fossa 129, 184, **212**, 213
Fox
Bat-eared 79, 103, 231, 240, 312
Cape 103, 240, 296, 304, 312
Fennec 75, 86, **86**, 97, 124
Pale 124, 233
Red **60**, 351
Rüppell's 75, 97
Francolin
Chestnut-naped 357
Crested 249
Gray-winged 313, 318
Hartlaub's 273
Jackson's 299, **356**, 357
Moorland 348, 357
Swierstra's 179
Freshwater Habitats (Africa) 358–385
Afrotropical Deep Freshwater Marsh 214, 244, **358**, 358–364, **360**, **361**, **362**, **363**, **364**, 365, 368, 374, 378, 379, 380, 382, 384, 390, 393, 395, 397, 400
Afrotropical Shallow Freshwater Marsh 244, 255, 358, 360, **365**, 365–369, **366**, **367**, **368**, **369**, 374, 378, 379, 382, 390, 393, 395, 397, 400
Freshwater Lakes and Ponds 358, 365, 370, 374, **379**, 379–381, **380**, **381**, 382
North African Temperate Wetland **370**, 370–372, **371**, **372**, 374, 379, 382, 390, 393, 395

Rivers 358, 365, 370, 374, 379, **382**, 382–385, **383**, 384, **385**
South African Temperate Wetland 358, 364, 365, 369, 370, **374**, 374–378, **375**, **376**, **377**, **378**, 379, 382, 390, 393, 395, 395
Freshwater Lakes, Dams, and Ponds 379
Freshwater Lakes and Ponds 358, 365, 370, 374, **379**, 379–381, **380**, **381**, 382
Frigatebird
Great 404, 408
Lesser 404, 408
Frog
Cape Mountain Rain **307**
Cape Sand 114
Desert Rain 92
Drakensberg 348
Ice 348
Long-toed Tree 377
Madagascar Bright-eyed 192
Madagascar Jumping 192
Marbled Rain 187
Marbled Rubber 81
Mascarene Ridged 292
Mediterranean Tree 353
Micro 378
Montane Marsh 378
Moroccan Painted 353
Mozambique Forest Tree **167**
Namaqua Rain 114, **114**
Namaqua Stream 114
Sand Rain 314
Starry-night Reed 187, **187**
Stripeless Tree 324, 331
Tomato **187**
frogs
bright-eyed 187, 192
ghost- 308
Madagascar fringed 187
moss 308
Fruit Bat
Madagascar 290
Seychelles 200
Straw-colored 188, 425
fruit bats, epauletted **188**
fruit crops 415, 417, 418, 419
Fynbos 63, 101, 103, 106, 107, 114, 286, 300–308, **301**, **302**, **303**, **304**, **305**, **306**, **307**, **308**, 309, 312, 314, 315, 318, 319, 321, 326, 332, 354, 378, 418

G

Galago
Southern Lesser 173
Thomas's 148
galagos 173
Galápagos Lowland Desert 133
Gallinule, Allen's 376
Gannet, Cape 400, **403**, 404, **404**
Garrigue 326
Asian 326
Maghreb 49, 59, 118, 321, 322, **324**, 326, 326–327, **327**, 328, 329, 330, 419

Gazelle
Cuvier's 69, 330
Dama 73, 124, **124**, 233
Dorcas 74, 85, 97, 124, 233
Red 56
Red-fronted 124, 233
Rhim 74, 85, 97, 124
Thomson's 283, 425
gazelles 283
Gecko
African Wall 99
Agadir 75
Algerian 99
Angola Dwarf 257
Barbour's Day 273
Böhme's 119
Chobe Dwarf 279
Common Giant Ground **106**
Common House 426
Common Leaf-tailed 187
Drakensberg Rock 348
Dune 86, **87**
Hawequa Flat **307**
Helmethead 120
Lined Leaf-tailed 187
Madagascan Ground **131**
Morocco Lizard-fingered 120
Mossy Leaf-tailed **195**
Namib Dune 91
Namib Sand 91, **92**
Oudri's Fan-footed 75, 120
Round Island Day 405
Satanic Leaf-tailed 187
Socotran Giant **137**
Tropical House 426
Wahlberg's Velvet **168**
White-spotted Wall 99
geckos 136, 179, 426
dwarf 179, 273
ground 187
leaf-toed 308
thick-toed 106
in urban habitats 426
Gelada 298, **298**, **356**, 357
Gemsbok (Southern Oryx) 79, **79**, 89, 103, 225, **237**, 240, **282**, 312
Genet
Cape 304, 312, 317
Common 425
Giant 147
Large-spotted **155**
Miombo 256
Rusty-spotted **155**, 277
geographical regions, habitats of 33–53
geophytes 109, 301, 303, 309
Gerbil
Dune Hairy-footed 90
Hoogstraal's 327
Lataste's 56, 69
Occidental 327
Western 327
Gerenuk **231**, 233
ghost-frogs 308

INDEX

Giant Gecko, Socotran **137**
giant-petrels 408
Giant Tortoise, Aldabra 200, **201**, 201–202, 403
Gibbaeum 317
Gibber Chenopodlands 71, 72, 101, 107
Gifboom 78
Giraffe 167, 225, 231, **238**, 239, **239**, 240, 249, **250**
    Angolan 79
Girdled Lizard
    Armadillo **114**
    Black **314**
    Large-scaled 314
    Western Dwarf 105
    Zimbabwe 296
Gnu, White-tailed (Black Wildebeest) 103, **295**, 296
Godwit, Bar-tailed 394
Golden Mole
    Cape 112, 312
    De Winton's 112
    Grant's 112, 312
    Van Zyl's 112
    Visagie's 104
Goldfinch, European 141
Gonolek
    Crimson-breasted 225, 241, **241**
    Papyrus 362, **362**, 364
Goose
    African Pygmy- 362
    Egyptian 425
    Spur-winged 381
Gorilla
    Eastern 147
    "Mountain" Eastern 178, **178**, 296, 417
    Western 147
Goshawk
    Dark Chanting- 277, 331
    Eastern Chanting- 234
    Gabar 241
    Pale Chanting- 234, 241
granite forest 197, 199, **199**
    see also Seychelles Granite Forest
Grass
    Gamba 229
    Hippo 359, 360
    Mountain Wire 346
    Red 237
    Red Oat 295
grass(es)
    bushman 78
    thatching 261
Grassbird, Cape 298, 304
Grasscutter 146
grasshopper (*Phymateus saxosus*) 292
Grassland 280–299
    Australasian Montane 293
    Australian Tropical Tussock 280
    Hawaiian 293
    Indian Tropical 287
    Mesoamerican Savanna and 280
    Shola 293

Terai Flooded 280
Grassland (Africa)
    Afrotropical 45, 121, 145, 150, 164, 167, 203, 227, 235, 256, 261, 280–286, **281**, **282**, **283**, **284**, **285**, 287, 293, 334, 358, 365, 374, 412, 415, 418, 419, 427
    Malagasy Grassland and Savanna 209, 211, 286, 287–292, **288**, **289**, **290**, **291**, **292**, 299, 338, 340, 409, 412, 415, 417
    Montane 34, 38, 40, 45, 61, 64, 101, 103, 106, 107, 175, 251, 280, 286, 293–299, **294**, **295**, **296**, **297**, **298**, 300, 315, 332, 343, 344, 347, 348, 354, 356, 378, 412, 415, 418, 419, 427
grasslands
    "derived" 282
    edaphic 280–281
Grass-Owl, African 362
Grassy Mulga 227, 260, 338
Grazing Land **427**, 427–428, **428**
Greater Southern Africa, habitats 44–46
Great Rift Valley 44, 349, **349**, 379
Grebe
    Eared 392
    Great Crested 381
    Little 381
Greenbul
    Black-browed Mountain 65
    Cabanis's 206
    Fischer's 161
    Gray-olive 155
    Pale-olive 219
    Red-tailed 148
    Simple 411
    Sombre 168, 337
    Yellow-bellied 206, 219
    Yellow-necked 219
    Yellow-throated 219
Greenfinch, Eurasian 421
Green-Pigeon
    Bruce's **188**, 417
    Madagascar 213
Greenshank, Common 377, 396
Grosbeak
    São Tomé 148
    Socotra 136, **137**
Ground Gecko, Common Giant **106**
Ground-Roller 32
    Long-tailed 129, 132, **132**
    Pitta-like 192, **194**
    Rufous-headed 32, 192
    Scaly 185, **185**
    Short-legged 185, **185**
groundsel, giant 346
Ground Squirrel, Barbary 118
Ground-Thrush, Spotted 168
Groundwater Forest 153, **156**
Grysbok
    Cape 304, 312, 317
    Sharpe's **224**, 225
"guelta" 95
Guineafowl 29

    Helmeted **29**, 249, 290, 313, 318, 340, 419, 425
    Southern Crested **167**, 168
    Western Crested 155, 206
Guinea Savanna 34, 36, 39, 51, 62, 85, 143, 145, 146, 150, 227, 230, 243, 248, 250, 252, 253, 260–267, **261**, **262**, **263**, **264**, **265**, **266**, 275, 281, 286, 382, 409, 411, 412
    South to North transition 267
Gull
    Gray-headed 381
    Hartlaub's 404
    Kelp 401, 404
    Lesser Black-backed 398
    Mediterranean 398
    Sabine's 408
    Sooty **403**
    White-eyed 405
    Yellow-legged 398
Gum, Southern Blue 140
Gum Acacia 117, 422, 423
Gundi
    Felou 98
    Mzab 98, **98**
    Val's 98
gundis 98, **98**
Gusu 45, 62, 204, 215, 221, 226, 235, 243, 244, 248, 251, 252, 253, 258, 260, 275–279, **276**, **277**, **278**, **279**, 412

H

habitat architects 337, 373
habitat fluctuations 234
Had (*Cornulaca monacantha*) 72
Hagenia (trees/bushes) 62, 299, 354
Hamada 49–50, 53, 71, 82, 94, **100**
Hamada Montane Woodland 95
Hamerkop 361, **382**, 385, 417
Hare
    Cape 283
    Ethiopian Highland **346**, 347
    Scrub 104
Harrier
    Black 313, 318, **320**
    Madagascar 290
harriers 362
Hartebeest 240, 368
    Bubal Red 56
    "Coke's" **426**
    Red 104, 283, 312, 336
Hawaiian Grasslands 293
Hawfinch 69, 331
Hawk, Bat 423
Heath, Tree 356
heathland
    African, dendrogram 342
    Afrotropical montane *see* Alpine Tundras and Montane Heaths
    Asian Temperate 300, 309, 315
    Australasian Montane 315
    Australian Southern Lowland 300, 309
    Fynbos *see* Fynbos

INDEX 437

Northwest European Coastal and
  Montane Heath 300, 309, 315
West European Coastal Heath 354
Hedgehog
  Algerian 323
  Desert 75, 97
  Long-eared 118, 323
Helmetshrike
  Angola **218**
  White 225, 278
  Yellow-crested 178
herbs 327
Heron
  Goliath 361, **361**
  Gray 381, 394
  Malagasy Pond- 364
  Rufous-bellied 362, 376
  Western Reef- 398
  White-backed Night- 171, **173**, 362
Heuweltjies **110**, 111–112, 114
High Atlas Alpine Meadow 55, 350–353, **351**, **352**, **353**
Highland Acacia Woodland 247, **247**
Highland Ouhout Shrubland 64, 299, **299**
Himalayan Pine Forest 67
Hippopotamus 284, 360, 273, **373**, 376, 380, **384**, 425
  Pygmy 171
Hog
  Giant Forest 147, 177
  Red River 147, **150**
Holly, Common 59
Honeyguide
  Least 207
  Scaly-throated 337
Hoodia **333**
Hoopoe
  Eurasian 414, 421
  Madagascar 290, 340
Hoopoe-Lark
  Greater **84**, 86
  Lesser 125
Hornbill
  African Gray 277
  Black-and-white casqued 178
  Black-casqued 148
  Bradfield's 277, **278**
  Crowned 65, 167, 337
  Ground **30**
  Pale-billed 257
  Silvery-cheeked 178
  Southern Ground **30**
  Trumpeter 167
  White-crested 148
  Yellow-casqued 148
hornbills 148, 428
"Horn of Africa" 33–34
Humid Broadleaf Forests *see* Warm Humid Broadleaf Forests
Humid Chaco 252, 260
Humid Lowland Cultivation **409**, 409–411, **410**, **411**, 412, 415, 417

Humid Puna 293, 343
Hyacinth, Common Water 359, 360, 384
Hyena 296
  Brown 79, 90, **91**, 225, 240, 398, 400
  Spotted 74, 79, 225, 231, 240, 249, 250, 336, 425
  Striped 97, 231
Hylia, Green 148
Hyliota **31**
  Southern **31**, 207
  Usambara 161
Hyphaene 247
Hyrax 272, **272**, 273
  Eastern Tree 177
  Rock 97, **268**, 272, **272**

I

Ibex
  Nubian 97, 118
  Walia 65, **294**, 347
Ibis
  Dwarf 148
  Hadaba 425
  Madagascan **185**
  Northern Bald **323**, 421
  Southern Bald **296**
  Spot-breasted 171
  Wattled 298, 348, 417
Ice-Rat, Sloggett's **347**
Igapó flooded forest 169
Impala 225, 240, 249, 425
  Black-faced 225
Indian Dry Deciduous Forest 209, 252
Indian Ocean, habitats 41–44
Indian Ocean Lowland Rainforest 43, 129, 132, 142, 143, 169, 180–188, **181**, **182**, **183**, **184**, **185**, **186**, **187**, 189, 197, 209, 211, 287, **398**, 409
Indian Ocean Montane Rainforest 43, 129, 132, 142, 174, 180, 182, 189–196, **190**, **191**, **192**, **193**, **194**, **195**, **196**, 197, 199, 211, 287, 409
Indian Tropical Grassland 287
Indigobird **31**
  Village 426
Indo-Malayan Limestone Forest 197, 268
Indo-Malayan Mangrove Forest 386, 388
Indo-Malayan Offshore Islands 403
Indri 183, 192, **193**
Inselbergs and Koppies 268, 268–274, **269**, **270**, **271**, **272**, **273**, **274**, 304, **366**
irises 316
Ironplum, Water 166
Island Arc Lowland Rainforest 180
Itigi-Sumbu Thicket 335
Itigi Thicket **153**, **153**, 156

J

Jacana
  African 362, **362**
  Lesser 369
  Madagascar 214, 364

Jackal
  Black-backed 79, 90, 103, 123, 225, 231, 240, 296, 391, 398, 400
  Golden 56
  Side-striped 249, 250, 361
jackfruit 410
Jack Pine Forest 67
Jaegers
  Long-tailed 408
  Parasitic 408
  Pomarine 408
Jay, Eurasian 330
Jellyfish Tree 199–200
Jerboa
  Four-toed 396
  Greater Egyptian 97
  Lesser Egyptian 74, 97
Jery
  Common 290, 340
  Stripe-throated 290
  "Subdesert" Stripe-throated 131
Juniper **58**, **63**
  African 62, **63**
  Cade 59
  Common 59, 350
  Greek 62
  Phoenician 55, 59, 60, 322
  Spanish 55, 56, 59

K

Kalahari Dry Thorn Savanna 77, 88, 101, 103, 105, 106, 107, 221, 225, 226, 227, 235–242, **236**, **237**, **238**, **240**, **241**, **242**, 243, 248, 250, 252, 258, 275, 277, 332, 390
kalanchoe 271
Kapok Bush 122
Karee 112
Karoo
  Nama Karoo 72, 77, 79, 88, 95, **101**, 101–106, **102**, **103**, **104**, **105**, **106**, 107, 112, 114, 115, 121, 235, 300, 309, 315, 319, 332, 336, 344, 354
  Succulent Karoo 77, 88, 101, 102, 106, 107–114, **108**, **109**, **110**, **111**, **112**, **113**, **114**, 121, 122, 133, 286, 300, 309, 311, 314, 315, 319, 336
Kashmir Tree 422
Kestrel
  Common 60, 69
  Eurasian 425
  Greater 105
  Lesser 421
  Madagascar 290
  Rock 425
  Seychelles 201
Kingfisher
  Collared 389
  Giant 361, 385
  Half-collared 385
  Madagascar 417
  Malachite 377, **387**
  Malagasy 389

INDEX

Kingfisher *continued*
   Mangrove 388
   Pied 377, **377**, 381, 385
Kipunji 177
Kite
   Black 60, 69, 290, 314, 327, 411, 414, 425
   Black-shouldered 417
   Black-winged 60, 69
   Scissor-tailed 233
Klipspringer 271, **271**
Knot, Red 394
Kob 264, 283
Kokerboom 112
Koppies *see* Inselbergs and Koppies
Korhaan, Southern Black **313**
"kraal" enclosure **427**
Kudu
   Greater 225, 240, 249, 250, 277, 336
   Lesser 233

L

Lala-Palm savanna 168
Lammergeier 353
"Landibe" (silk moth) 340–341
Lanner 241
Lapwing
   African Wattled 376–377
   Blacksmith 376
   Brown-chested **378**
   Crowned **282**
   Long-toed **363**
   Northern 421, **421**
   Spot-breasted 298, 348
lapwings 385
Lark 75
   "Agulhas" Long-billed 319, 419
   Ash's 125
   Barlow's 90, 91
   Bar-tailed 75
   "Beesley's" Spike-heeled **285**
   Benguela 79
   Botha's **298**
   Calandra 421
   Cape 91, 114
   Cape Clapper 307, 319
   Crested 124, 421
   Desert 75, 124
   Dune 90, 91, **91**
   Dupont's 327
   Fawn-colored 278
   Flappet 286
   Gray's 79
   Horned 351, **352**
   Karoo 104
   Large-billed 319, 419
   Madagascar 290, 340
   Masked **122**, 125
   Obbia 125
   Red 104, **104**
   Red-capped 417, 419
   Rufous-naped 417
   Sclater's 104
   Somali Long-billed 125

Stark's 79
Temmick's 75
Thekla 421
Thick-billed 75, **119**
William's **122**, 125
Lasimorpha senegalensis 360
Laurel
   Azores 140
   Bay 140
   Canary 140
   Portuguese Cherry 140
   Spurge 59
Laurel Forest 50, 138–141, **139**, **140**, **141**
Laurisilva 140
Leaf Chameleon, Elongate **195**
Leaf-tailed Gecko
   Common 187
   Lined 187
   Mossy **195**
   Satanic 187
Lechwe
   "Black" **284**
   Nile 283
   Southern 283
lechwes 361
Lemon-rope, Sand 166
Lemur 44
   Black-and-white Ruffed 183, **184**
   Cleese's Woolly 213
   Common 211
   Coquerel's Giant Mouse 213
   Crowned 211, **213**, 272
   Fat-tailed Dwarf 213
   Giant Mouse 411
   Gray Bamboo **183**
   Gray-brown 129
   Gray Mouse 129, 290
   Lac Alaotra Bamboo 361
   Madame Berthe's Mouse 213
   Mongoose 211
   Northern Bamboo **183**
   Northern Giant Mouse **411**
   Pale Fork-marked 213
   Pariente's Fork-marked 411
   Petter's Sportive 129
   Red-fronted Brown 211, **212**
   Red Ruffed 183
   Ring-tailed 129, 340, **340**
   Sanford's 211
   Tavaratra Mouse **212**
   Western Lesser Bamboo 211
   Western Woolly 213
   White-footed 129
lemurs 183, 192, 211, 213, 411
Lentisk 117
Leopard 65, 97, 147, 154, 167, 177, **224**, 240, 249, 250, 264, 265, 271, 277, 296, 425, 427
   "Barbary" 56, 351
   "Cape" 304, 312
lichens 78
   Spanish Moss-like old man's beard 255
   wandering 78

lilies 316
   Fire 303, **303**
   water 376
Limbali 145
Limestone Forest 211
   Indo-Malayan 197, 268
Linnet, Warsangli 65
Lion 74, 79, 177, 231, 240, 249, 250, 264, 265, 277, 296, 336, 427
   African 265, **265**
   Barbary 56
   "Cape" 304, 312
litchi 410
Lizard
   Armadillo Girdled **114**
   Black Girdled **314**
   Black-lined Plated 225, 278
   Bosc's Fringe-toed 120
   Bushveld 278
   Caprivi Rough-scaled 279
   Common Crag 348
   Cottrell's Mountain 348
   Desert Plated 91
   Dumeril's Fringe-toed 86
   Essex's Mountain 348
   Lang's Crag 348
   Large-scaled Girdled 314
   Long-fingered Fringe-toed 86
   Ornate Sandveld 296
   Small-spotted 120
   Snovel-snouted 91
   Three-eyed 132
   Tsingy Plated 214, **274**
   Wedge-snouted 91
   Western Dwarf Girdled 105
   Zimbabwe Girdled 296
lizards
   crag 106
   forest 179
   girdled 106
   grass 296
   plated 187
   sand 105–106
   sandveld 105
lobelias, giant 346, **346**
Lodgepole Pine Forest 67
Longclaw
   Abyssinian 298
   Cape **314**
   Orange-throated 377, 396
   Rosy-throated **283**
   Sharpe's 299
   Yellow-throated 286
Longtail, Green 179
Lovebird
   Black-winged 65, 250
   Fischer's 415
   Gray-headed **291**
Lowland Rainforest 142
   Afrotropical *see* Afrotropical Lowland Rainforest
   Australasian Tropical 143, 180
   Central American 143, 180

INDEX

Indian Ocean *see* Indian Ocean
Lowland Rainforest
Island Arc 180
Malabar 143, 180
Lowland Rivers 382

M

Macaque, Barbary 56, 323, 330
Madagascar Swift (reptiles)
Dumeril's 273
Grandidier's 273
Merrem's 132
Madwort, Spiny 350
Maghreb 47, **50**, 371
Maghreb Broadleaf Woodland 54, 56, 67, 321, 322, **328**, 328–331, **329**, **330**, **331**, 338
Maghreb Fir and Cedar Forest **54**, 54–57, 59, 60, 67, 69, 322, 323, 327, 328, 329, 330, 350
Maghreb Garrigue 49, 59, 118, 321, 322, **324**, **326**, 326–327, **327**, 328, 329, 330, 419
Maghreb Hot Shrub Desert 71, 73, 77, 82, 83, 94, 100, 101, 107, **115**, 115–120, **116**, **117**, **118**, 119, **119**, **120**, 121, 123, 124, 322
Maghreb Juniper Open Woodland 54, 55, **57**, 57–60, **58**, **60**, 61, 67, 321, 322, 350, 419
Maghreb Maquis 49, 55, 59, 64, 68, 138, 300, 309, 315, 321–325, **322**, **323**, **324**, **325**, 326, 327, 328, 356, 419
Maghreb Pine Forest 54, 57, **67**, 67–69, **68**, **69**, 138, 322, 422, 423
Magpie, Maghreb 69, 331
Magpie-Robin
Madagascar 290, 411
Maghreb **421**
Seychelles 201
Mahogany, Madeira 140
maize 410, 412, 415
Malabar Lowland Rainforest 143, 180
Malabar Semi-evergreen Forest 157, 164, 203
Malagasy anthropogenic savanna **292**
Malagasy Deciduous Forest 128, 129, 132, 180, 189, 209–214, **210**, **211**, **212**, **213**, **214**, 215, 271, 287, 338
Malagasy Grassland and Savanna 209, 211, 286, 287–292, **288**, **289**, **290**, **291**, **292**, 299, 338, 340, 409, 412, 415, 417
Malbrouck 217, 256
Malkoha, Green 168
Mamba
Black 241, 278
Eastern Green 168

Grey-cheeked 171
Sooty 388
mangellas (*Mantella*) 187
Mango 210, 211
Mangrove
Afrotropical 39, 143, 161, 167, 169, 386–389, **387**, **388**, **389**, 395, 396, 399, 403
Neotropical 386
Mangrove Forest
Australasian 386, 388
Indo-Malayan 386, 388
mangroves 386–388, **387**, **388**, **389**
manioc **414**
Manketti 276
Mannikin
Bronze 411, 417
Red-backed 168
Mantella, Baron's **195**
Maple, Montpellier 59
Maquis 300, 309, 315, 321
Maghreb *see* Maghreb Maquis
Middle Eastern 321
Maritime Pine 68
Marsh
Afrotropical Deep Freshwater *see* Afrotropical Deep Freshwater Marsh
Afrotropical Shallow Freshwater *see* Afrotropical Shallow Freshwater Marsh
salt *see* Salt Marsh
Martin
African River 385
Banded 286
Crag 69
House 69
Plain 69
Rock 99
Marula 247
Masked-Weaver
Southern 241, 362, 377, 417
Vitelline **163**
Mastic 322
Mastigure
Moroccan **97**, 99
North African 120
Sudan 99
Mavunda 142, 203–207, **204**, **205**, **206**, **207**, 215, 275
meadow, High Atlas Alpine 55, 350–353, **351**, **352**, **353**
Mediterranean Dry Pine Forest 67
Mediterranean Forests, Woodlands and Scrubs (Africa) 300–341
Albany Thicket 45, 106, 153, 164, 167, 243, 249, 332–337, **333**, **334**, **335**, **336**, **337**
Fynbos 63, 101, 103, 106, 107, 114, 286, 300–308, **301**, **302**, **303**, **304**, **305**, **306**, **307**, **308**, 309, 312, 314, 315, 318, 319, 321, 326, 332, 354, 378, 418
Maghreb Broadleaf Woodland 54, 56, 67, 321, 322, **328**, 328–331, **329**, **330**, **331**, 338

Maghreb Garrigue 49, 59, 118, 321, 322, **324**, **326**, 326–327, **327**, 328, 329, 330, 419
Maghreb Maquis 49, 55, 59, 64, 68, 138, 300, 309, 315, 321–325, **322**, **323**, **324**, **325**, 326, 327, 328, 356, 419
Renosterveld 101, 103, 106, 107, 300, 304, 308, 309, 313, 315–320, **316**, **317**, **318**, **319**, **320**, 354, 378, 418, 419
Strandveld 101, 107, 114, 164, 300, 308, 309–314, **310**, **311**, **312**, **313**, **314**, 315, 319, 336, 354, 418
Tapia 328, 338–341, **339**, **340**, **341**
Mediterranean Junipa and Cypress Forest 54
Mediterranean Oak Forest 57, 328
Mediterranean Pine Forest 54
Mediterranean to Caspian Pelagic Waters 406
Mediterranean to Caspian Tidal Flats 393
Meerkat **81**, 103, 240, 296
Meru 229
Mesite **32**
Brown 32
Subdesert 129, 132
White-breasted 213, **213**
Mesoamerican Cloudforest 174, 189
Mesoamerican Dry Deciduous Forest 209, 215
Mesoamerican Lowland Rainforest 143, 180
Mesoamerican Savanna and Grassland 280
Mesoamerican Semi-evergreen Forest 151
Middle Eastern Juniper Forest 57
Middle Eastern Maquis 321
migration, bird 56, 76, 136, 286, 322, 364, 372, 378, 399, 403, 421
intra-African 378
migration, mammal herds 36, 73, 103, 124, 265, **281**, 283, 384, 414
Milkwood 309, 313
Miombo 36, 39, 45, 61, 62, 63, 64, 66, 150, 151, 152, 154, 156, 157, 159, 161, 203, 204, 205, 219, 221, 225, 226, 227, 235, 243, 248, 251, 252–259, **253**, **254**, **255**, **256**, **257**, **258**, **259**, 260, 261, 267, 275, 276, 277, 279, 281, 286, 366, **366**, 368, 369, 382, 412
etymology 252
"Miombo flush" 253, **253**
Miombo trees (*Brachystegia*) 159, 204, 252, **255**, **258**, 261, 275
Mistigure, Moroccan 75
Modulatricidae **31**
Moist Mixed Savanna 33, 36, 39, 45, 51, 61, 62, 63, 151, 157, 159, 161, 164, 167, 176, 215, 217, 219, 221, 225, 226, 227, 229, 235, 239, 240, 243–251, **244**, **245**, **246**, **247**, **248**, **249**, **250**, **251**, 252, 253, 258, 260, 261, **263**, 266, 267, 275, 287, 332, 336, **349**, 380, 382, 409, 411, 412, 415, 424, 426

INDEX

Moist Montane Forest 33, 35, 38, 40, 45, 61, 62, 63, 63, 64, 66, 138, 142, 143, 144, 147, 151, 156, 157, 159, 164, 165, 167, 169, 174–179, **175**, **176**, **177**, **178**, **179**, 189, 247, 293, 299, 348, 354, 356, 357, **357**, 415, 422
  subhabitats 177
Mole, Golden *see* Golden Mole
Mole-rat
  Big-headed African 347
  Cape Dune 312, **312**
  Damaraland 312
  Naked 233, 312
  Namaqua Dune 90, 112, 114
mole-rats 312
Mongoose
  Black 271
  Cape Gray 103, 304, 317, 419
  Egyptian 317
  Jackson's 177
  Marsh 361
  Slender 65, 177
  Sokoke Dog 160
  Yellow 103, 312, 317, 419
Monitor
  Nile 361, 380
  Rock 279
Monkey
  Blue 206, **206**
  Campbell's 388
  De Brazza's **172**
  Dryas 171
  Green 264
  Kipunji 177
  L'Hoest's 177
  Mona 388
  Patas 98, **98**, 233, 264
  Preuss's 177
  Red-tailed 148
  Sykes' Blue 167
  Tantalus 264
  Vervet 167, 250, **259**, 336
Monsoon Forest (Afrotropical) 36, 142, 143, 146, 147, 151–156, **152**, **153**, **154**, **155**, **156**, 157, 161, 164, 165, 173, 174, 177, 203, 204, 205, 215, 219, 226, 244, 248, 250, 255, 256, 258, 276, 382, 424
Montane Forest
  Albertine Rift 41
  Moist (Afrotropical) *see* Moist Montane Forest
Montane Grassland 34, 38, 40, 45, 61, 64, 101, 103, 106, 107, 175, 251, 280, 286, 293–299, **294**, **295**, **296**, **297**, **298**, 300, 315, 332, 343, 344, 347, 348, 354, 356, 378, 412, 415, 418, 419, 427
Montane Heath
  Africa 38, 40, 46, 61, 62, 140, 141, 176, 196, 293, 299, 300, 309, 315, 343, 347, 348, 354–357, **355**, **356**, **357**
  Northwest European Coastal and 300, 309, 315

  see also Alpine Tundras and Montane Heaths
Montane Heathland, Australasian 315
Montane Rainforest *see* rainforest
Montane Renosterveld 317, 320
Moorhen, Lesser 369
Mopane (habitat) 45, 62, 215, 217, 221–226, **222**, **223**, **224**, **225**, 235, 240, 243, 244, 248, 252, 258, 260, 275, 277, 279, 390, 412
Mopane, Large False 204
Mopane Tree 221, 223, **224**, 225
Mopane worm 226
Moringa, Desert 78
Mountain-Babbler
  Red-collared 178
  White-throated 179
Mouse
  Barbary Striped Grass 327
  Cape Spiny 304
  Four-striped Grass **396**
  Rosevear's Striped Grass 205
  Verreaux's 304
  Wood 56
Mousebird
  Blue-naped 231
  Red-faced 91
  Speckled 312–313, 411, 417
  White-backed **29**, 313
mousebirds 414
Mouse Lemur
  Coquerel's Giant 213
  Giant 411
  Gray 129, 290
  Madame Berthe's 213
  Northern Giant **411**
  Tavaratra **212**
mouse lemurs 213
Muchesa 253
Munondo 253
Mushitu 152
mussels **402**
Mustard Tree 90
Myna, Common 200, 411, 425
Myrtle, Saharan 95–96

N

Nama Karoo 72, 77, 79, 88, 95, **101**, 101–106, **102**, **103**, **104**, **105**, **106**, 107, 112, 114, 115, 121, 235, 300, 309, 315, 319, 332, 336, 344, 354
Nama Padloper 81
"Namaqualand blooms" 109, **110**
Namib Rock Desert 71, 77–81, **78**, **79**, **80**, 81, 88, 90, 94, 101, 106, 107, 114, 123, 385
Namib Sand Desert 77, 81, 82, 88–93, **89**, **91**, **92**, 114, 123, 385, 397
Nara Melon 89
Nearctic Alpine Tundra 350
Nearctic Cropland 418
Nearctic Desolate Desert 82, 88
Nearctic Montane Mixed-Conifer Forest 54, 67

Nearctic Open Water 379
Nearctic Pastureland and Rangeland 427
Nearctic Pelagic Waters 406
Nearctic Reedbed Marshes 358
Nearctic Rocky Coastline 400
Nearctic Salt Marsh 395
Nearctic Sand Beach and Dunes 397
Nearctic Sedge and Grassland Marshes 365
Nearctic Tidal Mudflat 393
Nearctic Tree Plantations 422
Nearctic Urban 424
Neddicky 307
Neospiza finches 404
Neotropical Coastal Lagoons 395
Neotropical Cow Pasture 427
Neotropical Flowing Rivers 382
Neotropical Mangroves 386
Neotropical Oil Palm 409
Neotropical Rocky Coastline 400
Neotropical Semi-evergreen Forest 151, 157, 164
Neotropical Temperate Wetland 370, 374
Neotropical Tidal Mudflat 393
Neotropical Tropical Wetland 358, 365
Nesomys, Western 214
Newt, Sharp-ribbed 331
Nicator **31**
  Eastern 168
  Yellow-throated **31**
Night-Heron, White-backed 171, **173**, 362
Nightingale, Common 324, 330, 331
Nightjar
  Abyssinian 356
  Collared **186**
  Egyptian 99
  Fiery-necked 414
  Freckled 272
  Golden 233, **233**
  Montane 65
  Nubian 99
  Pennant-winged **259**
  Plain 414
  Red-necked 69, 324, 331
  Rwenzori 356
  Standard-winged **264**, 266
Nile River 383
Noddy
  Black 404
  Brown 404
  Lesser 405
North Africa, habitats 47–50
North African Temperate Cultivation 418, 419–421, **420**, **421**
North African Temperate Wetland 370, 370–372, **371**, **372**, 374, 379, 382, 390, 393, 395
North-Central Africa, habitats 39–41
Northeast Africa, habitats 33–34
Northern Acacia Shrubland 221, 235
Northern Dry Thorn Savanna 51, 87, 117, 121, 122, 124, 125, 133, 136, 227–233, **228**, **229**, **230**, **231**, **232**, **233**, 235, 240, 242, 243, 248, 252, 260, **263**, 266, 267, 390, 427

INDEX 441

Sahal 227, 229, 230, 233, 267
Somali–Masai zone 227, 231, 233, 241
Northwest European Coastal and
 Montane Heath 300, 309, 315
num-num trees 309, **335**
Nuthatch
 Algerian 56, 331
 Eurasian **330**, 331
Nyala 160, **160**, 167, 188, 249
 Mountain 65, 357, **357**

O

Oak
 Atlas 56, 330
 Cork 328, 329, 330, 331, 422, 423
 Holly 55, 59, 322, 328, 329, 330
 Holm 331
 Kermes 328, 329
Oak Dehesa 275, 338
oak forest
 Maghreb 54
 Mediterranean 49
oases 49, 76, 100, 421
ocean currents 45, 88, 93
octopus trees 127, 128, **128**
Oedera squarrosa **320**
Offshore Islands 393, 395, 397, 400, **403**,
 403–406, **404**, **405**, 406
 Indo-Malayan 403
oil palm 169, 410, 411
Okapi 147, 148
Olive, Wild 59, **60**, 322, 328
Olive Tree, Laperrine's 96
Open Eucalypt Savanna 243, 275
Orca **408**
orchids 306
Oribi 250, 264, 283, 295
Oriole
 African Black-headed 278
 African Golden 277
 Black-headed 337
 Eurasian Golden 421
Oryx 73
 Beisa 233
 Scimitar-horned 73, **123**, 123–124,
  233
 Southern (Gemsbok) 79, **79**, 89, 103,
  225, **237**, 240, **282**, 312
Ostrich 79, **279**, 313
 Common 123, 124, 249, 286, 392
 Somali 124
"Ostrich toes" **317**
Otter
 Cape Clawless 376, **376**
 Spotted-necked 376
Ouhout shrubland 64, 299, **299**
Ouzel, Ring 60
Owl
 Abyssinian 65
 African Bay- 178
 African Grass- 362
 African Wood- 65, 168
 Barn 414, 421

Cape Eagle- 65, 356
Comoro Scops- 192
Congo Bay 149, **149**
Little **322**, 324, 421
"MacKinder's" Cape Eagle- 273
Maghreb 69, 331
Maned 149
Marsh 369
Pallid Scops- 99
Pel's Fishing- 155, **155**
Pharaoh Eagle- **99**
Red 185, 192
Rufous Fishing- 149, 171
Seychelles Scops- 201
Shelley's Eagle- 148, 149
Socotra Scops- 136
Sokoke Scops- 161
Southern White-faced **241**
Usambara Eagle- 179
Vermiculated Fishing- 171, **171**
Owlet
 African Barred 154
 Albertine 149
 "Cape" Barred 337
 Pearl-spotted 225
Oxpecker 31, 249, 428
 Red-billed 31, **428**
 Yellow-billed **428**
Oxylabes, Yellow-browed 32, 192
Oystercatcher, African 398, 401, **401**

P

pachypodiums 127, 271
Pacific Chaparral 300, 309, 315, 321
Paddy Fields 409
Padloper
 Boulenger's 105
 Greater 105
 Parrot-beaked 105, 307, **313**, 314, 337
 Speckled 114, 314
Painted-Snipe, Greater 362
Palearctic 53, 76
paleoclimate 234
Palm **198**
 African Oil 169, 410, 411
 Coconut 199, 422
 Date 84, **420**, 422
 Doum 100
 Nipa 387
 Traveler's 182, 334
 Wild Date 247
palms, raffia 169
Palm Savanna **246**
Palm Swamp 169
pandanus trees 191, 287, **288**
Pangolin
 Giant 147
 Ground **237**, 240, 256, 284
 White-bellied Tree 205–206
Papyrus 38, 358, 359, **359**, 360, 364, **367**
Paradise-Flycatcher
 African 277, 417
 Seychelles 201, **202**

Paradise-Whydah, Eastern **31**
paradise-whydahs 415
Parakeet
 Echo 185
 Rose-ringed 415, 425
 Seychelles 200
Paramo 343
Parrot
 Gray 411, 417
 Greater Vasa 192
 Lesser Vasa 192
 Mascarene 186
 Seychelles Black 201
 Yellow-fronted 65
Partridge
 Barbary 56, 60, 69, 98
 Madagascar 290
 Sand **97**, 98
 Stone **262**
 Udzungwa 178–179
Patagonian Steppe 101
Peacock, Congo 148
Pectinator, Speke's 271
Pelagic Waters 393, 395, 397, 400, 403,
 406–408, **407**, **408**
Pelican
 Great White 381, 391
 Pink-backed 381
Penduline-Tit, Southern 241
Penguin, African **398**, 400, 404
Perch, Nile 380
Petrel 404, 405
 Band-rumped Storm- 405
 Barau's 405
 Black-winged 405
 Bulwer's 405
 Cape 408
 Cape Verde Storm- 405
 European Storm- 405, 406, 408
 Fea's 405
 Giant **407**
 Herald 405
 Kermadec 405
 Leach's Storm- 406
 Mascarene Black 405
 Monteiro's Storm- 405
 Trinidade 405
 White-faced Storm- 405
 Wilson's Storm- 406
 Zino's 141, 405
petrels
 giant- 408
 storm- 404
Philippine Montane Rainforest 189
photosynthesis 111
"Phrygana" 327
Phymateus saxosus grasshopper **292**
Piapiac 266, 415, 427
Picathartes 148
Picathartes (Rockfowl)
 Gray-necked 273
 White-necked **30**, **145**, 273
Pied-Babbler, Southern 241

Pigeon
  African Olive 167–168
  Bolle's 141
  Bruce's Green- **188**, 417
  Comoro 186
  Laurel 141
  Madagascar Blue- **181**, 192
  Madagascar Green- 213
  Rameron 178
  Rock 425
  Seychelles Blue- 201, **202**
  Trocaz 141, **141**
  White-collared 65, 269, 425
"pig's ear" (*Cotyledon*) **111**
Pine
  Aleppo 55, 67, 68, 69, 322, 329
  Canary Island 68, **68**, 141
  Maritime 55, 68, 140
  Stone 67, 68
  Umbrella 67, 68
Pine Forest
  Himalayan 67
  Jack 67
  Lodgepole 67
  Maghreb 54, 57, **67**, 67–69, **68**, **69**, 138, 322, 422, 423
  Mediterranean 54
  Mediterranean Dry 67
pines 288, **290**, 422, **422**
Pintail, Eaton's 404
Pinyon–Juniper Woodland 57
Pipit
  African 286, 417, 419
  Mountain 348
  Plain-backed 417
  Red-throated 421
  Sokoke 161
  Tree 65
  Yellow-breasted **297**, 299
Pitta
  African 155, 161, 219
  Green-breasted **145**
Plains Monte 107
Plantain-eater, Western Gray 411
Platanna, Cape 377
Plated Lizard
  Black-lined 225, 278
  Desert 91
  Tsingy 214, **274**
Platypelis olgae 195
Playas 390
Plover
  Chestnut-banded 392, **392**
  Crab- **394**, 394
  Egyptian 29, **384**, 385
  Forbes's 286, 414
  Kittlitz's **392**, 396
  Quail- 233
  Three-banded 376
  White-fronted **398**, 399, **399**
Plum, Guinea 145–146
Polecat, Saharan Striped 117
pollinators 306

Pond-Heron, Malagasy 364
Porcupine
  African Brush-tailed 147
  Cape 317
  Crested 323
  South African **159**
Porkbush 332–333, **333**
Potto 148
  West African **150**
Pratincole
  Gray 385
  Madagascar 385
  Rock 383, 385
Pre-Puna Semi-Desert Scrub 77
Prinia
  Black-chested 241
  Karoo 307, 318
  Tawny-flanked 336
prion 404
proteas 301, **301**, 303, **305**
Puffback
  Black-backed 155, **161**, 168, 278, 336
  Northern 411
  Pink-footed 178
Puku 256, 277, 283, 361
Purslane, Herero 89
Pygmy-Goose, African 362
Pygmy Toad
  Dombe 81
  Southern 106
Python, Anchieta's Dwarf 81

Q
Quagga 103, 317, 318
Quail
  Blue 286, 369
  Common 290, 414, 417, 419, 421
  Harlequin 290
Quailfinch, African 377, **377**, 381
Quail-Plover 233
Quelea, Red-billed 241
Quiver Tree 112

R
Rabbit
  Riverine 104
  Smith's Red Rock 336
Rail
  African 376
  Inaccessible Island 404
  Madagascar 367, 369
  Rouget's **357**
  Sakalava 364
  Tsingy Wood- 213, 214
  White-throated **387**, 389, 404
rainforest 40, 142, 143
  Afrotropical Lowland *see* Afrotropical Lowland Rainforest
  Afrotropical Montane *see* Moist Montane Forest
  Australasian Littoral 157, 164
  Australasian Tropical Lowland 143, 180
  Australasian Tropical Montane 174, 189

  Central American Lowland 143, 180
  Indian Ocean Lowland 43, 129, 132, 142, 143, 169, 180–188, **181**, **182**, **183**, **184**, **185**, **186**, **187**, 189, 197, 209, 211, 287, **398**, 409
  Indian Ocean Montane 43, 129, 132, 142, 174, 180, 182, 189–196, **190**, **191**, **192**, **193**, **194**, **195**, **196**, 197, 199, 211, 287, 409
  Island Arc Lowland 180
  Lowland 129, 132, 142
  Malabar Lowland 143, 180
  montane *see* Afroparamo
  Philippine Montane 189
  Sambirano 193, 210
  Sunda Montane 174, 189
Rain Frog
  Desert 92
  Marbled 187
  Namaqua 114, **114**
  Sand 314
rain frogs 308
Randonia 72
Rat
  Big-headed African mole- 347
  Cape Dune mole- 312, **312**
  Damaraland mole- 312
  Dassie 271
  Fat Sand 119
  Grant's Rock 104, 304
  Greater Cane 146
  Malagasy Giant Jumping 213, 214
  Naked mole- **233**, 312
  Namaqua Dune mole- 90, 112, 114
  Sloggett's Ice- **347**
rats
  mole- 312
  vlei 376
rattlepods 229
Raven
  Brown-necked 86
  Fan-tailed 86
Ravenala 182
Ravintsara 210
Red-hot poker **344**, 375
Redstart
  Black 331
  Moussier's **58**, 60, 331
Redwing 331
Reed, Giant 364
reedbeds 375, **375**
Reedbuck
  Bohor 283
  Mountain 295, 336
  Southern 283
reedbucks 361
Reed Frog, Starry-night 187, **187**
Reed Warbler
  African 362
  Clamorous 389
Reef-Heron, Western 398
refugia 234
Reg 49, 53, 71, 72, **72**, 73, 75, 76, 94

INDEX

Saharan Desert *see* Saharan Reg Desert
Turanian Reg and Hamada 71
Renosterbos 315, **316**
Renosterveld 101, 103, 106, 107, 300, 304, 308, 309, 313, 315–320, **316**, **317**, **318**, **319**, **320**, 354, 378, 418, 419
restio 301, 303, **310**, **311**
 sedge-like **301**
Rhebok, Gray 296, 304, 312, 317
Rhigozum somalense 122
Rhinoceros
 Black 79, 223, 225, 231, 240, 249, 264, 265, **265**, 317, 336, 356
 White 249, 264, 265, 283, **284**, 336
Rhombophryne tany 195
rice 411, 413, 417, **426**
Rivers 358, 365, 370, 374, 379, **382**, 382–385, **383**, 384, **385**
 Flowing 382
 Lowland 382
 Neotropical Flowing 382
 Upland 382
Robin
 Bearded Scrub- 249
 Black Scrub- 233
 Brown Scrub- 168
 European 141, 330
 Forest Scrub- 219
 Kalahari Scrub- 241
 Karoo Scrub- 105, 312, 318
 Miombo Scrub- 257
 Red-backed Scrub- 278
 Rufous-tailed Scrub- 324, **324**, 331
 White-browed Scrub- 336
 White-starred 178
Robin-Chat
 Cape 318
 Red-capped 206, **207**
 White-browed 417
 White-throated **166**, 250
Rock Agama, North African 75, 120
Rockfowl (Picathartes) 30
 Gray-necked 273
 White-necked 30, **145**, 273
Rockjumper **30**
 Cape 304, **305**
 Drakensberg **30**, 348
 Orange-breasted **347**
Rock Rabbit, Smith's Red 336
Rockrunner **31**, 273
Rock Sengi, Karoo 104
Rock-Thrush
 Blue 60, 331
 Cape 304
 Forest 340
 Littoral 129, **129**
 Miombo 257
 Rufous-tailed 60, 331
 Sentinel 304
rock-thrushes 272
Rocky Canyon 77
Rocky Cerrado 268

Rocky Coastline 390, 393, 403, 406
Rocky Hamada and Massif 62, 71, 73, 77, 82, 83, 94–99, **95**, **96**, **97**, **98**, **99**, 100, 115, 419
Rocky Shoreline 365, 370, 395, 397, **400**, 400–402, **401**, **402**
Rocky Spinifex 94
Roller
 Blue-bellied 266
 Blue-throated **144**
 Cuckoo- **32**, 44, 192
 Lilac-breasted 249
 Long-tailed Ground- 129, 132, **132**
 Pitta-like Ground- 192, **194**
 Rufous-headed Ground- **32**, 192
 Scaly Ground- 185, **185**
 Short-legged Ground- 185, **185**
Rubber Tree 410, 422, 423
Ruff 377, 392
Ruffed Lemur
 Black-and-white 183, **184**
 Red 183
Rush-Warbler, Grauer's 369

S

Sagebrush Shrubland 101, 107, 115, 121
Sahara 49, 53, 55, 71, 73, 85, **85**
Saharan Erg Desert 71, 73, 82–87, **83**, **85**, **86**, 88, 94, 100, 115, 121, 397
Saharan Reg Desert **71**, 71–76, **72**, **73**, **74**, **75**, **76**, 77, 82, 83, 84, 85, 94, 97, 100, 115, 118, 121, 322, 419
Sahel Thornscrub, transition to 267
Salamander, North African Fire 331
Salicornia spp. 395, **396**
Saline Andean Lakes 390
Saline Habitats (Africa) 386–408
 Afrotropical Mangrove 39, 143, 161, 167, 169, 386–389, **387**, **388**, **389**, 395, 396, 399, 403
 Offshore Islands 393, 395, 397, 400, 403, 403–406, **404**, **405**, 406
 Pelagic Waters 393, 395, 397, 400, 403, 406–408, **407**, **408**
 Rocky Shoreline 365, 370, 395, 397, **400**, 400–402, **401**, **402**
 Salt Marsh 382, 386, 390, 393, **395**, 395–396, **396**, 397, 399, 400, 403, 406
 Salt Pans and Lakes 379, **390**, 390–392, **391**, **392**, 393, 395, 396, 397, 400
 Sandy Beach and Dunes 386, 390, 393, **397**, 397–399, **398**, **399**, 400, 403, 406
 Tidal Mudflats and Estuaries 358, 365, 370, 374, 379, 380, 382, 386, 389, 390, **393**, 393–394, **394**, 395, 396, 397, 399, 400, 403, 406
Salt Marsh
 Afrotropical 382, 386, 390, 393, **395**, 395–396, **396**, 397, 399, 400, 403, 406
 Australian Coastal 395
 European Coastal 395
 Nearctic 395
 Soda Pans and Inland Salt Marshes 390

Salt Pans and Lakes 379, 389, **390**, 390–392, **391**, **392**, 393, 395, 396, 397, 400
Salt Steppe 287
Salvinia, Giant 359
Sambirano rainforest 193, 210
Sand Desert, Namib 77, 81, 82, 88–93, **89**, **91**, **92**, 114, 123, 385, 397
Sanderling 399
Sandfish, White-banded 86
Sand Forest 164, 165, 168
Sandgrouse
 Burchell's 241
 Crowned 74
 Madagascar **291**
 Namaqua 79, 105, **105**, 241
 Yellow-throated 286
Sandpiper
 Curlew 392
 Green **370**
 Marsh 377
 Terek **394**
 Wood 377
Sand Rat, Fat 119
sand seas 82, **83**, 88, 90
Sand Snake
 Cape 314
 Mahafaly 132
Sandy Beach and Dunes 386, 390, 393, **397**, 397–399, **398**, **399**, 400, 403, 406
Sarcocornia spp. 395
Savanna Cultivation 290, 409, 410, **412**, 412–415, **413**, **414**, 415, **415**
Savannas (Africa) 44, 51, 220–279, 280, 373
 Guinea 34, 36, 39, 51, 62, 85, 143, 145, 146, 150, 227, 230, 243, 248, 250, 252, 253, 260–267, **261**, **262**, **263**, **264**, **265**, **266**, 275, 281, 286, 382, 409, 411, 412
 Gusu 45, 62, 204, 215, 221, 226, 235, 243, 244, 248, 251, 252, 253, 258, 260, 275–279, **276**, **277**, **278**, **279**, 412
 Inselbergs and Koppies 268, 268–274, **269**, **270**, **271**, **272**, **273**, **274**, 304, 366
 Kalahari Dry Thorn 77, 88, 101, 103, 105, 106, 107, 221, 225, 226, 227, 235–242, **236**, **237**, **238**, **240**, **241**, **242**, 243, 248, 250, 252, 258, 275, 277, 332, 390
 Malagasy anthropogenic **292**
 Miombo 36, 39, 45, 61, 62, 63, 64, 66, 150, 151, 152, 154, 156, 157, 159, 161, 203, 204, 205, 219, 221, 225, 226, 227, 235, 243, 248, 251, 252–259, **253**, **254**, **255**, **256**, **257**, **258**, **259**, 260, 261, 267, 275, 276, 277, 279, 281, 286, 366, **366**, 368, 369, 382, 412
 Moist Mixed 33, 36, 39, 45, 51, 61, 62, 63, 151, 157, 159, 161, 164, 167, 176, 215, 217, 219, 221, 225, 226, 227, 229, 235, 239, 240, 243–251, **244**, **245**, **246**, **247**, **248**, **249**, **250**, **251**, 252, 253, 258, 260, 261, **263**, 266, 267, 275, 287, 332, 336, **349**, 380, 382, 409, 411, 412, 415, 424, 426

Savannas (Africa) *continued*
   Mopane 45, 62, 215, 217, 221–226, **222**, **223**, **224**, **225**, 235, 240, 243, 244, 248, 252, 258, 260, 275, 277, 279, 390, 412
   Northern Acacia 221, 235
   Northern Dry Thorn 51, 87, 117, 121, 122, 124, 125, 133, 136, 227–233, **228**, **229**, **230**, **231**, **232**, **233**, 235, 240, 242, 243, 248, 252, 260, **263**, 266, 267, 390, 427
savoka 182
Scaly-Tail, Cameroon 148
Schefflera 190
Scimitarbill **30**
   Common 241
Scops-Owl
   Comoro 192
   Pallid 99
   Seychelles 201
   Socotra 136
   Sokoke 161
   Southern White-faced **241**
Scrub-Robin
   Bearded 249
   Black 233
   Brown 168
   Forest 219
   Kalahari 241
   Karoo 105, 312, 318
   Miombo 257
   Red-backed 278
   Rufous-tailed 324, **324**, 331
   White-browed 336
scrubs *see* Mediterranean Forests, Woodlands and Scrubs
Sea Coconut **198**, 199
Seal
   Cape Fur 90, 398, 400, 404, **405**
   Southern Elephant 404
Sea Turtle
   Green 399, 404, 405
   Hawksbill 399, 405
   Leatherback 399, 405
   Loggerhead 399, 405
   Olive Ridley 399, 405
Secretarybird **29**, 105, **415**
seed dispersal 188
Seedeater, Protea 308
Semi-Desert, Dragon's Blood Tree 34, 133–137, **134**, **135**, **136**, **137**
Semi-Desert Scrub, Pre-Puna 77
Semi-evergreen Forest
   Central American 151
   Malabar 157, 164, 203
   Neotropical 151, 157, 164
   Southeast Asian 151
Sengi, Karoo Rock 104
Serin, European 421
Serpent-Eagle, Madagascar 185, 192
Sertao Caatinga 332
Serval 296, 361, **368**
Seychelles Granite Forest 143, 180, 189, 197–202, **198**, **199**, **200**, **201**, **202**

Shag
   European 403
   Imperial 404
   Kerguelen 404
"shamba"/shamba cultivation 416, 417
Shark, Great White **407**
Shearwater
   Balearic 408
   Barolo 405
   Boyd's 141, 405
   Cape Verde 405
   Cory's 405, 406
   Flesh-footed 408
   Manx 405
   Sooty 406
   Tropical 404, 408
   Wedge-tailed 404, 405, 408
   Yelkouan 408
shearwater species 404
Sheathbill, Black-faced 404
Sheep, Barbary 97, 118, 351
Shelduck
   Ruddy 348, **372**
   South African **374**, 376
She-oak, Coastal 422
Shoebill 38, 361, **362**
Shola Grasslands 293
Shoveler
   Cape 376
   Northern **370**
Shrew
   Golden-rumped Elephant **160**
   Mauritanian 74
   North African Elephant 118, 119
   Tarfaya 74
   Whittaker's 74
Shrike
   Great Gray 99
   Masked 327
   Red-backed **76**
   Souza's 278, **278**
   White-tailed **30**, 225
   Woodchat 324, 331, 421
   Yellow-billed 266
Shrub Desert
   Afrotropical Hot *see* Afrotropical Hot Shrub Desert
   Maghreb Hot *see* Maghreb Hot Shrub Desert
Shrubby Savanna 221
Shrubland
   Arid Tussock Acacia 121
   Highland Ouhout 64, 299, **299**
   Sagebrush 101, 107, 115, 121
Sifaka
   Coquerel's 211
   Crowned 211
   Golden-crowned 211
   Perrier's 211
   Verreaux's 129, **131**, 211
   Von der Decken's 211
sifakas 183, 211
silicrete 126

Silverbird **249**, 250
Siskin 348
   Cape 307
   Ethiopian 417
Sitatunga 206, 360
Skimmer, African 385
Skink
   Afro-Malagasy 187
   Bayon's 296
   Boulenger's Sand 86
   Coastal 331
   Grass 286, 296
   Manuel's 120
   Moorish 120
   Mountain 353
   Ocellated 120
   Ovambo Tree 226
   Round Island Ground 405
   Wahlberg's 278
   Wedge-snouted 86
Skua, Brown 404
Skylark, Eurasian 421
Snake
   Awl-headed 86
   Bernier's Striped 292
   Cape Sand 314
   Cream-spotted Mountain 296
   Madagascar Leafnose **214**
   Mahafaly Sand 132
   Mole 278
   North African Cat 76
   Savanna Vine 278
   Stilletto 241
   Western Bark 81
   Western Keeled 81
Snake-Eagle
   Black-chested **251**, 286
   Brown 277
   Fasciated 167
Snipe
   African 369, 376
   Greater Painted- 362
   Madagascar 369, 417
Socotra Island Blue Baboon (tarantula) 136
Soda Pans and Inland Salt Marshes 390
Soldier-in-a-Box **311**
Solitaire, Rodrigues 186
Sonoran Desert 133
South African Temperate Cultivation 316, 319, 415–416, **418**, 418–419, **419**
South African Temperate Wetland 358, 364, 365, 369, 370, **374**, 374–378, **375**, **376**, **377**, **378**, 379, 382, 390, 393, 395
South American Temperate Pelagic Waters 406
South Coast Forest Matrix 62, 151, 157, 159, 161, 162, 164–168, **165**, **166**, **167**, **168**, 174, 203, 243, 244, 248, 309, 332, 336, 382, 386, 388, 399
Southeast Asian Dry Deciduous Forest 209, 215, 260
Southeast Asian Semi-evergreen Forest 151

INDEX

Sparrow
  Abd al Kuri 136
  Desert **83**, 86
  House 425
  Northern Gray-headed 414–415
  Socotra 136
  Southern Gray-headed 414–415
  Spanish **60**, 421
  Swainson's 417
Sparrowhawk
  Eurasian 60, 69
  Ovambo 277
  Red-breasted 423
Sparrowlark
  Black-crowned 119
  Black-eared 81, 104
  Gray-backed 81, 104
Sparrow-Weaver, Chestnut-backed 257
Spinetail
  Bat-like 219
  Mottled 219, 414
Spinifex Eucalypt Savanna 268
Spiny Forest 44, 127–132, **128**, **129**, **130**, **131**, **132**, 133, 180, 189, 209, 211, 213, 214, 287, 332, 333, 338
  Limestone Plateau 128, 129, **130**
  Red-sand Lush 128, **131**
  White-sand Coastal 128, 129, **129**
Spot-throat **31**
Springbok 79, 104, **113**, 225, 240, 312
Springhare
  South African **279**
  Southern 231, 240
Spurfowl
  Cape 313, 318
  Mount Cameroon 179
  Red-necked 337
Spurge, Tenerife Milk 117
spurges (Euphorbiaceae) 109, 309, 333
  cactus-like 117
Squirrel
  African Pygmy 148
  Barbary Ground 118
  Carruther's Mountain 177
  Forest Giant 148
  Gambian Sun 264
  Ochre Bush 160
  Red Bush 167
  Red-legged Sun 148
  Slender-tailed 148
  Smith's Bush 160, 167, 277
  Tanganyika Mountain 177
squirrels
  palm 148
  rope 148
Starling
  Black-bellied 168
  Bristle-crowned 269
  Bronze-tailed 266
  Cape 336
  European 425
  Golden-breasted **232**, 233
  Greater Blue-eared 225

  Lesser Blue-eared 257, 266
  Meves's 225
  Neumann's 269
  Pied 312, 318, 337
  Purple 266
  Red-winged 269, 318, 418, 425
  Réunion 186
  Sharpe's 178
  Socotra 136
  Splendid 411
  Spotless 421
  Superb 415
  Violet-backed 411
  Wattled 427
Steenbok 104, 240, 277, 304, 312, 317, 419
Sterculia setigera 217
Stilt, Black-winged **370**, 376, 394, 396
Stinkwood 140
Stint, Little 377
Stipagrostis **104**
Stonechat
  African 290, 356, 417
  European 331
  Réunion 356
stonecrops (Crassulaceae) 109, 111, **111**, 309, 334
Stork
  Abdim's 286, 369
  Black 376
  Marabou 414, 425, **426**
  Saddle-billed 361, **361**, 385
  White 286, 369, 417, 421
  Yellow-billed 361, **369**
Storm-Petrel 404
  Band-rumped 405
  Cape Verde 405
  European 405, 406, 408
  Leach's 406
  Monteiro's 405
  White-faced 405
  Wilson's 406
storm-petrels 404
Strandveld 101, 107, 114, 164, 300, 308, 309–314, **310**, **311**, **312**, **313**, **314**, 315, 319, 336, 354, 418
Strelitzia 333–334
Striped Grass Mouse
  Barbary 327
  Rosevear's 205
Subhumid Yungas 61
Succulent Karoo 77, 88, 101, 102, 106, 107–114, **108**, **109**, **110**, **111**, **112**, **113**, **114**, 121, 122, 133, 286, 300, 309, 311, 314, 315, 319, 336
Succulent Puna 72, 115, 121
Sugarbird **31**
  Cape **31**, 304, **305**
sugarcane 411, 414
Sunbird
  Amethyst 337
  Anchieta's 257
  Bannerman's 155
  Carmelite 388

  Collared 411, 417
  Congo 171
  Copper 411
  Eastern Miombo 257
  Eastern Violet-backed 233
  Giant 148
  Golden-winged **306**
  Madagascar **341**, 356
  Malachite 298, 304, 313, 337, 356
  Malagasy 340
  Mariqua 241
  Mouse-brown 388
  Mouse-colored 168
  Neergaard's 168
  Olive 168
  Olive-bellied 411
  Orange-breasted 304, **304**
  Plain-backed 168
  Purple-banded 168
  Red-chested 362
  Scarlet-chested 417
  Scarlet-tufted 347
  Seychelles 201
  Shelley's 155
  Socotra 136, **136**
  Souimanga 340
  Southern Double-collared 304, 313
  Splendid 411
  Tacazze 356
  Variable 250, 417
  Western Miombo 257
  Western Violet-backed 266
  White-breasted 250
Sunbird-Asity
  Common **32**
  Yellow-bellied 192
sunbirds 306
"Sundaic savanna" 251
Sunda Montane Rainforest 174, 189
Sungazer 296
Suni 65, 167
Swallow
  Blue **297**, 423
  European 69
  Gray-rumped 286
  Montane Blue 298
  Pearl-breasted 319
  Red-rumped 69
  White-tailed 428
Swamp Forest 39, 142, 143, 147, 152, 159, 169–173, **170**, **171**, **172**, **173**, 180, 358, 360, 364, 380, 382, 386, 388
  Asian Peat 169
  Australasian 169
  dendrogram 142
Swamphen, African 362
Swamp Warbler
  Cape Verde 364
  Grauer's 365
  Lesser 377
  Madagascar 364, **364**, 389
  White-winged 362
Sweet Thorn 102

INDEX

Swift (bird)
   Alpine 69
   Common 69
   Forbes-Watson's 136
   Little 69, 425–426
   Malagasy Palm 290
   White-rumped 426
Swift (reptile)
   Dumeril's Madagascar 273
   Grandidier's Madagascar 273
   Merrem's Madagascar 132
Swiftlet, Seychelles 201
Symphonia 169

T

Table Mountain Pride Butterfly 303
Tahara 122
Tailorbird
   African 179
   Dapple-throat 179
   Long-billed 179
   Spot-throat 179
Talapoin
   Angolan 217, 388
   Gabon 388
Taman 100
Tamarind 210
Tamarisk 117, 229, 239
   Nile 84
   Wild 90, 102, 112
tamarisk shrubs 100
Tamarix 84
Tamboti, Bastard 166
Tapia 328, 338–341, **339**, **340**, **341**
   Malagasy, classification 341
Tapia tree 338, **339**, 339–340, **341**
tarantula, Socotra Island Blue Baboon 136
Tchagra
   Black-crowned 278, 331, 336
   Southern 314, **335**, 337
Teak 422
   Angolan 275, **276**
   Zambezi 275, 276, **277**
Teak Forest 275
Teal
   Blue-billed 362
   Cape 392, **392**, 396
Tea Plantation 415
Temperate Broadleaf Forests (Africa) 138–141
   Laurel Forest 50, 138–141, **139**, **140**, 141
Temperate Cropland
   Asian 419
   Australian 418
Temperate Cultivation
   North African 418, 419–421, **420**, 421
   South African 316, 319, 415–416, **418**, 418–419, **419**
Temperate Deciduous Forest, Asian
   Hyrcanian 138
Temperate Freshwater Wetland,
   Australasian 370, 374

Temperate Heathland, Asian 300, 309, 315
Temperate Heath Thicket, Australian 321, 338
Temperate Pelagic Waters
   Asian 406
   Australian 406
   South American 406
Temperate Rocky Coastline and Sandy Beach, Asian 400
Temperate Tidal Mudflat, Asian 393
Temperate Wetland
   Asian 370, 374
   Neotropical 370, 374
   North African 370, 370–372, **371**, **372**, 374, 379, 382, 390, 393, 395
   South African 358, 364, 365, 369, 370, **374**, 374–378, **375**, **376**, **377**, **378**, 379, 382, 390, 393, 395
Ténéré Sand Sea **83**
Tenrec
   Common 183
   Lesser Hedgehog 129, **132**
   Tailless 183, 184, 200
   Web-footed 184
tenrecs 183–184
Terai Flooded Grassland 280
termite mounds 163, **205**, 259, **259**
termites 373
Tern
   Common 401
   Great Crested 398, 401
   Gull-billed 381
   Kerguelen 404
   Lesser Crested 398, 401, 405
   Roseate 401, 405
   Sandwich 398, 405
   Sooty 404, 405
   Subantarctic 401, 404
   Whiskered 377, 381
   White-cheeked 405
   White-winged 377, 381
terns, marsh 377
Terrapin, Serrated Hinged **381**
terrapins 370
Tethys flora 135
Tetradonta Woodland Savanna 252
Tetraka
   Appert's 213
   Dusky 185
tetrakas **32**
Thamnornis 129
Thickets 335
   Albany 45, 106, 153, 164, 167, 243, 249, 332–337, **333**, **334**, **335**, **336**, **337**
   Australian Temperate Heath 321, 338
   Itigi 153, **153**, 156
   Itigi-Sumbu 335
thick-knees (birds) 385
Thorn
   Camel 237
   Sweet 102, 112
   Umbrella 83, 100, 122, 228, 237
   Whistling 228, **230**, 246

thorn savanna
   Kalahari *see* Kalahari Dry Thorn Savanna
   Northern *see* Northern Dry Thorn Savanna
Thrush
   African 411, 417
   Blue Rock- 60, 331
   Cape Rock- 304
   Common Rock- 60, 331
   Forest Rock- 340
   Kurrichane 278
   Littoral Rock- 129, **129**
   Miombo Rock- 257
   Mistle 330
   Sentinel Rock- 304
   Somali 65
   Song 324, 331
   Spotted Ground- 168
   Tristan 404
thrushes, rock- 272
Tidal Mudflats and Estuaries 358, 365, 370, 374, 379, 380, 382, 386, 389, 390, **393**, 393–394, **394**, 395, 396, 397, 399, 400, 403, 406
Tinkerbird
   Green 161
   Moustached 178
   Red-fronted 167, **336**, 337
   Western 178
   White-chested 206
   Yellow-fronted 266
   Yellow-necked 277
   Yellow-rumped 167, 206, 207
Tit
   African Blue 69, **69**
   Ashy 241
   Carp's 225
   Coal 331
   Gray 312, **345**
   Great 331
   Miombo 257
   Rufous-bellied 207, 278
   Southern Black- 278, 336
   Southern Penduline- 241
   White-backed 65
Toad
   African Green 86
   Berber 324, 331
   Cape Sand 314
   Common 353
   Dombe Pygmy 81
   Karoo 106, 114
   Paradise 81
   Southern Pygmy 106
   Spiny 353
   Western Leopard 377
Toothbrush Tree 100
Torchwood 166
Tortoise
   Aldabra Giant 200, **201**, 201–202, 403
   Angulate 307, 314, **337**
   Geometric 319

Kleinmann's 75
Leopard 105
Radiated **130**, 131
Spider 131
Tent 105, 114
Treecreeper, Short-toed 330
tree ferns 179, **192**, **196**
Tree Frog
  Mozambique Forest **167**
  Stripeless 324, 331
Tree Pangolin, White-bellied 205–206
Tree Plantations 414, **422**, 422–423
trees *see individual types/genera*
Trogon
  Bar-tailed 178
  Narina 155
Tropical Deciduous Forests 209–219
  Angolan Deciduous Forest 151, 203, 209, 215–219, **216**, **217**, **218**, **219**
  Malagasy Deciduous Forest 128, 129, 132, 180, 189, 209–214, **210**, **211**, **212**, **213**, **214**, 215, 271, 287, 338
Tropical Montane Cultivation 409, 414, 415–417, **416**, 418, 419
Tropicbird
  Red-billed 405
  Red-tailed 404
  White-tailed 401
Tsessebe, Common 283
Tsingy (limestone karst) 211, 269, 271, 273, **274**
Turaco **29**
  Bannerman's 179
  Fischer's 161
  Knysna **29**, 337
  Livingstone's 167
  Prince Ruspoli's 65, 250
  Red-crested 179, **219**, 417
  Ross's 207, 417
  Schalow's **155**, 207
  White-cheeked **64**, 65
Turanian Cold Erg Desert 82, 88
Turanian Reg and Hamada 71
Turnstone, Ruddy 401
Turtle, Sea *see* Sea Turtle
Turtle-Dove, Malagasy 201, 340
turtles, marine 388
Twinspot
  Green 168, 207
  Peter's 155
  Pink-throated 168
  Red-throated 206

U

Uapaca bojeri (Tapia) tree 338, **339**, 339–340, **341**
Umbrella Pine 68
Umbrella Thorn 83, 100, 122, 228, 237
Upland Rivers 382
Usnea lichen **257**

V

Vachellia abyssinica 247, **247**

Vachellia spp. 83, 90, 100, 122, 228, **237**, **238**, 246, 247
Vanga
  Archbold's 129
  Bernier's 185
  Blue 213
  Chabert 290
  Helmet 185, **186**
  Hook-billed 213
  Lafresnaye's 129
  Madagascar Blue **210**
  Red-shouldered 129
  Red-tailed 185
  Rufous 213
  Sickle-billed 213
  Van Dam's 213
  White-headed 213
Várzea flooded forest 169
vineforest
  Australasian Monsoon 151, 157, 164
  Australian Dry 203
Viper
  Gaboon 162, 168
  Mountain 353
  Saharan Horned **87**
  Saharan Sand 86
Vleis (Dambos) 366, **366**, 368, 369
Vontsira
  Grandidier's 129
  Narrow-striped 129, 214
  Ring-tailed 185, **196**
Vulture
  Egyptian 421
  Hooded 414, 425
  Palm-nut 155

W

wadis 50, **83**, 96, **96**, 100, **136**
Wagtail
  African Pied 385, 396, 411
  Cape 377
  Madagascar 385
  Yellow 421
Wallum (Australian Southern Lowland Heathland) 309
Wanderflechten 78
Warbler
  African Desert 119
  African Reed 362
  Barred Wren- 241
  Cape Verde Swamp 364
  Chestnut-vented 91, 312
  Cinnamon Bracken- 356
  Cinnamon-breasted **112**
  Clamorous Reed 389
  Cryptic 192
  Cyprus 119
  Dartford 323
  Grande Comoro Brush- 356
  Grauer's 178
  Grauer's Rush- 369
  Grauer's Swamp 365
  Gray-capped 362, 417

Kopje 104, 273
Laura's Woodland- 155
Layard's 104, **345**
Lesser Swamp 377
Little Rush 377
Madagascar Swamp 364, **364**, 389
Marmora's 324
Marsh 377
Melodious 323, 331
Miombo Wren- 225
Namaqua 105
Neumann's 178
Papyrus Yellow- 362, 364
Rufous-eared 104, **113**
Sardinian 323, 331
Scrub 119, **120**
Seychelles 201
Socotra 136
Spectacled 323, 327
Subdesert Brush- 129
Tristram's **53**, 60, 119, 323, **323**
Victorin's 304, **308**
Western Bonelli's 331
Western Olivaceous 323, 324, 331
Western Orphean 324, 331
Western Subalpine 323, **327**
White-tailed 179
White-winged Swamp 362
Wood 76
warblers, African **31**, **53**
Warm Humid Broadleaf Forests (Africa) 142, 143–207
  Afrotropical Lowland Rainforest 62, 142, 143–150, **144**, **145**, 146, **147**, **148**, **149**, **150**, 151, 154, 156, 157, 160, 169, 170, 173, 174, 177, 180, 181, 197, 203, 205, 215, 244, 251, 258, 260, 261, 266, 267, 273, 364, 382, 386, 388, 409, 411
  East Coast Forest Matrix 61, 62, 147, 151, 156, 157–163, **158**, **159**, **160**, **161**, **162**, **163**, 164, 168, 173, 174, 203, 219, 243, 244, 248, 252, 258, **335**, 382, 386, 388
  Indian Ocean Lowland Rainforest 43, 142, 143, 169, 180–188, **181**, **182**, **183**, **184**, **185**, **186**, **187**, 189, 197, 209, 211, 287, **398**, 409
  Indian Ocean Montane Rainforest 43, 129, 132, 142, 174, 180, 182, 189–196, **190**, **191**, **192**, **193**, **194**, **195**, **196**, 197, 199, 211, 287, 409
  Mavunda 152, 203–207, **204**, **205**, **206**, **207**, 215, 275
  Moist Montane Forest 33, 35, 38, 40, 45, 61, 62, 63, **63**, 64, 66, 138, 142, 143, 144, 147, 151, 156, 157, 159, 164, 165, 167, 169, 174–179, **175**, **176**, **177**, **178**, **179**, 189, 247, 293, 299, 348, 354, 356, 357, **367**, 415, 422
  Monsoon Forest 36, 142, 143, 146, 147, 151–156, **152**, **153**, **154**, **155**, **156**, 157, 161, 164, 165, 173, 174, 177, 203, 204, 205, 215, 219, 226, 244, 248, 250, 255, 256, 258, 276, 382, 424

Warm Humid Broadleaf Forests (Africa) *continued*
  Seychelles Granite Forest 143, 180, 189, 197–202, **198, 199, 200, 201, 202**
  South Coast Forest Matrix 62, 151, 157, 159, 161, 162, 164–168, **165, 166, 167, 168**, 174, 203, 243, 244, 248, 309, 332, 336, 382, 386, 388, 399
  Swamp Forest 39, 142, 143, 147, 152, 159, 169–173, **170, 171, 172, 173**, 180, 358, 360, 364, 380, 382, 386, 388
Warthog, Common 167, 240, 264, 336
Waterbuck 249, 264, 361
water features/sources 236
waterholes 236
water lilies 376
Wattle
  Australian **375**, 376
  Black 422
  Coastal 311
Wattle-eye **30**
  Black-throated **207**
  Yellow-bellied **149**
Waxbill
  Black-faced 241
  Black-tailed 155, 168, 206
  Swee 337
  Violet-eared **236**, 241
Weaver
  Baglafecht 417
  Bar-winged 155, 207, 257
  Brown-capped 178
  Cape 362, 377
  Chestnut-backed Sparrow- 257
  Clarke's 161
  Forest 168, 337
  Grosbeak 155, **360**
  Northern Brown-throated 362
  Olive-headed 257
  Orange 171
  Sakalava 213
  Scaly **236**
  Southern Brown-throated 362
  Southern Masked- 241, 362, 377, 417
  Thick-billed 362
  Village 426
  Vitelline Masked- **163**
weavers 163
Weevil, Giraffe-necked **195**
Weinmannia 190
Welwitschia **78**, **78**, 223
West Africa, habitats 51–53
West Asian Hamada 94
Western Desert Steppe 101, 107
Western Riparian Woodlands 78, 102, 112
West European Coastal Heath 354
wetlands 370
  Asian Temperate 370, 374
  Asian Tropical 358, 365
  Australasian Temperate Freshwater 370, 374
  Australian Tropical 358, 365

Neotropical Temperate 370, 374
Neotropical Tropical 358, 365
North African Temperate **370**, 370–372, **371, 372**, 374, 379, 382, 390, 393, 395
South African Temperate 358, 364, 365, 369, 370, **374**, 374–378, **375, 376, 377, 378**, 379, 382, 390, 393, 395
Whale
  Blue 406
  Humpback 406
  Orca 406
wheat 418
Wheatear
  Atlas 351, **353**
  Black 327, **327**
  Black-eared 60, 327, 331
  Capped 312, 417
  Desert 75, **118**
  Hooded 75, 99
  Mourning 75
  Northern 421
  Red-rumped 119, 331
  White-crowned 75, 99
Whimbrel 394
Whipsnake, Madagascan 292
Whistling Thorn 228, **230**, 246
White-eye
  Cape 318
  Comoro 356
  Madagascar 192, 356, 411
  Malagasy 340
  Marianne 200
  Northern Yellow 411
  Orange River 91
  Réunion 192
  Seychelles 201
  Southern Yellow 278
Whitethroat, Greater 60, 421
Whydah **31**
  Eastern Paradise- **31**
  Shaft-tailed 241
whydahs, paradise- 415
Widdringtonia cedars 62
Widowbird
  Fan-tailed 337, 377
  Red-collared 337, 377, 417
  Yellow-mantled 369
Wigeon, Eurasian **372**
Wild Ass, African 97, 124
Wildcat, African 97, 312, 323, 425
Wild Dog, African 74, 225, **242**, 249, 256, 264, 265, 277, 317
Wildebeest
  Black 103, **295**, 296
  Blue 36, 240, 251, 281, 283, 384, 414
Wild Service Tree 59, 225
Willow, Port Jackson 311
Witels Tree 177
Wolf
  African 56, 118, 330
  Ethiopian 65, 298, **345**, 347, 357

Wood-Dove, Emerald-spotted 336
Woodhoopoo **30**
  Green **30**, 225
  Violet 225, **225**
Woodland
  Afrotropical Montane Dry Mixed *see* Afrotropical Montane Dry Mixed Woodland
  Argan 329, **329**
  dry montane 61–62, **63**, 64, 65, 66
  Hamada Montane 95
  Highland Acacia 247, **247**
  Maghreb Broadleaf 54, 56, 67, 321, 322, **328**, 328–331, **329, 330, 331**, 338
  Maghreb Juniper Open 54, 55, **57**, 57–60, **58, 60**, 61, 67, 321, 322, 350, 419
  Mediterranean *see* Mediterranean Forests, Woodlands and Scrubs
  Pinyon–Juniper 57
  Western Riparian 78, 102, 112
  see also Forest
Woodland-Warbler, Laura's 155
Wood-Owl, African 65, 168
Woodpecker
  Abyssinian 65
  Cardinal 241
  Fine-spotted 266
  Great Spotted 330, **331**
  Ground **105**, 273, 304
  Knysna 314, 337
  Lesser Spotted **331**
  Levaillant's 56, **68**, 69, 331
  Mombasa 161
  Olive 337
  Reichenow's 225
Woodpigeon 60
  Common 141
Wood-Rail, Tsingy 213, 214
Woolly Lemur
  Cleese's 213
  Western 213
Wormwood, White 59, 117
Wren-Warbler
  Barred 241
  Miombo 225
Wryneck
  Eurasian 421
  Rufous-necked 417

Y
Yellow-Warbler, Papyrus 362, 364
yellowwoods 62, 176

Z
Zebra
  "Cape" Mountain 103, 304, 312, 336
  Grévy's **232**, 283
  "Hartmann's" Mountain 79, 225
  Plains 36, 225, 240, 249, **272**, 283, **380**, 384, 414, 425
Zitting 286
Ziziphus lotus 117